LIFE'S

SPLENDID

DRAMA

SCIENCE AND ITS CONCEPTUAL FOUNDATIONS

A SERIES EDITED BY DAVID L. HULL

EVOLUTIONARY
BIOLOGY AND THE
RECONSTRUCTION OF
LIFE'S ANCESTRY
1860-1940

LIFE'S
SPLENDID
DRAMA

PETER J.
BOWLER

THE UNIVERSITY OF CHICAGO PRESS
CHICAGO & LONDON

PETER J. BOWLER is professor of history and philosophy of science at
the Queen's University of Belfast. He is the author of eleven books
including *Fossils and Progress: Paleontology and the Idea of Progressive
Evolution in the Nineteenth Century* (1976); *The Eclipse of
Darwinism: Anti-Darwinian Evolution Theories in the Decades around
1900* (1985); *Evolution: The History of an Idea* (1984); *Theories of
Human Evolution: A Century of Debate, 1844–1944* (1988); *The
Mendelian Revolution: The Emergence of Hereditarian Concepts in
Modern Science and Society* (1989); and *The Fontana/Norton History
of the Environmental Sciences* (1992).

The University of Chicago Press, Chicago 60637
The University of Chicago Press, Ltd., London
© 1996 by The University of Chicago
All rights reserved. Published 1996
Printed in the United States of America
05 04 03 02 01 00 99 98 97 96   1 2 3 4 5

ISBN: 0-226-06921-4 (cloth)
0-226-06922-2 (paper)

Library of Congress Cataloging-in-Publication Data

Bowler, Peter J.
    Life's splendid drama : evolutionary biology and the
reconstruction of life's ancestry, 1860–1940 / Peter J. Bowler.
        p.   cm. — (Science and its conceptual foundations)
    Includes bibliographical references and index.
    1. Evolution
(Biology)—History. 2. Biology—History. I. Title.
II. Series.
QH361.B685 1996
575'.009—dc20   95-25394
                                                        CIP

In memory of my father

WALLACE BOWLER

1913–1995

CONTENTS

LIST OF FIGURES   *ix*

PREFACE   *xi*

TABLE OF GEOLOGICAL PERIODS AND ERAS   *xiii*

1   THE FIRST EVOLUTIONARY BIOLOGY   *1*
A New Biology   *6*
A Revolution in Science?   *15*
Transforming Traditions   *18*
The Professional Framework   *25*

2   THE TREE OF LIFE   *40*
Relationships Redefined   *45*
Form and Function   *62*
Convergence and Parallelism   *67*
Ontogeny and Phylogeny   *74*
The Base of the Tree   *83*

3   ARE THE ARTHROPODA A NATURAL GROUP?   *97*
The Problem of Arthropod Origins   *103*
The Genealogy of the Crustacea   *106*
Peripatus and the Origin of the Tracheata   *120*
Limulus an Arachnid   *123*
The Debate Widens   *127*
The Fossil Record   *132*

4   VERTEBRATE ORIGINS   141
The Ascidian Theory   147
The Annelid Theory   154
The Arthropod Theories   171
Nemertines and the Actinozoa   180
Balanoglossus and the Echinoderms   185
The Environmental Trigger   192
Later Developments   196

5   FROM FISH TO AMPHIBIAN   203
The Origin of Fish   206
The Fin Problem   219
The Origin of the Amphibians   229
From Water to Land   244

6   THE ORIGIN OF BIRDS AND MAMMALS   259
From Reptile to Bird   261
Taking to the Air   273
Monotremes, Marsupials, and Mammals   280
The Mammal-like Reptiles   297

7   PATTERNS IN THE PAST   313
Putting Things Together   320
Adaptive Radiation   328
Laws and Trends   339
Rise and Fall   352
Mass Extinctions   366

8   THE GEOGRAPHY OF LIFE   371
Zoological Provinces   379
Lost Worlds   389
Northern Origins   394
Southern Continents   407

9   THE METAPHORS OF EVOLUTION   419
Trees and Ladders   424
The Biology of Imperialism?   435
Phylogeny and Modern Darwinism   441

BIOGRAPHICAL APPENDIX   447

BIBLIOGRAPHY   461

INDEX   511

FIGURES

1.1   Anton Dohrn working at his microscope   29
1.2   Henry Fairfield Osborn   30
1.3   Cartoon of E. Ray Lankester   36

2.1   Homologies of the hand in mammals   47
2.2   Part of Darwin's evolutionary tree   50
2.3   Gastrula of various animals   86
2.4   Monophyletic pedigree of animals   90
2.5   Pedigree of man   92
2.6   Lankester's schematic mollusc   93

3.1   A eurypterid, or sea scorpion   105
3.2   Larvae of the shrimp *Peneus*   108
3.3   The phyllopod *Apus*   111
3.4   *Peripatus capensis*   121
3.5   *Limulus,* the horseshoe crab   123
3.6   Trilobites, showing leg structure   134

4.1   Segmentation of head of a selachian   145
4.2   Development of an ascidian and *Amphioxus*   149
4.3   Internal structure of an ascidian, *Amphioxus,*
       and *Petromyzon*   151
4.4   Tadpole of a frog and of an ascidian   152
4.5   Transformation of an annelid worm into a vertebrate   158
4.6   Phylogeny of the arachnid, ostracoderm, and vertebrate
       stock   174
4.7   Internal structure of *Balanoglossus.*   186

5.1   *Palaeospondylus gunni*   211
5.2   *Cephalaspis murchisoni*   212

5.3   Gegenbaur's theory of origin of fin skeleton   *221*
5.4   The living lungfishes   *221*
5.5   Formation of fins from a hypothetical ancestor   *222*
5.6   Pelvis and pelvic appendages   *249*
5.7   Protetrapod   *251*

6.1   First specimen of *Archaeopteryx*   *264*
6.2   Restoration of *Compsognathus longipes*   *266*
6.3   Hypothetical four-winged bird   *276*
6.4   Restoration of hypothetical proavian   *279*
6.5   Transition from mammal-like reptile to mammal   *304–5*
6.6   Origin of auditory ossicles   *306*
6.7   Knight's restoration of mammal-like reptiles   *309*

7.1   Four-toed *Orohippus* to modern horse   *332*
7.2   Ancestral tree of the mammals   *337*
7.3   Brain proportions in archaic and modern mammals   *341*
7.4   Development of spinose condition   *348*

8.1   Zoo-geographical map of the world   *383*
8.2   Migration routes to the British Isles   *399*
8.3   Distribution of the primates   *404*
8.4   Map of continents in the Jurassic   *411*
8.5   Map showing the extent of Antarctica if the sea level were
      to drop   *415*

9.1   Monophyletic pedigree of the back-boned animals   *427*
9.2   Phylogenetic tree of the animal kingdom   *430*

PREFACE

The circumstances leading to the writing of this book are outlined in chapter 1 and need not be repeated here. I would, however, like to repeat the warning that I am writing as a historian, not as a scientist. Some of the issues raised by the earlier generation of scientists discussed in this book are still active today, and I am aware—in part thanks to comments by David Hull—that what I say will be perceived as relevant to the modern debates. But my primary purpose as a historian is to provide an overview of the techniques and theories used by those biologists who first tried to build upon the basic concept of evolution. The results will be of interest to modern biologists—but also to historians and others who do not have technical training in biology. I have tried to give some impression of where the debates are still relevant today, but it is not my primary intention to evaluate past theories by modern standards. Nor do I want to translate past discourse into modern terminology, in part for fear of imposing modern preconceptions onto the past, but also because that terminology would be unintelligible to readers with no biological training.

I am grateful to the Queen's University of Belfast for sabbatical leave in the Michaelmas terms of 1991 and 1994, which greatly speeded up work on the project. The library staff at Queen's have been of immense help during the course of the research. Additional work has been done at Cambridge University Library, the libraries and archives of the Natural History Museum, the Science Museum, and the Linnean Society, London, and the Stazione Zoologica, Naples. Thanks to the staff of these institutions, including John Thackray, Gina Douglas, and Christiane Groeben. The School of Philosophical and Anthropological Studies at Queen's University provided funds for assistance with the preparation of the illustrations and with translation from German. Thanks to Nora Lynch for help with the pictures and Valeria Lima Passos for the translations. Final stages in the preparation of the book were smoothed by a period of leave at the Rockefeller Study Center at Bellagio, Italy. Finally I should like to thank my wife Sheila, who helped both with the printing of the final manuscript and by taking over several tasks (including conference administration) which would otherwise have held up the project.

The manuscript has been read by Michael Ghiselin, Ernst Mayr, James Secord, Michael Benton, and an anonymous referee. I have greatly benefited from their advice (although I must record the ususal warning that any errors and omissions remaining are my own responsibility). David Hull read the first draft of the whole manuscript; he provided valuable advice and pointed out many stylistic problems. Some of my early thoughts on evolutionary morphology were published in an article, "Development and Adaptation: Evolutionary Concepts in British Morphology, 1870–1914," in the *British Journal for the History of Science* 22 (1989): 283–297. Chapter 3 is based on an article published under the title "Are the Arthropoda a Natural Group?" in the *Journal of the History of Biology*, 27 (1994): 177–213, reproduced by permission of Kluwer Academic Publishers. Some of the points made, especially in chapter 7, are included in an article, "Darwinism and Modernism: Genetics, Paleontology, and the Challenge to Progressionism, 1880–1930," published in Dorothy Ross, ed., *Modernist Impulses in the Human Sciences, 1870–1930* (Johns Hopkins University Press, 1994). The material on T. H. Huxley scattered through this book has been collected in an article prepared for a volume on Huxley being edited by Alan Barr for the University of Georgia Press.

TABLE OF GEOLOGICAL PERIODS AND ERAS

Cenozoic ero ("Age of Mammals")	Holocene (recent) Pleistocene	Quaternary	
	Pliocene Miocene Oligocene Eocene Paleocene	Tertiary series	
Mesozoic era ("Age of Reptiles")	Cretaceous Jurassic Triassic	Secondary series	
Paleozoic era ("Age of ancient life")	Permian Carboniferous Devonian Silurian Ordovician Cambrian	Transition series	
	Precambrian	Primary series	

Note that estimates of the absolute ages of geological periods changed dramatically during the period covered by this book. Although Darwin accepted a bast, but unspecified, age for the earth, late nineteenth-century geologist (under the influence of physicist such a Lord Kelven) reduced this to approximately 100 million years. Primitive methods of radioactive dating greatly extended the age of the earth in the early twentieth century, and by the 1930s dates similar to those now accepted had become available. The beginning of the Cenozoic is now dated at 65 million years (m.y.) before present, of the Mesozoic at 225 m.y., and of the Paleozoic at 570 m.y.

1

THE FIRST EVOLUTIONARY BIOLOGY

It is widely assumed that the publication of Charles Darwin's *Origin of Species* in 1859 triggered a scientific revolution with immense consequences both for biology itself and for Western culture as a whole. Darwin forced biologists to confront the possibility that the natural world was a product not of divine wisdom but of a historical process driven by purely material forces. Exploration of this possibility generated new initiatives in science as biologists tried to understand the nature of the evolutionary process. Widespread acceptance of a natural origin for the human species also sparked a cultural revolution in which the foundations of moral values had to be reassessed, with major consequences for religious beliefs, philosophical positions, and social thought.

Historical literature dealing with this revolution abounds; indeed the phrase 'Darwinian revolution' has become part of the common language of historical discourse. Historians of science, in particular, have focused on the process by which Darwin was driven to accept the basic idea of evolution, and his discovery of the theory of natural selection. The scientific community's reaction to the *Origin* has been extensively studied. We now know a great deal about the debates over the plausibility of evolutionism and even more about the controversies centered on the mechanism of natural selection. But it remains an open question whether we have a fully balanced understanding of Darwin's impact on science and hence of his impact on wider debates. I have complained elsewhere about the limitations of the traditional historiography of the Darwinian revolution.[1] So much attention has been given to the discovery and reception

1. Bowler, *The Non-Darwinian Revolution;* see also Bowler, *The Eclipse of Darwinism.* Literature on the Darwinian revolution includes Loren Eiseley, *Darwin's Century;* John C. Greene, *The Death of Adam;* Gertrude Himmelfarb, *Darwin and the Darwinian Revolution;* and Michael Ruse, *The Darwinian Revolution.*

of the selection theory that we tend to forget the extent to which many biologists—and many social thinkers—preferred non-Darwinian mechanisms of evolution such as the Lamarckian theory of the inheritance of acquired characteristics. If Darwin converted the scientific world to evolutionism despite being unable to convince his contemporaries that he had solved the problem of how evolution worked, we need to reassess our understanding of his impact on both science and society.

One product of the traditional historiography has been a concentration on the later development of the theory of natural selection. In particular we now know a great deal about the process by which concerns over the mechanism of variation and heredity led to the emergence of Mendelian genetics in the early years of the twentieth century, and the eventual synthesis of genetics with the selection theory.[2] But it was precisely this concentration on the fate of Darwin's proposed mechanism that was responsible for historians' lack of interest in the non-Darwinian theories of evolution that flourished in the late nineteenth and early twentieth centuries. We knew that controversy surrounded the selection theory, but we were more concerned with how the debates were resolved in the theory's favor than with the alternatives that were preferred by a whole generation of biologists.

My intention is now to present a further modification of the traditional historiography. The 'selection-centered' viewpoint of many earlier studies was whiggish in that it assumed a straight line of development leading from Darwin to the modern genetical theory of natural selection. All else was a sideline of no real interest to the modern world—even if a whole generation of scientists and other thinkers was profoundly interested in the alternatives. I now admit that my own attempts to draw attention to the alternatives were still influenced by the traditional interpretation. I was still focusing on the *mechanism* of evolution, even though I wanted to explore the non-Darwinian ideas that flourished during the 'eclipse of Darwinism.' But my own original interest in those mechanisms had been aroused when working on late nineteenth-century paleontolgists such as Edward Drinker Cope. These 'neo-Lamarckians' were attracted to a non-Darwinian version of the evolutionary mechanism because they thought that the selection theory was unable to explain the patterns they were uncovering in the history of life on earth.

I now want to suggest that our standard histories of the impact of Darwinism have been skewed by a concentration on the debate over mechanisms at the expense of the debates that arose over how to interpret the

2. See William Provine, *The Origins of Theoretical Population Genetics;* Ernst Mayr and William Provine, eds., *The Evolutionary Synthesis;* and Bowler, *The Mendelian Revolution.*

course of life's evolution. The first generation of evolutionary biologists were primarily concerned to reconstruct the history of life on earth. They debated mechanisms only because they had an interest in the processes which might have produced the patterns of development they were uncovering. Of course, some evolutionists concentrated on the mechanism—August Weismann would be a good example—but only in the last decades of the nineteenth century did this concern become paramount as the crisis built up toward the emergence of Mendelian genetics. Even then, work on reconstructing the history of life on earth continued, although it was rapidly overtaken in influence by the newly expanding experimental biology. The paleontologist William Diller Matthew wrote about his work in reconstructing the 'splendid drama' of life on earth—a phrase borrowed for the title of this study because it encapsulates the fascination of this area of science for a whole generation of biologists.[3]

The present work is thus both descriptive and polemical. It is descriptive in the sense that I want to provide an outline of evolutionary debates that have been left largely unrecorded by historians of science. This is not a complete survey of the attempts that were made to reconstruct the history of life on earth during the post-Darwinian era. But the book does attempt to provide a comprehensive sample of such debates, and to outline the evolutionary principles that were accepted by the participants. The only restriction I have allowed myself is a concentration on zoology. The evolution of plants has been left out of the story except where it was thought to have directly influenced the history of animal groups.

In providing this description, however, I have a polemical purpose because I want to encourage historians to confront the reasons why these extensive debates have been ignored in most surveys of the Darwinian revolution. We need to recognize that a whole generation of evolutionary morphologists, paleontologists, and biogeographers engaged in scientific research that was both of permanent value (because we are still interested in the course of life's evolution) and largely unrelated to the subsequent crisis over the mechanism of heredity. In so doing, we are led yet again to reassess the criteria by which we selected those topics that were to be included in the traditional story of the Darwinian revolution. We need to think more carefully about the impact of evolutionism on the still embryonic professional biology of the time. We shall also see how many of the cultural influences traditionally ascribed to the selection theory—concepts of 'social Darwinism' and the like—may equally well have had their

---

3. In his *Outline and General Principles of the History of Life*, Matthew wrote, "The story of life on the earth is a splendid drama" (p. 6). I am indebted to Edwin H. Colbert's biography, *William Diller Matthew: Paleontologist: The Splendid Drama Observed*, for suggesting this title.

origins in applications of lessons derived from the history of life on earth to human affairs.

The bulk of this study offers a series of chapters devoted to major episodes in the reconstruction of life's ancestry in the post-Darwinian era. The survey begins with the more fundamental question of how the main types of animal life had evolved, and then moves on to more detailed work on the internal history of particular groups. The first two chapters serve as a general introduction to the whole study. The remainder of this first chapter assesses the impact of Darwin's work on contemporary biology, identifying those areas which were transformed by the new approach to the study of life's history. I ask how significant the new approach was: Did it constitute a revolution, or merely a transformation of existing disciplines? My argument is that the attempts to reconstruct life's ancestry fall into the latter category—but that in minimizing the 'revolutionary' impact of Darwinism, we should not lose sight of the cumulative transformation of biologists' attitudes brought about by decades of effort. In the end, I hope to show that the emergence of twentieth-century Darwinism was encouraged by profound modifications that had taken place in evolutionists' views on the history of life in the period from the 1870s to the 1930s. The concluding sections of the chapter illustrate the professional framework of those parts of biology that underwent this massive, if fairly gradual, transformation.

Chapter 2 outlines some of the theoretical issues encountered by the biologists who sought to use the structural resemblances between living and fossil animals as clues to their evolutionary relationships. In effect, it asks what effect evolutionism had on the concept of 'relationship,' especially in view of the fact that it soon became necessary to recognize that some structural resemblances were not signs of an evolutionary link. The possibility that similar structures could be developed independently in several different branches of the tree of life was to bedevil the whole program of phylogenetic research. Evolutionists also had to ask whether the history of life was shaped primarily by forces internal to the organisms themselves, or by the less predictable effects of an ever-changing environment. My long-term argument is that the growing willingness of evolutionists to consider a role for the external environment represents a belated acceptance of principles that would become enshrined in modern Darwinism. This chapter ends with a brief look at some of the problems encountered by biologists seeking to understand the earliest steps in the evolution of life on earth, beginning with the origin of life itself. It thus serves as a prelude to the chapters dealing with the origin of certain groups of animals.

Taken together, the chapters making up this book offer a cross-section of the debates over the history of life in the period before the emergence of modern Darwinism in the 1940s. The chapters do not attempt to carry the story of each debate through to the present, although I have made some references to later developments where it seems appropriate. My aim is to capture the spirit of an age unified by acceptance of a constellation of ideas and attitudes that we no longer take for granted, and to chart significant developments that took place even before the consolidation of twentieth-century Darwinism. The impact of the 'modern synthesis' of Darwinism and genetics on these areas of research is well understood—it is most visible, perhaps, in the work of George Gaylord Simpson in paleontology. A number of non-Darwinian mechanisms of evolution were abandoned, with important consequences for ideas on the pattern of evolutionary change. But the rejection of these non-Darwinian ideas should not be seen as a sudden revolution precipitated by events outside the realm of paleontology. As suggested above, this study will show that the path toward modern Darwinism had already been broken by an earlier generation of evolutionists asking new questions about the forces that shape the history of life on earth.

A number of factors contributed to this steady transformation. Priorities had changed among the disciplines that had been affected by the advent of evolutionism. As morphology turned away from phylogenetic reconstruction, paleontology and biogeography became more active. We shall see that this transition promoted a greater awareness of the extent to which the history of life on earth could be explained in terms of changing environments through geological time. If Simpson was able to make important contributions to the modern synthesis, he did so because he could draw upon important innovations, both technical and conceptual, made by the previous generation of paleontologists. They had, in effect, become more 'Darwinian' in their overall view of the history of life, even though many still accepted a role for mechanisms other than natural selection. The impact of the modern synthesis of genetics and selectionism on the techniques of paleontology was profound, but we should not ignore the steady move toward a more Darwinian way of thinking about the history of life on earth among the previous generation of biologists.

The advent of the modern synthetic theory of evolution marks the end-point of this study. The introduction of later techniques such as cladistic taxonomy and the use of molecular biology to assess degrees of relationship has dramatically influenced modern ideas and modern attitudes toward the whole question of reconstructing ancestries. To trace the impact of these later developments would take another book, one that I would not be qualified to write. Scientists involved in current debates on these issues will, I hope, take an interest in what their predecessors did, but they should bear in mind that my purpose is to study an earlier period, not to decide which were the 'right' and 'wrong' theories as judged by modern standards.

Some of the earlier debates have been resolved quite recently by the techniques of molecular biology. Others linger on, or reemerge in new guises despite the greater sophistication of modern investigations. Some systematists in the 1980s rejected the use of fossils in the determination of relationships, although cladistic techniques are now being successsfully extended to include fossil species.[4] But many cladists are still unwilling to discuss the 'adaptive scenarios' that had become a standard component of evolutionary explanations even before the advent of modern Darwinism. Other scientists have become deeply embroiled in a renewed debate on the causes of mass extinctions. The potential for evolutionary convergence to produce similarities that will confuse the reconstruction of ancestries is still debated. Vicariance biogeographers still challenge the orthodox theory of the dispersal of each group from a point of origin. Some of the issues that emerge from this study are thus still of interest today, but the story has not been told from the viewpoint of hindsight. As a historian, my purpose is to understand a particular episode in the past, not to convert earlier ideas into their closest modern equivalent. I have also refrained from translating the terminology of the period I am dealing with into the language of modern biology; to do so would, I believe, impose an illusion of modernity onto biologists whose way of thinking may have been significantly different from ours.

A NEW BIOLOGY

What was the impact of Darwin's theory on biology? Traditionally we have simply assumed that biologists must have been profoundly affected by the transition from seeing living things as the creations of a wise and benevolent God to seeing them as the products of a ruthless struggle for existence within a purely material universe. In fact most late nineteenth-century biologists continued to hope that the development of life on earth was a purposeful process. They responded to the more general possibility that natural processes could transform one species into another. Critics fulminated against the materialism of Darwin's theory, but how exactly did a working biologist apply the idea of evolution in the study of living things? Darwin certainly outlined a program, or several programs, of re-

4. On the impact of cladism and the reemergence of an interest in fossils, see Andrew B. Smith, *Systematics and the Fossil Record.* A useful survey of modern work is K. C. Allen and D. E. G. Briggs, eds., *Evolution and the Fossil Record.* For an account of the debates in modern systematics, see David L. Hull, *Science and Progress.* A brief outline of modern debates in evolution theory is given in Bowler, *Evolution: The History of an Idea,* chap. 12.

search in the *Origin of Species*. But just as we have been forced to reassess the success of his arguments for the selection theory, so we need to think more carefully about how his fellow biologists responded to the challenge offered by the general theory of evolution.

Ernst Mayr has argued that the *Origin* contains five different but interrelated theories, each of which would suggest different modes of application.[5] For our purposes we can identify three components of Darwin's work which could have inspired potential applications in research. First, Darwin began his book with an extended account of the theory of natural selection, and clearly one could study the various processes which go to make up this mode of transforming populations through time. Second, he made it clear that his theory had implications for the way new species are produced: they 'bud off' from existing species either when sections of the population become isolated from the rest in a different environment or (as Darwin later came to believe) when different ecological pressures at either end of a species' range effectively pull it apart. Again, field naturalists ought to be able to investigate the process of speciation to see if this is how it works. They also ought to be able to show that the existing geographical distribution of species is consistent with this mode of production. Finally, Darwin outlined the consequences of a theory of common descent for traditional ideas about the relationships between species. Related species must be the descendants of a single common ancestor. Thus classification becomes based on a genealogical tree, and it should be possible to reconstruct the branches of the tree from a knowledge of each species' structure. If sufficient fossils were available, it ought to be possible to reconstruct the actual course of a group's evolution, but Darwin warned about the "imperfection of the fossil record" and believed that it would be difficult to obtain enough reliable information to make this kind of detailed reconstruction possible. He ventured none.

Historians have devoted a great deal of effort to the impact of the selection theory upon research into the problems of heredity, variation, and speciation. Important developments in these areas certainly paved the way for the emergence of modern genetics and modern Darwinism. But the biologists of the late nineteenth century devoted at least as much attention to the exploration of a quite different aspect of evolutionism. They took it for granted that one of the most exciting ways of building upon Darwin's initiative would be to venture into the territory he had largely ignored: reconstructing the course of life's evolution through geological time. The German evolutionist Ernst Haeckel proclaimed the importance of studying what he called "phylogeny"—the evolutionary history of each known form of life. The program he founded was to transform the

5. See, for instance, Ernst Mayr, *One Long Argument,* chap. 4.

sciences of morphology (comparative anatomy and embryology), paleontology, and biogeography. Let us look first at the selection theory itself,
and then at the emergence of phylogenetic research.

The theory of natural selection assumes that a species consists of a
population of individuals subject to random (or at least undirected, purposeless) variation, and that variant characters can be transmitted to the
offspring by inheritance. Since in each generation more individuals are
born than can survive, a struggle for existence will ensue in which those
individuals best adapted to the local environment survive at the expense
of those less well adapted and pass their advantageous characters on
to the next generation. Over many generations, the addition of minute
adaptive variations will accumulate to produce a sigificant change in the
population.

Darwin himself conducted an extensive study of the variability of populations and the process of heredity, outlined in his *Variation of Animals
and Plants under Domestication* (1868). His assumption that variation
was largely undirected and purposeless was based on extensive studies of
domesticated animals, but for purely theoretical reasons many biologists
found this the most difficult part of the theory to accept. They preferred
to believe that variation was somehow directed along predetermined channels, thus imposing a direction on the course of evolution. Darwin himself
accepted that at least some variation arose from the animals' interaction
with the environment. He would not rule out the Lamarckian effect in
which characteristics acquired by the organism through change of habit
are transmitted to the offspring. Many late nineteenth-century evolutionists preferred Lamarckism to the selection of 'random' variation because
it allowed evolution to seem a more purposeful process and eliminated the
need for the harshness of the struggle for existence. A somewhat desultory
program of experimental work was undertaken to provide evidence for
the Lamarckian effect, culminating in the early twentieth century with the
notorious 'case of the midwife toad.'[6]

Others supposed that variation was directed by forces internal to the
organism, perhaps under the control of the process of embryological development. This would impose a direction on evolution that allowed it to
be seen as a more orderly process than would be possible if it were directed
merely by response to changes in the local environment. The resulting
theory of 'orthogenesis' was, like Lamarckism, supported mainly on the
grounds of indirect evidence.[7] It did not inspire a program of experimental

6. See Arthur Koestler, *The Case of the Midwife Toad,* and Bowler, *The Eclipse of Darwinism,* chap. 4.

7. On orthogenesis, see Bowler, *The Eclipse of Darwinism,* chap. 7.

work, unlike the related belief that variation might occur in sudden leaps, or saltations, which might establish a new species more or less overnight. When this theory became the subject of serious experimental study in the last decade of the century, it created the climate of opinion which paved the way for the emergence of Mendelian genetics.[8]

Widespread suspicion greeted Darwin's own theory of heredity, pangenesis, which was published in the *Variation of Animals and Plants under Domestication*. The most notable product of this suspicion was August Weismann's theory of the germ plasm, developed during the 1880s. Although highly controversial at first, this theory eventually helped to eliminate Lamarckism by creating the hereditarian viewpoint characteristic of early genetics. Weismann himself became an outspoken selectionist (a 'neo-Darwinist' to use G. J. Romanes's phrase). But it is significant that the first serious efforts to test the effect of selection on wild populations were not made until the 1890s. At this point members of the biometrical school, including Karl Pearson and W. F. R. Weldon, tested the effect of changed environments on the range of natural variation.[9]

For all the controversy triggered by the publication of the selection theory, little scientific work was inspired by Darwin's proposed mechanism until the last decades of the century. The debate over selection was conducted more on the grounds of its conceptual plausibility than on the basis of serious empirical studies. Darwin's own later work on the structure of flowers, on insectivorous plants and on climbing plants, was designed to provide indirect evidence that evolution was driven by a process such as natural selection, but this did not stimulate his contemporaries to research these areas. The theory of natural selection failed not only to gain the acceptance of the scientific community, but even to inspire any serious research programs during Darwin's own lifetime. If we hope to detect a revolutionary effect of the theory on late nineteenth-century biology, we shall have to look for applications of the wider principles of evolutionism that were also proposed in the *Origin of Species*.

Certain kinds of field study were inspired by Darwin's broader view of how evolution worked. Henry Walter Bates's studies of mimicry were seen as an important application of the idea of adaptive evolution.[10] Alfred Russel Wallace, the codiscoverer of natural selection, worked on local

8. On saltations and discontinuity, see Bowler, *The Eclipse of Darwinism*, chap. 8, and *The Mendelian Revolution*.

9. On biometry, see Provine, *The Origins of Theoretical Population Genetics;* and Bernard Norton, "The Biometric Defence of Darwinism."

10. Henry Walter Bates, "Contributions to an Insect Fauna of the Amazon Valley." See Barbra G. Beddall, *Wallace and Bates in the Tropics*.

variability and geographical distribution.[11] But there is little evidence that this kind of work was part of a really substantial research tradition in the years following the publication of Darwin's theory. Wallace later undertook an extensive program of research on geographical distribution, designed to show how the existing faunal provinces had been formed. He was exploiting the concept of evolution because he was anxious to trace the point of origin and the subsequent migrations of each family of animals. He was certainly following in the footsteps of Darwin himself, who saw the study of migration and distribution as an important component of his theory. Yet even here the application of the theory was delayed. The flow of publications on the topic remained a mere trickle until the appearance of Wallace's *Geographical Distribution of Animals* in 1876 triggered an explosion of research in the last decades of the century.

The example of geographical distribution allows us to make some general points that are relevant to the evaluation of Darwinism's impact. First, the theory's influence came about through the transformation of an existing field of study, not through the creation of an entirely new scientific discipline. Janet Brown has described a long tradition of debate over the distribution of animals and plants into which Darwin and his botanist ally Joseph Dalton Hooker inserted their program of seeking to explain distribution in terms of evolutionary origins and subsequent migrations.[12] Second, the transformation came about through the injection of a historical element into the explanatory scheme used to account for the observations. One did not have to accept the theory of natural selection to see how evolutionism could help to explain the present state of affairs as the result of a long historical process. Reconstructing the history of life on earth became a central feature of the explanatory technique adopted by evolutionary biogeographers.

In this area, Darwinism came perhaps closest to creating an entirely new area of debate, transforming the identification of static geographical provinces into a genuinely historical research program. Yet—although this program was closer to Darwin's own interests than any of the others discussed below—it was slow to emerge as a really substantial topic of debate. As with all of Darwin's more radical insights, it took time for the naturalists of the late nineteenth century to realize that evolutionism offered them an entirely new research field in this area. One could certainly argue that the advent of evolutionism transformed the study of geographi-

---

11. See the papers collected in Alfred Russel Wallace, *Contributions to the Theory of Natural Selection* and *Natural Selection and Tropical Nature*. See also Wallace's *Island Life* and his *The Geographical Distribution of Animals*.

12. Janet Browne, *The Secular Ark*.

cal distribution, whether or not it is legitimate to use the term *revolution* in this context. No doubt this is not as revolutionary a change as that which led to the emegence of Mendelian genetics—but it was substantial nonetheless. The striking point is that, even if less dramatic in scope than the 'Mendelian revolution,' the transformation was a relatively slow one which did not get fully underway until the last decades of the nineteenth century.

One final point about the emergence of evolutionary biogeography as a major research program also attracts my attention. We have no detailed survey of the immense outburst of activity in this area during the last decades of the nineteenth century. When compared with the extensive body of historical literature on the debates over natural selection and the emergence of genetics, this lack is particularly striking. Our understanding of Darwin's impact on science has been distorted by a relative lack of information about those areas where evolutionism transformed the explanatory scheme used in traditional areas of research. If there was a revolution—or even a substantial transformation of research—in these areas, we have bypassed it because our attention has been fixed on the more glamorous developments leading to the new experimental science of genetics.

The same arguments can be made in the case of the other two areas which form the subject of the present study—paleontology and evolutionary morphology. The fossil record had been the subject of extensive study since the early decades of the nineteenth century, both as a stratigraphical tool for geologists and as a source of information about the earth's past inhabitants. Some highly controversial attempts to link the progress of life on earth with the idea of evolution were made, as in Robert Chambers's *Vestiges of the Natural History of Creation* (1844). Chambers's hypothetical phylogenies were bizarre by modern standards, and the critical reaction by paleontologists convinced Darwin of the need to invoke the imperfection of the record to explain the apparently sudden leaps in the evolution of life on earth.[13] By converting the scientific world to evolutionism, Darwin stimulated paleontologists to look out for 'missing links' among their new discoveries. His theory also suggested that one might look for patterns of evolutionary development among known fossil sequences, allowing for the inevitable gaps. Here was another existing field of study that could be transformed by evolutionism—in fact, paleontology was the only direct source of information about the history of life on

13. On Chambers's theory, see M. J. S. Hodge, "The Universal Gestation of Nature"; Milton Millhauser, *Just Before Darwin;* and James Secord, "Behind the Veil." On Darwin's response to the attacks on *Vestiges,* see Frank Egerton, "Refutations and Conjectures."

earth by which different hypotheses about the course of evolution could be tested.

In the case of paleontology, evolutionism had a fairly rapid impact— although some students of the fossil record, including T. H. Huxley, were still rather slow to appreciate its significance. Important discoveries were made during the 1860s and 1870s which helped to fill in some of the gaps in the record of evolution. Fossil sequences were put together which seemed complete enough to be presented as evidence for evolution, especially the ancestry of the modern horse. Most accounts of the Darwinian revolution pay at least some attention to the debates over the fossil record in the 1860s and 1870s. But for details we must turn to specialist histories of paleontology, and even these limit themselves to the decades following immediately after the publication of the *Origin of Species*.[14] Historians have paid comparatively little attention to the extensive work done by paleontologists trying to reconstruct the history of life on earth in the decades around 1900, although recent work on American paleontology has begun to fill in part of the gap.[15] Several of the chapters below have been written to demonstrate that this was an active field of study at the time. I shall argue that paleontology became the prime focus of attention for this aspect of evolutionary biology, taking over from an increasingly discredited evolutionary morphology. Paleontology may have been eclipsed in relative terms by the rise of experimental biology, but it remained an active field—and is still the area of science most closely associated with evolutionism in the mind of the general public. The paleontologists were also responsible for important transformations in the way we think about the history of life on earth. Although often wedded to non-Darwinian ideas, they became increasingly aware of the possibility that the history of life has been punctuated by phases of rapid change that may have been triggered by geological forces.

Morphology forms the third branch of our study, and here we see an existing area of science which was incorporated rapidly into the evolutionary paradigm of the 1860s. Morphology is the study of animal structure, and as an extension of comparative anatomy, it had been used throughout the early nineteenth century to throw light on the relationships between species. Anatomists working in university-based laboratories were involved—but so were systematists working in museums, since they too needed information on the detailed structure of organisms if

---

14. See M. J. S. Rudwick, *The Meaning of Fossils;* Adrian Desmond, *Archetypes and Ancestors.* See also Bowler, *Fossils and Progress,* and Eric Buffetaut, *A Short History of Vertebrate Paleontology.*

15. See Ronald Rainger, *An Agenda for Antiquity.*

they were to classify them accurately. Darwin realized that taxonomy and hence morphology would be transformed by evolutionism once it was accepted that the 'relationships' between species should be seen as genealogical. Morphology would allow one to reconstruct the branchings in the tree of life. Ernst Haeckel set out most enthusiastically to use morphology as a guide to reconstructing the history of life on earth, and his work inspired a generation of biologists who saw their research in comparative anatomy and embryology as a means of extending the evolutionary paradigm. Evolutionary morphology was the centerpiece of scientific evolutionism until enthusiasm for it faded during the last decade of the nineteenth century.

Historians have debated the extent to which evolutionism transformed the morphological research tradition. I shall argue that, as in biogeography and paleontology, the injection of a historical element into its explanatory scheme forced this area of biology to undergo a significant transformation, if not a revolution. Darwin himself had done important morphological work on the barnacles, but he did not participate directly in the creation of Haeckel's evolutionary morphology. In this case, the use of structural relationships to reconstruct the history of life on earth extended evolutionism into an area that Darwin himself did not regard as central. Precisely for this reason, the rise and fall of evolutionary morphology has gone largely unrecorded by historians. A few studies mention the introduction of the soon-discredited 'recapitulation theory,' but one is left with the impression that the whole episode should be regarded as something of an embarrassment. The classic survey of the field is still E. S. Russell's *Form and Function* of 1915, little having been written since that time.[16] A recent study by Jane Maienschein looks at American morphologists at the turn of the century but emphasizes how they turned away from the evolutionary paradigm to other concerns.[17]

The fact that morphologists such as William Bateson and Thomas Hunt Morgan abandoned the field for genetics has helped to create the impression that evolutionary morphology was a sideline in the development of modern biology. Bateson was openly scornful about the claims that had been made for morphology's ability to reconstruct evolutionary ancestries. But the evolutionary questions did not go away just because some biologists wanted to move toward the study of more easily controlled topics. My argument is that, whatever one thinks about its quality as science, evolutionary morphology provided the conceptual framework

16. See E. S. Russell, *Form and Function,* chaps. 14–17. See also William Coleman, "Morphology between Type Concept and Descent Theory."

17. Jane Maienschein, *Transforming Traditions in American Biology.*

within which a whole generation of post-Darwinian biologists worked. In the early twentieth century, those morphologists who retained an interest in evolutionary questions—E. S. Goodrich is a good example—were able not only to interact with paleontologists but also to play a role in influencing some of the founders of modern Darwinism.

The morphologists addressed serious questions about evolutionary relationships, and their work dovetailed with that of systematists who approached the same questions from a different angle. Systematics could still be done without appeal to evolution theory, and in some areas the leading figures continued to chart 'relationships' without interpreting them in evolutionary terms. But by the turn of the century, systematists in the great natural history museums were routinely incorporating evolutionary models into their thinking, and where their studies appealed to the internal structure of organisms, they interacted with the laboratory-based morphologists. Their work became highly controversial, but this was because—like the morphologists—their efforts had exposed complexities in the nature of the evolutionary process that interfered with the practical effort to reconstruct phylogenetic relationships.

It is an oversimplification to suggest that evolutionary morphology disappeared overnight when faced with the defection of Bateson, Morgan, and their supporters. And while it is true that many later morphologists became less concerned with evolutionary questions, those questions did not drop out of the agenda altogether, and in some cases reemerged as active controversies in later decades. Nor did the systematists join their laboratory-based cousins who turned the spotlight elsewhere. Systematics still required the careful study of organic structures to assess relationships, and morphology thus continued to play a signficant role in this very different professional environment.

If we bypass evolutionary morphology because it does not lie on the main line leading to the genetical theory of natural selection, we cannot pretend to have a proper understanding of the impact of evolutionism on nineteenth-century science. If we wish to tell our story as a triumphant advance toward modern evolutionism, we may be justified in ignoring the role of morphology in the Darwinian revolution. But if we want to understand what evolutionism actually meant to late nineteenth-century biologists—and to those twentieth-century ones who did not join the rush toward experimentalism—we ignore it at our peril. Paleontology and biogeography similarly need to be brought into the story if we are to understand the state of affairs in biology preceding the emergence of modern synthetic Darwinism. Here, at least, we can see important developments still taking place into the twentieth century.

A REVOLUTION IN SCIENCE?

Evolutionism clearly had some impact on biology in the post-Darwinian era, even if the selection theory was not accepted. But was this a sudden revolution, or a gradual transformation in those sciences which found that they could exploit the basic idea of evolution? One thesis of this book is that in the late nineteenth century, three areas—evolutionary morphology, paleontology, and biogeography—interacted to create a network of debates about the course of life's evolution on earth. These were not new areas of biological science, but each was significantly transformed by the appearance of the evolutionary world view. They were forced to interact with one another to create a network of explanatory schemes based on the idea of a historical process in the development of life on earth. Priorities shifted as morphology declined in influence and the other two areas took up the slack, but we can recognize a coherent set of interests and problems addressed by the exponents of each of these three fields. An interdisciplinary program of evolutionary biology had come into operation, although its primary concern was the course of evolution in the past rather than the mechanism of change.

I have suggested above that this research program has been neglected by historians of the Darwinian revolution because the traditional historiography has focused attention on the mechanism of evolution. Darwin's most radical insights were concentrated here, and by focusing on this area we are encouraged to form the impression that the *Origin of Species* contained the basis for an entirely new world view. The assumption that there was a Darwinian revolution in biology was based on an approach to the history of science which stressed the role of theories and theory-change. T. S. Kuhn's classic *Structure of Scientific Revolutions* mentions the introduction of Darwinism as an example of a revolutionary paradigm change, although Kuhn was primarily concerned with revolutions in the physical sciences and astronomy.[18] The transition from a static world view based on design by God to a dynamic or historical one based on material processes is clearly a major conceptual revolution. To the extent that those world views were represented by the theoretical frameworks accepted by the scientists of the time, a scientific revolution must have been associated with the impact of evolutionism.

Understanding the role of nonselectionist evolution theories in late nineteenth-century biology has modified our perception of Darwin's impact. The full implications of the theory of natural selection did not sink in until the early twentieth century. The immediately post-Darwinian decades were dominated by ideas of change that were far less radical than

---

18. See Thomas S. Kuhn, *The Structure of Scientific Revolutions,* pp. 171–172.

those proposed in the *Origin*. It was for this reason that I entitled an earlier study *The Non-Darwinian Revolution;* the new evolutionism was in many respects a far less radical challenge to traditional ideas and values than the full Darwinian theory. But was there a *revolution* in biology at all, or merely a minor transformation of existing modes of thought? If the latter is the case, in what sense can we talk about a genuinely evolutionary biology in the late nineteenth century? I want to argue that, although the initial impact of Darwinism was limited, continued efforts to reconstruct the history of life on earth brought about a gradual transformation in scientists' attitudes which allowed them to appreciate at least some of the more innovative aspects of Darwin's thinking.

An important point is at issue here concerning our understanding of how we reconstruct the past. The various nonselectionist ideas mentioned above are characteristic of a 'developmental' or historicist way of thought which assumes that change is directed along predictable channels towards a meaningful goal. The recapitulation theory assumed that the evolution of life on earth was equivalent to the embryo's development towards maturity. Both Lamarckism and orthogenesis were linked to this embryological analogy. Both allowed their supporters to assume that evolution was directed towards a goal, either by the animals' choice of habits or by some inbuilt trend forcing variation in a predictable direction. The most extreme version of this model is the image of evolution as a ladder leading step-by-step to the human species as the triumph of creation. An exact parallel is the nineteenth-century anthropologists' belief in a ladder of cultural stages leading toward European-style industrial civilization. To the extent that both the biological and social sciences of the late nineteenth century were dominated by this way of thinking, they reconstructed the past by creating a preconceived model of the stages through which life or society must inevitably mount toward their goal.

The alternative is a more genuinely historical approach to the reconstruction of the past which does not seek to uncover a predictable trend. Darwin's theory fitted this model because he saw no predetermined trend in variation. The evolution of each population was driven solely by the demands of the local environment, which changed in an unpredictable way—especially for small numbers of individuals who migrated, perhaps unwittingly, to some new locality. On this model, evolution is not a ladder but a tree or bush with ever-diverging branches. There is no predictable direction of change and no goal toward which everything is tending. If this is how life has evolved, then one cannot hope to identify a 'main line' of evolution with steps which succeed one another inevitably through time. Each episode and incident in the history of life on earth was unpredictable, and the resulting pattern (if one can call it a pattern at all) is a haphazard, open-ended one. If this is the case, then the best that the historian, or the evolutionist, can do is to look at the evidence for each individual step in the past development of culture or living things and try to reconstruct what actually happened.

Much of the phylogenetic research conducted in the immediately post-Darwinian era was developmental rather than truly historical in approach. Morphology became the first evolutionary science, and it was the least likely branch of biology to concern itself with field studies revealing the effect of an ever-changing environment, or the hazards of migration. The study of organic form, especially through embryology, encouraged an abstract view of the evolutionary process in which the developmental mode of thought could flourish. This is not to say that evolutionism had only a trivial impact on morphology. I shall argue in the following chapter that the post-Darwinian morphologists had to face up to a number of challenges that would simply not have been apparent to their predecessors. Even so, their initial reliance on the recapitulation theory was characteristic of the developmental way of thinking—but faith in the reliability of that theory declined some time before morphologists began to lose interest in evolutionary questions. Morphologists became increasingly aware of the possibility of links with other areas, such as paleontology, that might provide information about the actual conditions under which evolutionary transformations had taken place.

I shall argue that phylogenetic research underwent significant changes in the late nineteenth century, moving it from a purely developmental to a more historical view of how to reconstruct the past. If the initial impact of Darwinism was somewhat less than revolutionary, the potential implications of evolutionism gradually became more apparent as biologists grappled with the problems of reconstructing the history of life. We should not think of late nineteenth-century evolutionism as a static discipline simply marking time until it was overtaken by the sudden emergence of new experimental sciences such as genetics. Early twentieth-century accounts of the history of life on earth are quite different from those of the 1860s and 1870s. To compare Ernst Haeckel's *History of Creation* of 1873 with Henry Fairfield Osborn's *Origin and Evolution of Life* of 1917 is to compare the product of an obviously outdated evolutionism with something which is recognizably part of the same genre as modern books on the topic. If we concentrate on those branches of evolutionary biology devoted to reconstructing the past, we see not stagnation and revolution, but a steady transformation brought about by increasing recognition of evolutionism's potential.

The most obvious sign of this transformation is the growing emphasis placed by evolutionary texts on paleontology and biogeography at the expense of pure morphology. Morphology did not disappear in the 1890s, but it was certainly overshadowed in in general surveys of evolution by the study of fossils and of geographical distribution. The result was, inevitably, a shift of attention to the more recent areas of the history of life where evidence from these fields was more appropriate. But this did not represent a sudden change of direction—more a gradual change of emphasis within a coherent network of debate. Morphologists became more aware of fossil evidence, while paleontologists and biogeographers were inevitably brought face to face with lines of evidence that required a more historical mode of enquiry. They could no longer treat the study of the history of life on earth as the unfolding of a developmental trend.

Developmental thinking was not eliminated from paleontology overnight. Osborn himself continued to champion orthogenesis into the 1930s, and many German paleontologists remained enthusiastic about the theory. But even Osborn had to deal with the evidence for migration and adaptation to new environments in the history of life. Paleontogists also interacted with geologists who were promoting the claim that relatively sudden bursts of continental elevation were responsible for triggering new initiatives in the evolution of life on earth. Evidence from a variety of different backgrounds thus forced the evolutionary biologists of the late nineteenth and early twentieth centuries to confront the fact that the history of life on earth was not based on a predictable trend. Elements of the old way of thinking remained, especially when it came to evaluating the origin of the human species itself. But on average we can see a steady transformation in the style of explanations offered which represents a move towards a more historical, and in that sense a more Darwinian, view of the past. Even those biologists who retained an interest in non-Darwinian ideas responded to this new awareness that the history of life contained at least some unpredictable episodes. When compared to the revolutionary impact of new sciences such as genetics, this transformation may look a good deal less than revolutionary. But in another sense it represents the culmination of the impact of Darwin's original proposals, and an important aspect of the process by which early twentieth-century biology was primed to accept the synthesis of Darwinism and genetics.

TRANSFORMING TRADITIONS

The suggestion that evolutionism effected a gradual transformation of these areas of biology, rather than a sudden revolution, fits in with other developments in the history of science. We have become more aware of the extent to which the development of science is influenced by the existence of networks of researchers unified by their adherence to particular fields of study and techniques, often defined by their professional location and identification. New research traditions are founded from time to time, but the emergence of these new disciplines does not always coincide with the great theoretical revolutions recognized by the old historiography. Darwinism had an impact on science because it gradually transformed existing research programs in systematics, morphology, and paleontology. By recognizing this continuity we are forced to see the impact of Darwinism as a less revolutionary process. The danger is that by concentrating exclusively on research traditions, and on the creation of new traditions such as classical genetics, we cease to acknowledge the power of important theories such as evolutionism to transform research over a period of time.

It has long been recognized that the morphological research tradition played an important role in nineteenth-century biology, even if it did not figure prominently in accounts of the Darwinian revolution. Some time ago, Garland Allen argued that there was a "revolt against morphology" at the end of the century, leading to the emergence of new experimental sciences such as genetics.[19] A number of historians protested that the morphological research tradition survived into the twentieth century in areas such as paleontology.[20] More recently Jane Maienschein has argued strongly for a gradual transformation of the morphological research program at the turn of the century.[21] But this concentration on research programs has had the effect of diverting attention away from initiatives at the theoretical level. Maienschein insists that it is pointless to concentrate on the impact of evolutionism, because morphologists gradually abandoned their interest in this topic without ceasing to be morphologists.[22]

The new historiography based on research traditions and professional disciplines tends to have the same effect as the old when it comes to evaluating the impact of evolutionism. Where the old interpretation of the Darwinian revolution ignored evolutionary morphology because it did not accommodate Darwin's most revolutionary insights, the new historiography sidelines the impact of evolutionism because it did not form the basis of coherent professional groupings among biologists. In a recently circulated manuscript, Michael Ruse has gone so far as to claim that there was no evolutionary biology in the late nineteenth century.[23] Ronald Rainger's

---

19. See Garland E. Allen, *Life Science in the Twentieth Century.*

20. See, for instance, Ronald Rainger, "The Continuation of the Morphological Tradition in American Paleontology."

21. Maienschein, *Transforming Traditions;* on the role of research traditions in science, see Larry Laudan, *Progress and its Problems.*

22. Maienschein, *Transforming Traditions,* esp. p. 299. Curiously, Maienschein herself has noted the dangers of highlighting certain areas of science as being at the "cutting edge" of progress; see her "Cutting Edges Cut Both Ways." By implication, other areas are marginalized even when they are still active—which is precisely what happened to phylogenetic studies.

23. Michael Ruse, "Molecules to Men: The Concept of Progress in Evolutionary Biology," in preparation for Harvard University Press, chap. 7

studies of American paleontology accept that evolutionary concerns were central to the work of Henry Fairfield Osborn and others, but present paleontology as an increasingly stagnant discipline, marginalized in the early twentieth century because it was excluded from the new universities by the rise of experimental biology.[24] Ernst Mayr had earlier claimed that fields such as biological taxonomy were reduced to second-rank status because they were practiced in museums rather than universities.[25]

By concentrating on the professionalization of biology, the creation of coherent research traditions, and the institutionalization of those traditions in the new universities, especially in America, the new historiography thus threatens once again to marginalize those branches of late nineteenth and early twentieth century biology concerned with reconstructing the history of life on earth. Even when exploring the evolutionary work of his paleontologists, Rainger implies that Osborn's group at the American Museum of Natural History survived in part because of his political and social influence. My attempt to identify a significant and coherent evolutionary biology in this period must obviously confront these efforts to deflect attention away from evolutionism. I certainly do not want to deny the importance of the great explosion of interest in experimental biology that began in the decades around 1900. But I do want to argue that too exclusive a focus on these disciplines, especially as they developed in America, has blinded us to the existence of a coherent program of evolutionary biology that continued to flourish throughout this period. This program may not have been the focus for professional societies, journals, or academic departments, and its level of activity may seem stagnant when compared to the explosion of interest in the new experimental disciplines, but it continued to function and produced work of lasting value. Scientists' understanding of the evolution of life on earth was different in the 1930s from what it had been in the 1870s precisely because the activities of morphologists, paleontologists, and biogeographers had led to an increased awareness of how evolutionism could be applied in their areas.

Let us deal first with the impact of evolutionism on morphology. Maienschein accepts that the morphologists of the 1870s and 1880s did have as one of their major concerns the use of comparative anatomical and embryological evidence to reconstruct evolutionary relationships. Unlike some commentators, she also acknowledges that this project had an effect on the kind of work they did. But the influence of evolutionism was temporary, and by the 1890s the more innovative morphologists had moved on to become more interested in the actual process of individual development—from which position some moved even further to become the

24. See Rainger, *An Agenda for Antiquity.*
25. Ernst Mayr, "The Role of Systematics in Biology."

founders of genetics. The advent of evolutionism was thus a relatively temporary episode in the history of this research tradition. Ruse goes even
further; he accepts the claims of earlier historians of morphology such as
E. S. Russell that the impact of evolutionism on morphology was trivial at
the conceptual level.[26] But he goes beyond this to argue that morphologists
such as T. H. Huxley and E. Ray Lankester actively excluded evolutionary
considerations from most of their scientific work. Evolutionism was popular science to be discussed only in the same breath as ideas of social progress. Hard scientific morphology was descriptive and eschewed theoretical
concerns, especially problems of ancestry. While acknowledging that morphology was the centerpiece of the new scientific biology that Huxley and
his followers introduced into the university system, Ruse argues that there
was little evolutionism involved. He concedes that in Germany the research schools founded by Ernst Haeckel and Carl Gegenbaur were genuinely concerned with evolutionary questions. But he regards their work as
second-rate and points out that in any case morphology collapsed within
the German educational system for reasons explored in detail by Lynn
Nyhart.[27] For Ruse, Maienschein's claim that American morphologists
soon turned their backs on evolutionism is a sign of how trivial the theory's impact was.

The impact of evolutionism on morphological techniques will be explored in the following chapters. More significant is Ruse's claim that Huxley, despite being Darwin's champion, refused to allow evolutionary considerations to intrude into his scientific work or his teaching. To support
this latter point, Ruse cites the absence of evolutionary questions from
university examinations and the reminiscences of Huxley's students. Now
it has to be conceded that much morphological work was purely descriptive in character. But the whole point of morphology was to understand
the relationships between different organic structures, and it is precisely
the understanding of these relationships that was transformed by evolutionism. The term *evolution* may not occur in the discussion of an organism's significance, but any trained reader would know that the question
of relationships was to be understood in terms of evolutionary origins.
Huxley also worked on fossils, where evolutionary problems were even
more apparent.

As evidence of the extent to which morphological research was per-
vaded by the question of evolutionary origins, I quote the invertebrate
specialist Henry M. Bernard, trained in Germany but now, in the 1890s,
working at the Huxley Laboratory of Imperial College, London: "[M]ost

26. Russell, *Form and Function*, p. 247.

27. Lynn Nyhart, *Biology Takes Form: Animal Morphology and the German Universities,
1800–1900*.

modern works dealing with the morphology of any Arthropod form contain discussions as to the probable affinity of the family described with other members of the group, and its bearing on the ancestry of the Arthropods in general."[28] To say that the term *evolution* does not occur in these discussions, or to note that the actual mechanism of change is not at issue, is to miss the point that the whole question of relationships was to be understood in evolutionary terms.

To confirm that perceptive students realized the evolutionary implications of morphology, we may note H. G. Wells's recollections of the year he spent training under Huxley at the Normal School of Science (1884–85):

> Our chief discipline was a rigorous analysis of vertebrate structure, vertebrate embryology and the succession of vertebrate forms in time. We felt our particular task was the determination of the relationship of groups by the acutest possible criticism of structure. The available fossil evidence was not a tithe of what has been unearthed today; the embryological material also fell far short of contemporary resources; but we had the same excitement of continual discovery or correcting our conclusions, widening our outlook and filling up new patches of the great jig-saw puzzle that the biological student still experiences.[29]

The fact that Wells subsequently incorporated evolutionary themes into many of his science-fiction stories suggests that he was well aware of the underlying message of the biology he was taught. The same was almost certainly true for any of the zoological courses being taught by E. Ray Lankester, E. S. Goodrich, or their American and European counterparts. Evolutionism as understood by a modern Darwinian may not have been mentioned, but everyone knew that the search for structural relationships was now to be understood in terms of the history of life on earth.

Huxley, Lankester, and the others would not have defined themselves as professional evolutionists. They were zoologists or morphologists, who would have seen evolutionism as merely part of their explanatory scheme rather than as the center of their professional life. They founded no Society for the Study of Evolution, no *Journal of Evolutionary Biology,* and no university departments of evolutionary studies. The professional locations of these biologists are the subject of the concluding section of this chapter. But just because biologists do not define themselves by their acceptance of evolutionism does not mean that the theory had no effect on their work. Evolutionary morphology may not have had the same revolutionary impact as experimental disciplines like genetics, but it shaped a whole gen-

28. H. M. Bernard, "The Comparative Morphology of the Galeodidae," p. 305.
29. H. G. Wells, *An Experiment in Autobiography,* II, pp. 200–201.

eration of detailed scientific work. Hindsight tells us that the techniques it brought to bear on the question of origins were to prove less than adequate to solve some of the more fundamental problems involved. Problems such as the origins of the major animal phyla are not always resolved even when tackled by the techniques of modern molecular biology. But the evolutionary morphologists set out in good faith to reconstruct the genealogical tree of life, and they cannot altogether be blamed if it turned out that the real world was more complex than they had hoped. Evolutionists such as Lankester also played a role in defending the Darwinian view of evolution when it was under fire, and the next generation helped to lay some of the foundations for the modern evolutionary synthesis.

If pure morphology was unable to solve some problems, it was by no means powerless to throw light on others. Some debates, such as the one centered on the question of arthropod polyphyly, were rekindled in the mid-twentieth century after decades of relative neglect (chapter 3). Other problems continued to attract the attention of systematists, even if the laboratory-based anatomists and embryologists had lost interest. Morphologists also realized that they could turn to the ever-growing fund of fossil evidence to clear up disputes concerning at least the later phases in the history of life on earth. The paleontologists themselves remained steadily active as fossils flooded in from increasingly remote localities, and as links with geology provided evidence of the climates and environments in which many extinct animals had lived. Biogeographers studied collections of plants and animals from all over the world and sought to explain how the flora and fauna of particular areas had been built up through migration and local evolution. They too were able to use the increasingly more substantial fossil record as a clue to the past.

Systematists, paleontologists, and biogeographers were generally museum workers rather than academics; they were interested in evolutionary issues because these arose naturally out of their efforts to explain the data they worked with. Their professional work was defined by description, taxonomy, and (for the paleontolgists) stratigraphy, but again this did not prevent them from taking evolutionary questions very seriously indeed. Far from declining in significance during the early twentieth century, as Rainger and Mayr imply, these areas kept up a steady level of activity that continued to solve increasingly complex questions about the history of life on earth. The impression that they declined is an illusion created by the relative expansion of experimental biology within the university system, especially in America. In absolute terms, the level of activity by museum workers was maintained, and their ability to tackle evolutionary questions remained intact. The chapters below should serve as an adequate confirmation of this claim. If evolutionary morphology declined in the early years of the twentieth century, systematists, paleontologists, and biogeographers continued to debate the network of questions established by their predecessors and to come up with important new insights. The fact that field naturalists such as Ernst Mayr and paleontologists such as George Gaylord Simpson were able to contribute to the emergence of the modern evolutionary synthesis was a tribute to decades of steady activity on evolutionary problems which had paved the way to recognition of a more 'Darwinian' way of understanding the history of life on earth.

Evolutionism did not provide the basis for a professional biologist's identity in the late nineteenth century, but it should not thereby be excluded from the attention of historians of the Darwinian revolution, or of biology itself. The reconstruction of life's ancestry was so extensive a project that it necessarily affected a wide range of existing disciplines. Here was a theme so broad that it could not be made the basis of a single research tradition: it had consequences for any biologist who wanted to know why living things are structured, related, and distributed as they are. Specialists in many different areas were forced to interact because they were all concerned with similar questions of origins. Morphologists studied comparative anatomy and embryology with a view to establishing relationships—but they could not afford to ignore fossil and even biogeographical evidence when it related to the origins of the groups they established. Systematists wanted to classify the species in their collections, but in some cases they had to appeal to comparative anatomy to establish true relationships. Paleontologists needed morphology to help reconstruct extinct animals, and some used embryology as a guide to evolutionary trends. But they were also aware of the changing geographical distribution of groups through time, and of geological evidence for past climates. Biogeographers needed morphology to be certain of the identification of species from remote locations, and also looked at the fossil record to compare the past distribution of organisms with modern faunal groupings.

I argue that evolutionism should be retained as a historiographical theme precisely because it served to connect a wide range of existing— and some newly emerging—disciplines. Research traditions may have their own internal coherence defined by shared techniques and professional interests, but they do not exist in isolation, and many important developments in science occur because of cross-fertilization between them. If we want to understand the interactions, both conceptual and professional, among research traditions in biology at the turn of the century, the network of debates on reconstructing life's ancestry serves as a valuable tool.

Perhaps the clearest evidence that scientists saw phylogeny as a central theme is the increasing number of popular surveys they wrote to outline the history of life on earth. Making use not only of new evidence, but also new techniques for illustrating books with photographs of fossils, artists' reconstructions of extinct animals, and so forth, they produced a new genre of literature that has a distinctly modern appearance. Darwin distrusted Chambers's *Vestiges of Creation* partly because he wanted to deflect attention away from such broad surveys—yet one effect of the Darwinian revolution was to legitimize scientists' efforts to reconstruct the evolution of life on earth. If pioneering efforts such as Haeckel's *History of Creation* look primitive by modern standards, popular surveys by early twentieth-century biologists such as Osborn would look quite familiar to a modern nonspecialist reader. The fact that many biologists thought it appropriate to write such surveys suggests that they took their evolutionism seriously.

THE PROFESSIONAL FRAMEWORK

Who, then, were the members of this first generation of evolutionary biologists? In the 1860s and 1870s it was still possible for wealthy 'amateurs' such as Darwin and Sir John Lubbock to make important contributions. Others existed at the margins of organized science: A. R. Wallace pieced together a living from writing, supplemented later by a pension obtained for him by Darwin's influence. Fritz Müller had to leave Europe after the revolutions of 1848, and eked out an existence from schoolteaching in South America. Yet Wallace and Müller both had a major impact on the thinking of biologists concerned with reconstructing the history of life on earth. We still cannot rule out the impact of these 'amateurs' during the early part of the post-Darwinian era. The mammal-expert Richard Lydekker, for instance, had no official position, but wrote many descriptions of specimens for the British Museum (Natural History) in the 1890s. Increasingly, however, the important work was being done by a new generation of biologists who were 'professionals' in the literal sense of that term: they gained their livelihood from their science by working in museums, geological surveys, or universities.

The rate of expansion in professional opportunities was painfully slow at first. Even T. H. Huxley found it difficult to get his foot on the first rung of the ladder of professional appointments in the 1850s. In European countries such as Britain and Germany, the expansion of the educational system was comparatively slow—although Germany certainly had a head start and was seen as the leading scientific nation in the late nineteenth century. Huxley and his followers worked tirelessly to encourage the expansion of British science so that it would catch up with Germany's, and they were equally tireless in their efforts to gain control of the slowly expanding network of professional influence and patronage. The staff of the natural history museums expanded slowly, but by far the greatest source of new opportunities came from the new universities and colleges, both in Britain and her empire. In America, there was an even greater explosion in the university system in the last decades of the nineteenth century, although this tended to benefit the new experimental sciences rather than the more traditional areas where reconstructing the history of life was an active concern.

Although opportunities for professional biologists increased, the system through which they disseminated their results remained fairly traditional. Many of the societies and journals in which results were described and controversies fought out were the long-established institutions dating from the first half of the century or even earlier. To use Britain as an example, papers on topics relevant to the reconstruction of life's ancestry appeared in the *Annals and Magazine of Natural History,* the *Journal of the Linnean Society,* and the *Proceedings of the Zoological Society.* Fossils were described and interpreted in the *Quarterly Journal of the Geological Society of London.* Some substantial contributions were published in the *Philosophical Transactions of the Royal Society.* Since much zoological and paleontological work was descriptive, a new kind of journal was not needed, the evolutionary commentary often appearing only in the conclusion of an otherwise traditional essay. Tracing the debates is rendered more complicated precisely because the relevant papers are scattered through a wide range of periodicals devoted primarily to the publication of descriptive material. Yet we should not thereby conclude that evolutionism had no impact; the new generation of evolutionists were often employed to describe new species, new fossils, and new observations on the internal structure of organisms. It was their interpretation of the observations that was being gradually transformed by the decision to evaluate their findings in evolutionary terms.

The morphologists did, at least, have their own organ, the *Quarterly Journal of Microscopical Science,* founded by Edwin Lankester and edited subsequently by his son, E. Ray Lankester, and then by the latter's disciple, E. S. Goodrich. Charles Otis Whitman founded the American *Journal of Morphology* in 1887. Although these journals were devoted to the new laboratory-based morphology, they too published articles that were mainly descriptive, the evolutionary implications coming out only when the significance of new observations was discussed. In Britain, Huxley helped to found the new semipopular journal *Nature* in 1869, at least in part as a vehicle for publishing shorter reports that might more directly address theoretical issues.

Huxley and his younger colleagues presented morphology and physiology as the twin pillars of a new biology that would form the basis for a program of scientific education suited to a world power in the age of modern industry.[30] In a sense, they deliberately wanted to separate their laboratory-based science from the image of the old-fashioned field naturalist. To some extent this could be sustained through the traditional link between comparative anatomy and medicine. Anatomists had engaged in evolutionary debates during the pre-Darwinian period, and many retained an interest in the relationships between different organic forms. University anatomists often taught medical students, and some used evolution to provide a conceptual framework for their surveys of the animal kingdom. A few anatomists teaching in the medical schools made contributions to phylogenetic debates.

The preference for laboratory-based biology was also noticable in the new research universities that were founded in America in the later decades of the century. But we should be careful not to accept the clear-cut distinction between field and laboratory work at its face value. Museum-based naturalists could do morphology as part of their taxonomic work just as easily as the new generation of university lecturers. Some individuals moved between the two areas: Lankester was professor of zoology at University College, London, and then at Oxford, but became a highly controversial director of the British Museum (Natural History) at the end of his career.[31] Both museums and universities continued to send expeditions out into the field, and the old individual explorer and collector was supplanted by organized expeditions aimed at producing new discoveries in areas of known scientific value. Although some museum and university workers never went out into the field, others did participate in this kind of activity, thereby bridging the gap between the new and the old approaches. Henry Moseley, who preceded Lankester at Oxford, went on the expedition of H. M. S. *Challenger* and sent back important reports on newly discovered organisms.[32] Museum paleontologists such as William Diller Matthew went on annual field trips.[33]

The opportunities for exploration increased because governments, museums, and universities were now willing to fund expeditions. Science was being incorporated into the network of imperial power that the Western nations were establishing over the globe. Scientific activity in collect-

30. See Joseph A. Caron, " 'Biology' in the Life Sciences." The most complete account of the reorganization of science in Britain is D. S. L. Cardwell, *The Organization of Science in England*, although this concentrates mainly on the physical sciences.

31. See Joseph Lester, *E. Ray Lankester and the Making of Modern British Biology*.

32. On Moseley's career, see the biographical memoir in H. N. Moseley, *Notes by a Naturalist*, pp. v–xvi. More generally, see Roy MacLeod, "Embryology and Empire."

33. For details of Matthew's field trips, see Colbert, *William Diller Matthew*. On Matthew's relationship to Osborn and the latter's support for—but very superficial participation in—fieldwork, see Rainger, *An Agenda for Antiquity*.

ing and describing specimens symbolized Europeans' domination of the wild and primitive parts of the earth—and could also have direct value in showing how the resources of conquered areas could be better exploited. The four-year voyage of H. M. S. *Challenger* (1872–1876) set the scene for nationally financed voyages of exploration. The British government funded this expedition under the leadership of the Edinburgh naturalist John Murray. It returned with a wealth of new species, many of which were distributed to experts around the world for description.[34] Other countries followed suit, and indeed began to eclipse the British in this area of activity.[35] Governments also funded geological surveys which provided for the discovery and description of new fossils. The reduction in the U.S. Geological Survey's willingness to support fossil-hunting in the West had major consequences for American paleontology in the late nineteenth century.[36] But other surveys did not necessarily suffer the same fate: the Geological Survey of India, for instance, supported major discoveries of mammal fossils in the Sewalik hills.

At the same time, natural history museums and universities began to finance expeditions of their own to remote areas of the world, where important discoveries were made. At first these were centered on individual scientists, as when W. H. Caldwell went to Australia to study the reproduction of the platypus with a view to determining its relationship to the placental mammals and reptiles (see chap. 6). Soon much larger expeditions were setting out. The Natural History Museum in London sent out expeditions to East Africa and other parts of the British empire during Lankester's tenure as director. In the 1880s the British Association for the Advancement of Science began to set up committees supporting biogeographical studies, including surveys of the faunas of remote islands. In seeking support for field trips run by the American Museum of Natural History, Henry Fairfield Osborn complained that three separate expeditions were setting out from England in 1895 alone.[37] J. P. Morgan financed the Princeton expeditions to Patagonia (1896–99), which gener-

34. For a popular account of the *Challenger* expedition, see Eric Linklater, *The Voyage of the Challenger;* see also Moseley, *Notes by a Naturalist.* The specialist publications arising from the voyage were substantial; for a survey, see John Murray, *Report of the Scientific Results of the Voyage of H. M. S. Challenger . . . A Summary of the Scientific Reports.*

35. On later nineteenth-century developments in oceanography, see Eric L. Mills, *Biological Oceanography: An Early History.*

36. On the collapse of the U.S. Geological Survey's support for paleontology, see Rainger, *An Agenda for Antiquity.*

37. H. F. Osborn, Report of the Department of Vertebrate Paleontology, American Museum of Natural History, 1895, quoted in Rainger, *An Agenda for Antiquity,* p. 93. On the Princeton expeditions, see W. B. Scott, ed., *Reports of the Princeton University Expeditions to Patagonia, 1896–1899,* especially the survey by J. B. Hatcher which forms vol. 1.

FIGURE 1.1   Anton Dohrn working at his microscope at the Stazione Zoologica, Naples, in 1889. Reproduced with the permission of the archives, Stazione Zoologica "Anton Dorhn," Naples.

ated masses of fossils designed to test theories on the origin of South American mammals. Admittedly, few universities could support this level of activity without donations from wealthy patrons, but that support was forthcoming and helped to produce the evidence upon which paleontologists and biogeographers relied.

Another source of research opportunities was offered by the network of marine biological stations opened up in the late nineteenth century. Anton Dohrn founded the first of these at Naples in 1872, with support from the Prussian government (see fig. 1.1).[38] The station gave Dorhn himself a professional base and offered temporary research opportunities for many academic biologists, often at an early stage in their careers. It also provided a journal and a monograph series for the publication of results. Lankester had been with Dohrn at an early stage in the station's history; later

38. On Dohrn's career, see Theodor Heuss, *Anton Dohrn: A Life for Science*. On the history of the Naples Zoological Station, see the special supplement of the *Biological Bulletin*, Christiane Groeben et al., "The Naples Zoological Station and the Marine Biological Laboratory: One Hundred Years of Biology."

FIGURE 1.2   Henry Fairfield Osborn. Negative No. 36106 (photo by J. Kirschner). Courtesy Department of Library Services, American Museum of Natural History.

he led the campaign which founded the Marine Biological Association's laboratory at Plymouth in 1888.[39] This had a small staff of professional biologists. In America, the Marine Biological Laboratory was founded at Woods Hole in the same year.[40]

The natural history museums provided a small but slowly expanding source of employment for professional biologists. Precisely because the total staff in any one museum was small, these institutions were often strongly influenced by their directors. Louis Agassiz kept evolutionism out of Harvard's Museum of Comparative Zoology until the 1870s, and his son, Alexander, was skeptical about the possibility of reconstructing the history of life from morphological evidence.[41] Even so, some significant evolutionary work was done, including Carl Eigenmann's studies of evolution in fish during the 1920s. Osborn used his links with New York's social elite to boost funding for the Department of Vertebrate Paleon-

39. See the anonymous "The History of the Foundation of the Marine Biological Association," and E. S. Russell, "The Plymouth Laboratory of the Marine Biological Association."

40. On Woods Hole, see Groeben et al., "The Naples Zoological Station and the Marine Biological Laboratory"; Keith R. Benson, "From Museum Research to Laboratory Research"; and Philip J. Pauly, "Summer Resort and Scientific Discipline."

41. On the Museum of Comparative Zoology, see Mary P. Winsor, *Reading the Shape of Nature*.

tology at the American Museum of Natural History.[42] His subordinates, including W. D. Matthew and W. K. Gregory, were able to make use of the resources provided by Osborn's manipulation of the patronage network, even though they disagreed with his views on evolution. Private sponsorship also created the University of California's Museum of Vertebrate Zoology in 1908.[43]

Ronald Rainger's study of Osborn's career suggests that the American Museum's work was sustained by his social contacts at a time when the real driving force in biology was switching to the universities. But American institutions drew an unusually large proportion of their funding from private sources and were thus vulnerable to the changing interests of individual donors. There is, in any case, evidence that American museums were by no means the professional backwaters implied by this account. European museums were largely state-financed as symbols of imperial power; their total funding may have been less than that available to the larger American museums, but it was at least more secure. While museums did not enjoy any massive expansion of resources in the early twentieth century, they were able to sustain a steady level of activity, much of which could be related to evolutionary questions. In addition, the various countries of the British empire themselves set up universities and natural history museums, which provided a steady source of employment for biologists trained in England.[44] Thus Arthur Willey, trained by E. Ray Lankester, was both director of the Colombo Museum in Ceylon (Sri Lanka) and professor of zoology at McGill University, Montreal (he also taught for a while at Columbia University in New York).

The development of the British Museum (Natural History) in London gives a good indication both of the restrictions and opportunities that such institutions offered for evolutionary biologists.[45] Richard Owen was appointed superintendant in 1856, and was largely responsible for the move to the present buildings in South Kensington in 1880. Owen was an opponent of the Darwinians—indeed Huxley tried to block the building of a separate natural history museum because it increased Owen's power.[46] But in fact Owen suported the general idea of evolution and used it in his

42. On the American Museum of Natural History, see Rainger, *An Agenda for Antiquity.*

43. See James R. Griesemer and Elihu M. Gerson, "Collaboration at the Museum of Vertebrate Zoology."

44. On colonial natural history museums, see Susan Sheets-Pyenson, *Cathedrals of Science.*

45. See Albert E. Gunther, *A Century of Zoology at the British Museum;* and William T. Stearn, *The Natural History Museum at South Kensington.* Also of interest is P. Chalmers Mitchell, *Centenary History of the Zoological Society of London.*

46. On the controversy surrounding the creation of a separate Natural History Museum, see Nicolaas Rupke, "The Road to Albertopolis."

own studies of fossils. If Darwinism made only limited inroads into the museum at first, it was because the keeper of zoology, Albert Gunther, was simply indifferent to it. Even so, some of Gunther's staff (which numbered seven by the 1890s) managed to publish on evolutionary questions, especially the arachnid specialist R. I. Pocock. William Henry Flower, who took over the directorship after Owen's death, was certainly not hostile to evolutionism. When Lankester succeeded Flower in 1895, the museum came under the directorship of one of the country's leading evolutionists. The primary concern of the museum's specialists was always the description and classification of new specimens. But they were always at liberty to apply their expertise to evolutionary questions, and by the end of the century would have been actively encouraged to do so. It is also worth noting that some of their work had economic value. Thus Lankester supervised the first systematic study of the mosquitoes by F. V. Theobald just when it was being recognized that one particular species was the carrier of malaria.[47]

Museums were not static organizations determined to preserve the status quo. On the contrary, there was a good deal of concern over the best way to ensure that a museum served its scientific function in the best possible way. W. H. Flower wrote extensively on the nature and purpose of museum work, and Lankester introduced many (often controversial) innovations during his tenure at the Natural History Museum.[48] In America, George Brown Goode, Joseph Grinnell, and others debated how best to organize a museum so that it served as the basis for serious research.[49] The image of Osborn as the outdated patriarch dominating an increasingly marginalized professional locus does not do justice to the dynamism of museum work at the time. To be sure, much of the work under discussion concerned questions of systematics, biogeography, and ecology, but these often required input from evolutionary topics.

This being said, it would be wrong to deny that the greatest expansion of opportunities for trained biologists occurred in the university system. By the mid-nineteenth century Germany already had a network of universities employing trained researchers as teachers. Expansion of the system was steady rather than spectacular, but even so, British biologists such as Lankester viewed it with envy and saw it as a model that would have to be emulated if Britain was to preseve its role as a world power. We have already noted how Huxley and his protégés, including Lankester, were

47. See F. V. Theobald, *A Monograph on the Culcidae, or Mosquitoes.*

48. See W. H. Flower, *Essays on Museums,* and on Lankester's work at the museum, see Joseph Lester, *E. Ray Lankester,* chap. 11.

49. See George Brown Goode, "The Principles of Museum Administration," and Joseph Grinnell, "The Uses and Methods of a Research Museum." More generally see Griesemer and Gerson, "Collaboration at the Museum of Vertebrate Zoology."

gradually able to convince the government that the system of scientific education would have to be reformed and expanded, and were able to incorporate biology into the system. In America there was an even greater expansion of science in the university system toward the end of the century. The founding of research-based universities such as Johns Hopkins and the University of Chicago allowed America quite rapidly to assume a major position in the world of international science, and biology was certainly part of this process.

Because American historians of science have been particularly active in this area, we know more about the expansion of biology here than anywhere else.[50] At Johns Hopkins, William Keith Brooks trained a whole generation of biologists who spread out into the rapidly expanding university network. Brooks himself was certainly interested in evolution, and many of his students began their careers working in evolutionary morphology. As Maienschein and others have shown, however, some of the most active graduates of the Hopkins school lost interest in the evolutionary implications of morphology and led the way towards a more experimental form of biology. T. H. Morgan in particular went on to become one of the founding fathers of genetics. From the American historians, we get a strong sense of evolutionary issues being marginalized as they were excluded from the universities in favor of experimental biology, and increasingly restricted to the less active museum network.

I have already suggested that we should be careful not to get swept away by this image of evolutionism in decline. Evolutionary morphology became less active everywhere in the early decades of the twentieth century, but systematics, paleontology, and biogeographical studies remained active both in the museums and in some universities. Geology departments provided additional centers of interest in the history of life on earth. Thomas C. Chamberlin at Chicago certainly incorporated evolutionary topics into both his research and his teaching. Charles Schuchert brought the history of life on earth into his teaching of historical geology at Yale's Sheffield Scientific School. Admittedly, paleontologists were often at a disadvantage because their work fell between the two stools of biology and geology. Alfred Sherwood Romer faced this problem at Chicago, while Matthew, after his move to the University of California, compared the paleontologist to a man with two wives, "perfectly happy with either, were t'other dear charmer away."[51] Yet both Romer and Matthew continued to do important work in reconstructing ancestries from the fossil record.

50. See, for instance, Rainger, Benson, and Maienschein, eds., *The American Development of Biology;* and Maienschein, *Transforming Traditions in American Biology.*
51. Matthew, *Outline and General Principles of the History of Life,* p. 8. On Romer, see Ronald Rainger, "Biology, Geology, or Neither, or Both: Vertebrate Paleontology at the University of Chicago, 1892–1950."

Evolutionism was also seen as an important conceptual framework by some anatomy teachers in the medical schools. Texts such as Herbert V. Neal and Herbert W. Rand's *Comparative Anatomy* and Harris H. Wilder's *History of the Human Body* show that evolutionary problems were seen as a legitimate topic for medical students. The network of debate on evolutionary issues may seem fragmentary when compared to the specialized discipline of genetics, but the sheer breadth of subjects which took an interest in evolution shows that it was never marginalized completely.

Moreover, we must be careful not to take the American experience as typical of what happened elsewhere. The rapid expansion of the American university system provided a perfect environment for the establishment of new disciplines based on laboratory work. This was not the case in Europe, where the pioneers of new research programs such as genetics found it more difficult to create professional niches for themselves. Thus William Bateson struggled hard to establish genetics at Cambridge. He was eventually successful in the sense that he got a chair (professorship) in genetics in 1907, but he left soon afterwards for the privately funded John Innes Horticultural Institution because he still could not get enough research funding at Cambridge.[52] Genetics did not establish itself as a separate discipline in France or Germany during the early decades of the century, and as a consequence there was nothing resembling the clear-cut hereditarianism of Morgan's classical genetics in those countries.[53] Pragmatic German biologists wanted to specialize in the new experimental studies, but others retained the traditional respect for a comprehensive view of nature which included evolutionary problems. Active schools of research on evolutionary questions emerged in other countries such as Russia, where Nikolaevitch Svertsov did important work at the University of Moscow in the 1920s.[54]

In the middle decades of the nineteenth century, English-speaking biologists had looked to Germany for a model of how scientific education and research should develop. The Germans pioneered the system in which the university was a center for both research and teaching, and there was a steady increase in the number of universities with chairs (professorships) in biological subjects. It was a very rigid system in the sense that the professor had considerable powers, and younger scientists often had a hard struggle to reach professorial rank. Active professors such as Carl Gegen-

52. See Robert C. Olby, "Scientists and Bureaucrats in the Establishment of the John Innes Horticultural Institution under William Bateson."

53. See R. M. Burian, J. Gayon, and D. Zallen, "The Singular Fate of Genetics in the History of French Biology"; and on German genetics, see Jonathan Harwood, *Styles of Scientific Thought.*

54. See Mark B. Adams, "Severtsov and Schmalhausen: Russian Morphology and the Evolutionary Synthesis."

baur and Ernst Haeckel could attract students and create centers of research based on new ideas such as evolutionism. The use of morphology to trace evolutionary ancestries became a central plank in the program of German academic biology in the 1870s and 1880s.

There were problems, however, as Lynn Nyhart's analysis of German morphology has shown.[55] Morphology was divided between professors of zoology, located in philosophy faculties, and professors of anatomy, which belonged to medicine. When the two got on well, as did Haeckel and Gegenbaur at Jena, much was achieved. But there was often a conflict of interest, and when Gegenbaur moved to Heidelberg in 1873, he found it much less easy to function effectively. A conflict arose within morphology between comparative anatomists and embryologists. Anton Dohrn and Gegenbaur became bitter enemies following a disagreement over the relative status of these two areas in morphological research. By the end of the century, embryology was increasingly seen as a separate field, paving the way for the elimination of evolutionary research in favor of the study of individual development for its own sake. German paleontologists were largely confined to geology departments, where it was more difficult for them to interact with biologists.

Evolutionary morphology thus underwent a cycle of expansion and decline in the German universities. We should remember, however, that some aspects of biology in the German-speaking world are not covered in Nyhart's account. Thus Carl Claus headed an influential school of invertebrate morphology at Vienna until his retirement in 1896. He trained a new generation of invertebrate zoologists, including Berthold Hatschek and Karl Heider, and also taught the young Sigmund Freud.[56] Important work was still being done by vertebrate morphologists in the early twentieth century, including Ernst Gaupp's analysis of the homologies between the bones of the reptilian jaw and mammalian ear (see chap. 6). The real collapse in German influence came with World War I, as the financial difficulties following Germany's defeat made it impossible for many scientists to function adequately. The isolation that followed the war allowed some German biologists and paleontologists to rediscover the idealist philosophies of an earlier period, promoting theories of saltative and internally directed evolution of the kind that was becoming unfashionable in the English-speaking world.

The situation in Britain offers a useful model, falling somewhere in between Germany and America. In the middle decades of the century, the traditional universities of Oxford and Cambridge were still largely mori-

---

55. Nyhart, *Biology Takes Form;* on the careers of the Hertwig brothers within the German system, see also Paul Weindling, *Darwinism and Social Darwinism in Imperial Germany.*
56. On Claus, see Lucille B. Ritvo, *Darwin's Influence on Freud,* chap. 9.

FIGURE 1.3   Cartoon of E. Ray Lankester from *Vanity Fair,* supplement, 12 January 1905. The caption reads, "His religion is the worship of all sorts of winged and finny freaks."

bund. T. H. Huxley struggled to get a modern system of biological education under way in London, and pioneered an influential summer school in South Kensington in the 1870s, using Lankester and Michael Foster as assistants. This initiative led indirectly to the creation of the Normal School of Science (where H. G. Wells studied) and the Imperial College of Science and Technology. Lankester was soon elected to the chair of zoology at University College, London, succeeding the venerable Robert Grant, whose radical transformism had shocked the medical establishment of the 1830s and 1840s. Here Lankester built a major research school, training a host of biologists who spread out to occupy positions throughout the British empire. His successor at University College was the paleontologist D. M. S. Watson.

Lankester began his career as a fellow of Exeter College, Oxford, where he struggled to create a place for science against the barriers of tradition. The Linacre professor of anatomy and physiology was George Rolleston, who had been supported by Huxley and who made some efforts to modernize the system. But Rolleston was not active enough for Lankester, who criticized him severely before departing for London. When Rolleston died in 1881 his chair was divided into two, and the responsibility for anatomy was taken up by Henry Moseley. Moseley was a close friend of Lankester and a zoologist of the modern type, but he soon became ill and died shortly afterwards. When Lankester returned to occupy the Linacre chair, he still found himself struggling against vested interests suspicious of science. Oxford was thus slow to generate a research school on the same scale as London and Cambridge, although Lankester and his successors, W. F. R. Weldon and G. C. Bourne, were more successful. Lankester's student, Edwin S. Goodrich, worked under Bourne and had a chair of comparative embryology created for him in 1919. He succeeded Bourne to the Linacre chair in 1921 and was still teaching morphology in the early 1940s. Like Lankester, Goodrich retained an interest in broader evolutionary questions, and his generally Darwinian viewpoint helped to shape the views of biologists such as Julian Huxley and Gavin De Beer, who played a major role in the creation of the evolutionary synthesis.[57]

The professor of zoology at Cambridge in Darwin's time was the ornithologist Alfred Newton.[58] Newton has been largely ignored by historians, but he was by no means a cipher—although he hated lecturing and was hardly likely to create a research school. He was one of the first zoologists to support Darwin openly, and later did important work in biogeography.[59] More important, he made no effort to block the appointment of one of Huxley's disciples, Francis Balfour, to a lectureship (eventually a readership and then a chair) in morphology.[60] Balfour was both a leading embryologist and a man who inspired a new generation of students. His death in a climbing accident in 1882 was widely regarded as a disaster for British biology. He was succeeded by one of his students, Adam Sedgwick, who later moved to the chair of zoology at Imperial College. One of

57. See Janet Howarth, "Science Education in Late-Victorian Oxford: A Curious Case of Failure?" On Rolleston, see the "Introduction" to vol. 1 of his *Scientific Papers and Addresses*. On Lankester at Oxford, see Lester, *E. Ray Lankester,* chaps. 5 and 10. On Moseley, see the biographical introduction to his *Notes by a Naturalist.* On Goodrich's career, see the obituary by A. C. Hardy; the fact that he was still teaching biology through morphology in the 1940s is confirmed by one of his students (A. J. Cain, personal communication). On Goodrich's influence, see Steven James Waisbren, "The Importance of Morphology in the Evolutionary Synthesis."

58. See A. F. R. Wollaston, *Life of Alfred Newton.*

59. See I. Bernard Cohen, "Three Notes on the Reception of Darwin's Ideas on Natural Selection."

60. On Balfour's career, see the biographical introduction to *The Works of Francis Maitland Balfour,* I, pp. 1–24. Balfour worked closely with Michael Foster, who transformed the teaching of physiology at Cambridge; see Gerald Geison, *Michael Foster and the Cambridge School of Physiology.*

Sedgwick's students was E. W. MacBride, who took the chair of zoology at McGill University in Montreal before returning to Britain as Sedgwick's assistant, and later successor, at Imperial College.[61]

London, Cambridge, and later Oxford thus developed as important centers of research, training zoologists who moved out to take up positions in the universities now opening up throughout Britain and her empire. Although many of the morphologists followed Sedgwick in turning away from evolutionary questions, others, including MacBride and Goodrich, did not. Important positions were still occupied by morphologists, or by paleontologists such as Watson, well into the twentieth century. The British university system did not enjoy the same massive expansion as the American. It did continue to expand at a slower rate, however, and precisely because the rate was slower, it was more difficult for new fields like genetics to take over existing departments or create new specialist departments.

In 1909 the British scientific community marked the fiftieth anniversary of the publication of Darwin's *Origin of Species*. Stephen Jay Gould has argued that paleontology was already being marginalized within the framework of what was perceived as Darwinian research. There were, he notes, only two "insipid" articles on the theme, by W. B. Scott on the record for fossil animals, and D. H. Scott on the plants.[62] He might have added that the article on embryology was by Sedgwick, who insisted on the outdated character of phylogenetic research based on this line of evidence.[63] The focus of evolutionary research seemed to have passed to the experimental disciplines represented by Bateson and De Vries. Here we see an early phase in the process by which modern historians were encouraged to see the history of evolutionism largely in terms of debates over the mechanism of change. Gould is surely correct to suggest that phylogenetic research had been an integral component of the original Darwinian debates. But the elimination of this branch of evolutionism from the development of twentieth-century Darwinism may have been more apparent than real. Evolutionary morphology did indeed decline rapidly in influence, but paleontology and biogeography remained active disciplines in the early twentieth century, even if they could not compete in terms of resources with the new experimental disciplines. It may have been no ac-

---

61. On MacBride's career, see the obituary by W. T. Calman, and Bowler, "E. W. MacBride's Lamarckian Eugenics."

62. See Gould, "Irrelevance, Submission, and Partnership: The Changing Role of Paleontology in Darwin's Three Centennials." See W. B. Scott, "The Palaeontological Record. I. Animals."

63. Sedgwick, "The Influence of Darwin on the Study of Animal Embryology," pp. 173–175.

cident that the contribution on animal paleontolgy in the 1909 anniversary volume was by an American. British paleontologists remained active in the study of phylogenetic topics, but it was the Americans who were most innnovative in developing new approaches to the history of life based on correlations with changes in the physical environment. If, as Gould suggests, paleontology reemerged as an active participant in the Darwinian synthesis of the 1940s, it did so because early twentieth-century paleontologists had pioneered new avenues of research on evolutionary issues. The chapters which follow will demonstrate how themes centered on the reconstruction of life's ancestry continued as an active focus of research despite the rapid expansion of rival fields.

2

THE TREE OF LIFE

For many late nineteenth-century biologists, the great challenge offered by the advent of evolutionism was the boost it gave to their efforts to reconstruct the history of life on earth. We have already noted a group of disciplines—systematics, morphology, and paleontology—that were able to reinterpret their research programs in terms of phylogenetic study. Biologists in all of these areas began to ask new kinds of questions and encounter new kinds of problems. This chapter seeks to outline some of those questions and problems, and will end with an outline of evolutionists' ideas concerning the very earliest phases in the emergence of life on earth. The chapter will thus serve as a guide to the underlying issues that run through the phylogenetic debates forming the first half of this study. Later chapters will reveal how these issues were gradually supplemented by additional questions arising from geologists' studies of past environments and from biogeography.

In the earliest phase of phylogenetic research, the critical questions focused on the use of structural resemblances as a clue to evolutionary relationships. Where fossils were available, they could be arranged into plausible phylogenetic sequences, and 'missing links' could be sought among new discoveries. But the structure of living animals could also be used as a guide to their relationships. The theory of common descent encouraged the hope that biologists would be able to provide a complete reconstruction of the tree of life, determining where each branch of the animal kingdom separated off. Major problems were encountered as biologists began to realize that structural resemblances were not always a reliable guide. Debates erupted over the extent to which separate branches of the tree might independently acquire similar characteristics through parallel evolution. Evolutionists also debated the extent to which the development of each group was shaped by the animals' response to the external environment.

By the middle of the nineteenth century an outline of the history of life had become available from the fossil record. But as Darwin himself had explained in the *Origin of Species,* the record was so incomplete that it offered at best only imperfect evidence for the evolutionary history of most groups. Some important fossil discoveries were made in the 1860s and 1870s, but even so the record was not necessarily seen as the best source of evidence for tracing evolutionary relationships. The situation would change by the end of the century, when the flood of new discoveries began to make paleontology the chief center of attention for reconstructing the history of life. Incorporating fossils into the sequence would pose its own problems, some of which still rear their heads today. But the origins of the major animal phyla lay lost in the mists of pre-Cambrian time, so the fossil record offered little guidance on the most fundamental questions.

It was thus to morphology that many biologists turned in their efforts to reconstruct the tree of life. Darwin had argued that a true system of classification should reflect evolutionary origins: two species were 'related' because they shared a common ancestry at some point in the past. By assessing degrees of relationship, biologists could thus reconstruct the branching-points in the tree of evolution. The tree would lack an absolute time dimension (except where this could be inferred from fossil evidence), but it would link the whole series of living forms into an overall pattern defined by their evolutionary past. The *relative* sequence of the nodes in the tree of life (the points from which branches diverged) could be determined. Morphology, the detailed study of animal form or structure, would provide the information from which the degrees of relationship would be assessed. Morphology thus became the first center of evolutionary biology. The rise and fall of evolutionary morphology constitutes the main theme in the early history of scientific evolutionism, with paleontology gaining influence only as the morphologists began to realize that their science could not provide unambiguous answers to evolutionary questions.

An active science of morphology already existed. Indeed, the study of animal form had made great strides in the early nineteenth century, as both comparative anatomy and embryology were used to throw light on the relationships between the various types of animal structure. Historians such as E. S. Russell and William Coleman have noted the continuity between pre- and post-Darwinian morphology and have implied that the impact of evolutionism was a good deal less than revolutionary. Their position echoes a long-standing tradition in the historiography of German biology represented by Adolph Naef.[1] These writers believed that the old idealist view of 'relationship' could be translated quite easily into the materialist concept of forms related by descent from a common ancestor. In this model of the history of evolutionism, the new theory had only mimimal conceptual impact on what was proclaimed as one of its most important areas of application.

This claim is paralleled in a related area, when it is suggested that evolutionism had little or no real impact on systematic biology. We shall return to this claim in chapter 7, but for the time being it is important to note the similarities and the differences between the responses of systematists and morphologists to the challenge of evolutionism. Taxonomists try to erect a workable system for representing the relationships between species, and some have claimed that it is virtually impossible to use the concept of branching evolution as a guide. They continue to express relationships in terms of degrees of similarity, without worrying about whether the similarities are the products of common descent. The claim that evolution has had no useful effect is, however, based on grounds that are exactly opposed to those used by the morphologists. Instead of arguing that evolutionism simply reinterpreted the old idea of relationship without requiring any significant modification, the critics argue that evolution theory has implictions for systematics that make it quite unusable. When a theory's impact on two integrated areas is dismissed on such divergent grounds, we may be sure that something interesting was really going on. Darwin forced biologists to rethink what they meant by the concept of 'relationship,' and even those who rejected a role for evolutionism had to do so by defining which aspects of the evolutionary process made it so difficult to apply as a guide.

It is important to note that the systematist and the evolutionary morphologist have different goals. Both are interested in the relationships between species, but for different purposes. In some respects, pre-evolutionary morphology was far more directly compatible with the aims of biological taxonomy. Evolution theory had an impact on morphologists' work precisely because it forced them to adopt a new way of understanding what the foundation for a group of related organisms might be. The evolutionary notion of a common ancestor derived from another group is incompatible with the pre-Darwinian view of relationship, which was purely formal in nature; that is, it did not presuppose that the basic character of the group was imposed by its descent from a real animal.

---

1. See E. S. Russell, *Form and Function*, p. 247; William Coleman, "Morphology between Type Concept and Descent Theory"; and Adolf Naef, *Idealistische Morphologie und Systematik*. On the parallel claim that evolution theory had no value in systematics, see Mary P. Winsor, "The Lessons of History."

Systematists had problems with evolutionism precisely because their goals turned out to be in some ways more compatible with a nonevolutionary view of relationships. Common ancestors are a problem for practical taxonomy, even if one believes, in theory, that the species have been generated by a process of evolution.

To begin with morphology, this chapter will argue that the interpretation cited above does not do justice to the morphologists who saw themselves as the spearhead of the new evolutionism. The history of the relationship between morphology and evolutionism has all too often been written by those in whose interest it was to minimize its significance. Morphology began from a standpoint which encouraged the search for timeless unities among animal forms. A group of related organisms was seen as a collection of superficial modifications imposed upon an archetype, an idealized essence of the group's basic character, which existed at the level of pure thought, perhaps in the mind of the Creator. Naef seemed explicitly to endorse a return to this idealist view of organic form, and Russell himself was no friend to Darwinism. For these writers of the early twentieth century, the advent of evolutionism did not affect the development of their science because the only aspects of the theory acceptable to them were those that could be translated into their own world view. This does not mean that they were unable to address any evolutionary topics—indeed we shall see that Naef published contributions to what he regarded as phylogenetic questions. But at the same time, the idealist view of evolution disregarded what many Darwinians regard as the most important issues generated by the need to see transformations as the result of natural processes.

Coleman's position accepts the same view of the relationship between morphology and evolutionism, but possibly for different reasons. In the period following the emergence of modern synthetic Darwinism, the historiographical trend favored a stress upon the mechanism of evolution. Developments in the application of the general concept of evolutionism were of no significance because they did not contribute to the eventual emergence of the genetical theory of natural selection. Here again there was a reluctance to admit that the more general search for natural modes of evolutionary change might affect the kind of questions a morphologist asks about organic relationships.

Darwinism certainly did not create a new science of morphology, but it dramatically transformed the existing science. New questions had to be asked, and new techniques sought by which these questions could be answered. This was perhaps most obvious in the search for relationships between the major animal phyla. For the idealist, the archetype was a timeless essence which could not, by definition, be derived by modifying the essence of another group. Evolutionists had to think about how the common ancestor from which a new group was derived could have been produced by modifying a real species belonging to a previously existing group. Embryological evidence took on a new level of significance when Ernst Haeckel began to argue that individual development recapitulated the course of the species' evolutionary history, thus illustrating these major transformations. New conceptual problems emerged that would have been inconceivable in the old days of idealist morphology. This is especially true for the problem of convergence, where the evolutionists came face to face with the prospect that common descent might not, after all, be the only source of the structural similarities by which living forms seem to be 'related.' The fact that the new techniques were unable to justify the faith initially placed in them helps to explain why evolutionary morphology eventually slipped into the background. But it does not justify the historian ignoring a major episode in the history of biology.

Convergence was equally problematic for those systematists who tried to use evolution as a guide to relationships. The increasingly obvious fact that characters do not map unambiguously onto a treelike pattern made it clear that some similarities are not relationships in the evolutionary sense, but have been independently acquired by several different groups. Those systematists who suspected that there was a great deal of convergent evolution despaired of ever being able to reconstruct the ancestral tree. Modern cladistic taxonomy has pointed out further problems that were seldom realized by the first generation of evolutionists. Precisely because the common ancestor of a group is itself a real species, it has to be fitted into the taxonomic system along with its descendants: but should the common ancestor be counted as a member of the same taxon as its descendants? How would one confirm that a particular species was the actual ancestor of others? The taxonomic status of ancestors has proved to be so problematic that the modern cladists refuse to acknowledge ancestor-descendant relationships in their taxonomies. Their aim is to identify monophyletic groups, that is, groups of species identified by possession of a unique character derived from a common ancestor—but without seeking to identify the ancestor itself. At this level of analysis even the fossils do not help, since a fossil species is just another species to include within the taxonomy. The early evolutionists certainly hoped to find fossils that might be ancestral to living species, and even thought that some ancestral species might have survived almost unchanged through to the present. Later generations of biologists have found that—even if one starts out from the assumption that the evolutionary tree is the source of the relationships we seek—things are not so simple. Some would argue that that in the circumstances it is better to forget all about evolution and simply compare characters.

Those systematists who reject evolution as a guide to taxonomy thus come back to a view of relationship which is more compatible with the idealist view of morphology. The idealist seeks to identify the underlying essence of a group without trying to pin that essence down to a particular species, living or fossil. This approach does not raise awkward questions about how one might recognize an ancestor, and how one might classify it even if one were certain that it was the ancestor. The evolutionary morphologist is less interested in classifying all the members of a group; indeed he or she may cheerfully take a single species as representative of the group in order to determine its derivability from others. The evolutionist seeks to visualize the common ancestor of a group not only to define the starting point from which all the members of that group have radiated, but also to see which previously existing group might have a member that could have transformed itself into the founder of the new one. Evolutionism thus had a profound effect on morphology—even if it turned out to be difficult to apply in the area of systematics.

RELATIONSHIPS REDEFINED

The science of biological taxonomy, which seeks to produce a rational classification of living things, had been created in something like its modern form in the eighteenth century. The Swedish naturalist Carolus Linnaeus established the basis of the modern system of classifying plants and animals and had also founded the binomial system of nomenclature in which each species is identified by a double-barreled name. Species which show a very close degree of similarity are grouped into a genus, and each species is known by its generic name and its individual specific name. Thus the dog is *Canis familiaris,* while the wolf (obviously a close relative) is *Canis lupus.* All the genera of doglike animals are then grouped into the family Canidae, which in turn is grouped along with a number of other flesh-eating families into the order Carnivora. (Not all flesh-eating vertebrates belong to the Carnivora—think of killer whales, for instance.) The Carnivora are one order of the class Mammalia, the class of all those animals which are warm-blooded and suckle their young (the Cetacea, whales and dolphins, are another). The taxonomic hierarchy thus rests on the identification of varying levels of relationship, the most basic or fundamental characteristics defining the higher-level categories.

In the early decades of the nineteenth century, Georges Cuvier put taxonomy on a new footing by making extensive use of comparative anatomy, the study of the internal structure of animals, to analyze their relationships.[2] Many previous attempts to create an overall arrangement

2. On Cuvier's work, see Russell, *Form and Function,* chap. 3; and William Coleman, *Georges Cuvier.* See also Toby Appel, *The Cuvier-Geoffroy Debate.*

of the animal species had been based on the 'chain of being,' the assumption that nature is based on a single hierarchy, or ladder, from the simplest to the most complex. Although linear models continued to resurface even in post-Darwinian evolutionism, Cuvier showed that they are in principle inadequate. The natural system is branching, not linear. He argued that there were four "embranchements" of the animal kingdom, four basic types of structure to which all individual species belong: the vertebrates, molluscs, articulates, and radiates. Each type was, in effect, an underlying ground plan which nature had modified in different ways to produce animals adapted to various ways of life. The relationships used to classify animals depended upon varying degrees of modification to the ground plan of the type. Thus the mammals represented a group of vertebrates defined by being warm-blooded, and so on, the carnivores a group of mammals defined by certain structural peculiarities (teeth, claws, etc.), and so on. At each successive level of the taxonomic hierarchy, the group is defined by an additional set of characters imposed upon those already possessed by virtue of belonging to the more basic categories.

The term *morphology* was coined by J. W. von Goethe to denote the search for underlying similarities between different organic forms. It was, in effect, the science lying behind the study of comparative anatomy. This branch of biology became intimately related to taxonomy, because morphology revealed the degrees of relationship upon which classification would be based. The morphologists of the mid-nineteenth century used the microscope to probe ever deeper into the internal structure of animals in their efforts to identify significant relationships. Karl Ernst von Baer and others studied the development of the individual organism to define its fundamental character. It was von Baer who showed that the most basic characteristics defining the type appear first in the organism's development: the embryo is recognizable as a vertebrate before it can be identified as a mammal, and so on.

In the 1840s Richard Owen clarified the two kinds of relationship known as *homology* and *analogy*. Homologies are fundamental similarities which underly superficial adaptive modifications. Thus the bones of the human arm are homologous with those in the forelimb of any other mammal; each bone in the human arm has an equivalent in the foreleg of the horse, for instance, despite the two limbs being adapted to very different purposes (see also fig. 2.1). The skilled morphologist can recognize such homologies or equivalences in very different animals. The main purpose of training in morphology was to develop the intuitive sense of which relationships are genuine homologies. The most important thing for the morphologist to watch out for was those alternative similarities which Owen called analogies. These were superifical adaptive relationships which were of no value in classification. Thus the fishlike shape of

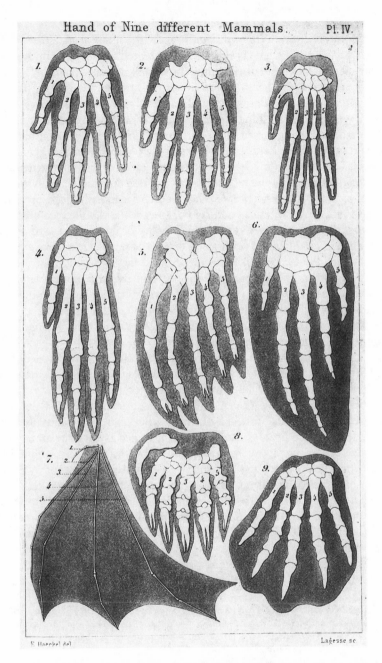

F. Haeckel del   Lagesse sc

FIGURE 2.1 Homologies of the hand in mammals. 1: a man; 2: a gorilla; 3: an orang; 4: a dog; 5: a seal; 6: a porpoise; 7: a bat; 8: a mole; and 9: a duck-bill. From Ernst Haeckel, *The History of Creation* (New York, 1883), II, plate IV.

the dolphin is of no taxonomic significance: the internal structure identifies it as a mammal which is superficially adapted to the same mode of life as the members of that entirely different vertebrate class, the fish.

In the pre-evolutionary world view, it was by no means easy to define why certain relationships were fundamental, and others merely superficial. Owen, who shared the idealist philosophy of many German naturalists, introduced the concept of the vertebrate "archetype" in an attempt to explain why certain kinds of relationship were fundamental.[3] The archetype was an idealized model of the basic structure of the type, stripped by the imagination of all the specialized adaptations possessed by living species of animals. The vertebrate archetype was the most basic, most generalized form of vertebrate structure. For Owen, the significance of the archetype lay in its ability to unify the diversity of forms possessed by real animals. He modified the traditional version of natural theology to argue that God's influence could be seen most clearly in the fact that He had designed nature to a unified plan or pattern. By stripping the adaptive modifications away to reveal the archetypical form of each class, the morphologist was revealing the underlying unity of God's creation.

By the 1850s Owen had already realized that his view of organic relationships could be translated into a model for the history of each class. The fossil record showed that the earliest mammals, for instance, were highly generalized forms, closest to the idealized archetypical mammal. From these radiated out lines of increased specialization leading toward the modern species. For Owen, this pattern of development in the fossil record constituted the unfolding of the divine plan. He eventually came to accept that the process of development might involve the transmutation (evolution) of one form into another, but he never abandoned his belief that the process must be controlled by a divine plan, not by purely material forces. Owen thus came to accept what has been called *theistic evolutionism*—the belief that evolution is controlled by nonphysical forces that can be explained only by an appeal to supernatural guidance.[4] The history of life was not a contingent process, but the unfolding of a preordained plan.

We shall see that Owen made some cogent suggestions about how evolution might have proceeded in groups such as the crocodiles, but he

3. Richard Owen, *On the Archetype and Homologies of the Vertebrate Skeleton* and *On the Nature of Limbs.* See Adrian Desmond, *The Politics of Evolution,* and Nicolaas Rupke, "Richard Owen's Vertebrate Archetype." On German morphology before 1850, see Lynn Nyhart, *Biology Takes Form,* chap. 3.

4. See Bowler, *Fossils and Progress,* chap. 5; and Dov Ospovat, "The Influence of Karl Ernst von Baer's Embryology." On Owen's theistic evolutionism, see Bowler, *The Eclipse of Darwinism,* chap. 3.

retained one fundamental principle of the idealist viewpoint. He was often unwilling to accept that the earliest, most archetypical, member of a group could have been derived by gradual modification from some previously existing type. The main taxa were defined by archetypes, and modifications could take place within each group thus defined, but there could be no links between the fundamental essence of one group and that of another. Some later evolutionists would postulate major saltations to explain how a new type could come into being. For the idealists, nature consisted of a fixed number of groups, each defined by its archetype. The archetypes were eternal components of the universal plan; any changes that took place within the physical manifestations of the archetype (species of living animals) were relatively trivial and predetermined by the possibilities lying latent within the original pattern. The morphologists' job was to identify how many fundamental taxa there were, and define their archetypes. The fact that they could never agree on just how many basic types there are illustrates (to the Darwinian mind) the artificiality of their world view. Nevertheless, we shall see how powerful this position was in limiting many biologists' ability to accept, for instance, that the vertebrate type might have originated from a previously existing invertebrate (chapter 4).

Darwin accepted Owen's ideas about the process of specialization revealed by the fossil record, but dismissed his appeal to a divine plan as mere verbiage. For him, specialization was an inevitable consequence of natural selection. In the penultimate chapter of the *Origin of Species,* he argued that the taxonomists' intuitive sense of a 'natural' order arose from the fact that a true classification must reflect genealogy. Species are 'related' because they share features which have been inherited from a common ancestor. The most basic characteristics of the ancestral form persist while superficial adaptive specializations are imposed upon them in the descendants. The morphologists' efforts to recognize true homologies and to reconstruct 'hypothetical' archetypes for each class could be seen as unwitting efforts to identify those features which were inherited in each group from its common ancestor. Darwin believed that

> the natural system of classification is founded on descent with modification; that the characters which naturalists consider as showing true affinity between any two or more species, are those which have been inherited from a common parent, and, in so far, all true classification is genealogical; that community of descent is the hidden bond which naturalists have been unconsciously seeking, and not some unknown plan of creation . . . [5]

5. Charles Darwin, *On the Origin of Species* (1859 ed.), p. 420.

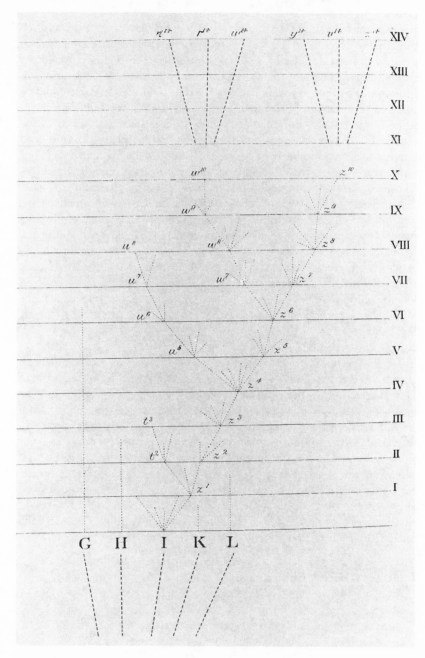

FIGURE 2.2   Part of the evolutionary tree. From Charles Darwin, *On the Origin of Species* (London, 1859), facing p. 117.

In effect, the archetype for the group could be seen as an idealized reconstruction of the real common ancestor which, in theory at least, might eventually be revealed by the fossil record. Since evolution was a process of divergence, the tree of life should be represented as having fan-like branches, just as in the diagram that Darwin included with the *Origin* (fig. 2.2).

For Darwin, then, the systematists' and morphologists' efforts to identify real relationships were an indirect way of trying to reconstruct the history of life. His theory would use history to give a materialistic explanation of organic relationships that had hitherto been ascribed to the mind of the Creator. Stated so baldly, it all sounds terribly simple—and there would indeed be no very revolutionary impact of evolution theory on the actual practice of classification. But in fact the advent of evolutionism offered a potential transformation of both taxonomy and morphology, although many morphologists failed to accept the full challenge offered to their assumptions by Darwin's theory. If one really thought about what it would mean to explain the form of animals as the outcome of a contingent historical process, a whole set of new questions and techniques would be required.

The assumption that taxonomic relationships reflect genealogy does not entail that genealogies can be translated directly into classification systems—even if it is possible to reconstruct the actual pattern of a group's evolution. Darwin, as an experienced naturalist, knew that it is in fact very difficult to decide in some cases which are the most useful characters upon which to base the classification of a particular group. Some characters reveal common descent more reliably than others, and a character which is useful in classifying one group might be unreliable in classifying another. For the evolutionist, reliability would depend upon the exact circumstances of the history of each group, that is, which characters had varied most to produce adaptive modifications, and which had been conservative in retaining the ancestral form. Some new developments in the history of life seemed to have constituted major new 'initiatives' or 'breakthroughs' which had defined successful new types of structure whose basic form was then adapted to a whole range of habitats. The most extreme version of the attempt to classify solely in terms of branching-points in the tree of life is the modern taxonomic school known as *cladism* (after the Greek for "branch"). In the language of cladism, a taxon is defined by shared derived characters, that is, by those characters all its members share because they were developed only in the founder members of the group from which all others radiated. Shared primitive characters, which may also be passed on to other branches of the tree, do not count.

The cladistic approach certainly originated from the belief that taxa are monophyletic in the evolutionary sense, but some of its more extreme advocates reject all efforts to infer genealogies from their trees. They recognize that it is extremely difficult to incorporate the notion of 'ancestry' into a workable taxonomy, because in a tree defined solely by relationships among known species, the nodes, or branching-points, are purely formal: they do not and cannot correspond to real species. If anyone suggests that a certain species (a fossil, or a supposedly primitive living species) is ancestral to others, that species nevertheless has to be classified like any other. It cannot be inserted into the tree that represents the relationships between known species except as another terminal branch of that tree. For this reason it becomes impossible to translate taxonomic relationships directly into information about evolutionary sequences—even though the grouping was originally meant to represent evolutionary divergences. From the systematists' perspective, the idealist morphologists had the more useful technique, since their notion of the archetype for each major group was not meant to represent a living species (even if some species might be further removed from the archetypical pattern than others).

The impact of evolutionism on systematics thus turned out to be rather complex. We shall return to this problem in chapter 7, but for the time being it is more important to explore the response of the morphologists. They did not have to worry about the problem of classifying every individual species; indeed when membership of a group was not in doubt, they routinely used a single species (generally one that was assumed to be 'primitive') to illustrate the character of the whole group. For those who took evolutionism seriously, the claim that relationships are based on genealogy was an invitation to trace the actual sequence of evolutionary steps by which the various modern groups had differentiated themselves from one another. The basic character of the group was defined not by an abstract archetype, but by descent from a common ancestor, and they hoped that it might be possible to reconstruct what that ancestor had looked like. In some cases there might still be animals alive today whose structure was a relatively unchanged relic of the ancestral form. More important, it should be possible to derive the common ancestor of each group from a previously existing type. The main branches of the tree must link up if traced far enough back in time, and it should thus be possible to show that now-distinct types such as the vertebrates had an ancestry in some more primitive form. Showing that the vertebrates were more closely related to the tunicates, or the annelids, was tantamount to arguing that at some point the vertebrates had split off from the appropriate antecedent group.

The metaphor of the branching tree as the basis for classification had already been developed by philologists seeking to explain the evolution of languages.[6] In animal evolution it could be assumed that divergent evolution was the only source of new characters. This cannot be assumed in philology, where elements can be transferred from one language to another, nor even in plant evolution, where new species can be produced by hybridization. But on the assumption that transfers from one branch to another cannot occur, morphologistss dealing with the animal kingdom could seek to identify the crucial 'inventions' that had founded each clade, or branch. As we shall see below, the only major problem was the possibility of 'independent invention'—the parallel evolution of identical characters in two independent branches. Most Darwinists were reluctant to admit this possibility, but there were non-Darwinian concepts of evolution which positively relished it.

Darwin seems to have believed that experienced naturalists developed an intuitive sense of which characters were more reliable for the purposes of classification. Morphology was the science which offered to put this intuitive sense on a more systematic foundation. By looking in ever more detail at the internal structure of organisms, the morphologist could determine which similarities were genuine homologies, and which were mere adaptive analogies of no taxonomic value. In effect, the morphologist would determine what the definitive characters of each group should be, which evolutionary initiatives had played the most important role in defining its future range of structural modification.

Darwin's theory supplied a new explanation of homologies. In idealist morphology, the homology of, say, the human arm and horse forelimb was explained in terms of both being adaptive modifications of the same basic part in the vertebrate archetype. For Darwin, the relationship existed because both were derived from the forelimb of the primitive tetrapod ancestor. The recognition of genealogical relationships would depend upon correctly identifying the true homologies which indicated descent from a common ancestor. This was by no means a straightforward process, especially when embryology was brought in to supplement comparative anatomy. It was widely assumed that in segmented animals such as the vertebrates, parts could only be homologous if they were derived from the same segment in the embryo. But as E. S. Goodrich pointed out in 1913, the limbs of vertebrates are derived from different segments in different species—yet they must be homologous to make any sense of evolutionary relationships. Goodrich appealed to Bateson's work on discontinuous variations to argue that changes of this nature can take place in the course of evolution without negating the concept of genealogical relationships.

6. On the tree metaphor in linguistics and biology, see Henry M. Hoenigswald and Linda F. Wiener, eds., *Biological Metaphor and Cladistic Classification*.

The limbs can jump as units from one segment to another in the course of evolution.[7]

By recognizing the significance of morphological relationships, revisions of traditional groupings still went on in the post-Darwinian era. When E. Ray Lankester confirmed in 1880 that *Limulus,* the horseshoe crab, was not a crustacean but an arachnid, he used morphology to challenge the traditional definition of a crustacean. It had been assumed that the Crustacea was the class of arthropods which breath through gills. Hence, since *Limulus* had gills, it must be a crustacean. By showing that it had a number of important morphological features identifying it as an arachnid, Lankester forced naturalists to realize that breathing through gills was an adaptive character that might be developed by any arthropod (or indeed any other animal living in water). This changed perceptions of the basic nature of the arthropod classes—something that could have been appreciated by a pre-Darwinian morphologist. But it also helped to open up a new problem centered on how the arthropod classes might be related. As we shall see in chapter 3, Lankester's discovery posed major questions for evolutionists seeking to reconstruct the evolutionary origins of the Arthropoda. Morphological questions had consequential effects for evolutionists that would simply not have been problematic for biologists of the pre-Darwinian era.

In Darwinian terms, the morphologist's job was to identify those characters which allowed the taxonomist correctly to group organisms according to their common ancestry. The branches were then linked together to reconstruct the evolutionary tree by tracing each species back to a series of hypothetical ancestral forms—the common ancestors of its genus, family, order, class, and type. Lankester showed that *Limulus* belonged to a side branch of the stem leading to the spiders, not the one leading to the true crabs. But by altering morphologists' perception of what the early arachnids must have been like, his discovery impinged on the problem of how to reconstruct the hypothetical common ancestor of the whole arthropod phylum, the proto-arthropod from which the insects, arachnids, and crustaceans had diverged.

The idealist morphologists' archetype was a pure abstraction, a compound of the group's defining features that could not exist as a real animal. It was not, indeed could not be, derived from any other animal or any other archetype. The evolutionists' hypothetical common ancestor for a group had to be a real organism, one that could have functioned in the real world. It could not be a merely idealized abstract of the common features of the group. It had to have all the organs needed to live, and had to be adapted to some way of gaining a livelihood. Owen's ver-

7. E. S. Goodrich, "Metameric Segmentation and Homology."

tebrate archetype looks something like the small marine creature *Amphioxus*, which was indeed widely taken to be the best living representative of the earliest vertebrate form. But *Amphioxus* is a real organism with special adaptations of its own. The ancestral vertebrate may have had its own adaptations, possibly to a different way of life. The morphologist had to study modern forms and try to determine which were the ancestral characters—and then try to understand how the ancestral form itself might have lived in its own environment, and how it might have responded to changes in the environment in a way that would generate new evolutionary developments.

This transition to thinking in terms of common ancestors that are real organisms illustrates the crux of the debate between idealist morphologists and those who adopted a more materialistic approach under the influence of evolutionism. Later idealists such as Naef have been dismissed by modern Darwinians as mere throwbacks to a pre-evolutionary worldview. They are thought to have discouraged the search for evolutionary relationships by stressing the existence of distinct archetypes, or *Baupläne,* which could not be transformed into one another (except by some mysterious saltation). This accusation is not quite fair: Naef called his morphology "idealist," yet he subsequently produced a whole series of studies devoted to phylogenetic topics, including a contribution to the debate over the origin of the vertebrates and a study of mammalian evolution that made extensive use of fossil sequences. The paleontologist Otto Schindewolf accepted a link between certain kinds of idealist morphology and phylogenetic research, although his evolutionism was clearly non-Darwinian in character.[8] Idealists did not ignore evolutionism—they merely incorporated what they could understand of it into their own version of morphology. It is thus important for us to understand how the idealists could contribute to certain levels of evolutionary debate without accepting what the Darwinists would regard as a truly evolutionary perspective.

Naef's work on vertebrate origins shows that the concept of an underlying type could be made flexible enough to allow the search for connections between quite widely divergent forms. It was simply a matter of defining a sufficiently small number of basic types to allow many superfi-

8. Naef, "Notizen zur Morphologie und Stammesgeschichte der Wirbeltiere. 6" and "Notizen . . . 7"; Otto H. Schindewolf, *Basic Questions in Paleontology,* pp. 409–416. The influence of Naef's idealism is discussed in Wolf-Ernst Reif, "The Search for a Macroevolutionary Theory in German Paleontology" and the same author's "Afterword" to the translation of Schindewolf's *Basic Questions in Paleontology.* On Naef's contribution to the debate over vertebrate origins, see K. Nübler-Jung and D. Arendt, "Is Ventral in Insects Dorsal in Vertebrates?" This article also contains a protest against the Darwinists' assumption that Naef's idealism was anti-evolutionary.

cially different animal forms to be included in each type. Chordates and annelids could then be accepted as modifications of an even more fundamental pattern. It would also be possible to look for the sequence of developmental changes involved in the creation of the two major phyla—and of any particular group within each phylum. The problem for the idealists was precisely the difficulty of deciding how many basic types there were. Nor can it be denied that they instinctively tended to prefer a larger number of more rigidly defined archetypes, and only reluctantly conceded to the evolutionists that what they had at first thought to be distinct forms should perhaps be seen as modifications of something more basic. Owen's early efforts to claim that humans were distinct from apes is a manifestation of this reluctance—although he later accepted many evolutionary links including the derivation of the vertebrates from an invertebrate type.

The other major difference between idealist and materialist morphologists concerns the nature of the underlying pattern from which the structure of real organisms is derived. Naef proposed a hypothetical developmental pattern which could be modified to produce both annelids and chordates, but this was idealized in the sense that it was never manifested in the ontogeny of a real organism: it represented the primitive features of the developmental process stripped of any specialized details that would have had to exist in the ontogeny of a real organism. Using such a model, the idealist could look for ways in which the hypothetical original pattern could be modified to produce the organisms we actually see in the real world, and could even incorporate fossil evidence illustrating at least the later phases of the sequence of modification. But since there was no need to think about the starting point as a real organism, there was no incentive to search for the adaptive modifications that might have indicated how the early members of the group had lived, or the kinds of adaptive pressures that might have forced their transformation into the divergent forms we know today.

It was by no means easy for the materialistic evolutionist to work with the assumption that the common ancestor was a real organism. As we shall see, evolutionists such as Haeckel opened up a wide range of problems when they tried to use living organisms as illustrations of 'primitive' evolutionary stages, precisely because it was difficult to determine what was primitive, and what adaptive. But by taking this step they opened up the way for a later generation of evolutionists to consider questions about the actual causes of evolutionary transformations, questions of a kind that were all too often marginalized in the idealists' more abstract search for underlying relationships. Idealists accepted some aspects of the evolutionary program, but only some—and then with a reluctance that delayed their participation in the evolutionary debates.

Once defined by the evolutionist, the common ancestor of a group dictated the structure of the resulting hypothesis of common descent, and the group's taxonomy. Problems resulted when, as happened on all too many occasions, morphologists disagreed as to the exact nature of the ancestral form. Advocates of some new theory would complain that the previously accepted common ancestor had become so firmly embedded in the textbooks that it was impossible for anyone trained in the old interpretation to visualize the newly proposed relationships. There were immense debates over the true nature of the common ancestor for the molluscs, the arthropods, and the vertebrates, each position reflecting a particular interpretation of the group's origin. (Idealist morphologists were no more successful: they disagreed bitterly over the number and nature of the basic archetypes, with less possibility of reaching a consensus on what would end the disputes.)

Perhaps the most important characteristic of the ancestor was that it must be derivable from some previously existing form, which must itself be a viable organism. In a Darwinian scheme, furthermore, the whole process of transformation must be explicable in materialistic terms. Darwin rejected the claim by the Duke of Argyll that rudimentary organs might be structures undergoing development for future use: in a materialistic system future goals do not count, and rudiments can be explained only as the relics of once-useful structures.[9] Each step in the process must represent a viable organism—and critics of Darwinism such as St. George Jackson Mivart insisted that many hypothetical intermediates were nonfunctional. Of what use is a partially formed wing?[10] The Darwinian evolutionist had to explain how a leg could be transformed into a wing via a series of intermediates, each of which could still serve some purpose as a means of locomotion. For this reason, Darwinians such as Anton Dohrn proposed that organs can have more than one function: evolution often consists in the shift of priorities from one function to another, not the creation of a new structure with an entirely new function (see chapter 4).

The search for common ancestors fitted exactly into the program outlined in Ernst Haeckel's *Generelle Morphologie* of 1866 and his more popular *Natürliche Schöpfungsgeschichte* of 1868, translated as *The History of Creation*. Haeckel more or less explicitly set out to convert the idealist morphology of Goethe into something compatible with Darwinian materialism. It was he who coined the term *phylogeny* to denote the evolutionary history of the species. Where Darwin merely discussed

---

9. See George Douglas Campbell, Duke of Argyll, *The Reign of Law*, p. 213; see also Argyll's letters to Sir William Flower reproduced in the Dowager Duchess of Argyll, ed., *George Douglas Campbell . . . Autobiography and Memoirs*, II, p. 483.

10. St. George Jackson Mivart, *On the Genesis of Species*, pp. 26–70.

the general principles upon which a genealogical classification would be based, Haeckel set out to reconstruct the whole history of life on earth as the basis for a classification for the plant and animal kingdoms. Like Darwin, he declared that the natural system of classification was based on the pedigree of the various forms:

> For the true cause of the intimate agreement in structure can only be the actual blood relationship. Hence we may, without further discussion, lay down the important proposition that all animals belonging to one and the same circle or type must be descended from one and the same original primary form.[11]

Haeckel stressed the need to provide a 'monophyletic' outline of the tree of life, an outline in which each natural group was traced back to a single original ancestor.

The unity of the group would be maintained only by the preservation of basic features derived from the common ancestor. There was no mystical archetype guaranteeing the fixity of the type. Yet, as Mario Di Gregorio has shown, Haeckel and his colleague Carl Gegenbaur found it difficult to throw off the legacy of the old typological viewpoint.[12] Gegenbaur in particular was unwilling to accept that one major type could be linked to another via evolutionary intermediates (as we shall see in chapter 4). The founders of evolutionary morphology could accept the notion of relatively limited divergence from a common ancestor, but were sometimes reluctant to extend the more materialistic way of thinking to its ultimate conclusion. This illustrates the limited impact of Darwin's new way of thinking, and in so doing helps to explain some of the confusion among modern commentators on the role of idealism in evolutionary biology. If we suppose that a 'true' evolutionist has to go the whole way in accepting Darwinism, then any biologist retaining a vestige of the idealist or typological way of thinking cannot be an evolutionist. Later idealists such as Naef can be dismissed as throwbacks to the pre-evolutionary worldview. The case of Gegenbaur shows us the dangers of setting up this kind of rigid polarization. Evolutionary morphology was founded by biologists who were often unable to accept the full materialism of Darwin's theory. They were by no means prevented from working out more limited phylogenetic connections—and neither were later idealists such as Naef. Gegenbaur figures prominently in what follows because in the 1870s even his limited evolutionism could break new ground by leading him to address important phylogenetic questions. He also used the language of the new mate-

---

11. Ernst Haeckel, *History of Creation*, II, p. 122; see Russell, *Form and Function*, chap. 13.

12. Mario Di Gregorio, "A Wolf in Sheep's Clothing: Carl Gegenbaur, Ernst Haeckel, the Vertebrate Theory of the Skull, and the Survival of Richard Owen."

rialism, even if he failed to adopt the full Darwinian way of thinking. By the early twentieth century, Naef's idealism—although in practice not very different from Gegenbaur's position—was so obviously an attempt to turn the philosophical clock back that we tend to forget that he too could work out phylogenetic connections.

Inevitably, our story tends to focus on those morphologists who went further into the logic of the evolutionary worldview, and who became willing to contemplate the origin of one type from another by natural processes. Haeckel, at least, realized that to make his evolutionary system consistent, he must show how the common ancestor of each group could, in turn, have evolved from a member of a previously existing group. Even without fossil evidence, it was obvious on structural grounds that the earliest mammal, for instance, must have evolved from a branch of the reptiles. Having reconstructed the original mammal, the evolutionist could seek to identify which reptile group was the most likely source of the higher class, by trying to identify the kind of reptile structure that could have been most easily transformed into that of the mammal.

Even more revolutionary was the evolutionists' assertion that the vertebrate type itself must have had its origins in a previously existing, presumably invertebrate, form. Cuvier and von Baer had insisted that the types were absolutely distinct: there could be no homologies between types. But in the 1820s Geoffroy Saint Hilaire had developed a "transcendental anatomy" in which there could indeed be homologies between the types, and had linked this to a controversial transmutationist theory.[13] In the post-Darwinian era the prospect of searching for an evolutionary origin of the main phyla of the animal kingdom opened up entirely new problems for the morphologist. How would one identify homologies between structures which differed so fundamentally? Evolutionary morphologists began to appeal to new kinds of evidence, including histological similarities (similarities of tissues rather than of organ structures) and the origins of organ systems in the germ-layers of the earliest embryonic structures. It was certainly possible for later, more flexible, idealists like Naef to accept this evidence, but it seems unlikely that the idealists would have moved in this direction by themselves.

In the evolutionary perspective, the original form of each group could have existed only in the past. Strictly speaking, one would have to turn to paleontology to have any hope of providing concrete evidence for its existence. Owen had noted that the earliest mammals were more generalized than their modern descendants, but given the incomplete state of the fossil record one would have to be very lucky to discover the common ancestor

13. On Geoffroy's transcendental anatomy, see Appel, *The Cuvier-Geoffroy Debate;* and Desmond, *The Politics of Evolution.*

from which all the mammals had diverged. Haeckel was certainly willing to use fossil evidence when it was available, but he believed that morphology would allow one to reconstruct the ancestral form even when fossil evidence was completely lacking.

Problems arose even when fossils were available. In theory, the fossil record might help us to fill in the whole history of life. But—as Darwin himself warned—its imperfection made it a very dangerous tool. The first known fossil member of a group might nevertheless represent a form that had evolved some time after its first appearance. One might find a fossil which threw light on the ancestry of a group, without being the actual common ancestor. The fossil form might be a relative of the common ancestor of later forms, but would have modifications that were not characteristic of the ancestor itself. The well-known *Archaeopteryx* was widely seen as a primitive bird, but many authorities believed that it occurred too late in the fossil record to be the actual ancestor of all modern birds. It was a side branch from the main tree of bird evolution which had preserved some characters of the common ancestor from which the whole class had diverged. In this case, the morphologist had to ask which were the truly primitive characters, and which the secondary modifications— and expert opinions differed widely. It is for this reason that modern cladists treat fossils exactly the same as living species for the purposes of classification. Coming earlier in time is no reliable indication of ancestral status.

One potentially dangerous aspect of Haeckel's reconstructions was his tendency to look for living equivalents of the ancestral form of each class. In theory it was most unlikely that the common ancestor could have survived unchanged while its descendants had radiated out into various lines of specialization. But some species do not specialize for a particular way of life, and there are certainly some living forms which seem to be relatively little-changed relics of earlier periods in the earth's history. *Limulus* was seen as the closest living analog of the trilobites which flourished in the Palaeozoic seas. The monotremes and marsupials were widely regarded as relics of early stages in the evolution of the mammals, predating the appearance of the placental mammals. These 'living fossils' might throw light on the ancestral stages even though true fossils were not available. Haeckel seems to have gone out of his way to provide a living analog of each node in the tree of life. Only when absolutely necessary would he invoke a totally hypothetical ancestral form. In part he did this to distance himself from the idealists: by identifying a living animal that could at least illustrate the character of the ancestor, he made it clear that the evolutionists were not dealing with abstractions. But it is difficult to avoid the impression that he was also succumbing to the lure of the old chain of being. He wanted to believe that all stages in the ascent of life were still represented in the modern world. Many of his contemporaries shared this assumption: biologists such as Richard Semon and W. H. Caldwell undertook expeditions to Australia in search of the platypus and other living fossils.

The dangers of the assumption that ancestral forms could survive unchanged were highlighted by E. Ray Lankester. He warned against

> a very common tendency of the mind, against which the naturalist has to guard himself,—a tendency which finds expression in the very widespread notion that the existing anthropoid apes, and more especially the gorilla, must be looked upon as the ancestors of mankind. A little reflexion suffices to show that any given living form, such as the gorilla, cannot possibly be the ancestral form from which man was derived, since *ex hypothesi* that ancestral form underwent modification and development, and in so doing ceased to exist.[14]

The problem was that even if one assumed that a particular living form was a *relatively* unchanged relic of the past, one would have to accept that in the course of time it had aquired at least some new characters. The morphologist would have to decide which were truly primitive characters, and which were later or secondary ones. This ought to be easy in principle, but the determination of which were truly primitive characters might depend upon the hypothesis being adopted as to the evolutionary relationships involved. One morphologist's primitive character often turned out to be another's secondary modification. Modern paleontologists tend to be very suspicous of the claim that any primitive form can have survived unchanged for vast periods of time.

Another difficulty arising from the attempt to use living species as models for the primitive form of each class was the possibility of evolutionary degeneration. If all evolution were progressive, that is, consisted of the addition of new characters to existing ones, then any form which retained a low level of organization could be assumed to be an unchanged relic of an earlier stage in the evolution of higher animals. Lankester again pointed out the dangers:

> We have no sufficient ground for assuming that, even in respect of the simplicity of their structures, any given animal forms at present existing exhibit a mere survival of a corresponding degree of simplicity in their remote ancestors. Such an assumption was almost universally made, until a more

---

14. E. Ray Lankester, article "Vertebrata" in Lankester, ed., *Zoological Articles Contributed to the Encyclopaedia Britannica*, p. 175. Note, however, that later in the same article Lankester himself refers to *Balanoglossus* as very close to the hypothetical vertebrate ancestor, see pp. 182–183. For a survey of modern paleontologists' views, see Niles Eldredge and Stephen M. Stanley, eds., *Living Fossils*.

correct view was pressed on the attention of naturalists by Dr. Anton Dohrn. . . . So far from its being the case that simplicity of organization necessarily implies the continuous hereditary transmission of a low stage of structural development from remote ancestors, there are numerous instances in which it is certain that the existing simplicity of structure is due to a process of degeneration, and that an existing form of simple structure is thus descended from ancestors of far higher complexity of structure than itself.[15]

Dohrn had introduced this principle as a corollary of his theory of vertebrate origins (see chapter 4). Unfortunately the theory was highly controversial, and opinions differed as to which were degenerate forms, and which were truly primitive.

FORM AND FUNCTION

Haeckel's efforts to apply the general principles of evolutionism to morphology thus raised a number of problems. Reconstructing common ancestors was not as easy as it might appear at first sight, especially when it was possible to reconstruct two quite different phylogenies to explain the origin of a particular group. Underlying many of the problems was the continuation of a long-standing debate within morphology over the relative significance of form and function in the shaping of organic structures. Strictly speaking, Goethe's 'morphology' had been a science of pure form: the morphologist studied the relationships between structures without taking into account the functions which those structures performed in the real world. Many morphologists actually followed this program in their research, ignoring functional questions altogether. Others followed Cuvier in arguing that a science of pure morphology was a meaningless abstraction. Real organisms had to function in the real world, and their structure was shaped by adaptation to a particular way of life.

    Those who ignored function tended to assume that it was, in any case, a trivial factor which exerted only minimal contol over the organism's structure. The structure of real animals was determined mainly by purely formal constraints, with only superficial features being determined by adaptation. In embryological terms, this position translated into the view that adult form was shaped by forces deriving from the developmental process itself. The 'laws of growth' determined form, not the demands of the environment within which the organism would have to function. Needless to say, the advocates of this view were mainly laboratory biolo-

15. Lankester, "Vertebrata," p. 175; see also Lankester, *Degeneration: A Chapter in Darwinism.*

gists rather than field naturalists: they studied organisms in a totally arti-
ficial situation and had no interest in the way they interacted with one
another or with the physical environment in the real world. Even today
there are some biologists studying the developmental process who insist
that form must be explained in terms of ontogenetic laws. They find it
difficult to believe that the genetic program which unfolds through the
developmental process can have been constructed by anything so haphazard as a historical process dependent upon the contingencies of local
adaptation.[16]

The nineteenth-century debate over the relative significance of form
and function certainly continued within an evolutionary context. At one
level, Darwin's theory stresses the role of function. Natural selection can
modify a species only by adapting it to changes in its way of life. But this
does not mean that species are infinitely plastic in the sense that they can
be rapidly remolded into any new configuration by a change in the environment. Heredity is a conservative factor, and the genetic program that
shapes the organism's development is itself the product of history. Existing
structure can only be modified within certain limits, especially if the modifications are to be viable in the struggle for existence. Moreover, Darwin
did not believe that variation is totally 'random' in the sense that an infinite range of variability is available to every species. Many variations are
possible, none of which are actually generated by the needs of the organism, but there are also constraints imposed by the existing constitution of
the organism. In modern terminology, not all conceivable mutations are
equally likely to occur, and some are quite impossible. Thus, although the
demands of function represent the driving force of evolution, the existing
form exerts strong constraints on the evolutionary process. Groups retain
basic features derived from their common ancestor, underlying each member's recent adaptive modifications.

Darwin used history to work out a compromise between the advocates of pure form and pure function. Species are modified by adaptation,
but they cannot be modified to an unlimited extent because heredity dic-
tates the starting point for each subsequent modification. Each species is
a product of a unique sequence of events in its past history. The deep structures once assigned to the archetype were in reality the constraints of history imposed on the modern organism by the fact that it was the descendant of a particular sequence of ancestral forms.

Many evolutionists, however, found it difficult to see the modern
world as a product of historical contingency. Even Haeckel seems to have
paid little attention in practice to the environmental factor in evolution.

16. See, for instance, Gerry Webster, "The Relations of Natural Forms," and Webster's introduction to the reprinted edition of Bateson's *Materials for the Study of Variation.*

His search for the ancestral form of each group was conducted largely in terms of pure morphology, looking for structural resemblances without worrying too much about the adaptive processes that could have converted one form into another. In this sense, Haeckel's program did indeed look very much like a continuation of the old idealist morphology. And, as we shall see below, his attempt to use embryology as a guide to reconstructing phylogenies tended to emphasize the internal control of form rather than the demands of function in the external world.

To some explicitly anti-Darwinian evolutionists, stressing the role of internal, purely form-driven processes was a way of undermining the element of historical contingency which they found so disturbing in the *Origin of Species*. Many morphologists, and some palaeontologists, rejected the claim that function, that is, adaptation, is the major driving force of evolution. They argued that the environment places relatively few constraints on the organism, leaving internally programmed forces derived from the developmental process free to drive evolution along a predetermined path. For theistic evolutionists such as Owen and Mivart, evolution was the unfolding of a preordained pattern of development that could not be explained in terms of how the species adapted to an ever-changing environment. Major aspects of evolution were thus nonadaptive, opening up an entirely different set of explanatory techniques that evaded many of the constraints imposed by Darwin's reliance on adaptation as the driving force. Naef's brand of idealism had similar consquences for a later generation of German evolutionists.

The paleontologists who supported the theory of 'orthogenesis' also believed that evolution was driven in predictable directions by nonadaptive forces arising from within the organism. They even believed that in some cases the resulting trends created structures so maladaptive that they eventually caused the species' extinction. In such a theory, internal forces did not merely constrain or limit the range of possible variations—they generated a single trend in individual variation which imposed a rigid direction on the evolution of the species.

The tendency for many morphologists—Haeckel included—to think in terms of purely formal, almost geometrical, relationships was pointed out by Dohrn in the course of his attempt to reconstruct the origins of the vertebrate type. Dorhn adapted a more 'Darwinian' view by insisting that phylogenies should be reconstructed by taking into account the functional changes that could have provided a reason for one structure to be changed into another. His theory of *Functionswechsel* assumed a succession of functions in the history of an organ. His emphasis on degeneration was a by-product of his increased willingness to see function as the driving force: if a structure ceased to have a function, it would degenerate, something that was most likely to happen when species adopted a less mobile way of life.

Some morphologists agreed with Dohrn, although their efforts to relate changes in structure to new habits in new environments often seemed to pay only lip service to the functional approach. It was relatively easy to speculate vaguely about an organism's habits when dissecting it in the laboratory without any reference to how it functioned in the wild. Thus the embryologist E. W. MacBride took up the theme of degeneration in his theory explaining the origin of the various invertebrate types (discussed below). Each type was a degenerate offshoot from the vertebrate stock, adapting to a sessile or burrowing lifestyle, while the more progressive line continued to evolve in the open sea. MacBride could certainly speculate about the habits of the common ancestor of each group, but his purely morphological approach offered little hope of tracing any environmental trigger for the change of habits. Only when geology and paleontology began to offer better information about the climates and environments of the distant past did it become possible to speculate with any plausibility about the role of external factors in the direction of evolution. In modern terminology, morphologists found it difficult to offer adaptive scenarios to explain precisely why a population found it advantageous to change in a particular way to adapt to changes in its local environment.

In the early twentieth century morphologists became increasingly concerned about function. They began to ask exactly how the structures they were observing fulfilled their roles in the animal's lifestyle. In America, William King Gregory developed a functional morphology for the vertebrate limbs which had important implications for many of the major steps in evolution.[17] H. Graham Cannon in England did major work on the functional morphology of arthropod feeding mechanisms. For Gregory, at least, the study of function led to a more 'Darwinian' view of evolution, in the sense that he began to oppose the orthogenetic explanations offered by his mentor, Henry Fairfield Osborn. Gregory was no field naturalist and offered no adaptive scenarios to explain major evolutionary breakthroughs. But the more he studied the way the organism functioned, the less he believed that evolution was driven by trends that were totally unrelated to the organism's adaptive needs.

If Gregory's functional morphology led him toward a more Darwinian viewpoint, Cannon retained an interest in the increasingly discredited Lamarckian theory.[18] Lamarckism, the inheritance of acquired character-

17. See Ronald Rainger, *An Agenda for Antiquity*, chap. 9.
18. H. G. Cannon, *Lamarck and Modern Genetics*.

istics, was traditionally a theory which supposed that evolution was governed by habit. In a new environment, the organism adopted new habits and used its body in new ways (the classic example is the ancestral giraffe stretching its neck to reach the leaves of trees). If such acquired characters were inherited, then the effects of new habits shaped the evolution of the species. Lamarckism came increasingly under attack from the geneticists of the early twentieth century, but some morphologists and paleontologists refused to accept the findings of the new experimental biology. Cannon's increased sense of how habit determined structure seems to have made it difficult for him to abandon the belief in a direct relationship between them via the Lamarckian effect. He assumed that the organism was much more plastic under the influence of the environment than most Darwinians would allow. Cannon's student Sidnie Manton went on to argue that the arthropods were polyphyletic (chapter 3). She believed that structure was so intimately dependent on habit that several lines of evolution had independently 'invented' the same kind of structure. Those stuctures were not homologous and were not a sign of common ancestry. The Lamarckian Frederic Wood Jones insisted that the similarities between humans and apes were independently acquired as both groups adapted to the same way of life. Lamarckism thus survived as an extreme version of the belief that form is shaped by function.

Paradoxically, Lamarckism and orthogenesis were originally products of the same anti-Darwinian viewpoint. American neo-Lamarckians such as Edward Drinker Cope and Alpheus Hyatt certainly argued that orthogenetic trends existed alongside adaptive trends directed by habit. Theodor Eimer, who popularized the term *orthogenesis* to denote nonadaptive evolutionary trends, was also a Lamarckian. Both theories arose out of the refusal to accept that evolution was a haphazard process driven by the accidents of history.[19] Variation was supposed to be directed along predictable channels either by habit or by internally programmed trends. Yet in the end Lamarckism and orthogenesis moved to opposite extremes in the debate over the relative importance of form and function, leaving Darwinism in the middle. Where orthogenesis rejected any role for adaptation, Lamarckians began to suppose that function could create almost any structure the organism needed to suit its habits. What makes this distinction all the more crucial is that both theories tended to promote the view that similar structures could appear in different lines of evolution, thus opening up a major problem for the use of morphological evidence in reconstructing the history of life.

19. On the origins of orthogenesis and neo-Lamarckism, see Bowler, *Eclipse of Darwinism,* chaps. 4–7.

The most intractable problem for the morphologists was that posed by the possibility that the same structure could evolve independently in more than one group. Darwin and Haeckel both assumed that if one could identify really fundamental characters uniting a group of organisms, then those characters must have been inherited from a single common ancestor of the whole group. Superficial adaptive analogies could always be recognized and eliminated from the analysis. This assumption that all natural groups must be monophyletic was not a necessary consequence of the basic postulate of evolution, and it is important for us to recognize why the Darwinians were so committed to it. In part, of course, the monophyletic assumption is a direct consequence of parsimony in the construction of hypotheses.[20] If evolution is to be any guide to classification, then most shared characters must be a consequence of common ancestry, and it is simplest to assume that they all are. The idealists' archetype was a single imaginary form encapsulating the essence of a group's common structure, and the simplest possible translation of this into materialistic terms was to postulate a single common ancestor as the source of structural affinities.

Darwin had other reasons for advocating common descent as a guide to taxonomic relationships. He believed that newly acquired characters were generally imposed on the organism at a superficial level; that is, they were produced by modifications of only the later stages of the process of individual development. The more fundamental characters were produced at an earlier stage in ontogeny and were less likely to be modified by natural selection. Thus the fundamental characters which unite a group are likely to have been inherited from a common ancestor. Moreover, Darwin's whole approach to evolution stressed that it was an essentially undirected process, except for the requirements of adaptation to the local environment. The variations from which nature selects are undirected and cannot by themselves create evolutionary trends. Thus it is most unlikely that an apparently identical structure will be independently developed in two unconnected lines of evolution. Each adaptive innovation is unique, and hence each fundamental evolutionary initiative must be unique too. Any major new character can be developed only once, and all subsequent species sharing that character must do so because they have inherited it from the common ancestor in which it was first developed. (Characters

20. For a modern discussion of parsimony as a guide to taxonomy, see Elliot Sober, *Reconstructing the Past.* For a biologist's protest that the multiple origin of similar characters occurs more often than most Darwinists are prepared to allow, see A. J. Cain, "On Homology and Convergence."

can be lost by degeneration, as in the case of the snakes' legs, but a complex of other characters forces us to believe that the snakes have evolved from tetrapods.) Modern cladists adopt exactly the same point of view; one of the main purposes of morphology is to detect convergences, so that they can be eliminated from the list of characters used for classification.

Unfortunately, not everyone accepted the logic of this argument. Many naturalists were desperately unhappy with the essentially open-ended character of natural selection. They wanted to believe that evolution was somehow directed along predetermined channels. In the case of theistic evolutionists such as Owen and Mivart, evolution was supposed to have advanced along channels preordained by a divine plan of creation. Mivart's *Genesis of Species* (1870) argued strongly against Darwinism on the grounds that predetermined evolution would generate relationships quite different to those predicted by the theory of common descent. For Mivart, two species might share a character because both had developed in response to a trend imposed on them by the divine plan.[21] The shared character would be something more than a mere analogy, and would represent a unifying factor of genuine taxonomic value, but it would not be the product of common descent. For Mivart, similar characters could appear in many different lines of evolution, and the tree of life would look more like a tangled bush than a simple, ever-diverging fan.[22]

Mivart was ostracized by the Darwinians, but he continued to work in zoology and to challenge the credibility of the monophyletic hypothesis. It is probably fair to say that the theory of common descent was the major unifying factor accepted by all Darwinians. No member of the Darwinian camp went out of his way to stress the possibility of the independent evolution of similar characters. But in the end even good Darwinians were forced to admit that in some circumstances the simple hypothesis of common descent might have to be abandoned. Lankester's repositioning of *Limulus,* for instance, encouraged some biologists to argue that the arthropods were polyphyletic. They believed that one could reconstruct the evolutionary tree only by deriving the arachnids, insects, and crustaceans from separate ancestors which had independently acquired the distinctive arthropod structure (the view later revived by Manton). To those biologists who retained doubts about the adequacy of natural selection, the possibility that the evolution of life on earth might be riddled with examples of parallel evolution remained attractive well into the twentieth century.

21. Mivart, *Genesis of Species,* pp. 76–87.
22. On Mivart's view of the history of life, see Bowler, *Eclipse of Darwinism,* chap. 3; and Desmond, *Archetypes and Ancestors,* chap. 6.

Ernst Mayr has argued that the success of the theory of common descent was one of Darwin's great triumphs.[23] Our survey of the debates over the history of life on earth will show that this assessment cannot be accepted uncritically. Many morphologists and paleontologists retained doubts about the universal validity of the theory. While never doubting that, sooner or later, all branches could be traced back to a common origin, they were convinced that many species shared basic characters that were not inherited from a common ancestor. The multiple origin of an apparently identical structure can occur in three ways: by independent invention, by the convergence of two lines of evolution toward the same goal, and by parallel evolution. Independent invention can occur when two groups solve the same adaptive problem by developing the same kind of structure. Vertebrates and cephalopods have developed very simliar kinds of eyes, which no one would suggest were inherited from a common ancestor. Convergence occurs when two originally very different forms move into the same environment and gradually acquire similar characters. In superficial examples the result is obviously a mere analogy, as with the fishlike form of dolphins, but there are other cases where it is by no means so easy to decide which are the homologies, and which the characters independently acquired. In cases of parallelism, whole groups of species were supposed to have advanced independently through the same sequence of developmental stages, each acquiring new characters that were certainly a sign of relationship, but which were not inherited directly from a single common ancestor.

The problem of multiple origins was first addressed from a Darwinian perspective by Lankester himself in 1870. Lankester complained that the term *homology* was ambiguous because certain fundamental similarities of structure, far more significant than mere adaptive analogies, were not a sign of common descent. He proposed to call this phenomenon *homoplasy* (*convergence* became the more popular term for most examples of this, but *homoplasy* has recently come back into vogue). Homoplastic relationships occurred when two species independently solved the same adaptive problem by developing identical structural modifications. The results might be so similar that taxonomists were deceived into thinking that they were a sign of common descent. In fact, such cases of 'independent invention' had to be identified and eliminated from the list of true homologies used to assess genealogical relationships. Lankester compared the process to two tribes of humans independently inventing the same tools or weapons.[24] Even a good Darwinian might have to concede that in

23. Ernst Mayr, *One Long Argument*, pp. 21–24.
24. Lankester, "On the Use of the Term Homology in Modern Zoology," p. 41.

some cases there was only one good engineering solution to a particular mechanical problem, so it was not unlikely that two species had developed identical structures when placed in identical circumstances.

Independent invention was often seen as merely a special case of convergence, the acquisition of a similar structure by two groups. Warnings against the misleading effects of convergence abound in the literature of the late nineteenth and early twentieth centuries. I have explored in another study the many arguments supporting the idea that the similarities between humans and apes were examples of convergence rather than indications of close evolutionary relationship.[25] As an example of this kind of warning, one quotation may suffice, from G. B. Howes's preface to the English translation of Robert Wiedersheim's *The Structure of Man* (1895).

> It is now becoming evident that an essentially similar definitive condition may be independently reached, under advancing modification along parallel lines, by members of independent groups of animals; and there is reason to suppose that some of our classificatory systems are unnatural and erroneous from want of appreciation of this principle of 'convergence.' We must, therefore, not lose sight of the possibility that some of the characters which modern man and the higher Apes have in common may have been independently acquired.[26]

For Frederick Wood Jones, at least, this element of convergence was clear evidence of Lamarckian evolution.[27]

In paleontology, parallelism loomed even larger in the list of dangers to be taken into account when assessing evolutionary relationships. Paleontologists such as Edward Drinker Cope and Henry Fairfield Osborn believed that each class divided into a number of lines of evolutionary development at an early stage in its history, a phenomenon that Osborn called "adaptive radiation." But once the lines were established, each developed in parallel through a more or less identical sequence of morphological stages, driven by some internally programmed force, inherited from the common ancestor, which governed the future variation of all the descendant species. Similarities among the later descendants were thus independently acquired, and were not a direct product of common ancestry (although they were an indirect product in the sense that all the species had inherited the same tendency to vary in that particular direction). In a 1902 article, "Homoplasy as a Law of Latent or Potential Homology," Osborn drew a distinction between convergences produced by similar ad-

25. See Bowler, *Theories of Human Evolution.*

26. G. B. Howes, preface to the translation of Robert Wiedersheim, *The Structure of Man,* p. vi.

27. Frederick Wood Jones, *Man's Place Among the Mammals;* and Bowler, *Theories of Human Evolution,* chap. 8.

aptation to external conditions and parallelisms brought about by internally controlled evolution in related species. Both were major sources of difficulty when reconstructing phylogenies, although they were produced by different mechanisms. He later called the law of analogous evolution the "will o' the whisp" that tended to lead the student of descent astray.[28]

A study by Hermann Friedmann in 1904 presented convergence as an alternative to Darwinism, but concentrated more on the mechanisms likely to produce it.[29] The most complete discussion of the problem in English was Arthur Willey's *Convergence in Evolution* of 1911. Like Lankester, Willey had a career as both an academic morphologist and as a museum zoologist—he was for a time the director of the Colombo Museum in Ceylon (Sri Lanka). His own earlier study on the origin of the vertebrates (discussed in chapter 4) had warned against the dangers of convergent evolution, but Willey thought morphologists were reluctant to acknowledge that convergence was a major feature of the evolutionary process. If characters were discovered to be independently acquired, this was dismissed as "mere convergence," and "there is more joy amongst morphologists over one attempted genealogy than over ninety and nine demonstrations of convergence."[30] Willey gave numerous examples of convergence, including even mimicry among insect species as one aspect of the phenomenon.

Willey distrusted Osborn's neat distinction between convergence and parallelism, asking what degree of relationship between the superficially similar forms was sufficient to make the process count as parallelism.[31] In this, I suspect, he was trying to confuse an issue that most authorities were relatively clear about. Convergence was always a phenomenon supposedly produced as a consequence of adaptation to similar habits in a similar environment. Parallelism was produced by internally programmed trends that had no necessary connection with the demands of the environment. These trends might have *begun* as a consequence of adaptation, but their continuation was automatic and might generate even nonadaptive structures. Thus it was widely believed that the formation of horns in several mammal families acquired a kind of 'momentum' which subsequently developed the structures to a size that might become positively harmful to the species, contributing to their eventual extinction.[32]

28. Henry Fairfield Osborn, "Homoplasy as a Law of Latent or Potential Homology"; and Osborn, *The Age of Mammals*, p. 34. On parallelism in paleontology, see Bowler, *Eclipse of Darwinism*, chap. 6.

29. Hermann Friedmann, *Die Konvergenz der Organismen.*

30. Arthur Willey, *Convergence in Evolution*, p. 53.

31. Ibid., preface, pp. x–xi.

32. F. B. Loomis, "Momentum in Variation."

Osborn's definition was misleading only in that it excluded certain very basic cases of parallelism caused not by the inheritance of a common trend, but by even more fundamental biological phenomena. The debates over the origin of the major phyla (chapters 3 and 4) had to confront the possibility that the metamerism, or segmentation, of the body structure characteristic of the annelid worms, the arthropods, and the vertebrates might have been acquired independently. A basic tendency in the process of embryogenesis might lead to the duplication or repetition of existing parts. If this was at work in several different phyla, each would independently acquire a body consisting of multiple segments.

It was hardly surprising that, as Willey claimed, convergence and parallelism were treated as secondary phenomena which tended to interfere with the search for genuine homologies. The whole purpose of evolutionary morphology had been to identify such homologies as a guide to reconstructing the tree of life. The point of detecting convergences was to eliminate them from the calculation of true (i.e., genealogical) relationships. The main difficulty confronting morphologists was that, while everyone admitted that convergence and parallelism occurred, individuals differed over which structures were produced in this way and which were genuine homologies. Thus in Dorhn's theory deriving the vertebrates from the annelids, segmentation was a homologous characteristic uniting the two types. For the majority of morphologists who rejected the theory, segmentation was independently acquired through parallel evolution in the two phyla. It was the hypothetical phylogenies which determined their supporters' opinions on which were homologies. One could not detect 'obvious' convergences, so that they could be eliminated, because what counted as a convergent character depended upon preexisting ideas about how the groups of animals fitted together in the evolutionary tree.

This point helps us to see how the search for hypothetical pedigrees had become the central research program for morphologists and paleontologists. There was no purely morphological technique by which they could detect homologies, although each side in a debate would insist that the convergent characters it had detected were obvious. Morphologists trusted their instincts to judge which were true homologies, and their instincts were, in fact, formed by the hypotheses they constructed about evolutionary relationships. What was obvious to one morphologist was not necessarily obvious to all his contemporaries, a situation which led to acrimonious debates in which each side accused the other of failing to realize how it had been deceived by superficial resemblances. In some cases a consensus was reached in which the vast majority accepted a particular interpretation of the facts, but in others it proved impossible to reach agreement over the true significance of the structures under study.

Evolutionary morphology will provide much evidence for those historians seeking to argue that scientific knowledge is socially constructed.

The late nineteenth century saw an explosion of interest in non-Darwinian theories of the evolutionary mechanism, sparked in part by the effort to reconstruct phylogenies. Unfortunately for the historian searching for neat generalizations, rival phylogenetic theories cannot be mapped directly onto the well-known debates over the mechanism of change. In general, it is probably fair to say that the anti-Darwinians were more likely to invoke convergence or parallelism. The distinction is not clear-cut: as we have seen in the case of Lankester, even Darwinians might be forced to admit that in some cases adaptive convergences had taken place. But the Darwinians' preference for monophyletic evolutionary trees was fueled by the general assumption that evolutionary novelties would normally (if not inevitably) be unique. The history of life was not predetermined, and most major innovations were likely to have occurred only once. Convergence and independent invention were thus rather unlikely, to be admitted only as a last resort.

Lamarckians seem to have been far more likely to invoke convergence. As noted above, one of the most extreme exponents of the view that humans and apes were only distantly related was Wood Jones, a committed Lamarckian. Jones was prepared to accept that many of the characters common to humans and apes were independently acquired because he saw the organism as far more plastic, far more responsive to the demands of the environment, than the Darwinians. Lamarckism seems to have encouraged the belief that, when placed in similar environments, species would acquire similar habits and hence very similar physical self-adaptations. Morphological structure would be determined far more closely by the environment than was possible in Darwin's theory. It seemed more likely that different species would converge on the same structural adaptation, thus producing similarities so perfect that they might at first be taken as homologies.

Lamarckians shared the widespread distrust of the supposedly undirected nature of Darwinian evolution and were thus predisposed to see many species being driven toward a similar goal by similar adaptive pressures. But Lamarckism was not the only source of the belief that evolution was a more directed process than Darwinism would allow. The anti-Darwinism pioneered by Mivart was characteristic of a widespread distrust of the claim that all evolution was driven by the demands of adaptation—whether the mechanism of adaptation was Darwinian or Lamarckian. Many morphologists and paleontologists were convinced that species possess signficant nonadaptive characters, and they were prepared to admit orthogenetic mechanisms of directed variation to explain the production of such characters. It was precisely this kind of mechanism that was supposed to generate parallel evolution in related species. While the distiction between Lamarckism and orthogenesis seems clear enough in principle, it was seldom all that clear in practice. A similar character described by one anti-Darwinian as an adaptive product of convergence might be seen by another biologist as a nonadaptive feature produced by parallel evolution and othogenesis.

Morphologists could thus disagree profoundly over whether or not evolution was driven by functional adaptations or by purely formal constraints arising from developmental forces internal to the organism. But in practice, they might unite to support the polyphyletic origin of a group of 'related' organisms through either convergence or parallelism. If, as suggested above, Darwinism can be seen as offering a compromise between the rival demands of form and function, perhaps it is not so surprising that its reinvigoration via the synthesis with genetics was ultimately found acceptable by a majority of biologists. Nevertheless, there are some modern biologists who still suspect that the multiple origin of similar characters occurs far more often that most Darwinists are prepared to allow.

ONTOGENY AND PHYLOGENY

For many biologists, the stongest reason for believing that variation was directed in a predetermined manner was the assumed parallelism between the development of the individual and the evolutionary history of its species. The recapitulation theory, or "biogenetic law" as Haeckel called it, was developed by Fritz Müller and by Haeckel himself, originally as a means of using ontogeny (the development of the individual) to throw further light on the course of phylogeny (the evolutionary history of the species). Ontogeny would supplement comparative anatomy, offering new clues as to the course of evolution. Its use in this context was to become increasingly controversial among morphologists, as the phylogenies reconstructed using embryology as a model came into conflict with alternative schemes based on the comparative anatomy of adult forms. This problem led many morphologists to abandon the use of embryology in this context, and led some—of whom William Bateson is the best example—to reject the whole project in which morphology was used to reconstruct the course of evolution. The paleontologists were eventually left as the main supporters of recapitulationism, since some fossil sequences were widely regarded as paralleling individual development.

Although the original purpose of the recapitulation theory was to aid the reconstuction of genealogies, the theory had definite implications for the mechanism of evolution. Individual development would only repeat the sequence of adult ancestral forms if the individual variations which accumulate to give evolutionary change in the species are *additions* to the developmental process, not mere *deviations* from the original course of development. As Stephen Gould's study of the theory shows, this assumption fitted in much more naturally with the Lamarckian theory than with selectionism.[33] Orthogenesis also fitted the model, if one assumed that the developmental process had a kind of inner momentum which constantly tended to push variation further along a course already marked out by the existing pattern of ontogeny.

The recapitulation theory thus tended to reinforce the preference for non-Darwinian mechanisms of change. It also encouraged a temptation to see evolution as an inherently progressive force pushing life towards a preordained goal, just as ontogeny leads toward the mature organism. Haeckel knew that the phylogeny of the animal kingdom must be represented as a branching tree, but when he used human ontogeny as the basis for a popular exposition of evolutionism, he gave the tree a trunk defining a main line of development leading toward the human race. Small wonder that some commentators saw Haeckel's evolutionism as little more than a superficially materialist gloss on the old idealism or transcendentalism of German *Naturphilosophie*. This claim has important implications for our understanding of the whole enterprise of evolutionary morphology. The old transcendentalism had been associated with a linear model of the development of life on earth in which all lower animals were equivalent to immature stages in human ontogeny. It thus promoted the 'ladder' rather than the 'tree' or 'bush' model of evolution. We certainly need to reflect on the extent to which Haeckel and other evolutionists found it difficult to break away from the inherently progressionist image of a ladder of developmental stages.

The relationship between Darwinism and the recapitulation theory has now become a source of some controversy among historians. Robert Richards has claimed that, in their efforts to 'modernize' Darwin's own scientific work, historians such as Gould and myself have exaggerated the gulf between his thinking on embryology and Haeckel's recapitulationism.[34] According to Richards, Darwin has been presented as the pioneer of the branching model of evolution and hence as someone who had no interest in the ladder of developmental stages as a model for the history of life on earth. Richards claims that Darwin did accept recapitulation, and made it the basis of a progressionist view of evolution little different from Haeckel's. The first of these claims is plausible, but the effort to turn Darwin into a model Haeckelian goes too far when it portrays him as having

---

33. See Stephen Jay Gould, *Ontogeny and Phylogeny*, chap. 4. On embryology, see also Mark Ridley, "Embryology and Classical Zoology in Britain."

34. Robert Richards, *The Meaning of Evolution*.

a vision of evolution as the unfolding of a progressive trend aimed at the production of more 'mature' organisms.

To understand the significance of this debate, we must be clear about the various possible relationships between ontogeny and phylogeny. Darwin argued that ontogeny helped to reveal taxonomic relationships, because some forms which are widely different as adults pass through similar developmental stages which can only be explained on the assumption that they shared a common ancestry. Thus the barnacles he studied so extensively, despite looking nothing like crustaceans as adults, revealed their origins as crustaceans in their larval stages. This observation was no more than an extension of von Baer's law that specialized adult features are added later in the developmental process—a law that was introduced to emphasize the falsity of the ladder model of development. The critical question is whether early stages in ontogeny recapitulate the adult forms of ancestral stages in the species' history. Darwin did believe that this might happen in some cases, but his thinking on this question was fully compatible with the branching model of relationships, and shows little compromise with the old idea of a main line of development leading towards humankind.

The point is that even in the branching model, embryos may resemble ancestors because both lack the later specializations developed by modern adult forms. If—as Darwin certainly believed—the early members of any group are very generalized in structure, then the embryo of a specialized modern species may indeed resemble that generalized ancestor just before it acquires its final specializations. The Darwinian version of recapitulationism is fully compatible with the emphasis on evolution as a process driven by adaptation and specialization. To the extent that Haeckel still tended to present evolution as a tree with a trunk defining a main line leading towards humankind, he encouraged the belief that the process of individual development somehow drove evolution towards a predetermined goal. Elements of the ladder model are incorporated into the branching tree, because the status or grade of the 'lower' animals is defined by the point at which their phylogeny branches off from the main line leading toward the goal of the human form. The evolutionary bush becomes a Christmas tree with the human race as the angel on the top of the trunk.

At a more restricted level, the history of any particular group could be represented as a predictable ascent through a series of stages leading toward 'maturity' and—to carry the analogy to its ultimate conclusion— to senility and death (extinction). Here the analogy between ontogeny and phylogeny led not to a progressionism based on humankind as the goal of evolution, but to a more pessimistic image of internal forces pushing each group blindly on toward the overdevelopment of once-useful structures, generating bizarre and maladaptive forms that were seen as expressions of racial senescence or senility. The recapitulationism favored by many paleontologists thus led to support for internally directed mechanisms of evolution, orthogenesis, and parallelism.

The principles underlying the biogenetic law were outlined in Haeckel's *Generelle Morphologie,* although sheer lack of embryological information prevented him from applying it in some areas where it would later prove useful. Here he stated his conviction that there was a causal link between the evolutionary history of the species and the developmental process of the modern individual:

> Ontogeny is the short and rapid recapitulation of phylogeny, conditioned by the physiological functions of heredity (reproduction) and adaptation (nutrition). The organic individual . . . repeats during the rapid and short course of its individual development the most important of the form-changes which its ancestors traversed during the long and slow course of their palaeontological evolution according to the laws of heredity and adaptation.[35]

The same thesis was developed in chapter 12 of Haeckel's *History of Creation* and became the central theme of his *Evolution of Man.* Embryology provided a direct window looking back into the past. Instead of merely identifying groups united by common descent, one could actually reconstruct the sequence of ancestral forms through which each modern species had passed—although it was expected that the two lines of evidence would support one another.

In the *Generelle Morphologie* and more fully in the *History of Creation,*[36] Haeckel referred to the pioneering application of the recapitulation theory to the phylogeny of the Crustacea described in Fritz Müller's *Für Darwin* of 1861. The details of Müller's attempt to reconstruct the ancestry of the Crustacea are outlined in chapter 3. But we shall also see how Müller was conscious of the limitations under which the theory operated: one could only expect it to be valid if the variations accumulated by evolution were created by addition to the growth process. If the new character was formed by adding on a new stage to ontogeny, leaving the original growth pattern intact, then the old adult form became a stage in the ontogeny of the new individual. If the old pattern of growth were merely distorted, then the old adult form would disappear, and the old and the new species (assuming the former survived) would merely share

---

35. Haeckel, *Generelle Morphologie,* II, p. 300, translation from Russell, *Form and Function,* p. 253.

36. Haeckel, *Generelle Morphologie,* II, p. lxxxvi, and *History of Creation,* II, pp. 174–178.

the same pattern of development up to the point where the deviation occurred.

The assumption that variation occurred by the addition of developmental stages fitted in far more comfortably with the Lamarckian than with the Darwinian version of the evolutionary mechanism. The purposeless, multidirectional variations postulated by the selection theory are easily understood as distortions of the existing developmental process. But if deliberately acquired characters are inherited, it is easier to imagine them as additions to the original growth pattern which are then incorporated into ontogeny by an acceleration of the developmental process. Lamarckism thus turned out to be by far the best conceptual foundation upon which to build the recapitulation theory, and Haeckel had no qualms about endorsing it openly. The claim that growth could be accelerated to incorporate old adult forms into the ontogeny of new species was explored in more detail by the American school of neo-Lamarckism. Paleontologists such as E. D. Cope and Alpheus Hyatt invoked the acceleration of growth and the inheritance of acquired characters to explain what they saw as parallels between the fossil history of certain groups and the ontogeny of later individuals.[37]

Whether openly supportive of Lamarckism or not, many morphologists set out to explore the possibility that ontogeny could be used in the reconstruction of phylogeny. At the start of a major project to study the embryology of the Mollusca, Lankester wrote the following in 1874:

> The success which had attended Fritz Müller's investigation of the Crustacea, and his celebrated 'recapitulation hypothesis,' according to which we have, in the development of every individual organism, a more or less complete epitome of the development of the species, so that the series of changing forms passed through between ovum and adult form are but a series of dissolving views or portraits (often very much marred) of its line of ancestors—this, I say, led me to hope that materials might be found in the developmental history of the Mollusca for constructing their genealogical tree.[38]

The evolutionary significance of embryology was also stressed by Francis Balfour in the introduction to his "Monograph on the Development of the Elasmobranch Fishes" of 1876–78: "My object will have been fully attained if I have succeeded in adding a few stones to the edifice, the

37. See Edward Drinker Cope, *The Origin of the Fittest* and *The Primary Factors of Organic Evolution;* and Alpheus Hyatt, "On the Parallelism between the Different Stages of Life in the Individual and Those in the Entire Group of the Molluscous Order Tetrabranchiata." On the American school, see Bowler, *Eclipse of Darwinism,* chap. 6.
38. Lankester, "Observations on the Development of the Pond-Snail," p. 365.

foundations of which were laid by Mr Darwin in his work on the *Origin of Species.*"[39]

By the standards of the time, the techniques developed for studying embryos represented a high level of technological sophistication. As much evidence as possible was gleaned from observation of the living embryo, but internal structures could only be studied by complex processes designed to allow visible cross sections to be cut. Minute embryos were dehydrated in alcohol, and the protoplasm converted into a solid form by the use of reagents such as 'osmic acid' (a solution of osmium tetroxide in water). Since osmic acid blackened the specimen with metallic osmium, various techniques were used to render it more transparent. Specimens were then embedded in wax, sectioned with a microtome, and studied under a high-powered microscope.[40]

The information provided by these observations then had to be interpreted to ascertain its phylogenetic significance. Most embryologists realized very quickly that the initial expectation that ontogeny would reveal the direct course of phylogeny was unlikely to be fulfilled. There were too many processes that might interfere with the direct preservation of ancestral stages, even if this preservation were to be expected from the theory of heredity. In his *Evolution of Man* Haeckel distinguished between "palingenetic processes" in which ancestral forms were preserved, and "kenogenetic processes" which distorted the phylogenetic record.[41] In a paper of 1875 Balfour warned against one such distorting agency:

> If the genealogical relationships of animals are to be mainly or largely determined by embryological evidence, it becomes a matter of great importance to know how far evidence of this kind is trustworthy.
>
> The dependence to be placed on it has been generally assumed to be nearly complete. Yet there appears to be no *a priori* reason why natural selection should not act during the embryonic as well as the adult period of life; and there is no question that during their embryonic existence animals are more susceptible to external forces than after they have become full grown: indeed, an immense mass of evidence could be brought to show that these forces do act upon embryos, and produce in them great alterations tending to obscure their developmental histories.[42]

39. Francis Balfour, "A Monograph on the Development of the Elasmobranch Fishes," reprinted in Balfour, *Works,* I, pp. 203–520, see p. 206 (originally published 1876–78).

40. For description of the techniques used by late nineteenth-century embryologists, see E. W. McBride, *Textbook of Embryology (Invertebrata),* chap. 2; and B. Hatschek, *Amphioxus,* chap. 1.

41. Haeckel, *Evolution of Man,* I, pp. 10–11.

42. Balfour, "A Comparison of the Early Stages in the Development of Vertebrates," p. 198; reprinted in Balfour, *Works,* I, 112–133, see p. 112.

More generally, Balfour's *Treatise on Comparative Embryology* commented on the different effects that natural selection could be expected to have on fetal and larval developments. Fetal growth would always tend to become simpler or more direct, thus obscuring ancestral lines of development. Larval forms, which had their own active lifestyle, could acquire secondary adaptations of no phylogenetic significance, although they would preserve those ancestral characters that were actually needed by the larvae. The result was that ancestral characteristics were more likely to be lost in a fetus, but masked in a larva.[43] Balfour's work on the elasmobranches, like Lankester's on the molluscs, was also plagued by the problem of large yolks in the eggs, which substantially distorted the structure of the embryos in a way that again concealed ancestral characters.

Balfour and Lankester never abandoned the hope that their techniques would allow them to identify and eliminate the kenogenetic characters, leaving a valuable residue that would still throw important light on phylogenetic issues. But as the nineteenth century drew to a close, the majority of morphologists became suspicious of the claims originally made on behalf of the recapitulation theory. Many processes obscured the repetition of ancestral stages, and in some areas the phylogenies worked out on the basis of embryology conflicted with those based on the comparative anatomy of adult forms. E. S. Goodrich, who certainly retained his interest in the reconstruction of phylogenies, called Haeckel's biogenetic law an "incautious generalization" which had "done more to delay the progress of sound views on phylogeny than any other modern speculation." [44] The details of this growing disillusionment with the recapitulation theory are explored in the next two chapters of this study. But even more disturbing for the evolutionists was the growing chorus of opinion that morphology was simply incapabable of answering phylogenetic questions.

We can see this suspicion developing in the work of Adam Sedgwick, who was Balfour's favorite student and who replaced him as reader in animal morphology at Cambridge. In 1894 Sedgwick published an article attacking the credibility of embryological evidence as a guide to ancestral stages in evolution. He now insisted that neither the recapitulation theory nor even von Baer's law of increasing specialization represented a valid link between individual development and evolutionary history. The claim that the early stages of development helped to reveal ancestral relationships was empirically false: "Embryos of different members of the same group often resemble one another in points in which the adults differ, and differ from one another in points in which the adults resemble; and it

---

43. Balfour, *A Treatise on Comparative Embryology,* reprinted in Balfour, *Works,* II and III, see III, pp. 360–363 (originally published 1881–82).

44. E. S. Goodrich, *Living Organisms,* p. 146.

is difficult, even if possible, to say whether the differences or the resemblances have the greater zoological value (because we have no clearly defined standard of zoological value)."[45] Sedgwick argued that the earliest stages of development did not remain unchanged during evolution because "variations are . . . inherent in the germ and affect more or less profoundly the whole of development."[46] As a result, development was invariably altered in a way that eliminated ancestral stages. Only in the case of larvae, where developmental stages remained useful during the active larval stage, would a characteristic be preserved—and then only if the environment and habits of the larvae remained unchanged.

Sedgwick remained a morphologist, but he can no longer be considered an *evolutionary* morphologist. His widely praised *Student's Textbook of Zoology* (1898–1905) was a massive compilation of morphological detail but seldom offered any comment on phylogenetic issues. Another product of the Cambridge school, William Bateson, also lost interest in evolutionary morphology and went on to become one of the founders of the new science of genetics. The preface to his 1886 article on the ancestry of the Chordata already notes that the morphologists' search for genealogical trees had become "the subject for some ridicule, perhaps deserved."[47] The introduction to Bateson's *Materials for the Study of Variation* of 1894 repudiates the whole enterprise of evolutionary morphology, along with the assumption that evolution is a continuous process shaped by the demands of the environment. Bateson's transition from the study of saltations to the study of heredity, and his subsequent career as a geneticist, lie outside the scope of the present study.[48]

In America, doubts about the ability of morphology to reconstruct the past were expressed by Alexander Agassiz in an address to the American Association for the Advancement of Science delivered as early as 1880. Declaring that where the fossil record was lacking, morphology led to the unrestricted play of the imagination, Agassiz insisted that "the time for genealogical trees is past."[49] In his own work on the Echini, Agassiz found it possible to construct a number of equally plausible genealogical trees, which could not be tested in the absence of fossil evidence. Some characters must be homologous, others the products of convergence: but how

45. Adam Sedgwick, "On the Law of Development Commonly Known as Von Baer's Law," p. 38.

46. Ibid., p. 41.

47. William Bateson, "The Ancestry of the Chordata," reprinted in Bateson, *Scientific Papers*, I, 1–31, see p. 1 (originally published 1886).

48. See Peter J. Bowler, introduction to the reprinted edition of Bateson's *Materials for the Study of Variation*. On Bateson's later career, see B. Bateson, *William Bateson: Naturalist;* and R. C. Olby, *Origins of Mendelism.*

49. Alexander Agassiz, "Paleontological and Embryological Development," pp. 388–389.

could the morphologist decide which was which? Darwin himself wrote to Agassiz expressing the hope that he had "over-estimated the difficulties to be encountered in the future."[50] It was not, in fact, until the late 1890s that Agassiz's predictions were vindicated by the start of a more general move against evolutionary morphology. Thomas Hunt Morgan provides perhaps the best American equivalent of Bateson: a biologist who began his career as an evolutionary morphologist but who abandoned the project in the course of the 1890s to concentrate first on a saltative view of evolution and then on the new science of genetics.[51]

Bateson and Morgan provide classic illustrations of what Garland Allen has called the "revolt against morphology" at the end of the nineteenth century.[52] Their influence should not, perhaps, be exaggerated; the following chapters will show that evolutionary morphology remained an active field in the 1890s and went into decline only in the early years of the new century. Morphology itself survived, and major figures such as Goodrich continued to believe that it could be used to answer phylogenetic questions. By the early years of the new century, however, E. W. MacBride was perhaps the only embryologist who still saw the recapitulation theory as a guide to phylogeny. In 1895 he had responded to Sedgwick by defending the use of embryology to reconstruct phylogenies.[53] MacBride's *Textbook of Embryology (Invertebrata)* of 1914 admitted that the recapitulation theory was now considered out of date but insisted that new theories were not necessarily better than old ones. Recapitulation certainly happened in some cases, and might be a more general guide to phylogeny. Where the fossils were lacking, it was still the only guide to phylogeny.[54] MacBride remained one of the last outspoken supporters of Lamarckism.

By the turn of the century, MacBride's was a voice crying in the wilderness. Curiously, it was the paleontologists now who remained the chief advocates of the recapitulation theory. In the 1860s and 1870s the work of American paleontologists such as Cope and Hyatt had uncovered apparent parallels between the evolutionary history of certain groups and the ontogeny of later individuals. Hyatt's studies of fossil cephalopods

50. Darwin to Alexander Agassiz, 5 May 1881; *Life and Letters of Charles Darwin,* III, 245.

51. Morgan's evolutionary morphology will be mentioned in chapter 3; see, for example, his "Contribution to the Embryology and Phylogeny of the Pycnogonids." On his subsequent career, see Garland Allen, *Thomas Hunt Morgan.*

52. On the revolt against morphology, see Allen, *Life Science in the Twentieth Century.* For qualifications of this position, see, for example, Jane Maienschein, *Transforming Traditions in American Biology.*

53. MacBride, "Sedgwick's Theory of the Embryonic Phase of Ontogeny as a Guide to Phylogenetic Theory," p. 327.

54. MacBride, *Textbook of Embryology (Invertebrata),* p. 22.

such as the ammonites revealed patterns of progress and degeneration in the course of geological time. The same pattern seemed to be present in the shells of individual animals recovered from the later geological periods (the ammonite shell preserves the whole sequence of chambers occupied by the animal in the course of its life). Hyatt was thus led to claim an exact parallel between the life cycle of the individual and the evolutionary history of the group, including the degeneration into senility and death.[55] When the recapitulation theory came under strong attack in the 1890s, it was defended by paleontologists such as Francis Bather and James Perrin Smith, while Hyatt's concept of racial senility was upheld by H. W. Shimer as late as 1906.[56]

By the 1920s some paleontologists were beginning to look for more naturalistic explanations of the decline of the ammonites, although in 1932 J. B. S. Haldane still thought that that Hyatt's evidence was difficult to explain in Darwinian terms.[57] Most modern Darwinians followed the argument put forward in Gavin De Beer's Embryology and Evolution of 1930: there is simply no reason to suppose that the effect of mutation in producing new characters will work by adding those characters onto the existing adult form. Old adult characters are destroyed by the modifications of ontogeny needed to produce new ones. In recent years, however, molecular biologists have begun to suspect that genetic mechanisms might sometimes be additive in a way that allows the manifestation of phylogenetically ancient characters before recent ones in ontogeny.

THE BASE OF THE TREE

Whatever the problems encountered by the evolutionary morphologists in their efforts to define how structure could be interpreted as a clue to ancestry, their work remained the only avenue by which evolutionists could approach some of the deepest questions concerning the origins of phyla. Morphologists could certainly address problems such as the origin of the mammals or of particular mammalian groups, but here their work was increasingly overshadowed by the discoveries of the paleontologists. But the origin of the major phyla, including the vertebrates themselves, lay

55. See Hyatt as cited in note 37.

56. See Francis Bather, "The Recapitulation Theory in Palaeontology"; James Perrin Smith, "The Biogenetic Law from the Standpoint of Paleontology": H. W. Shimer, "Old Age in Brachiopoda." See also Ronald Rainger, "The Continuation of the Morphological Tradition in American Paleontology." On the collapse of recapitulationism in other areas, see Nicolas Rasmussen, "The Decline of Recapitulationism in Early Twentieth-Century Biology."

57. J. B. S. Haldane, The Causes of Evolution, pp. 23–28; for paleontologists' reevaluation of Hyatt's evidence, see for, instance, Edward W. Berry, "Cephalopod Adaptations"; and Carl Owen Dunbar, "Phases of Cephalopod Adaptation."

shrouded in the mists of pre-Cambrian time where the fossil record was nonexistent. The one major discovery of a pre-Cambrian 'fossil'—the so-called *Eozoön canadense* (supposedly a giant foraminiferan from the Laurentian rocks of Canada)—was soon discredited.[58] If evolutionists were to reconstruct the earliest stages in the history of life on earth, they would have to do it from the indirect evidence provided by morphology. When interest in evolutionary morphology diminished in the early twentieth century, those biologists who turned away were effectively abandoning all hope of dealing with certain fundamental evolutionary questions. Only with the emergence of techniques of molecular biology, along with the discovery of genuine (if still puzzling) pre-Cambrian fossils, would these questions be reopened in the later twentieth century.

The next two chapters of this study explore two episodes in the attempt to reconstruct the origins of major phyla. Of greatest general interest, perhaps, was the origin of the vertebrates. But no study of the reconstruction of life's ancestry should be confined to the vertebrates, lest we convey the impression that attention was solely centered on the 'main stem' of evolution leading toward humankind. Many biologists spent their whole careers studying the structure and origin of invertebrate groups, and their work encountered all the problems that arose in the study of vertebrate origins, even though they were dealing with groups whose evolution led off in an entirely different direction. The study of arthropod origins in chapter 3 is an attempt to redress the balance of a historiography which all too often ignores the 'side branches' of evolution.

Yet to use terms such as *main stem* and *side branch* reminds us that evolutionists had to confront the question of how the major phyla fitted together to form the overall branching tree of relationships. The debates studied in chapters 3 and 4 are just two examples chosen from the many controversies which erupted among evolutionary morphologists attempting to explain the origin of the major phyla. To understand their significance, we need to know a little about the general problem encountered by biologists seeking to reconstruct the earliest phases of the history of life on earth. How exactly were the major invertebrate phyla related to each other, and to the vertebrates? What were the major steps leading from the earliest origins of life to the first metazoans and the wormlike organisms from which it was believed all the higher phyla had diverged? These were the questions that a generation of evolutionary morphologists set out to answer. In effect, they were trying to determine the basic structure of the tree of life. Did it have a single root and a main stem—and if so, how were the side branches related to this stem?

The starting point for the tree of relationships was the origin of life

58. See Charles F. O'Brien, "*Eozoön canadense:* The Dawn Animal of Canada."

itself. Darwin had evaded this issue by writing in the conclusion of the *Origin of Species* of life being "breathed" into "some one primordial form."[59] The topic was certainly a controversial one, and would remain so through the rest of the century.[60] In the 1870s, the spontaneous generation of life from inanimate matter was defended by the professor of pathological anatomy at University College, London, Henry Charles Bastian. In his books *The Beginnings of Life* (1872) and *Evolution and the Origin of Life* (1874), Bastian insisted that to be consistent, the evolutionists had to accept the natural origin of life itself. Haeckel was certainly willing to endorse this proposition, and in his *History of Creation* he openly supported the spontaneous generation of life at some early stage in the earth's history.[61]

Spontaneous generation was supposed to produce the simplest level of life, forms of undifferentiated protoplasm that Haeckel called the "Monera." There was an embarassing episode in which T. H. Huxley was led to believe that apparently organic matter dredged up from the sea bed might represent this simplest original form of life and named it *Bathybius haeckelii*.[62] Note again the tendency to assume that an early stage in evolution, in this case the *Urschleim* from which all life began, still existed in the world today. Unfortunately the biologists on the *Challenger* expedition proved that the substance was an inorganic precipitate formed when chemicals were added to the sea water. Undaunted by this, Haeckel continued to argue that from the undifferentiated Monera had evolved the first single-celled organisms, coining the name "Protista" for the kingdom of single-celled organisms that represented the base of the tree of life.

The next step was the evolution of the first multicelluar organisms, and here Haeckel's application of the recapitulation theory achieved one of its most enduring successes. In his monograph on the calcareous sponges of 1872 Haeckel announced that the most important embryonic form in the whole animal kingdom was that of the *gastrula*, a simple sphere consisting of two concentric layers of cells, which appeared at an early stage in the ontogeny of all the metazoa (fig. 2.3).[63] The gastrula offered a clue to the common ancestor of the whole animal kingdom: a similar form, as an adult organism, had lived on the earth in the distant past and had served as the ancestor for all the higher animals. To this common ancestor

59. Darwin, *Origin of Species*, p. 484.

60. See John Farley, *The Spontaneous Generation Controversy*, chap. 7.

61. Haeckel, *History of Creation*, I, 327–342.

62. See Philip F. Rehbock, "Huxley, Haeckel and the Oceanographers: The Case of *Bathybius haeckelii*."

63. Parts of Haeckel's monograph were translated by W. S. Dallas as "On the Calcispongiae"; on the gastrea theory, see p. 254. See also Russell, *Form and Function*, chap. 16.

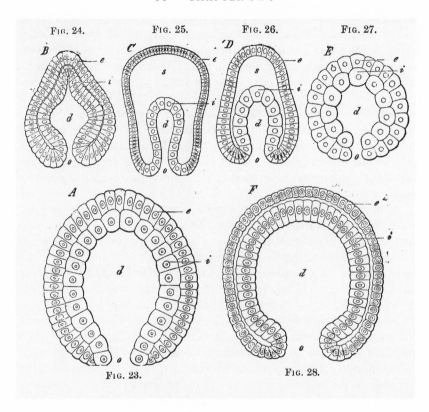

FIGURE 2.3   Gastrula of A: a zoophyte; B: a worm; C: an echinoderm; D: an arthropod; E: a mollusc; and F: a vertebrate. From Ernst Haeckel, *The Evolution of Man* (London, 1879), I, p. 193.

Haeckel gave the name "Gastraea." Lankester had made a similar deduction independently at about the same time, calling the two-layered sphere
the "planula." [64] There was much controversy over the mode of formation
of the Gastraea: were the two layers formed by invagination (an infolding
of an originally single-celled spherical layer) or by delamination (a division of each cell in a single-layered sphere into two)? The Russian embryologist Elie Metchnikoff, originally an enthusiast for Haeckel's approach,
eventually turned against the Gastraea theory. He suggested that the original multicellular creature, the parenchymella, had possessed no central

64. Lankester, "On the Primitive Cell-Layers of the Embryo as the Basis of a Genealogical
Classification of Animals," p. 327.

cavity, but had—like sponges today—cells capable of digestion through-out its body. He then abandoned phylogenetic research and used his views on the origin of life to suggest how the immune system functions.[65] Most late nineteenth-century morphologists, however, accepted that in this case the biogenetic law was valid at least as far as it elucidated the basic form of the Gastrea as a critical stage in the phylogeny of the animal kingdom.

The fate of the original opening into the interior of the Gastraea (which Lankester called the "blastopore") was much debated: did it be-come the mouth or the anus of the higher animals? Berthold Hatschek proposed that the originally round opening became a slit and then fused in the middle to create two openings.[66] The mode of formation of the mouth subsequently became a key feature in the division of the animal kingdom into two major groups, the Protostomia and Deuterostomia. As defined by Karl Grobben in 1908, the Protostomia retained the original blastopore as the mouth and subsequently developed a larval form known as the *trochophore*.[67] From this stem have sprung the annelid worms, the arthropods, and the molluscs, Hatschek himself having shown that the molluscs pass through the trochophore stage characteristic of the annelids (see below). The Deuterostomia form a secondary mouth distinct from the blastopore and then pass through a dipleurula larval stage; they constitute the echinoderms and the chordates. The division of the animal kingdom into these two major groups was a feature of most German teaching on morphology in the early twentieth century.[68]

English-speaking biologists developed a different basis for classifica-tion which held back recognition of the extent to which the higher animals had diverged into these two great branches. Lankester proposed a three-stage classication of the animal kingdom based on the germ layers of the early embryo. Single-celled organisms were the most basic form of life. Above them as the next higher stage of development were the "Diploblas-tica," those organisms which never advanced beyond the two-layered stage characteristic of the Gastraea. These consituted the phylum Coelen-terata, including the jellyfish, corals, and sea anemones. All the higher ani-mals constituted the "Triploblastica," whose early embryo developed a third layer, the mesoderm, in addition to the two layers characteristic of the gastrula stage.

---

65. For a detailed discussion of Metchnikoff's early embryological work, see Alfred I. Tauber and Leon Chernyak, *Metchnikoff and the Origins of Immunology*, chaps. 2–4.

66. Berthold Hatschek, "Beiträge zur Entwicklungsgeschichte und Morphologie der Anneliden."

67. Karl Grobben, "Die systematische Einteilung des Tierreiches."

68. A point confirmed from experience by Ernst Mayr (personal communication).

Lankester was prepared to use this method of classification to define a progressive sequence in evolution:

> The general doctrine of evolution justifies us in assuming, at one period or another, a progression from the simplest to the most complicated grades of structure; that we are warranted in assuming at least one progressive series leading from the monoplast to man; and that *until we have special reason* to take a different view of any particular case we are bound to make the smallest amount of assumption by assigning to the various groups of organisms the place which they will fit on the assumption that they do represent in reality the original progressive series.[69]

In effect, the central trunk of the tree of life would be used to define the various grades of organization. Divergent branches were unified in the sense that their grade was determined by the point at which they separated from the main stem. In a popular article on zoology, Lankester argued that Darwin's theory had introduced the model of a main stem of evolution with numerous diverging branches, the main stem being represented by the line leading toward the vertebrates.[70] All other evolutionary developments—the side branches—were of lesser significance.

The germ-layer theory was to become an important aspect of late nineteenth-century morphology. Tracing the various organs of the adult body back to one of the three original germ layers was central to much embryological research, although the phylogenetic implications of this work were controversial. Much attention also focused on the origins of the coelom, the body cavity within which the viscera of the vertebrates lie. Hatschek and others showed that the coelom developed from the genital follicles of lower forms and that metameric sementation originated as the repetition of the gonads.[71] The brothers Oscar and Richard Hertwig wrote extensively on the phylogenetic significance of the coelom.[72] Since all the higher animals (Lankester's Triploblastica) have at least a rudimentary coelom, they were often called the Coelomata. E. S. Goodrich expounded this theory for English-speaking biologists, and sought to trace the sequence of evolution from the original cavity containing the gonads to the

69. Lankester, "Notes on the Embryology and Classification of the Animal Kingdom," p. 440.

70. Lankester, "Zoology," p. 802.

71. For example, Hatschek, "Studien über Entwicklungsgeschichte der Anneliden."

72. For a translation of part of the Hertwigs's work, see O. and R. Hertwig, *The Chaetognatha;* see also Paul J. Weindling, *Darwinism and Social Darwinism in Imperial Germany,* chap. 4. For other surveys on the origin of the metazoa, see Balfour, *A Treatise on Comparative Embryology,* reprinted in Balfour, *Works,* II and III, chapter 13; Sedgwick, "The Development of the Cape Species of Peripatus," part 3 (1887), pp. 522–538.

coelom of the higher animals.[73] Lankester and others warned that the main body cavity of arthropods and molluscs was not the true coelom but a swollen vascular system known as the *haemocoel*.[74]

My concern in this study lies with the later stages in the history of life, the origins and subsequent evolution of the five major phyla, the annelid worms, the arthropods, the molluscs, the echinoderms, and the vertebrates, or chordates. Haeckel argued that at an early stage in the evolution of what Lankester called the Triploblastica, radial symmetry had been lost as the animals became mobile and developed an axis with a distinct head. The nonsegmented coelomate worms were thus the foundation from which the five major phyla diverged. The debates outlined in the following two chapters represent but a sample of those which raged through the late nineeenth and into the early twentieth centuries over how the genealogical tree of these five phyla should be constructed.[75]

Haeckel's use of progressionist language has led to the criticism that his evolutionism tended all too often to degenerate into a revived idealism in which the human form was seen as the goal of creation.[76] Like Lankester, he wanted to modify the tree metaphor to preserve the image of a main line of development. Yet we cannot accuse him of constructing a ladderlike arrangement of the major phyla (see fig. 2.4). He insisted that each was an independent offshoot from the primitive worms, developing in its own direction: "The Molluscs, Star-fishes, Articulated animals, and Vertebrate animals, do not stand in any close blood relationship to one another, but have originated independently in four different places out of the tribe of Worms."[77] There were, in fact, certain basic groupings. In the diagrammatic phylogenies given in his popular works, Haeckel represented the Arthropoda (crustaceans, insects, and spiders) as an offshoot of the Annelida, or segmented worms, a view accepted by almost everyone. He also included the Echinodermata (starfish and sea urchins) as a more distant offshoot of the same group. It was widely accepted that the echinoderms had once been bilaterally symmetrical animals which had relapsed secondarily into a radial symmetry, although Haeckel's view of their relationship to the annelids would not stand the test of time.

73. E. S. Goodrich, "On the Coelom, Genital Ducts and Nephridia."

74. Lankester, "The Coelom and the Vascular System of Mollusca and Arthropoda." On the rival theories of the coelom's origin and significance at the turn of the century, see Heinrich E. Zeigler, "Über den derzeitigen Stand der Cölomfrage."

75. For a modern approach to the issues discussed in the rest of this chapter, see Pat Willmer, *Invertebrate Relationships*.

76. For example, Russell, *Form and Function*, p. 253; and Bowler, *Non-Darwinian Revolution*, p. 85.

77. Haeckel, *History of Creation*, II, 130.

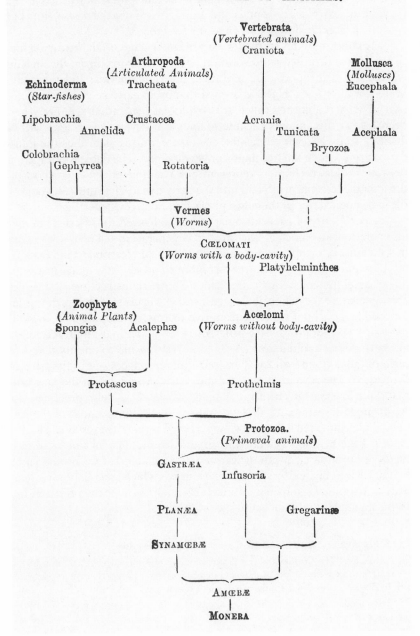

MONOPHYLETIC PEDIGREE OF ANIMALS.

FIGURE 2.4 Monophyletic Pedigree of Animals. From Ernst Haeckel, *The History of Creation* (New York, 1883), II, p. 133.

Curiously, Haeckel's diagrams imply a very loose relationship between the molluscs and the vertebrates—although in his account of the molluscs he went out of his way to insist that they were the most imperfectly organized of the major phyla, since they lacked even the metamerism, or segmentation, of the body which is characteristic of the vertebrates and articulated animals.[78] Here we see the true source of Haeckel's tendency to reintroduce the ladder model into his phylogeny of the animal kingdom. The major phyla are equally independent, but (to borrow Orwell's cynical motto) some are more equal than others. The vertebrate phylum stands above all the others not because these 'lower' animals are immature versions of the vertebrate form, but because the vertebrates have reached a higher level of organization leading to increased mental powers. In the diagram given in his *Evolution of Man,* the vertebrate phylum forms the whole top part of the tree, with the others relegated to insignificant side branches from the lower part of the trunk (fig. 2.5).Although popular opinion relegated the invertebrate phyla to a low status, many biologists spent their whole careers studying the structure of these animals and the relationships between them. To illustrate this, chapter 2 outlines the debate over whether or not the arthropods are polyphyletic. Was there a common ancestor with basic arthropod characteristics from which the crustaceans, insects, and arachnids had radiated out, or had several different annelid worms independently evolved the basic arthropod structure? In the latter case, the Arthropoda would represent a grade of organization achieved by several different phyla, and would not be a true phylum in the evolutionary or genealogical sense. But this was by no means the only major debate over invertebrate phylogeny. Equally controversial was the dispute over the relationship betwen the molluscs and the other invertebrates.

Haeckel saw the molluscs as independently evolved from the unsegmented worms; in effect, they were a group of organisms that had never had a segmented body. This position was later endorsed by Johannes Thiele,[79] and was indirectly supported by Lankester when he constructed a common ancestor for the molluscs which showed no sign of segmentation (fig. 2.6). But some morphologists wondered if the molluscs might have evolved from the annelid worms by loss of segmentation, in which case the earliest molluscs would have been expected to show some traces of the original segments. In 1877, Hatschek pointed out that some molluscs have a larval stage resembling the trochophore, the larval form

---

78. Ibid., p. 155.

79. Johannes Thiele, "Die Stammesverwandschaft der Mollusken." For a modern approach to this debate, see Michael Ghiselin, "The Origin of Molluscs in the Light of Molecular Evidence."

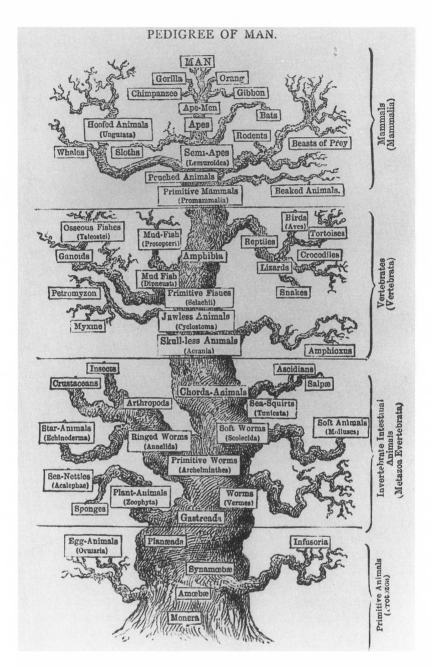

FIGURE 2.5   The Pedigree of Man. From Ernst Haeckel, *The History of Creation* (New York, 1883), II, facing p. 188.

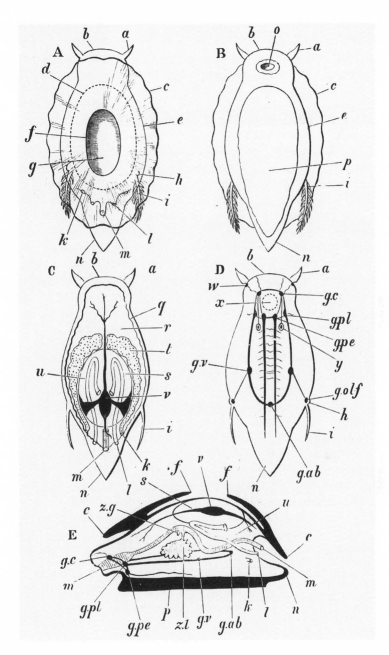

FIGURE 2.6   Lankester's schematic mollusc, the hypothetical common ancestor for the phylum. A: the dorsal aspect; B: the ventral aspect; C: the main internal organs; D: the nervous system and a cross-section cut in the median antero-posterior plane. From E. Ray Lankester et al., *Zoological Articles* (London, 1891), p. 98.

characteristic of the annelids.[80] He proposed his "trochophore theory," in which annelids and molluscs are descendents of a rotiferlike organism, the trochophore larva being a relic of this stage of evolution preserved in their ontogeny. Many authorities, including Karl Heider and Paul Pelseneer, supported this theory, accepting that the molluscs have evolved via loss of segmentation from the annelids and must be included within the grouping that became known as the Protostomia.[81] There were thus two divergent opinions on the origin of the molluscs, and to complicate matters further, Herman von Ihering tried to synthesize the two positions by suggesting that the molluscs might be diphyletic, some having evolved from the annelids, while others, including the cephalopods, had come from the flatworms. Lankester rejected this position on the grounds that it was unlikely that an organ as specialized as the lingual ribbon (a rasping tongue) could have been independently invented in two distinct phyla.[82] The debate over the origin of the molluscs remained unresolved in the early twentieth century and offered yet another example of evolutionary morphology's inability to provide unambiguous answers to such fundamental questions. Molecular biology now suggests that the molluscs are indeed derived from segmented animals.

Most attention, however, focussed on the origin of the vertebrates (chapter 4). Haeckel saw this, the 'highest' phylum, as a distinct line of specialization leading from the unsegmented worms. Eventually he accepted embryological evidence suggesting that the ascidians, or sea squirts, were the most likely intermediates. For Haeckel and those who followed him on this issue, the metamerism, or segmentation, of the vertebrate body was not homologous with that of the annelids and arthropods; each phylum had independently developed this characteristic. But Dohrn soon began to argue that vertebrate metamerism *was* homologous with that of the annelids. In effect, a vertebrate was an annelid which had turned itself upside down and acquired a new mouth. A more radical version of the same position, advocated by William Patten, derived the vertebrates from the arthropods. This theory came closest to presenting the history of life on earth as a ladder, with a main line leading from the unsegmented worms

80. Hatschek, "Beiträge zur Entwicklungsgeschichte und Morphologie der Anneliden" and "Studien zur Entwicklungsgeschichte der Anneliden."

81. See Karl Heider, "Phylogenie der Wirbellosen," pp. 504–511; and Paul Pelseneer, *Mollusca.*

82. Hermann von Ihering, *Vergleichenden Anatomie des Nervensystems und Phylogenie der Mollusken;* Lankester, "Notes on the Embryology and Classification of the Animal Kingdom," pp. 438–439. Note that in this early publication von Ihering's name is given as "von Jhering"; his later publications on biogeography (discussed in chap. 8) were issued under the name of von Ihering, and his book has been listed in the bibliography under this spelling to avoid confusion.

through the annelids to the arthropods and then on to the vertebrates. Only the unsegmented molluscs and echinoderms led off as side branches.

This was a highly controversial theory; most morphologists insisted that the similarities between vertebrates and arthropods were purely superificial. In a detailed consideration of the origin of metameric segmentation, Sedgwick argued for a deep division between the vertebrates and the other phyla. The vertebrate type was of immense antiquity, its ancestor having diverged at a very early stage from the common ancestor of the invertebrate types.[83] His position lined up with that proposed by Hatschek and Grobben: the division of the animal kingdom into the two major groups, Protostomia and Deuterostomia (although Sedgwick did not use these terms). Embryological evidence had led most biologists to suspect that of all the invertebrate types, the lowly echinoderms had the closest link with the vertebrates. The echinoderm larva revealed that the ancestors of this type had been bilaterally symmetrical forms very similar to some of the most primitive living chordates which (to those who rejected Dohrn's thesis) gave the best indication of vertebrate origins. All other invertebrate phyla lay on a different stem.

The tree of life thus had several major branching points at its base, with the molluscs, annelids, and arthropods representing types which had developed entirely separately from the line leading to the echinoderms and vertebrates. Even so, some biologists managed to see a 'main line' in the history of life defined by the ascent toward the vertebrates. We have already noted this tendency in the writings of Haeckel, Dohrn, Lankester, and Patten. But there were significant developments in thinking about the mechanism of progress. Patten still postulated a kind of internal driving force, akin to the French philosopher Henri Bergson's *élan vital* (chapter 4). But by the early twentieth century many biologists had rejected the notion of an inbuilt perfecting principle in favor of the belief that progress occurred when living organisms were stimulated by a challenging environment. Already in the 1870s, Dohrn and Lankester emphasized how degeneration could occur when organisms took up a less stimulating lifestyle such as crawling on the seabed or parasitism. This made the response to the environnment the key to evolutionary development: an active way of life guaranteed future progress; a passive one led to degeneration.

The Lamarckian embryologist E. W. MacBride also believed that active habits were the key to progress, and used this to justify the model of a main line of evolution. From the earliest free-swimming creatures, two lines had developed. One arose from creatures which had adapted to a relatively sedentary life on the seabed; this had given rise to the "creeping and gliding molluscs," the "burrowing annelids" and the arthropods, that

83. Sedgwick, "The Origin of Metameric Segmentation," pp. 69–70.

is, the Protostomia. The other line arose from a form called the dipleurula (still visible as a larval stage in many groups), which had retained an active, free-swimming lifestyle and had continued to progress until it gave rise to the earliest fish. The echinoderms and the lowest chordates represented side branches from this stem which had, like the other major branches of the tree of life, adapted to a life on the seabed. All these side branches represented the "weaker brethren [which] have given up the struggle and sought safety in a bottom life." MacBride concluded, "It is, therefore, broadly speaking, true that the Invertebrates collectively represent those branches of the Vertebrate stock which, at various times, have deserted their high vocation and fallen into lowlier habits of life."[84]

MacBride's language here is a fairly blatant example of the way in which the basic message of Darwinism—that the evolution of life must be visualized as an open-ended, ever-diverging tree—could be modified or perverted in order to retain the view that the vertebrates, and hence the human species itself, represent the goal toward which the main line of evolution has been striving. His claim effectively ignored the existence of the other great line of animal evolution, which also led to some highly organized and active creatures such as the cephalopod molluscs. The belief that constant struggle against a challenging environment guarantees progress was shared by many evolutionists, Darwinians and non-Darwinians alike. Whatever the details of the theories outlined in the following chapters, we must remain alert to the value-laden terms in which the relationships between the various groups could be described. The concluding chapter of this study will survey the constant interchange of metaphors and models between evolutionary biologists and writers seeking to derive a moral message from the idea of progress.

84. MacBride, *Textbook of Embryology (Invertebrata)*, p. 662.

3

ARE THE ARTHROPODA A NATURAL GROUP?

By the end of the 1860s Haeckel had established the principles upon which the search for a monophyletic pedigree of the animal kingdom should be established. He had defined the methods of comparative anatomy and embryology that would be used to identify relationships and had offered hypotheses as to how the major phyla fitted together to form a branching tree with a single root in the Gastraea. For Haeckel and many others, the branch leading toward the vertebrates was the most important, defining the main stem or trunk of the tree leading toward humankind. Others were merely side branches, leading to different forms of organization unrelated to our own. Of these, however, the branch leading to the Arthropoda (crustaceans, insects, and spiders) was the most significant, since it generated a major group of animals which paralleled the vertebrates in that their bodies were characterized by metamerism, or segmentation.

The topic of vertebrate origins itself became a major source of controversy (see chapter 4). But it soon became clear that the question of arthropod origins was also problematic. By the 1880s a major debate was underway among invertebrate morphologists, centered on whether or not the arthropods were polyphyletic. No one doubted that the arthropods had evolved from the annelid worms, and the most economical assumption was that the peculiar arthropod structure (segmented body with a hard exoskeleton and hollow, jointed legs) had evolved once as an evolutionary novelty allowing the organism more protection and better means of locomotion. The primitive arthropod line had then diversified into subbranches leading to the crustaceans, insects, and spiders. The type was thus monophyletic: it had a single origin, and its most basic structural characters were homologous in the three subgroups, preserved as an inheritance from the common ancestor.

But doubts soon began to arise based on the suspicion that the differences between the various orders were so great that it might be better to asssume that the Arthropoda were polyphyletic, that is, that several different lines striking out from the annelid worms had independently 'discovered' the arthropod type of structure. In this case, the underlying similarity was not the product of common descent. Parallel or convergent evolution had produced results in several phyla that were so similar that naturalists had been misled into thinking that they formed a unified group. If this were the case, it would force morphologists to reassess the criteria they were using to identify relationships. It would also challenge the assumptions underlying the search for the simplest possible arrangement of the branches in the phylogenetic tree, and would force evolutionists to think about the underlying processes responsible for generating major structural innovations.

The debate over arthropod origins thus raised fundamental questions about the nature of evolutionary biology. What is the relationship between classification and phylogeny, and how does one define a 'natural' group? How does the biologist adjudicate between different lines of evidence bearing on the question of origins, when the implications seem to contradict one another? To what extent must theories on the nature of the evolutionary mechanism influence our ideas about the possible relationships between groups? These were all issues that could not have been raised during the old days of idealist morphology. They were new problems created by the advent of evolutionism and which defined the area of debate for much post-Darwinian biology.

Evolutionism forced taxonomists to think more carefully about what they meant by a natural group. The very concept of the Arthropoda had been created on the assumption that the fundamental similarities of structure between crustaceans, insects, and spiders allowed them to be unified into a single natural group. One could postulate an arthropod archetype, an idealized representation of the most basic arthropod structure. Darwin's claim that classification would prove to represent degrees of evolutionary affinity translated the notion of the archetype into the most economical assumption about the group's origin: such a basic similarity must be derived from common descent. Fundamental similarities of form were homologies indicating a unitary origin that might be overlaid by superficial modifications, but was derived from the common ancestor's basic structure. The phylum Arthropoda was thus a natural group in both the taxonomic and phylogenetic senses.

The rival hypothesis forced biologists to confront the possibility that an apparently natural grouping reflected not common descent but the existence of similarities derived independently in separate groups. One could defend the existence of a monophyletic arthropod taxon by redefining the criteria for which groups were to be included, arguing that anything not derived from the common ancestor should not be included in the Arthropoda. More robustly, the phylum Arthropoda would have to be broken up into a number of equivalent phyla—but did this necessarily mean that the term *arthropod* was meaningless? Many biologists who thought that the arthropods were polyphyletic nevertheless accepted that the apparent similarity of structure was significant enough to make the concept of an arthropod a useful heuristic device. One might, for instance, still want to describe the arthropod phyla together in an introductory textbook. The term *arthropod* would denote a grade of organization reached by several phyla. But in terms of what would today be called a cladistic approach to classification, the phyla would be distinct, the similarities purely superficial analogies rather than evolutionarily significant homologies. Evolutionary morphologists thus had to face the very real question of redefining the criteria for a natural group. Similarities that an idealist morphologist could quite easily accept as significant enough to warrant the creation of a natural group were problematic to the evolutionist who believed those similarities could have arisen in two different ways.

Morphologists also had to beware of traditional assumptions about the definitive characteristics of a group. Since all living crustaceans breathe through gills, there was a tendency to regard this as a character unique to the Crustacea. Thus the horseshoe crab, *Limulus,* was automatically treated as a crustacean. E. Ray Lankester's demonstration that it had many features identifying it as an arachnid forced morphologists to realize that breathing through gills is an adaptation that might occur in any arthropodan group. Classes such as the Crustacea and Arachnida had to be identified by a complex of structural features, not by oversimplified assumptions about their preferred habitats.

The claim that a basic type of structure could have arisen independently forced evolutionists to think more carefully about the nature of the evolutionary process. They were not necessarily concerned about the actual mechanism involved (e.g., natural selection or Lamarckism), but they did have to think about how evolutionary novelties might be generated. The Darwin-Haeckel assumption that similarity indicates common descent implied that evolutionary novelties were unique; they could arise only once, and every form displaying that structure must have inherited it from the ancestor that first made the breakthrough. The rival claim that something as fundamental as the arthropod ground plan could have been developed several times in different lines meant that novelties were not unique. They could be seen as independent inventions, good engineering solutions to the problem of how to construct a workable organism that might have been discovered by several different forms faced with similar adaptive problems. This was the process usually called convergent evolution. Or they might be the product of orthogenetic trends that could drive several lines of evolution independently in the same direction (parallelism).

These alternatives certainly had implications for ideas about the evolutionary mechanism: Lamarckians were generally more inclined to accept convergence than Darwinians—although even Darwinians were forced to admit the possibility of independent invention when the evidence seemed strong enough. The theory of internally programmed trends was inherently non-Darwinian in that it postulated a nonadaptive driving force for evolution. But the fact that independently acquired similarities might be explained in terms of adaptive or nonadaptive trends means that it is impossible to identify a single position in the debate over evolutionary mechanisms that was most likely to generate support for the polyphyletic theory of arthropod origins.

Which was the most important line of evidence for assessing whether a character was so fundamental that it had to be regarded as a sign of real (i.e., phylogenetic) relationship, and how did one recognize that some similarities were independently acquired? Haeckel had stressed the role of embryology as the best source of evidence for recognizing ancestral relationships, and Fritz Müller's reconstruction of crustacean phylogeny was one of the first great triumphs of the recapitulation theory. But the relationship between embryology and comparative anatomy was never a simple one, and invertebrate anatomists soon began to challenge the adequacy of the embryological evidence, arguing that differences in adult structure were so great that they made the hypothetical common ancestor derived from embryology seem quite implausible. By the end of the century the recapitulation theory was in retreat, and the conviction that the arthropods were polyphyletic grew in proportion.

Haeckel had intended to include fossil evidence in the reconstruction of phylogenies, but many morphologists at first assumed that they were free to fight their battles without reference to the fossil evidence. The ultimate origin of the major phyla had occurred at such an early stage in the history of life that there was no hope that fossil evidence could throw direct light on any disputed question. The arthropods were already well defined when the fossil record began in the Cambrian. Yet by the end of the century, paleontology began to make a significant contribution to the debate. There were still no pre-Cambrian fossils, but better knowledge of early arthropods such as the trilobites now offered another source of indirect evidence bearing on the group's origin. Stephen Jay Gould has drawn attention to Charles Doolittle Walcott's work on the beautifully preserved Cambrian fossils from the Burgess shale.[1] In fact, better fossils of trilobites had already become available some time before Walcott dis-

1. Stephen Jay Gould, *Wonderful Life: The Burgess Shale and the Nature of History.*

covered the Burgess shale in 1909, marking the advent of a significant input from paleontology into what had begun as a purely morphological debate.

The geological record also allowed some attempts to reconstruct the environmental conditions under which the earliest phases of evolution had taken place. Most morphologists paid at best only lip service to the idea that evolution was driven by the interaction between the organism and its environment. They either postulated directly nonadaptive mechanisms of evolution, or speculated vaguely about the lifestyle to which a structural innovation was adapted. Even in so remote an area as the origin of the major phyla, a more realistic approach had begun in the late nineteenth century. The morphologists' search for abstract structural relationship was being replaced by a new emphasis on the conditions under which evolution took place.

Gould has criticized Walcott for 'shoehorning' the diverse Cambrian fauna into known phyla. Whether or not we accept Gould's own theory about the highly experimental nature of the first diversification of the higher animals, we must be careful not to read modern debates back into the past. Walcott's decision to classify many problematic types as arthropods must be understood in the context of a widespread conviction that the arthropods themselves constituted several distinct phyla. Gould argues that the early evolutionists were obsessed with the need to construct the simplest possible tree of relationships in which diversity constantly increased through time. But the debate over arthropod origins suggests that even by the 1880s significant efforts were being made to challenge the most economical way of reconstructing the tree. No one as yet was suggesting the kind of major burst of organic experimentation postulated by Gould, but it was by no means taken for granted that the tree of life was the product of a steady evolutionary trend. The claim that the arthropods were polyphyletic required its supporters to postulate much more branching at an early stage in the history of life than was required by the model that Gould criticizes. Some biologists, including William Keith Brooks, had also begun to think about the possibility of sudden bursts of adaptive radiation that would explain the abrupt appearance of the Cambrian fauna.

The debate over arthropod origins thus encapsulates many of the issues and trends highlighted in this study. It is important for us to recognize that the better-known debates over vertebrate evolution were not the only source of evolutionary biology in the late nineteenth and early twentieth centuries. There were many specialists in invertebrate anatomy and embryology who saw their branches of biology as fully integrated parts of the new evolutionism. We have already noted that Müller's phylogeny of the Crustacea helped to launch the recapitulation theory. Carl Claus' school at Vienna trained other invertebrate morphologists such as Berthold Hatschek and Karl Heider—and the young Sigmund Freud. E. Ray Lankester worked extensively on invertebrate morphology, and his classic paper proclaiming *Limulus,* the horseshoe crab (fig.3.5), to be an arachnid played a major role in the debate over arthropod phylogeny. In many cases, we have to look behind the scenes to realize that complex debates over arthropod relationships do, in fact, have an evolutionary agenda, but as H. M. Bernard, working at the Huxley laboratory of Imperial College, London, wrote in the mid-1890s, "most modern works dealing with the morphology of any Arthropod form contain discussions as to the probable affinity of the family described with other members of the group, and its bearing on the ancestry of the Arthropods in general."[2] During his early career as a morphologist, Thomas Hunt Morgan published an important paper on the structure and phylogeny of an unusual arthropod group, the sea spiders.[3]

Nor was the debate confined to the new morphological laboratories in the universities. Anton Dohrn made major contributions from his newly founded zoological station at Naples, and several later publications by the station also contributed. Newly discovered arthropod species from around the world were flooding into the great museums, and specialists such as R. I. Pocock of the Natural History Museum in London wrote at a theoretical as well as a descriptive level. Paleontologists from museums and geological surveys also became involved as the debate widened in the later years of the century.

Interest in these highly technical questions was not confined to professional scientists. The arthropods, especially the insects, had always figured prominently in traditional natural history, and there were still many well-informed amateurs who could appreciate what the experts were arguing about. The appearance of T. H. Huxley's *The Crayfish: An Introduction to the Study of Zoology* in 1879 suggests that arthropod anatomy was still seen as an important foundation for elementary biological education. Here Huxley stressed that the morphological relationships used to classify the various species of crayfish made sense only in the light of evolution. He also appealed to the evidence of geographical distribution to back up his proposed phylogeny for the group.[4] At this point, the debate over arthropod origins was only beginning, but when the popular British journal *Natural Science* published a survey of professional opinions on the question "Are the Arthropoda a Natural Group?" in 1897 (discussed below),

2. H. M. Bernard, "The Comparative Morphology of the Galeodidae," p. 305.

3. T. H. Morgan, "A Contribution to the Embryology and Phylogeny of the Pycnogonids."

4. T. H. Huxley, *The Crayfish,* esp. chap. 6.

it was clearly responding to general interest in what was perceived to be an important issue in taxonomy and evolutionary biology. The question still concerns modern biologists, and has become a source of controversy again in recent years following the work of Sidnie Manton (disussed below). Although the techniques used to answer it have changed considerably, the basic alternatives remain the same. The fact that Gould can revive interest in the origin of the arthropods as part of the popular reinterpretation of evolutionism offered in his *Wonderful Life* suggests that the general public can still be persuaded to take some interest in the topic.

THE PROBLEM OF ARTHROPOD ORIGINS

The arthropods provided one of the earliest success stories of evolutionary morphology. Inspired by the *Origin of Species,* Fritz Müller's *Für Darwin* of 1864 had explored many of the problems that would be encountered in the search for evolutionary relationships. Müller used embryology to trace the ancestry of the crustaceans back to the so-called Nauplius larva, which he now used as a model for the hypothetical ancestor of the group. Darwin himself applauded Müller's work, and his genealogy for the Crustacea was expounded at length in Haeckel's *Natürliche Schöpfungsgeschichte* (*History of Creation*). Most accounts of the Darwinian revolution mention Müller's work, but then drop the subject, leaving us to assume that his hypothetical genealogy became an integral part of late nineteenth-century evolutionism. In fact, Müller's triumph was short-lived, because by the late 1870s both Anton Dohrn and Berthold Hatschek had pointed out that if the Nauplius form did represent the common ancestor of the Crustacea, then it was most unlikely that the group was closely related to the other arthropods, which had very different embryonic histories. Dohrn and Hatschek's views were not accepted in their original form, but they fed into a growing debate over the unity of the arthropods which came to a head in the 1890s with the publication of a number of studies proclaiming that the group must be polyphyletic.

The arthropods are segmented animals with a chitinous exoskeleton, in which all or most segments possess paired, jointed appendages which may be variously modified. Traditionally, the Arthropoda were divided into a number of groups, although the relative status of these groups varied according to the opinions of each system's author on their relationship and origin. The five most common divisions follow:

1. Crustacea (crabs, shrimps, barnacles)
2. Arachnida (spiders, scorpions, mites, and ticks)
3. Insecta (insects)

4. Myriapoda (centipedes, millipedes)
5. Onychophora (*Peripatus*)

Some zoologists grouped the Onychophora, Insecta, and Arachnida into a unified phylum, the Tracheata, signifying that all breathed by tracheae, or pores penetrating into the body (many arachnids also breath by so-called book-lungs). The Crustacea formed an equivalent phylum, the Branchiata, or gill-breathing arthropods. This phylum was normally subdivided into the Malocostraca, the higher forms including the crabs and shrimps, and the Entomostraca, a diverse group of what were normally assumed to be more primitive forms.

The main debate over the polyphyletic character of the arthropods centered on whether the Tracheata and Branchiata had a common arthropod ancestry or were independently evolved from annelid worms. Müller's recapitulationist interpretation of crustacean evolution seemed at variance with evidence bearing on the origin of the Tracheata. Some of this evidence came from embryology, but more important was the role of the onychophoran *Peripatus,* several species of which were discovered and studied in the mid-nineteenth century (fig. 3.4). Here a newly discovered living form seemed to help resolve an evolutionary problem. *Peripatus* was a classic 'missing link'—it seemed to be a primitive form of the Tracheata, providing evidence of how this phylum had evolved from the annelids. By the turn of the century it was widely (although not universally) believed that the Tracheata formed an evolutionary sequence. The Onychophora represented the first stage in the evolution from the annelids; these evolved into the Myriapoda, which in turn evolved into the Insecta. Since *Peripatus* was an elongated form quite unlike the Nauplius larva of the crustaceans, the Tracheata and Branchiata must have had separate origins.

Another debate centered on the relationship between the Arachnida and the other Tracheata. When Lankester proclaimed that *Limulus,* the living member of an ancient group called the Xiphosura, was not a true crab and would have to be transferred from the Crustacea to the Arachnida, he effectively created a new category among arthropods, arachnids breathing through gills rather than tracheae. Redefining the Xiphosura in this way linked them to one of the most important groups of extinct arthropods known from the fossil record, the giant eurypterids of the Palaeozoic (fig. 3.1). These too had originally been treated as Crustacea because they breathed through gills, but had now been recognized as 'sea scorpions'—aquatic predecessors of the terrestrial arachnids. If the Arachnida had their origins in primitive aquatic forms unrelated to the ancestors of the other Tracheata, then they too might be an independently evolved phylum.

The other classic group of fossil arthropods is, of course, the trilo-

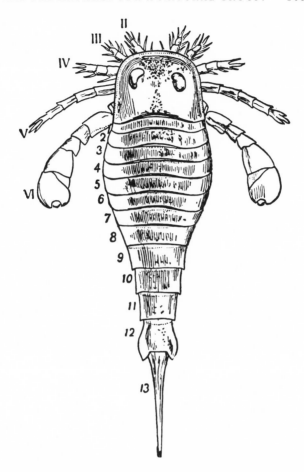

FIGURE 3.1   A eurypterid, or sea scorpion: *Eurypterus fischeri*. From E. Ray Lankester, *Extinct Animals* (London, 1905), p. 282.

bites. In the late nineteenth century these were widely treated as a primitive order of Crustacea, although the reclassification of *Limulus* (which seemed the closest living relative of the trilobites) threw this classification into doubt. As better information on the internal structure of trilobites came to light, their position in the early evolutionary tree of the arthropods came to be hotly debated. Without prejudicing the actual status of the groups, the accompanying table gives some idea of the subdivisions and relationships discussed below.

The number and diversity of insects is immense, and some entomologists vigorously debated their evolutionary relationships. These debates lie

TABLE 3.1.   PRINCIPAL SUBDIVISIONS OF THE ARTHROPODA

Crustacea	Entomostraca (including Phyllopoda or Branchiopoda)
	Malocostraca (including the Decapoda, i.e., crabs and shrimps)
	Trilobita ?
Arachnida	Eurarachnida (true spiders and scorpions)
	Pantopoda or Pycnogonidae (sea spiders)
	Merostomata (Eurypterida, sea scorpions, and Xiphosura, *Limulus*)
Insecta	
Myriapoda	
Onychophora	

outside the scope of the present study. The disagreements over the relative status and origins of the groups were highly technical, and what follows is an attempt to extract the gist of the evolutionary arguments from a welter of details which has driven one amateur (the author) almost to distraction.

THE GENEALOGY OF THE CRUSTACEA

In the *Origin of Species* Darwin had argued that a true classification should "reflect" genealogy. His own classification of the barnacles had been based on phylogenetic hypotheses, but in the *Origin* he refrained from constructing detailed phylogenies, in part because the fossil record was so incomplete. Evolutionary morphology developed as a means of bypassing this problem, especially in those areas where fossil evidence was unlikely to turn up. Inspired by Darwin's book, the German zoologist Fritz Müller set out to explore the possibility of constructing such an evolutionary tree for a single group, the Crustacea. Müller is a classic example of a biologist who significantly influenced the evolutionary debate despite having no professional position. He had been educated in medicine, including some work with the eminent physiologist and microscopist Johannes Müller, but had been forced to leave Germany for political reasons after the troubles of 1848. He moved to Brazil, where he supported himself by, among other things, teaching school. He continued to do important work on crustaceans, and it was from this work that he developed his application of the recapitulation theory, thereby inspiring Haeckel's enthusiasm for the link between ontogeny and phylogeny and providing a model for the evolutionary morphologists of the late nineteenth century.

Müller's conclusions were developed in his book, *Für Darwin* of 1864, translated five years later as *Facts and Arguments for Darwin*. Here he argued that

> if the establishment of such a genealogical tree, of a primitive history of the group under consideration, free from internal contradictions, was possible,—then this conception . . . must in the same proportion bear in itself the warrant of its truth, and the more convincingly prove that the foundation upon which it is built is no loose sand, and that it is more than merely "an intellectual dream."[5]

The first step in the creation of such a tree must be the analysis of relationships among known members of the group. It is necessary to identify homologies, which are by definition signs of common ancestry: two forms share the same basic structure because they have inherited it from a parent form, perhaps with superficial modifications imposed by later adaptive requirements. By extending this argument, it should be possible to reconstruct an outline of the way in which the branches of the tree fit together, even if the nodes from which they have diverged are unknown, that is, extinct. In the course of his analysis, Müller recognized and warned against many of the factors that would tend to mislead the morphologist. He noted that the crabs which are able to leave the water belong to widely different groups and have evolved very different adaptive solutions to this problem.[6] In this case the effects of adaptive convergence are obvious and cannot lead the evolutionist astray.

The real difficulty arose when two different groups had solved the same problem in the same way. Apparent contradictions would then emerge between relationships constructed using different characters. Müller did not address this problem directly, but he was certainly aware that some characters are more reliable guides to genealogical relationships than others. He constructed what look very much like modern cladograms to depict two possible relationships among the species of the genus *Melita*. One relationship was rejected on the grounds that the character upon which it is based is known to be highly variable in other genera.[7]

To reconstruct the original form from which the various kinds of crustaceans had diverged, Müller turned to embryology, thus pioneering the use of the 'recapitulation theory' later popularized by Haeckel. Crustaceans undergo substantial metamorphosis in the course of their individual development. Many Crustacea pass through a larval stage which was

5. Fritz Müller, *Facts and Arguments for Darwin*, p. 2.
6. Ibid., p. 8 and chap. 5.
7. Ibid., p. 11.

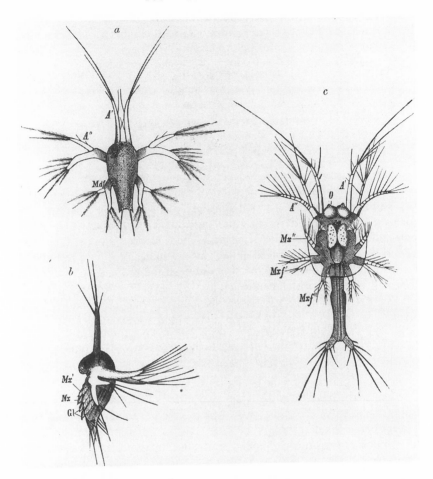

FIGURE 3.2   Larvae of the shrimp *Peneus*. Top left: the Nauplius stage; bottom left: the later metanauplius; right: the Zoea stage. From Adam Sedgwick, *Student's Textbook of Zoology* (London, 1898–1909), III, p. 528.

called the Nauplius (fig. 3.2). This form had first been noticed in the larva of the freshwater copepod *Cyclops* by O. F. Müller in the late eighteenth century and had been named under the mistaken impression that it was a distinct generic type. Fritz Müller now argued that since the Nauplius stage occurred in the development of many crustaceans, it must illustrate the ancestral form from which they had diverged. Indeed, for embryology to be a reliable guide to phylogeny, all crustacea should pass through the Nauplius stage, if the latter were derived from the ancestral form. Müller reported with some satisfaction his discovery of the Nauplius stage in shrimps, where it had hitherto been unnoticed.[8]

Müller also noticed another stage, the Zoea form, which is common in the later development of the higher Crustacea or Malocostraca. This structure must represent the common ancestor from which these higher groups had diverged, the lower forms splitting off from this main stem after the Nauplius stage. The assumption was that if all or most members of a group pass through a similar developmental stage, then this stage will represent the adult ancestor from which they have all diverged, as in the following diagram:

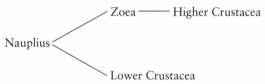

Müller stated the two possible modes of variation in the developmental process by which a species may change, and noted the different implications they have for the preservation of ancestral stages in ontogeny. *"Descendants therefore reach a new goal, either by deviating sooner or later whilst still on the way towards the form of their parents, or by passing along this course without deviation, but then, instead of standing still, advancing still further."*[9] In the former case, the development of the old and the new species will pass through similar stages only up to the point of deviation. Recognizing that ontogeny shares the same path up to a certain point will tell us that two different modern forms share an evolutionary common ancestor, but will not reveal the adult form of that ancestor.

In the latter case, however, the modern species will pass through all the earlier adult stages and will, in effect, recapitulate its whole evolutionary history. Variation by addition preserves old adult forms as stages in the ontogeny of later ones. Müller was aware that there could be no guarantee that the recapitulation theory was valid in the case of the Crustacea. Nevertheless, he assumed that the widespread existence of the Nauplius and Zoea stages in crustacean ontogeny meant that they probably represented the adult ancestors from which their later descendants had diverged. They were not specialized larval adaptations but genuine relics of past adult forms recapitulated in the ontogeny of their descendants.

Crustaceans, like all animals which pass through an active larval stage in their development, are subject to interference in the preservation of an-

8. Ibid., p. 13.

9. Ibid., p. 111. See Stephen Jay Gould, *Ontogeny and Phylogeny*, chap. 4, although Gould says surprisingly little about Müller's work.

cestral stages when the larvae are forced to adapt to different circumstances. Thus the development of the prawn through the Nauplius and
Zoea stages offers a better record than the ontogeny of crabs, since the
prawn maintains the same free-swimming habits throughout its life cycle.[10]
Müller concluded by noting that the Nauplius form represents the extreme outpost of the class, lost in "the gray mist of primitive time." He
did, however, suggest that the insects may have derived from the Zoea
stage of evolution, if some representatives of this stage could have raised
themselves to a life on the land.[11] Thus the other main line of arthropod
evolution had branched off from the crustaceans some time after the type
was established as one of the main animal phyla. For Müller, the arthropods could still be a monophyletic group, as shown in the following
diagram:

Darwin wrote to Müller saying that he found the latter's views on
classification and embryology "very good and original" and that "nothing
has convinced me so plainly what admirable results we shall arrive at in
Natural History in the course of a few years."[12] He also made arrangements for the book to be translated into English by W. S. Dallas.[13] Haeckel
took over the theory of the Nauplius and Zoea stages for his discussion of
the crustacean phylogeny in his *History of Creation* and saw it as a major
vindication for the recapitulation theory.[14] Thanks to the general neglect
of evolutionary morphology, no historian has taken the story any further,
and we have been left with the impression that Müller's work was simply
incorporated as dogma into the evolutionary canon of the post-Darwinian
era. The subsequent history of attempts to uncover the course of arthropod evolution was, however, much more complex.

Müller's view of the ancestral significance of the Nauplius form was
at first supported by the professor of zoology at Vienna, Carl Claus.[15] The

10. Müller, *Facts and Arguments for Darwin*, p. 125.

11. Ibid., pp. 140–141.

12. Darwin to Müller, 10 August 1865, in *The Life and Letters of Charles Darwin*, III,
pp. 37–38.

13. Darwin to Müller, 16 March 1868, ibid., pp. 86–87.

14. Ernst Haeckel, *The History of Creation*, II, pp. 174–178.

15. Carl Claus, *Untersuchungen zur Erforschung der genealogischen Grundlage der
Crustaceen-Systems*, pp. v–viii. On Claus, see Lucille B. Ritvo, *Darwin's Influence on Freud*,
chap. 9.

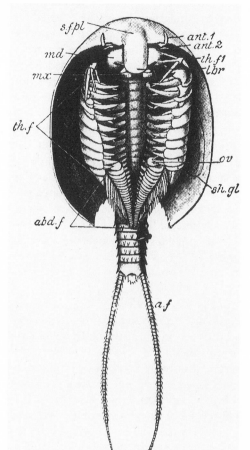

FIGURE 3.3   The phyllopod *Apus*.
From Adam Sedgwick, *Student's
Textbook of Zoology* (London,
1889–1909), III, p. 369.

Vienna school was one of the most influential centers of invertebrate morphology, Claus's students including important figures such as Berthold Hatschek and Karl Heider. Sigmund Freud received his early training in zoology under Claus, a significant point given the role played by recapitulationism in Freud's later psychological theories. Although known as an extremely touchy individual—we shall encounter this characteristic in a debate with Lankester discussed below—Claus was thus the center for an important network of zoologists attempting to understand the evolutionary significance of invertebrate structure.

Claus accepted the common view that the phyllopods were the most primitive crustaceans (fig. 3.3) and postulated that the Crustacea had evolved from a protophyllopod form resembling the modern Nauplius.[16] Yet although Claus supported the recapitulation theory, he rejected the claim that the Zoea form was a stage in the evolution of the Malocostraca.[17] He regarded its lack of segmentation as purely a secondary adaptation to the larva's swimming habits. Francis Balfour, the widely respected lecturer in morphology at Cambridge, defended Müller's position in his *Treatise on Comparative Embryology*. He argued that the same process of adaptation might have led the adult ancestors to lose their segmentation, only to regain it later in their evolution.[18] He believed that the sheer constancy of the Zoea form among the various Malocostracan groups made it unlikely that the same character could have been developed independently by the same secondary adaptation in each. But Balfour's suggestion that segmentation could be lost and then regained was regarded as implausible by most other zoologists, and the ancestral significance of the Zoea form was gradually rejected. Anton Dohrn led the attack and—as we shall see below—became an early convert to the view that the arthropods may be polyphyletic.[19] E. W. MacBride, however, remained loyal to Balfour's position as a means of preserving the recapitulation theory.[20]

Müller's view on the ancestral significance of the Nauplius form retained some support, but raised even more fundamental problems. As we shall see below, even Haeckel had warned that the arthropods might have to be considered a polyphyletic group, and by 1877 it was already becoming apparent that if Müller's interpretation were accepted, it would have major implications for this broader issue. An outline of the subsequent debates can be gleaned from surveys by later embryologists such as Balfour, Korschelt and Heider, and MacBride.[21] The basic problem was this: If Müller's neat embryological model for crustacean evolution was correct, then the Crustacea could no longer be regarded as closely related to

16. Claus, *Untersuchungen*, pp. 100–105.

17. Ibid., pp. 61–68.

18. Francis Balfour, *A Treatise on Comparative Embryology*, in *The Works of Francis Maitland Balfour*, II, pp. 506–507.

19. See Anton Dohrn, "Die Ueberreste des Zoëa-Stadiums in der ontogenetischen Entwickelung der verschiedenen Crustaceen-Familien." On the general acceptance of this position, see, for instance, W. T. Calman, *Crustacea* (E. Ray Lankester, ed., *A Treatise on Zoology*, vol. 7, 3d fascicule), p. 26.

20. See E. W. MacBride, *Textbook of Embryology (Invertebrata)*, pp. 210–211.

21. Balfour, *Treatise on Comparative Embryology*, *Works*, II, pp. 502–511; E. Korschelt and K. Heider, *Textbook of the Embryology of Invertebrates*, II, pp. 192–193 and 309–318; MacBride, *Textbook of Embryology*, pp. 203–220.

the other major grouping of arthropods, the Tracheata. As Müller himself had admitted, living species revealed no actual bridge between the Nauplius form of the Crustacea and the "nearer provinces of the Myriapoda and the Arachnida." [22] The Nauplius was an unsegmented form with three pairs of limbs. Yet the evidence from both embryology and comparative anatomy now suggested that the myriapods and insects had evolved from the (segmented) annelid worms, probably via a form similar to the onychophoran, *Peripatus*. If the recapitulationist history of the Crustacea was compared with the evidence on the origins of the Tracheata, then the unity of the Arthropoda was destroyed. The similarities which had led naturalists to link the Crustacea and the Tracheata in the phylum Arthopoda were independently acquired and were not a sign of a common evolutionary origin.

The debate was conducted at two levels. One was an internal dispute between embryologists dealing with two groups, the Crustacea and the Insecta. The other was a conflict between the two disciplines which contributed to the science of evolutionary morphology: comparative anatomy and embryology. Anatomists studying the relationships among adult forms came to conclusions about arthropod phylogeny that differed substantially from Müller's. Disagreements arose both within studies based on the Crustacea alone and, more important, within those using comparisons of the various arthropod groups. Embryologists were increasingly forced to concede that larvae did not throw direct light on the course of evolution. Some, like MacBride, fought to retain a limited role for recapitulation. Others, like Adam Sedgwick, eventually rejected the use of ontogeny for phylogenetic speculation altogether. More reasonably, Korschelt and Heider admitted that the Nauplius and Zoea had been "robbed of their glory as racial forms" but insisted that the study of crustacean metamorphosis still threw much indirect light on the group's origins. [23] The same authors conceded that comparative anatomy had replaced the study of larvae in the working out of genealogical relationships, but warned that some anatomists had gone too far in their efforts to break up the arthropod phylum. [24]

The inconsistency between the modes of development in the Crustacea and Insecta was pointed out in 1877 by Berthold Hatschek. In a study of the development of the Lepidoptera (butterflies and moths), Hatschek emphasized that in the insects and arachnids, the embryo is always long

22. Müller, *Facts and Arguments for Darwin*, p. 140.
23. Korschelt and Heider, *Textbook of the Embryology of Invertebrates*, II, p. 309.
24. Ibid., III, p. 426.

and distinctly segmented, quite unlike the unsegmented Nauplius.[25] He concluded that the Nauplius stage had no ancestral significance; in the process of segmentation, all embryos must pass through a stage with only three segments, and for some unexplained reason this stage had become modified into the Nauplius form in the course of crustacean development. Hatschek also used comparative anatomy to argue that the similarities between the nervous systems of annelids and crustaceans were so great that they could not have arisen by convergent evolution. The crustaceans must thus have originated from an annelid ancestor, and it was impossible for this ancestor to have resembled the Nauplius—it was more probably a segmented form similar to the primitive phyllopods. Hatschek thus preserved the unity of the Arthropoda at the expense of denying the significance of Müller's recapitulationist interpretation of crustacean evolution.

Hatschek's views were supported by Anton Dohrn, who conducted research on arthropod development at Naples while he was working to create the Zoological Station there. Originally trained as an entomologist, Dohrn began with the assumption that the arthropods were a natural group. He hoped to confirm Müller's theory that the insects had originated from a Zoea-like ancestor which had adapted to dry land, and began looking for traces of the Zoea in the early ontogeny of insects. He also appealed to some early insect fossils as evidence for this conclusion.[26] His initial efforts seemed successful, leading T. H. Huxley to write, "Thunder and lightning—this is the most important thing I have met with in a long time."[27]

By 1870, however, Dohrn's plan to write a book expounding this position had fallen apart, forcing him to set out with a new research program. He wrote the following to his father:

> I have become more and more convinced that the current opinion to combine crustaceans, spiders, millipedes, and insects in a genealogical unit like fishes, amphibians, reptiles, birds, and mammals, has to be abandoned. I had tried to prove the opposite, that is, the unity of the four arthropod classes, and all my initial publications maintained this position. Now I must refute myself. . . . If I have to formulate the result precisely, then I state: "I, Anton Dohrn, see in my embryological investigations no way by which a connection of arthropods and insects (or spiders) is possible."[28]

25. Berthold Hatschek, "Beiträge zur Entwicklungsgeschichte der Lepidopteren," see pp. 131–135.

26. Anton Dohrn, "On Eugereon Boeckingi and the Genealogy of the Arthropoda."

27. Huxley to Dohrn, quoted in Theodor Heuss, *Anton Dohrn: A Life for Science*, p. 63. On Dohrn's arthropod work, see Alfred Kühn, *Anton Dohrn und die Zoologie seiner Zeit*, chap. 4.

28. Dohrn to his father, June 1869, quoted in Heuss, *Anton Dohrn*, pp. 72–73.

Dohrn's rejection of the ancestral status of the Zoea form has already been noted. Indeed, by 1870 he was already beginning to have major doubts about the extent to which ontogeny could be trusted as a guide to phylogeny. He went on to become a strong exponent of the view that functional considerations should play a major role in evaluating the plausibility of hypothetical phylogenies.

Dohrn now recognized the possibility that the four arthropod phyla might have independently evolved their peculiar type of structure from annelid worms. He was more cautious in print, however, than the above-quoted letter might imply. Twelve of his papers, including his attack on the Zoea stage, were published in a collected form under the title *Untersuchungen über Bau und Entwickelung der Arthropoden* in 1870. The introduction to the second half of the collection merely states that the possibility of a polytypic origin must be taken seriously. A slightly later paper begins to move beyond the purely embryological approach, suggesting adaptive scenarios to explain the evolution of various groups of crustaceans. Dohrn also studied the Pycnogonidae, or sea spiders, which he regarded as modified Crustacea, although there was no sign of a Nauplius stage in their development.[29]

Dohrn was now prepared to consider the possibility that several lines of evolution led independently from the annelids towards the different arthropod forms. He went on to develop a theory in which the vertebrates had also evolved from the annelids. On this model, the various arthropod types and the vertebrates all possessed segmented bodies as an inheritance from an annelid ancestor. This hypothesis marked the completion of his transition from pure recapitulationism to the study of the functional or adaptive causes underlying major evolutionary transformations (see chapter 4).

The majority of embryologists who saw their work as a cornerstone of evolutionary morphology were naturally reluctant to accept this conclusion. Francis Balfour at Cambridge agreed that the Nauplius form was much modified from the ancestral state but insisted that it still had some significant ancestral features. He argued that the hinder part of the body had originally been segmented in the adult but that this had been lost secondarily when it became incorporated into the developmental process of the later Crustacea.[30] MacBride, himself a product of the Cambridge school, remained one of the last great exponents of the recapitulation theory and retained Balfour's conclusion as the only way of salvaging the

29. Dohrn, "Geschichte des Krebsstemmes" and *Die Pantopoden des Golfes von Neapel,* esp. pp. 87–88. See also the *Anhang* or appendix to Dohrn's *Ursprung der Wirbelthiere,* pp. 77–87.

30. Balfour, *Treatise on Embryology, Works,* II, pp. 502–503.

theory while retaining some unity for the Arthropoda.[31] He invoked the views of his own teacher, Adam Sedgwick, who argued that the embryonic stage (including the Nauplius in the Crustacea) was developed from what had once been a larval form. Since the larva was active, it was forced to adapt to its own evironment, a process which might distort any ancestral features before they could be incorporated into the developmental process of descendant forms. Sedgwick continued to support this position, although he became increasingly unhappy with such ad hoc assumptions about the distorting power of adaptation.[32] In the end, he became dissatisfied with the whole program in which embryology was used for phylogenetic research (see chapter 2). The intractability of the problems created by the attempt to understand arthropod origins thus contributed to the general decline of evolutionary morphology.

MacBride had preserved both the recapitulation theory and the unity of the Arthropoda by invoking massive secondary modification to explain the differences between the Nauplius form of the Crustacea and the segmented embryos of the other arthropod classes. MacBride also tried to introduce paleontology into his theory, arguing that the trilobites might represent the original form of the arthropods. He even indentified the trilobites as the Nauplius stage in the origin of the phylum. For MacBride, the Tracheata could be regarded as offshoots of the trilobite stock which had invaded the land, just as the higher Crustacea had replaced the trilobites in the oceans.[33] MacBride noted that even so, the fact that the trilobites had compound eyes generated a major difficulty. Crustaceans, the higher insects, and the arachnids also had compound eyes, but the lower Tracheata did not. The Tracheata thus could not have evolved from the known trilobites, and MacBride was forced to postulate a primitive trilobite with simple eyes which had invaded the land to give rise to early members of the Tracheata before the fossil record began. The compound eyes of insects would thus have had to evolve independently of those developed by crustaceans, the latter originating from appendages (since the eyes of modern crustaceans are often on 'stalks').

Here MacBride confronted a major problem for any hypothesis designed to explain the relationship between the arthropod groups. Given the absence of compound eyes in the more primitive Tracheata, it was necessary to postulate the independent evolution of compound eyes even if the Tracheata were derived from primitive Crustacea. If they were monophyletic, the Arthropoda still provided an important example of parallel evolution: two branches had independently solved the problem of

31. MacBride, *Textbook of Embryology*, p. 204.
32. Adam Sedgwick, *Student's Textbook of Zoology*, III, p. 452
33. MacBride, *Textbook of Embryology*, pp. 285–286.

developing a more sophisticated eye in the same way. Once this argument was accepted, then one might reasonably begin to wonder if the other similarities uniting the various arthhropod groups might also be the results of parallelism.

Other zoologists were less willing than MacBride to retain any vestige of the recapitulation theory, although they had to confront the same problem of parallel evolution. Arnold Lang, a former student and colleague of Haeckel, opted for the primacy of comparative anatomy and denied the ancestral significance of the Nauplius stage altogether. On anatomical grounds, one could reconstruct the probable ancestor of the Crustacea as a segmented form that would bring it more into line with the other arthropod classes. There had been no sexually mature Nauplius form in crustacean evolution, and the Nauplius stage was to be explained solely in terms of larval adaptation among later Crustacea.[34] Lang thus explicitly rejected Müller's once-popular recapitulationist interpretation of crustacean evolution.

Lang noted that the closest living analogs of the hypothetical crustacean ancestor were the phyllopods. The Phyllopoda were widely regarded as the most primitive Crustacea, and we have seen that Claus also regarded them as little-modified descendants of the ancestral form of the whole class. Korschelt and Heider, in their *Textbook of the Embryology of Invertebrates* adopted a similar view. Although writing from an embryological perspective, they abandoned Müller's Nauplius theory, postulating a hypothetical Protostracan form as the ancestor of the whole arthropod phylum. This hypothetical ancestor was a much simpler form than any living crustacean, simple enough to serve as a potential starting point for the phyllopod-crustacean line of development, the *Peripatus*-insect line, the trilobites, and the arachnids.[35] Like MacBride, they emphasized that many of the similarities between the various branches of the arthropods were developed independently by convergent evolution.[36]

An almost inevitable outcome of this growing awareness of the need for parallel or convergent evolution in the arthropods was the claim that the group was polyphyletic. Perhaps the Crustacea and the rest of the Arthropoda were the products of entirely separate lines of evolution starting out independently from the Annelida, the similarities all being due to parallel evolution in response to the problem of acquiring limbs and a more rigid body-structure. This position was implicit in a detailed study of crus-

---

34. Arnold Lang, *Textbook of Comparative Anatomy*, I, pp. 407–408. Lang also rejected the ancestral significance of the Zoea stage.

35. Korschelt and Heider, *Textbook of the Embryology of the Invertebrates*, II, pp. 315–316, 333–334, and 427.

36. Ibid., p. 415.

tacean evolution published in 1892 by Henry Meyners Bernard. Balfour had noted that the larvae of the phyllopod genus *Apus* at its Nauplius phase might be the best modern representative of the ancestral crustacean.[37] Bernard worked on *Apus* in Haeckel's laboratory and then extended his studies into a wholesale reappraisal of the origins of the Crustacea. His work is significant in that it represented a notable effort to synthesize morphological, embryological, and paleontological evidence, and in that he stressed the adaptive significance of the evolutionary modifications he postulated. He tried to show that *Apus,* a freshwater phyllopod which appears suddenly from dormant eggs in freshwater pools (fig. 3.3), was the ancestral form not only of the Crustacea, but also of the Trilobita and the Xiphosura (*Limulus*). In his own words, "Apus is perhaps the most perfect 'missing link' which zoology so far possesses, perfect, not only because its morphology is easily deducible from a carnivorous annelid, but also because the mechanical cause of the transformation is apparent." This was the first illustration of the rise of one class from another "by the simple and natural adaptation of the part of a single species of the latter to a new manner of life."[38]

Bernard derived the Apodidae (the ancestral group of which the modern *Apus* is the survivor) from a carnivorous annelid, the first five segments of which had been bent backwards so that the maxillae could push prey toward the mouth. This was an adaptation to browsing on the seabed. From such a hypothetical form, he claimed, the true Crustacea could easily be derived. A few members of this group had survived to the present with relatively little change because they had adapted to life in temporary freshwater pools, where there was little competition. The fact that they existed all over the world was a sign of their great antiquity, suggesting that their origin antedated the division of the globe into the modern biogeographical provinces.[39] This interpretation of crustacean ancestry was compatible with Müller's recapitulationist phylogeny because the Nauplius stage could be seen in the young *Apus*. Indeed, Bernard declared that the Nauplius form is the *Apus* stage in the ontogeny of the modern Crustacea.[40]

Bernard made a significant effort to incorporate paleontological evidence into his theory. His attempt is indicative of the growing importance attributed to fossil evidence in the later years of the nineteenth century,

37. Balfour, *Works,* II, p. 503.
38. H. M. Bernard, *The Apodidae: A Morphological Study,* p. 2.
39. Ibid., p. 9.
40. Ibid., p. 7 and chap. 11.

even among laboratory morphologists who did not normally study fossil forms. He revived the once-popular view that *Apus* was related to the trilobites, noting that if the one could be derived from a carnivorous annelid, then so could the other. An evolutionary tree was used to denote the relationship between the relevant groups.[41] The trilobites were a parallel and initially very successful modification of the annelid ancestor, of which the Ostracoda were a surviving remnant. But Bernard went even further in his effort to depict the aquatic arthropods as a group unified by their origin from a particular type of annelid. He saw *Limulus,* the horseshoe crab, as the survivor of another parallel development from the same ancestral 'bent' annelid, confirming the original view (now widely rejected, as we shall see below) that it was closely related to the Crustacea.[42] He made extensive use of Charles Doolittle Walcott's restorations of trilobites, complaining only that Walcott had ignored the antennae that must have existed even if they were not preserved.[43] Bernard then went on to include the Eurypterida, the Palaeozoic sea scorpions, as yet another branch from the same root. The Xiphosura (including *Limulus*) and the Eurypterida were trilobites modified for two different methods of feeding.[44]

Having unified the Crustacea, Xiphosura, and Eurypterida as descendants of a 'bent' annelid, Bernard ended his study with a chapter discussing the relationship of this group with the other great assemblage of arthropods, the Tracheata (Myriapoda, Insecta, and Arachnida.) Traditionally, all arthropods were derived from annelids, but Bernard now introduced a major division between the two groups. The Crustacea and the Tracheata were both derived from annelid ancestors which had developed their parapodia for mastication and locomotion, which explained the similarities between them. But the Tracheata were the descendants of an ancestor which had not bent over, thus avoiding the fusion of the five segments that form the crustacean head. The tracheae by which they breathe were developed after the ancestral member of the group had moved onto dry land.[45] They were not related to the Eurypterida or seascorpions, because the latter were too specialized to have developed the necessary adaptations to dry land.[46] Bernard subsequently challenged the increasingly popular view that the Arachnids had evolved from a form like

---

41. Ibid., p. 253.
42. Ibid., pp. 176–178.
43. Ibid., pp. 210 and 224. On the fossil evidence, see below.
44. Ibid., pp. 237–238 and 248.
45. Ibid., pp. 282–283.
46. Ibid., p. 285.

Limulus, arguing that the typical spider body was much modified by its blood-sucking habits, and that the most plausible reconstruction of a primitive Arachnid did not resemble *Limulus.*[47]

For Bernard, the evolutionary history of the Crustacea was so distinct from that of the Tracheata that there was no point in trying to defend the view that the Arthropoda as a whole were monophyletic. Two distinct phyla existed, their differences so great that they must have originated from distinct starting points among the annelids. The similarities were due both to the preservation of the deep ancestral characters, and to convergent evolution which had led two groups of annelids independently to evolve jointed appendages, a chitinous exoskeleton, and compound eyes. This view gained increasing support in the last decade of the century, and to understand why, we must look at a number of factors which shaped ideas about the evolution of the Tracheata.

PERIPATUS AND THE ORIGIN OF THE TRACHEATA

Müller's view that the Tracheata were derived from the Zoea stage in crustacean evolution was endorsed by Haeckel.[48] But while favoring the monophyletic interpretation of the arthropods, even Haeckel was forced to admit at least the possibility that the Crustacea and Tracheata had developed independently from the annelid worms.[49] The same caution can be seen in Carl Gegenbaur's comments on the topic. Gegenbaur advanced a theory to explain the origin of the tracheae based on the claim that the most primitive insects had resembled the larvae of Ephemeridae, or mayflies, which breath water through external tracheal gills.[50] The breathing function had gradually been internalized, while the external appendages became the wings. Similar external gills are found in many crabs, and also in the annelids. It was thus possible that the first Tracheata might have come either from the Crustacea or directly from the Annelida, and Gegenbaur admitted that there were many reasons for regarding the Tracheata as a separate stem form with independent origins.[51]

In Müller's view, the wormlike character of many insect larvae was of no phylogenetic significance, because the metamorphosis of modern insects was a form of development which had appeared only recently in the

47. Bernard, "Comparative Morphology of the Galeodidae," and "The Endosternite of *Scorpio* Compared with the Homologous Structures in Other Arachnida."

48. Haeckel, *History of Creation,* II, pp. 178–180.

49. Ibid., p. 173.

50. Carl Gegenbaur, *Elements of Comparative Anatomy,* p. 247.

51. Ibid., p. 228.

FIGURE 3.4 *Peripatus capensis.* From Adam Sedgwick, *Student's Textbook of Zoology* (London, 1889–1909), III, p. 552.

group's evolutionary history.[52] The argument for the independent origin of the Tracheata was based on the assumption that the segmented character of this group was preserved from that of the original annelid ancestor, thus eliminating anything resembling the Nauplius form from the ancestry of the group. Perhaps the strongest evidence for this interpretation came from the recognition that the Onychophora, represented by the modern genus *Peripatus,* could be used as a model for a plausible evolutionary bridge between the annelids and the myriapods. The latter were widely assumed to have given rise to the insects. The genus *Peripatus* had been established in 1826 for a specimen from the West Indies, which had originally been regarded as a kind of slug (fig. 3.4). More species were discovered throughout the tropical region, in Africa, South America, and Australasia. Later naturalists linked *Peripatus* either to the Annelida or Myriapoda, but it was Henry Moseley who conclusively demonstrated its primitive arthropodan character through research conducted while in South Africa during the celebrated voyage of oceanographical research by H.M.S. *Challenger.*

In his 1874 paper, "On the Structure and Developement of *Peripatus capensis,*" Moseley directly challenged Gegenbaur's theory of the origin of the Tracheata.[53] He showed that *Peripatus* could not be a degenerate myriapod. He also argued that although its mouth parts could be homologized with those of the higher annelids, its foot-claws showed that it could not be included in this phylum. Moseley demonstrated that the animal breathed through tracheae resembling those of the insects, but connected to openings distributed irregularly all over the body surface. He suggested

---

52. On this point, see John Lubbock, *On the Origin and Metamorphoses of Insects,* pp. 92–97.

53. H. N. Moseley, "On the Structure and Development of *Peripatus capensis,*" pp. 777.

that the tracheae probably originated from cutaneous glands which had gradually been adapted to the role of breathing. Although it had some specialized characters, *Peripatus* thus provided a clear indication of what Haeckel's hypothetical 'Protracheata' had looked like. It was, in effect, the stem form from which the rest of the Tracheata had diverged. In a more popular account of the discovery, Moseley wrote, "*Peripatus* is an animal of the very highest importance and antiquity, and I believe it to be a nearly related representative of the ancestors of all air-breathing Arthropoda, i.e., of all insects, spiders and Myriapods."[54] He also noted that the wide geographical distribution of the genus suggested that it was of high antiquity. According to this interpretation of the origins of the Tracheata, the group must have had an origin quite separate from the Crustacea, which seemed unlikely ever to have passed through a similar stage (see the accompanying diagram).

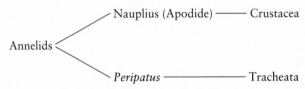

Moseley's discovery was soon recognized as a significant clue to the origin of the Tracheata. In 1876 Huxley accepted that *Peripatus* was a primitive arthropod and that it threw light on the relationship between the annelids and arthropods.[55] Balfour studied the development of specimens supplied by Moseley and wrote that the latter's discovery of the tracheal system "clearly proves that the genus Peripatus, which is widely distributed over the globe, is the persisting remnant of what was probably a large group of forms, from which the present tracheate Arthropoda are descended."[56] Balfour noted more annelid features, including glandular bodies resembling the segmental organs of annelids. Recognition that *Peripatus* was an arthropod, but with annelid characters that rendered it quite distinct from the other Tracheata, led to the creation of the separate class Onychophora. Many zoologists regarded the series Onychophora-Myriapoda-Insecta as an evolutionary one. As J. Arthur Thomson wrote in his popular *Outines of Zoology,* "These three classes form a series of which winged insects are the climax. The type *Peripatus* is archaic, and

---

54. Moseley, *Notes by a Naturalist,* p. 137.

55. Huxley, "On the Classification of the Animal Kingdom," see *The Scientific Memoirs of Thomas Henry Huxley,* IV, 35–60, p. 47.

56. Balfour, "On Certain Points in the Anatomy of Peripatus capensis," reprinted from *Proceedings of the Cambridge Philosophical Society* (1879) in Balfour, *Works,* III, 657–660, see p. 657. For a more detailed study, see Adam Sedgwick, "The Development of Peripatus Capensis."

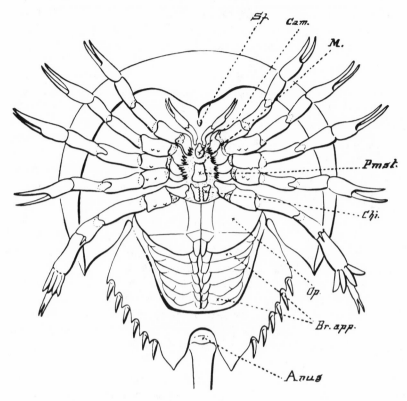

FIGURE 3.5 *Limulus,* the horseshoe crab (ventral surface). From E. Ray Lankester, *Extinct Animals* (London, 1905), p. 285.

links the series to the Annelids; the Myriopods lead on the the primitive wingless insects." [57] From such a model it was difficult to justify a close link with the Crustacea: whatever one's views on the Nauplius stage, it was increasingly difficult to believe that anything like *Peripatus* could be derived from even an early point in the evolution of the primitive crustaceans. This would imply that the Tracheata had an independent origin in the annelids, as Bernard had claimed.

LIMULUS AN ARACHNID

Another threat to the presumed unity of the Arthropoda came from E. Ray Lankester's demonstration that the horseshoe crab, *Limulus,* should be regarded not as a crustacean, but as a gill-breathing arachnid (fig. 3.5). Most

57. J. Arthur Thomson, *Outline of Zoology,* p. 318. The sequence Onychophora-Myriapoda-Insecta is still regarded as not only of phylogenetic significance, but also as illus-

authorities treated the Xiphosura, Trilobita, and Eurypterida as ancient forms of the Crustacea, of which only *Limulus* had survived to the present day. Alpheus Packard thought that his early studies of the embryology of *Limulus* would illustrate both crustacean metamorphosis and the nature of the trilobites.[58] Packard was a founding member of the American school of neo-Lamarckism and a prominent exponent of the recapitulation theory. By implication, Lankester's removal of *Limulus* to the Arachnida threatened the position of all the ancient types assigned to the Crustacea. The Eurypterida were now recognized as aquatic scorpions, and the possibility emerged that the whole Arachnid line might be derived from a Limuloid (*Limulus*-like) ancestor. Acceptance of this idea, however, would imply yet another line of arthropod evolution, because it was difficult to fit *Limulus* into either the crustacean or the *Peripatus*-insect lines of development. The Tracheata themselves became polyphyletic, with the Arachnids representing an offshoot of neither the crustacean nor the insect stems, but an entirely distinct line of evolution which had independently evolved the capacity to breath air.

As its popular name, horseshoe crab, signifies, *Limulus* looks like a crab with a large, horseshoe-shaped carapace. Lankester was by no means the first to suggest that its crustacean affinities are superficial. Eduard van Beneden and others had already noted this point and suggested that *Limulus* should be reclassified as an arachnid. But it was Lankester's detailed paper "Limulus an Arachnid" of 1881 which forced all zoologists to confront the arguments. Lankester enumerated many features which distanced *Limulus* from the true crabs and implied a closer link to the arachnids. In effect, he portrayed *Limulus* as a primitive gilled relative of the scorpions, similar to the Eurypterids of the Palaeozoic. It thus preserved many of the characters of the earliest arachnids, predating the point at which some members of the class moved onto the land.

In his conclusion, Lankester provided evolutionary trees to express his views on the relationships between the various arthropod groups.[59] At this point, he did not see the need to make the arthropods polyphyletic, postulating a Pro-arthropod form from which all the classes had diverged. *Peripatus* was a specialized terrestrial offshoot from low down on the

---

trating the order in which the genetic systems controlling the stages of the developmental process were established—thus reviving a form of the recapitulation theory; see R. Raff and T. C. Kaufman, *Embryos, Genes, and Evolution.*

58. Alpheus Packard, "The Embryology of Limulus Polyphemus." On Packard's neo-Lamarckism, see Bowler, *Eclipse of Darwinism,* pp. 134–136.

59. Lankester, "Limulus an Arachnid," p. 649. Note that in Britain, *Limulus* is sometimes colloquially called the "king crab," although in America this term is reserved for a true crustacean.

family tree of the Arthropoda, after which the two major branches of the Crustacea and the Arachnids had separated. For Lankester, the trilobites were primitive arachnids, branching off just before the Xiphosura and Eurypterida. He was uncertain as to the origin of the insects, suggesting three possible positions on the tree for them. They might be descendants of the *Peripatus* line, offshoots of the Crustacea, or the end-product of the arachnid line.

Lankester's reclassification of *Limulus* was controversial at first, although by the 1890s it had become widely accepted. Packard was one of the strongest opponents, although he eventually accepted *Limulus* as an arachnid in 1898.[60] In the late 1880s Lankester became embroiled in a vitriolic controversy with Carl Claus of Vienna, who had published a note on the classification of the Arthropoda in 1885.[61] Claus argued that the mites were degenerate arachnids and that the origin of this class lay in the Gigantostraca (Eurypterida and Xiphosura). Without mentioning Lankester, he argued that *Limulus* was a modern representative of the latter group. *Limulus* and the Gigantostraca had been treated as crustaceans because they breathed through gills, but Claus insisted that too much emphasis had been placed on this characteristic. The division of the Arthropoda into Branchiata and Tracheata had been made "without taking into consideration that breathing by air-spaces may have been developed in different ways at different times in the terrestrial forms, and that consequently no primarily decisive morphological value is to be ascribed even to the posession of *tracheae*."[62] While accepting that the Crustacea and Arachnida were related by descent from the Nauplius, Claus now regarded the Onychophora-Myriapoda-Insecta line as separate.

Lankester was incensed that Claus should advance the view that *Limulus* was an arachnid without acknowleging him as its originator and wrote a reply to Claus's note.[63] He now insisted that his reclassification of *Limulus* with the scorpions broke the unity of the Tracheata by showing that tracheae had developed independently in *Peripatus*, the Insecta, and the Arachnida. He admitted that he had hesitated to link the insects to the *Peripatus* stem but now regarded this as a plausible connection. Claus responded by insisting that his note merely repeated views he had published long before Lankester's article. The 1880 edition of his popular

60. Packard, "Hints on the Classification of the Arthropoda: The Group a Polyphyletic One," p. 145.

61. Claus, "On the Heart of *Ganasidae* and Its Significance in the Phylogenetic Consideration of the Acaridae and Arachnoidea," a translation of an article originally published in the *Anzeiger* of the Imperial Academy of Sciences, Vienna, in 1885.

62. Ibid., p. 170.

63. Lankester, "Professor Claus and the Classification of the Arthropoda."

textbook, *Grundlage der Zoologie,* for instance had treated the Tracheata as composed of two distinct groups.[64] He now insisted that Lankester had misunderstood his position: he (Claus) did not regard *Limulus* as a scorpion, and his division of the arachnid-crustacean phylum from the myriapod-insect phylum did not depend on the position of *Limulus.* Additional notes by Lankester and Claus contributed only invective to the debate and did nothing to resolve the confusion.[65]

Whatever their differences, Claus and Lankester had both moved towards a position in which the Arthropoda were polyphyletic. Both believed that air-breathing had developed independently in two distinct lines of evolution. Lankester continued to speculate on how the lung-books of the scorpions had evolved from the gill-books of *Limulus.* In 1884 he suggested that the gills had simply invaginated, that is, had begun to develop into the body instead of externally.[66] A rival hypothesis in which the lung-sac was developed by the gills being enclosed between the limbs and a depression in the body was proposed by J. MacLeod.[67] Whatever the actual course of evolution—and such hypotheses were always highly speculative when no fossils were available—the apparently clear demonstration that parallel evolution had occurred again threatened the unity of the arthropod phylum. As Korschelt and Heider put it, the apparent agreement of the Arachnida with the other Tracheata must be regarded as "nothing more than a similarity determined by their common Arthropodan nature and by a like development as a result of a similar manner of life."[68] Arthur Willey's 1911 survey of convergent evolution took the independent origin of the tracheae as a good example of the phenomenon.[69] R. I. Pocock of the British Museum (Natural History) thought that tracheae were so efficient a breathing device that they had been developed independently by different groups even within the Arachnida.[70]

64. Claus, "Professor E. Ray Lankester's Memoir 'Limulusy an Arachnid' and the Pretensions and Charges Founded upon It." See Claus, *Elementary Textbook of Zoology,* I, chap. 10. Here Claus treats the Arachnida, Onychophora, Myriapoda, and Insecta as classes equivalent to the Crustacea; *Limulus* and the eurypterida are included in the Crustacea.

65. Lankester, "Professor Claus: A Rejoinder," and Claus, "Reply to Professor E. Ray Lankester's 'Rejoinder.' "

66. Lankester, "New Hypothesis as to the Relation of the Lung-Book of Scorpio to the Gill-Book of Limulus."

67. J. MacLeod, "Recherches sur la structure et la signification de l'appareil respiratoire des Arachnides."

68. Korschelt and Heider, *Textbook of the Embryology of the Invertebrates,* III, p. 117.

69. Arthur Willey, *Convergence in Evolution,* pp. 148–149.

70. R. I. Pocock, "On Some Points in the Morphology of the Arachnida."

THE DEBATE WIDENS

The challenge to the presumed unity of the Arthropoda came to a head in the 1880s and 1890s as biologists began to survey the various lines of evidence. The American morphologist J. S. Kingsley first took up the theme in 1883.[71] Arguing that attempts to trace homologies between Crustacea and Tracheata had broken down, Kingsley insisted that the insects had evolved from a form like *Peripatus,* while the crustaceans originated from an organism resembling the Nauplius larva. He noted that the great problem with this interpretation was the fact that both groups contained forms with compound eyes, but insisted that this could be due to convergent evolution. To back up this point, he noted the similarity in the structure of the vertebrate and cephalopod eyes.

Kingsley extended his argument in a series of papers published in the early 1890s. He shared the view that the Tracheata were not a single phylum. Moseley's discovery of tracheae in *Peripatus* and Lankester's reclassification of *Limulus* had suggested two possible origins for the Tracheata. Either all the Tracheata had evolved from a Limuloid ancestry, or the similarities of the tracheae in the insects and arachnids were due to homoplasy (convergent evolution) not common descent.[72] The new discoveries had thrown the status of the whole Arthropod phylum open to debate. The two main divisions of the arthropods, the Tracheata and Branchiata (Crustacea), were of no morphological significance, since the differences between them were due merely to the physiological demands of breathing in air and water. For the time being most authorities retained the term Arthropoda to cover both, but thought it probable that they had no common ancestry closer than the annelids.[73]

An even more extreme position was argued by another American zoologist, H. T. Fernald.[74] He regarded the insects, arachnids, and crustaceans as independent phyla but insisted that the segmentation that supposedly linked them to an annelid ancestry might itself be the product of parallel evolution. The annelids and the arthropods represented two developments arising from the unsegmented worms. Each line of evolution had independently evolved segmentation, and the three main phyla of arthropods had independently evolved the other features that had traditionally been supposed to unite them. The search for homologies among the arthropods had ignored the warning (which Fernald attributed to Darwin)

71. J. S. Kingsley, "Is the Group Arthropoda a Valid One?"
72. Kingsley, "The Embryology of Limulus."
73. Kingsley, "The Classification of the Arthropoda."
74. H. T. Fernald, "The Relationship of Arthropods"; see especially pp. 499–501.

that groups may inherit from a distant common ancestor the tendency to vary in the same direction. His explanation of arthropod resemblances thus depended on parallel evolution driven by a common orthogenetic trend.

The conviction that the arthropods might be polyphyletic was clearly gaining support. The second edition of George Rolleston's classic text of comparative anatomy, *Forms of Animal Life* (1888), endorsed this position in its general introductory comments. Rolleston accepted two lines of evolution, one from the Nauplius form leading to the Crustacea and another from *Peripatus* to the Insecta.[75] By the 1890s, no arthropod expert could have been unaware of the growing debate over the possibility that the group might be polyphyletic.

In 1897 the New Zealand naturalist F. W. Hutton encouraged the semipopular British journal *Natural Science* to solicit a series of expert responses to the question "Are the Arthropoda a Natural Group?"[76] The editorial introduction to these responses noted that if *Peripatus* was a primitive tracheate arthropod, then its origins must lie in the annelids and be independent of the Crustacea. The characters that supposedly linked the crustaceans and insects, including the hollow-jointed limbs, could not be explained by common inheritance and must have evolved separately by convergence. The expert opinions opened with Bernard, who insisted that the arthropod characters had indeed been independently developed two or perhaps three times. He made it clear that his belief in convergence was inspired partly by a Lamarckian view of adaptive evolution:

> All these new forms of life arising out of that highly plastic form, the Chaetopod Annelid, were, I believe, due not to the natural selection of small chance congenital variations, but to definite responses to the environment. In each case, the adoption of a new method of feelings gave rise, slowly but surely, to a new organisation. All subsequent resemblance in the descendants of the annelid forms are purely convergent.[77]

Bernard's support for the polyphletic origin of the arthropods was endorsed by J. S. Kingsley, Malcolm Laurie, and R. I. Pocock. Kingsley differed from most other authorities in that he rejected the view that *Peripatus* was a primitive arthropod. Its primitive tracheae were acquired independently and did not link it to the myriapods and insects.[78] Within

---

75. George Rolleston, *Forms of Animal Life*, p. xxx.

76. Bernard et al., "Are the Arthropoda a Natural Group?" Hutton had played a prominent role in introducing Darwinism into New Zealand; see John Stenhouse, "Darwin's Captain."

77. Bernard, ibid., p. 100.

78. Kingsley, ibid., p. 108. George H. Carpenter also questioned the arthropod affinities of *Peripatus*, see ibid., p. 100.

the true arthropods, it was still necessary to ask whether this was a genuinely natural group, or rather "an assemblage of convergent forms with no common ancestor nearer than the annelids."[79] Kingsley favored the latter view, arguing for three main lines of development, the Crustacea, Insecta, and Arachnida. Laurie thought that Hutton's fears about the status of the Arthropoda were fully justified: "I have long felt that, while the term Arthropoda is a very convenient one as expressing a certain kind of structural modification, it does not imply a phylogenetic relationship among the forms included in it. In short, the Arthropoda had not an arthropod common ancestor, and jointed legs cannot be regarded as an absolute criterion of relationship."[80] The jointed leg was a product of convergent evolution, or homoplasy, because it was an obvious improvement in the means of locomotion, given the formation of an exoskeleton. Pocock also wrote of several parallel roads of evolution by which parapoda used in the same way for similar needs had independently evolved into feelers, jaws, and legs.[81]

Other views were expressed in the collection. Antoni Jaworowski argued for a close link between the Crustacea and Arachnida. He believed that the arthropod structure had evolved when a wormlike creature had developed tracheae and a hardened integument as a means of adapting to dry land. Following Heinrich Simroth's survey of the evolution of land animals, he claimed that the crustaceans had evolved from a land-dwelling form that had reentered the water in search of new sources of food.[82] Such views strongly opposed the normal ideas on arthropod relationships. Given the wide divergence of views expressed, it is hardly surprising that one authority, H. C. Hansen, expressed wholesale rejection of the search for ancestors and pedigrees.[83] The editors noted the differences among the expert opinions and somewhat dryly observed that they would like to be present when Hutton next lectured on the subject, "to hear how he has managed to reduce these varied views to dogmatic order."[84] It was this kind of indecisiveness which turned some morphologists against the search for phylogenies and encouraged experimental biology. But clearly the debate was still of interest to a general audience of naturalists, and there was certainly a predominance of opinion in favor of a polyphyletic interpretation of arthropod origins.

79. Ibid., p. 109.
80. Laurie, ibid., p. 111.
81. Pocock, ibid., pp. 113–114.
82. Jaworowski, ibid., p. 103–105. See Heinrich Simroth, *Die Entstehung der Landthiere,* p. 278, and on the arthropod form as a dry-land adaptation, pp. 255–256.
83. H. C. Hansen, in "Are the Arthropoda a Natural Group?" pp. 100–102.
84. Ibid., p. 117.

This latter position was supported in a 1903 article by Packard, which proclaimed in its title that the arthropods were polyphyletic.[85] Packard accepted the view of Kingsley and others that the development of jointed legs may have been due to convergence. He suggested that as many as five distinct phyla might be involved, all with separate annelid origins. Although once an opponent of the claim that *Limulus* was an arachnid, Packard now accepted that the arachnids had evolved from the eurypterids, which were in turn closely related to the trilobites.

Other authorities were having second thoughts, and the early twentieth century saw a gradual swing back to acceptance of a common ancestor for the arthropods. Lankester's reclassification of *Limulus* as an arachnid had certainly contributed to the original debate, but in his article on the Arthropoda for the eleventh edition of the *Encyclopaedia Britannica*, Lankester maintained that there were so many structural affinities between the various arthropods that it was impossible to regard them as polyphyletic.[86] He admitted a good deal of convergent evolution in the crustaceans and arachnids, but now seemed to believe that to explain all the similarities in this way would create more problems than it solved. The trilobites were presented as primitive arachnids along with *Limulus,* a position that gained little support as more fossil evidence became available.

In the early years of the twentieth century George Carpenter of the Science and Art Museum in Dublin launched a strong attack on the claim that the arthropods were polyphyletic. Surveying the debate up to that point, Carpenter noted that everyone accepted a substantial gap between the insect and crustacean lines, but differed over whether the gap was sufficient to render the postulation of a very primitive arthropodan common ancestor implausible. Authorities disagreed over how the arachnids fitted into the picture. Carpenter himself was inclined to minimize the differences, pointing out that H. C. Hansen had demonstrated exact correspondences between the segments of crustaceans and insects.[87] Carpenter's own approach was to reconstruct the ancestral form of each group, and then to show that the ancestral crustacean could also serve as a plausible common ancestor for the insects and arachnids.

Bernard had argued that the caterpillar might represent the ancestral form of the insects, which would support the view that the class had evolved from something like *Peripatus*. On this interpretation, the myria-

85. Packard, "Hints on the Classification of the Arthropoda: The Group a Polyphyletic One."

86. Lankester, "Arthropoda," p. 674.

87. G. H. Carpenter, "On the Relations of the Classes of Arthropoda," pp. 322–323. See H. C. Hansen, "A Contribution to the Morphology of the Limbs and Mouth Parts of Crustaceans and Insects."

pods (centipedes and millipedes) were the intermediate stage between *Peripatus* and the true insects. Carpenter attacked this position, pointing out that authorities such as John Lubbock, Packard, and more recently L. C. Miall believed that metamorphosis through a caterpillar stage was a comparatively recent insertion into insect ontogeny. The caterpillar was not a recapitulation of evolutionary ancestry.[88] Most entomologists believed that the wingless Campodea were primitive insects that could serve as a model for the class's origins. Kingsley and Pocock had also shown that the myriapods were unlikely to have been ancestral to the insects. The millipedes especially were only distantly related to the insects. They represented a distinct branch of evolution showing a marked increase in the number of segments. The insects and myriapods had thus evolved from a common ancestor with only a moderate number of segments.

Turning to the crustaceans, Carpenter noted that most authorities used highly segmented forms such as *Apus* as models for the group's ancestry. But the fossil record showed that the trilobites increased their number of segments in the course of time.[89] The evidence again suggested that in general the less segmented forms were more likely to be primitive. Carpenter believed that the trilobites were primitive crustaceans and that the arachnids were an early branch from the crustacean line. *Limulus* and the eurypterids were in turn specialized offshoots from the arachnids. He argued that the most primitive crustacean could serve as a common ancestor from which both the arachnid and insect lines had diverged. The most likely model for this primitive crustacean was the Nauplius form, leading Carpenter to predict that zoologists would soon be returning to Müller's position.[90] He also complained that many reconstructions of arthropod ancestry derived the type from the most highly evolved form of the annelid worms, the chaetopods. It would be more plausible to assume that both annelids and arthropods were divergent branches from a much more primitive common ancestor.

The new century thus began with no agreement among experts on the question of whether or not the arthropods were polyphyletic. The *impasse* here was even more fundamental than that over the origin of the vertebrates, and gives a better illustration of the frustrations building up within evolutionary morphology. The pace of morphological research bearing on this issue certainly slackened in the twentieth century, although some work continued. A survey by the American entomologist R. E. Snodgrass

88. See Lubbock, *On the Origin and Metamorphoses of Insects,* pp. 91–92; note that Lubbock himself was now tempted by the idea that the original insect might have had a wormlike structure: see ibid., pp. 95–97. See also L. C. Miall, "The Transformation of Insects."

89. Carpenter, "On the Relations of the Classes of Arthropoda," p. 332.

90. Ibid., p. 351.

in 1938 reveals that the issues were still alive, although the bibliography shows relatively few important twentieth-century contributions. Snodgrass himself did not accept that the arthropods were polyphyletic: he argued that the mandibles of insects and crustaceans were homologous and included both within the class "Mandibulata." He derived even the crustaceans from a centipede-like form, the "protomandibulata." Even so, he felt it necessary to warn against the possibility of convergent evolution, noting that this gave a speculative air to the reconstruction of phylogenies.[91] Snodgrass's position was probably typical of a growing skepticism toward arthropod polyphyly in the interwar years, although the issue was not hotly debated.

THE FOSSIL RECORD

Until the 1890s, the debate over arthropod origins had been conducted largely by morphologists and embryologists. Fossil groups such as the trilobites and eurypterids were noted, but their affinities were disputed. The origins of the phylum (or phyla) were lost in the mists of pre-Cambrian time, and the known fossils were thus not crucial to the debates. By the last decade of the century, however, better fossils were becoming available, and early stages of the fossil record began to figure more prominently in the debates. We have already noted Bernard's efforts to include the trilobites in his analysis of crustacean origins. Apart from helping to clarify the relationships between the arthropod groups, the fossil record also helped to throw new light on the conditions under which the early members of the groups had lived, opening up new questions about the environmental factors which might have induced key evolutionary developments.

The eurypterids were at first assumed to be primitive crustaceans, but their arachnid affinities had now been recognized. As Malcolm Laurie noted, it was difficult for many naturalists to free themselves from the assumption that any aquatic arthropod must be a crustacean.[92] Laurie accepted that the eurypterids must have evolved from primitive trilobites long before either appeared in the fossil record, but they were clearly linked with the arachnids along with *Limulus*. By the early twentieth century, the eurypterids were frequently described in colloquial terms as 'sea scorpions.'[93] The link between the aquatic and terrestrial scorpions was strengthened by new fossil discoveries of the latter type. Specimens had been found in the Silurian rocks of both Sweden and Scotland in the

91. R. E. Snodgrass, "Evolution of the Annelida, Onychophora, and Arthropoda."

92. Malcolm Laurie, "The Anatomy and Relations of the Eurypteridae," pp. 523–524.

93. For example, Henry Fairfield Osborn, *The Origin and Evolution of Life*, pp. 132–133.

1880s, but only with better descriptions in the following decade did the link become clear. As Pocock put it, the Scottish fossil scorpion "supplies a few more links to the chain of evidence pointing to the descent of the scorpions from marine Limuloid ancestors."[94]

Laurie argued that without knowing the structure of trilobite limbs, which are rarely fossilized, it was impossible to be sure about the relationship between this and the other arthropod groups. But better evidence of the structure of trilobites was now becoming available, as specimens from fine-grained rocks began to reveal details that were normally missing. Long before he discovered the Burgess shale, Charles Doolittle Walcott had found indirect evidence for the structure of trilobite limbs from the casts within the fossilized body. These discoveries, he argued, cleared up the doubts expressed by some earlier authors about the trilobites' crustacean affinities.[95] Walcott also linked the Trilobita, Xiphosura, and Eurypterida as orders within the "legion" Merostomata. He argued that the trilobites were bottom-crawling animals showing little evidence of swimming ability.

New discoveries in the 1890s began to distance the trilobites from *Limulus* and the eurypterids. In 1893 new specimens from the Heidelberg mountains near Albany, N. Y., revealed the structure of the antennae and led W. D. Matthew to doubt Walcott's link between the trilobites and *Limulus*.[96] This opinion was reinforced by studies of these and other specimens by Charles Emerson Beecher, professor of paleontology at Yale. A ten-millimeter-thick layer of black Utica shale, rich in the remains of the trilobite *Triarthus becki* was discovered near Rome, N.Y. by W. S. Valiant of the museum at Rutgers College. Beecher quarried two tons of this rock and studied the resulting trilobite specimens until his death in 1904. The internal structures of the trilobites had been mineralized as iron pyrites, leading Charles Schuchert to observe that in this case fool's gold had produced a "paleontological paradise."[97]

Beecher showed that the larvae of trilobites resembled the Nauplius stage of crustaceans, suggesting that the trilobites were ancient isopods, or proto-isopods.[98] Evidence from the limbs showed that the trilobites were crustaceans and that "no living type of crustacean more nearly con-

94. Pocock, "The Scottish Silurian Scorpion," p. 311. Lankester was by now director of the British Museum (Natural History) and had obtained the specimen for Pocock to study.

95. C. D. Walcott, "The Trilobite: New and Old Evidence Relating to Its Organization"; see p. 196 and for details pp. 201–208.

96. W. D. Matthew, "On the Antennae and Other Appendages of Triarthus."

97. Charles Schuchert, "Foreword," in Percy E. Raymond, *The Appendages, Anatomy, and Relationships of Trilobites*, p. 5.

98. C. E. Beecher, "Larval Forms of Trilobites from the Lower Heidelberg Group."

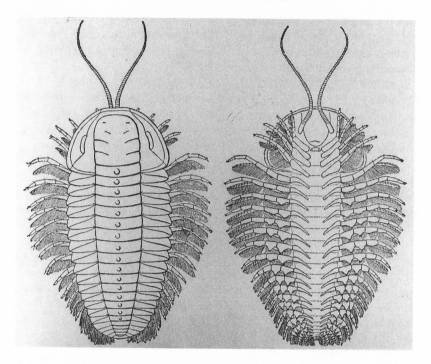

FIGURE 3.6   Trilobites, showing leg structure (after Beecher). From Thomas C. Chamberlin and Rollin D. Salisury, *Geology: Earth History* (London, 1908), II, p. 350.

forms to the theoretical archetype of the class than do the trilobites."[99] The trilobites did not, however, fit into the two existing subclasses of the Crustacea, the Entomostraca and Malocostraca; they were a third subclass equivalent to these.[100]

Beecher's view that the trilobites represented the ancestral form of crustacean paralleled that already expressed by Bernard, and the latter was not slow to draw upon the new fossil evidence to support his position. He now argued that the trilobites were an early specialization for bottom-crawling which branched off from the stem leading to the modern crustaceans. The American discoveries confirmed that the trilobites were "a fixed specialized stage in the evolution of the crustacea from an annelidan ancestor."[101] The similarities between trilobites and isopods were due to

99. Beecher, "The Morphology of Triarthus." This and a number of other papers on trilobites are reprinted in section 2 of Beecher, *Studies in Evolution*, pp. 213–219.

100. Beecher, "Outline of a Natural Classification of Trilobites." See *Studies in Evolution*, pp. 109–162.

101. Bernard, "The Systematic Position of the Trilobites," p. 432.

independent adaptation to the same way of life. Bernard predicted an annelid common ancestor of *Apus* and the trilobite *Triarthus*. The two branches had specialized in different directions: browsing, armored annelids had given rise to *Limulus* and the eurypterids, while a swimming form gave *Apus*, the trilobites, and crustaceans.[102]

It is in the light of the growing recognition that the arthropods might be polyphyletic that we must interpret Walcott's analysis of the beautifully preserved specimens from the Burgess shale. In his account of the shale fossils and their implications, Gould has implicity criticized Walcott for "shoehorning" the diverse organisms into existing arthropod groups in order to preserve the image of a "cone of diversity."[103] According to this model of evolution the Cambrian fauna could not contain forms more diverse in structure than those of today. I am not qualified to comment on Gould's views on the bizarre character of the shale organisms, but his interpretation of Walcott's position must be understood in a context in which the cone of arthropod diversity had already been greatly widened by the argument that the group comprised several independent lines of evolution. Walcott had five lines within the Crustacea alone, although one of these led to the eurypterids and arachnids.[104] Walcott supported Bernard's interpretation of arthropod origins and noted that the highly organized merostome *Sydneyia* showed that this group (which included the eurypterids) must date far back into pre-Cambrian time.[105] Later on, G. E. Hutchinson summed up the evidence suggesting that *Aysheaia* was an onychophoron, implying that the aquatic ancestors of *Peripatus* and the Tracheata were already in existence in the Cambrian.[106]

Walcott's belief in a long "Liparian" period of pre-Cambrian time, during which life had gradually evolved without leaving fossils, fitted a view of the history of life in which substantial arthropod diversity had already evolved by the Cambrian. But Walcott's work is also characteristic of an important trend in studies of the history of life pioneered by the paleontologists, who were now playing a much greater role. By studying the geological evidence for the conditions under which these early sediments were laid down, the paleontologists could offer information, rather than speculation, on the environment to which the newly evolved forms were adapting. Walcott supported W. K. Brooks's theory that life had

102. Bernard, "Supplementary Notes on the Systematic Position of the Trilobites," p. 359.

103. Gould, *Wonderful Life,* chap. 4.

104. Walcott, "Cambrian Geology and Paleontology," I, no. 6, "Middle Cambrian Branchiopoda, Malocostraca, Trilobita, and Merostoma," pp. 160–161.

105. Ibid., pp. 164–165.

106. G. Evelyn Hutchinson, "Restudy of Some Burgess Shale Fossils."

evolved first in the open ocean and then adapted to life on the bottom and the shorelines.[107]

Brooks' theory was a modification of the widely held view that the major phyla had first evolved in a marine environment. This had been supported by W. J. Sollas in 1884, whose argument was based on the comparatively small numbers of freshwater species in each class. Quoting Edmund Spenser's *Faerie Queene,* Sollas wrote, "That the sea is the fertile mother of all life was a poetic fancy which has now become a fair deduction from admitted facts." [108] He believed that marine forms rarely invaded fresh waters; the transition only occurred during epochs in which a general elevation of the land turned shallow seas into inland waters. Many authorities assumed that coastal marine waters had served as the most active location for the evolution of new types, which had then moved out into the open sea and into fresh waters.[109] Brooks challenged this view, arguing that the surface of the open ocean had formed the original home of all life.

Brooks developed his theory in the context of the debate over the origin of the vertebrates (chapter 4), but he also commented on the early evolution of the crustaceans. He was convinced that all the major phyla originated from a small number of pelagic organisms making up the earth's earliest population. The common view was that *Apus* and the highly segmented phyllopods were relics of the ancestral crustacean form, which had evolved on the sea bottom. Brooks insisted that the development of segmentation was a natural phenomenon within many evolutionary lines, so that highly segmented forms were never primitive. The ancestral crustacean was thus likely to have been unsegmented, and could well have resembled the Nauplius stage in the ontogeny of its modern descendants.[110] This ancestral form would have lived in the surface waters, not on the bottom, and the move to bottom-dwelling represented a crucial turning point in the history of life.

Brooks used the theory that there was an inbuilt tendency for many forms to increase their segmentation to comment on the origin of the arthropods. But his interest in the conditions under which these early stages of evolution had taken place represented a new departure. The fact that

---

107. Walcott, "Cambrian Geology and Paleontology," I, no. 1, "Abrupt Appearance of the Cambrian Fauna on the North American Continent," p. 14. See W. K. Brooks, "The Origin of the Oldest Fossils and the Discovery of the Bottom of the Ocean."

108. W. J. Sollas, "On the Origin of Freshwater Faunas," p. 88.

109. See J. Arthur Thomson, *Outlines of Zoology,* pp. 801–802. Thomson refers to the work of H. N. Moseley, Alexander Agassiz, and Sir John Murray, but gives no sources for the opinions quoted.

110. Brooks, "Salpa and Its Relation to the Evolution of Life," pp. 150–151.

Brooks, a morphologist, published an account of his idea in a geological journal suggests that he was aware of the growing interest of geologists such as T. C. Chamberlin in such matters.[111] A typical early twentieth-century survey of the history of life on earth was Henry Fairfield Osborn's *The Origin and Evolution of Life* of 1917, in which paleontology far outstripped morphology as the preferred source of information on this issue. Osborn stressed theories which invoked dramatic changes of climate and conditions to explain critical phases in the evolution of life. He accepted three lines of evolution leading from the annelid to the arthropods, thus supporting the polyphyletic theory.[112]

Further work on North American fossils continued to throw light on the ancestry of the arthropods. After Beecher's death in 1904, his fossils were studied by Percy E. Raymond of Harvard's Museum of Comparative Zoology. In 1920 Raymond summed up his work in a monograph which commented extensively on arthropod origins. He confirmed that the trilobites should be interpreted as primitive crustaceans, but challenged Bernard's view that they were closely related to the living *Apus*. The latter was specialized rather than primitive; its appendages differed considerably from those of trilobites, and any superficial resemblances were due to parallelism.[113] Raymond was particularly concerned to refute the position now being advocated by Lankester that the trilobites were arachnids. They were true, if primitive, crustaceans—although Lankester was quite right to suppose that the arachnids could have evolved from the trilobites. Raymond pointed to two of Walcott's Cambrian fossils, *Sidneyia inexpectans* and *Emeraldello brocki* as generalized trilobites with affinities to the eurypterids, or sea-scorpions.[114] He also approved of the conclusion developed in Anton Handlirsch's *Die Fossilen Insekten* of 1908, in which the insects were derived from trilobites which had adapted to terrestrial life. This position had also been defended by John D. Tothill in 1916.[115] Raymond was now prepared to use the possibility of a trilobite ancestry for all arthropods to challenge the polyphyletic theory.

According to Raymond, the ancestral trilobite—and hence the ancestral arthropod—was a pelagic animal with a short body divided into only a few segments, which settled onto the seabed and adapted to feeding there. As to its ultimate origins, he noted that the now widely accepted

111. See, for instance, Thomas C. Chamberlin, "On the Habitat of the Early Vertebrates," and more generally Chamberlin and R. D. Salisbury, *Geology: Earth History.*

112. Osborn, *Origin and Evolution of Life*, p. 132.

113. Percy E. Raymond, *The Appendages, Anatomy, and Relationships of Trilobites*, p. 107.

114. Ibid., p. 119.

115. John D. Tothill, "The Ancestry of Insects."

view deriving the arthropods from the annelid worms suffered from a number of problems, of which the most significant was "its total inhibition of the workings of that great talisman of the paleontologist, the law of recapitulation."[116] Paleontolgists were by now the major source of support for the recapitulation theory, the embryologists having largely abandoned it as a guide to phylogeny (see chapter 2). Raymond noted an important exception to this trend in W. T. Calman's 1909 volume on the Crustacea in the *Treatise on Zoology* edited by Lankester. True to Lankester's system of taxonomy, the trilobites were omitted from this volume and left to the arachnids. But Calman did protest that the ancestral significance of the Nauplius stage in crustacean ontogeny could not be ignored altogether.[117] Raymond thought that Calman's brief comments justified a reexamination of the old thesis that the Nauplius represented the ancestral form of the Crustacea and, by implication, of the whole arthropod type.

Although the question of arthropod origins remained a live issue, there was a marked retreat from the polyphyletic theory in the early decades of the twentieth century. The doubts of the 1890s were never overcome, but fewer zoologists were now prepared to accept that the differences between the arthropod classes were significant enough to warrant postulating an independent origin for each. Müller's thesis had regained some of its lost support, although paradoxically it was the paleontologists rather than the morphologists who were most keen to apply the recapitulation theory in this area. The most obvious trend visible in the literature, however, is the growing lack of interest in the whole debate. Precisely because the debate had proved so difficult to settle on morphological grounds, the morphologists themselves had retreated into a kind of agnosticism in which structures were described without any attempt being made to evaluate their implications for phylogeny.

Only in paleontology, where significant discoveries continued to be made, was there still some interest in the question. The tendency was still to stress homogeneity at the expense of diversity. Even those paleontologists who studied the bizarre creatures from the Burgess shale wanted to lump them into a single, highly diverse, arthropodan phylum, just as Walcott himself had done. Leif Stormer created the category "Trilobitomorpha" to include all the forms that have now been shown to represent a host of extinct phyla. Given the lack of support for arthropod polyphyly at the time, this attempt to create a very wide group based loosely on the trilobites (a "garbage-can" taxon, as its critics would say) was all the more remarkable.

The debate over arthropod polyphyly reopened in the late 1950s

---

116. Raymond, *The Appendages, Anatomy, and Relationships of Trilobites*, p. 140.
117. W. T. Calman, *Crustacea*, pp. 26–27.

when Sidnie Manton began to attack R. E. Snodgrass's claim that the mandibles of insects and crustaceans were homologous. Here morphology once again entered the fray at a time when the paleontologists were still inclined to lump everything of a remotely arthropodan nature together. Manton was a student of H. Graham Cannon, who in turn had begun his career working under MacBride. MacBride and Cannon were among the last British biologists to maintain support for Lamarckism. Cannon did extensive work on the functional morphology of arthropod feeding mechanisms in the 1920s, although his work was typical of the period in which phylogenetic considerations had dropped into the background even where the morphological tradition was continued. He did stress the role of habit in determining feeding structures, although his explicit support for Lamarckism came out only much later in his career.[118]

Manton shared Cannon's belief that habit shapes the structure of the organism, while never explicitly endorsing his unorthodox views on genetics. Through the 1960s and 70s she used a detailed morphological analysis to demonstrate that the mandibles were not homologous in the insects and crustaceans, and hence that the structural similarities were due to convergence. The possibility that the arthropod structure had been independently invented in two or more lines of evolution once again became credible. 'Arthropodization' had occurred twice and possibly three times in the course of life's history.[119] Manton specifically attacked the notion of a hypothetical ancestral arthropod from which all branches of the phylum could be derived.[120] Curiously, she was unaware that arthropod polyphyly had once been a popular thesis, imagining that her work was the first real challenge to the assumed monophyletic origin of the group.[121]

Manton was supported by the Australian embryologist Donald T. Anderson and by Harry B. Whittington, who went on to reveal the true diversity of Walcott's fossil 'arthropods' from the Burgess shale. Our story thus links up in the end with that told in Gould's *Wonderful Life*—although Gould himself rejects arthropod polyphyly while stressing the diverse and nonarthropodan nature of the Burgess shale creatures. The ma-

118. See, for instance, H. G. Cannon, "On the Feeding Mechanism of the Branchiopoda," and Cannon and S. M. Manton, "On the Feeding Mechanism of a Mysid Crustacean, Hemimysis lamornae." Cannon's Lamarckism came out in his *Lamarck and Modern Genetics;* on his enthusiasm for MacBride's views, see pp. 41–42.

119. Manton, *The Arthropoda,* p. 31. See S. M. Manton and D. T. Anderson, "Polyphyly and the Evolution of Arthropods," and Anderson, *Embryology and Phylogeny in Annelids and Arthropods.* For a more general survey, see A. P. Gupta, ed., *Arthropod Phylogeny.* I am grateful to Elihu Gerson for information on the postwar debates.

120. Manton, *The Arthropoda,* p. 291, attacking A. G. Sharov's *Basic Arthropodan Stock* of 1966.

121. Manton, *The Arthropoda,* p. 1.

jority of modern experts share Gould's lack of enthusiasm for Manton's thesis.

It is significant that the problem created by recognizing the possibility of the independent invention of arthropod characters in several evolutionary lines proved intractable for so long. The biologists who turned away from phylogeny reconstruction in disgust may have had a valid point: enthusiasm for evolutionism had indeed encouraged their predecessors to venture into a territory where the landmarks were too obscure to be identified with the techniques then available. Interest increasingly switched to those areas of the history of life where direct fossil evidence could be brought to bear. This inevitably focused attention on later episodes where the fossil record was more complete, and precipitated a change in the professional locus of debate from the morphological laboratory to the museum of paleontology. Molecular biology has reopened the old debates, and has gone some way toward resolving them. Recent studies suggest that the onychophorans cannot be regarded as the basis from which the insects evolved. They constitute an independent line sufficiently ancient that we must suppose multiple colonizations of the dry land by the various arthropod groups. The monophyletic status of the arthropods as a whole is maintained, undermining Manton's thesis with its reliance on massive convergence. The subject might seem one that few nonspecialists would find attractive, although the popularity of Gould's *Wonderful Life* shows that it is still possible for a skillful writer to arouse interest in the topic. A recent account of the latest molecular evidence for arthropod monophyly generated at least one newspaper headline about the "arthropod Eve." [122] The parallel between this issue and the debate over the unity of the human species had evidently not gone unnoticed.

122. Toronto *Globe and Mail*, 29 July 1995, referring to Jeffrey L. Boone et al., "Deducing the Pattern of Arthropod Phylogeny from Mitochondrial DNA Arrangements." Curiously, the main purpose of this article is to reassess the position of the Onychophora, and only incidentally to defend the monophyletic character of the arthropods as a whole.

4

VERTEBRATE ORIGINS

In the late nineteenth century, the debate over the origin of the vertebrate phylum became a centerpiece of evolutionary biology. The link between humans and apes may have been in the forefront of the drive to reassess 'man's place in nature,' but the origin of the vertebrate type represented the base of the human family tree. Here our ancestry was linked to the lowly invertebrates, thereby emphasizing our humble origins in the popular imagination. When Ernst Haeckel and Charles Darwin built the ascidian theory of vertebrate origins into their writings on the subject, the topic was highlighted in a way that raised it far above other areas of evolutionary morphology in the public consciousness.

Within the realm of biology itself, the idea of linking the vertebrates to an invertebrate type raised new questions. In the old transcendental morphology, homologies between organs were sought only within types. Many older morphologists—Karl Ernst von Baer is a good example—simply refused to accept the possibility of cross-type homologies. But evolutionism now made it possible—indeed inevitable, on the principle of a monophyletic origin for the animal kingdom—to seek links between the types. The vertebrates must have an origin somewhere, and the form from which they must have evolved must, by definition, have been an invertebrate. Yet even so, morphologists as sophisticated as Carl Gegenbaur sometimes found it difficult to adjust to the new way of thinking. New criteria for recognizing homologies would have to be worked out—a procedure which would turn out to be far from straightforward. Various hypothetical origins were suggested, each depending on one or another source of evidence for its chief means of support. The resulting debates dragged on into the early twentieth century and ended more through loss of interest by the majority of biologists than by any objective resolution.

Despite their importance for the early history of evolutionism, these debates have gone largely unrecorded by historians. The chapter in E. S. Russell's *Form and Function* is still the best published source, especially for the early period up to the formulation of Anton Dohrn's rival to the ascidian theory (the annelid theory).[1] An unpublished Ph.D. thesis by Roberta Beeson is the only modern analysis of the origins of the ascidian theory itself.[2] Beeson argues that the complexity of the debate forces us to reassess the claim by Russell and Coleman that evolutionary morphology was nothing more than a modified version of transcendentalism.[3]

In a superficial sense the old type concept could be translated fairly easily into the evolutionary notion of a common ancestor. And, as in the debates over arthropod origins, most morphologists made only superficial efforts to relate their ideas to the functions that new structures were evolved to perform. But the move into the world of cross-type homologies indicates that evolutionism had more than a trivial effect on morphology. New methods were introduced in an effort to seek links that derived from the darkest depths of the evolutionary past. The most obvious of these was Alexandr Kovalevskii's use of histological techniques in the study of the early embryo to determine homologies that might no longer be visible in the fully developed organism. Morphologists continued to disagree over the relative standing of anatomical and embryological evidence, and—as in the case of the arthropods—the gradual loss of faith in the recapitulation theory had a major effect on the course of the debate.

The debates also centered on differing concepts of how the genealogical tree linking the higher animal types should be structured. As explained in chapter 2, Haeckel had proclaimed that a whole series of phyla took their origins from the unsegmented worms. The vertebrates were the 'highest' phylum in the sense that they rose to the highest level of complexity, but Haeckel made no effort to construct a 'ladder' linking the major invertebrate types to the base of the vertebrate line. The echinoderms, the molluscs, the annelid-arthropod line, and the vertebrates were four independent lines radiating out from the unsegmented worms. The critical question for Haeckel was to identify a plausible link between those worms and the lowest vertebrate form.

The most popular alternative, Anton Dohrn's annelid theory, simplified the tree of life by deriving both the arthropods and the vertebrates from the annelid worms. Since both exhibited metamerism (segmentation), both should be seen as offshoots of the one group of worms that had

---

1. E. S. Russell, *Form and Function,* chap. 15.

2. Roberta Beeson, "Bridging the Gap."

3. Russell, *Form and Function,* p. 247; W. Coleman, "Morphology between Type Concept and Descent Theory."

been first to develop this character. An even more extreme version of the same argument was William Patten's claim that the vertebrates should be derived from the arthropods themselves. Patten created a single great phylum of metameric animals forming a ladder from the annelids through the arthropods to the vertebrates. The echinoderms and molluscs became mere offshoots from this one great line of progress. In the end, the evidence turned biologists' attention in an entirely different direction. Embryology suggested the surprising conclusion that the lowly echinoderms (sea urchins, etc.) were the invertebrate group most closely related to vertebrate origins. On this model, the molluscs and the annelid-arthropod lines were the side branches.

The problem of identifying convergent structures, or homoplasies, (identical structures produced independently in two separate lines) was critical. One of the most basic disagreements over the origin of the vertebrates centered on whether or not the metamerism, or segmentation, of the type was homologous with that of the annelid worms, or whether segmentation could have been developed independently in several unrelated phyla. The theories of Anton Dohrn and others which derived the vertebrates from the annelids or some other articulated types (e.g., the arthropods) were based on the assumption that so basic a similarity must be a genuine homology indicating common descent. In effect, Dohrn and his followers were adhering to the most simple intepretation of Darwin's claim that classification should reflect genealogy. They were also following the logic of the selection theory, in which evolution becomes an unpredictable and open-ended process. Each branching of the tree of life was based on the appearance of a unique new set of characters that was most unlikely ever to be duplicated in another line. The morphologist should not abandon the assumption that a basic similarity of structure must indicate community of descent unless absolutely forced to do so by evidence suggesting parallel development.

In this area, however, the majority of evolutionary biologists preferred to believe that the homology was only apparent, the similarities of structure being developed independently in all the metameric phyla. They felt that the additional problems created by the hypothetical transformation of an annelid into a vertebrate were so great that the assumption of common descent had to be abandoned. The major disagreement was thus over the criteria to be used in deciding when to abandon the simplest interpretation. This approach also required them to postulate evolutionary mechanisms that could produce such a close similarity. They could argue that the tendency to repetitive multiplication of segments was somehow a basic phenomenon of individual development which could easily be incorporated into the evolutionary history of many groups. Alternatively, it was suggested that the segmented form was such a good adaptation for mobility that it had been independently 'invented' by several different phyla. As in so many other areas of debate, both parallelism due to internally programmed trends and adaptive convergence could be invoked to explain identity of structure in supposedly unrelated forms.

The claim that metamerism was produced by internally programmed trends influencing variation was the more popular. William Bateson provides the clearest example of a biologist who, once convinced that metamerism had been developed independently, went on to assume that evolution in general was governed by processes internal to the organism. Far from being an open-ended tree shaped solely by pressures from the external environment, evolution would be directed along fixed lines by forces arising from within the process of individual development. We shall see other biologists proposing the same view of the evolutionary process—yet we must be cautious not to be trapped into the simple assumption that belief in internally directed evolution, or orthogenesis, automatically guarantees that a biologist will accept parallel evolution. William Patten opted for a strongly orthogenetic mechanism of development, yet derived vertebrate metamerism from that of the arthropods, thereby eliminating one of the most popular examples of parallelism.

What, exactly, was the most basic vertebrate form? In the era of transcendentalism, Richard Owen had produced his concept of the vertebrate "archetype"—the simplest possible form which still possessed the essential characteristics of the type.[4] In the age of evolutionism, the small marine creature *Amphioxus* was widely taken to be the closest living representative of the ancestral form from which all higher vertebrates had evolved. *Amphioxus* does, in fact, bear a resemblance to Owen's ideal archetype. It is a segmented animal with no obvious head, possessing a dorsal nerve cord and a notochord (or elastic rod) instead of a true vertebral column. It was to accomodate this primitive form that Balfour proposed the name "Chordate" for the phylum that would include it along with the true vertebrates (i.e., those animals with a vertebral column).

If a headless form such as *Amphioxus* was indeed the origin of the Chordate phylum, then the head of the true vertebrate became something of an evolutionary problem. Transcendentalists such as Lorenz Oken had proposed the vertebrate theory of the skull, in which the bones of the skull were supposed to be modified vertebrae.[5] This theory had been demolished by T. H. Huxley and others before the *Origin of Species* was published—but now the origin of the head from a headless form was implicit in the new evolutionism. In the course of the late nineteenth century, the

---

4. Richard Owen, *On the Archetype and Homologies of the Vetebrate Skeleton;* see Russell, *Form and Function,* chap. 8.

5. See Russell, *Form and Function,* chap. 10.

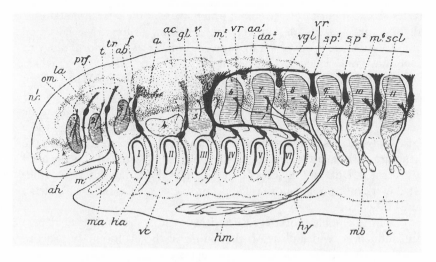

FIGURE 4.1   Segmentation of head of a selachian. From E. S. Goodrich, *Structure and Development of Vertebrates* (London, 1930), p. 221.

segmental theory of the skull was developed on the basis of embryological evidence to confirm that this evolutionary step had indeed taken place.

In his "Monograph on the Development of the Elasmobranch Fishes," originally published in 1876–78, Balfour identified eight segments coalescing to form the skull. Over the next few decades, numerous morphologists contributed to the theory, some—including Anton Dorhn—arguing that more segments were involved. E. S. Goodrich's 1918 paper on the skull of the dogfish *Scyllium* was perhaps the culminating point in the development of this theory, the history of which is outlined in the same author's monumental *Studies on the Structure and Development of Vertebrates*.[6] The evidence that the head region was a specialized development of the anterior segments of a headless form represents the kind of achievement by evolutionary morphology that has little chance of being recorded in histories of evolutionism. Far from the excitement of the debate over the invertebrate progenitor, morphologists such as Goodrich were actively putting together the evidence that would confirm the nature of the vertebrate type's ancestral form (fig. 4.1).

These solid achievements were far from incidental to the theoretical

6. For an account of this debate, see Goodrich's paper "On the Development of the Segments of the Head in *Scyllium*" and his *Studies on the Structure and Development of Vertebrates*, chap. 5. Another contemporary outline appears in chapter 16 of Herbert V. Neal and Herbert W. Rand's *Comparative Anatomy* of 1936. See also David Young, *The Discovery of Evolution*, pp. 180–181, and on Dohrn's contribution, Alfred Kühn, *Anton Dohrn und die Zoologie seiner Zeit*, pp. 67–133.

issues. Some theories of vertebrate origins denied the ancestral status of anything like *Amphioxus*. Patten referred to "that profound specialisation of the Vertebrate head which it is the goal of Vertebrate morphology to expound," and argued that the type must have evolved from a form which already possessed a well-differentiated head: the arachnids.[7] When Dohrn first suggested the origin of the vertebrates from annelid worms, he dismissed *Amphioxus* and the lower fish as degenerate offshoots from the main vertebrate stem. But for supporters of the ascidian theory, these forms were stages in the main line of ascent. Dohrn's later work on the structure of the head was intended to reinforce his theory of how the head was formed. Different theories on the origin of the vertebrates thus stimulated work on complex morphological problems such as the precise structure of the head.

Dohrn's emphasis on the role of degeneration shows that the debate also influenced wider ideas on the nature of the evolutionary process. Haeckel's theory required that the ascidians be regarded as a degenerate offshoot of the vertebrate stock. Alternatives such as Dorhn's required a different choice as to which forms were progressive and which degenerate. The theme of progress and degeneration thus permeated the debate over vertebrate origins. Ideas not only about progenitors but also about the relative status and taxonomic positions of a wide range of forms, both vertebrate and invertebrate, were influenced by intepretations of which were the main, and which the subsidiary, branches of the evolutionary tree leading ultimately toward humankind.

No fossils existed to guide the speculations. The debate over vertebrate origins was fought out almost exclusively within the realm of evolutionary morphology. Patten's arachnid theory can be seen as a belated and rather desperate attempt to make the fossils relevant: he postulated a direct link between the eurypterids (sea scorpions) and the armored fish of the Palaeozoic era. By the early twentieth century, attention was certainly switching to those areas of phylogenetic reconstruction where the fossil record offered some hope of finding direct evidence. In the English-speaking world, the defection of Bateson is often taken to highlight the growing distrust of evolutionary morphology and the emergence of a new emphasis on experimental biology. Coming from the originator of a theory which traced the vertebrates back to the enteropneust *Balano-glossus,* the ill-concealed contempt for the whole project expressed in Bateson's *Materials for the Study of Variation* of 1894 seems to encapsulate the frustrations of a generation which saw its hopes vanish in a cloud of untestable alternatives. In Germany, the bitter debate between Anton

7. William Patten, "On the Origin of Vertebrates from Arachnids," p. 318.

Dorhn and Carl Gegenbaur over the merits of the ascidian and annelid theories served the same negative purpose.

We should not, however, emphasize the speed of collapse. Goodrich's work confirms that important developments in morphology were still being made in the early twentieth century, although it is true that the phylogenetic element was becoming ever less visible in the mass of anatomical and embryological information. Important new ideas were still being introduced, including Walter Garstang's application of his theory of "neotony" to the problem of vertebrate ancestry. That Bateson defected to become one of the founders of genetics should not blind us to the fact that there was still some life left in the morphological tradition in the early decades of the new century.

Nor was morphology the only discipline brought to bear on the problem. As the search for purely morphological relationships declined in interest, some biologists and geologists began to develop hypotheses to explain the origin of the vertebrates in terms of climatic or environmental factors. As the emphasis on purely formal relationships began to seem ever more sterile, the possibility of approaching the question from a new direction was explored. William Keith Brooks introduced his theory of the "discovery of the sea bed" into a discussion of the ascidian *Salpa*. Thomas C. Chamberlin tried to explain the origin of the vertebrates in terms of his cyclic theory of geological change. While unable to halt the decline of interest in the problem, these new initiatives suggest that the factors which prompted a growing interest in the environmental dimension of evolution were able to influence even this remote and highly speculative area.

In the early decades of the twentieth century, the new experimental disciplines were brought to bear on the problem of vertebrate origins. Biochemical techniques were developed which allowed the various hypotheses to be tested in a new way. It became possible to identify degrees of relationships between phyla based on the evolution of different physiological processes in different branches of the tree of life. This new evidence came down firmly in favor of the theory which had become most popular during the heyday of evolutionary morphology, linking the vertebrates back through the enteropneusts to the echinoderms. Thus, paradoxically, new techniques allowed the resolution of old debates just when most biologists had become so enamored of the new approaches that they had lost interest in phylogenetic problems.

THE ASCIDIAN THEORY

In the years immediately following the publication of Darwin's *Origin of Species,* the problem of vertebrate origins seemed outside the realm of scientific hypothesis. No one could see a realistic way of detecting relation-ships that might link the supposedly distinct vertebrate type to an inver-tebrate ancestor. Convinced that the whole animal kingdom must have a monophyletic origin, Haeckel offered what even he admitted was an ex-tremely speculative way of linking the vertebrate type into the tree of life in his *Generelle Morphologie* of 1866. He suggested that the chaetognath, or arrowworm, *Sagitta* offered the closest living representative of what the vertebrate progenitor might have looked like.[8] Haeckel was depending on certain similarities between the musculature of these worms and that of the lowest vertebrates; so far, he had no useful embryological information that could be brought to bear on this question, despite his enthusiasm for the recapitulation theory.

The embryological work that filled in this gap and gave rise to the first serious theory of vertebrate origins was performed by the Russian biolo-gist Alexandr Onufrievich Kovalevskii (Kowalevsky in his original publi-cations). Some of Kovalevskii's earliest work was done at the Naples zoo-logical station; here he pioneered techniques for studying the early development of many invertebrate forms. He stressed the use of cellular methods of analysis which offered new insights into the mode of origin of fundamental structures in the early stages of development. By looking at the germ layers from which the organs differentiate, he established a new way of detecting homologies between organs in different forms. However different their final structures, it might be possible to see a common origin for organs in their mode of formation at this early stage.

Among a series of papers presented to the Academy of Sciences in St. Petersburg were two that would bear directly on the problem of vertebrate origins: one on *Amphioxus* and one on the simple ascidians.[9] *Amphioxus,* the lancelet (now known as *Branchiostoma*), is a small, almost transpar-ent creature that lives in shallow water in the Mediterranean and other seas. Kovalevskii studied *Amphioxus* on the assumption that it was the simplest representative of the vertebrate type, expecting to find that its early development would confirm the unique character of the type. The later stages of development certainly confirmed that it was a vertebrate (or chordate, in modern terminology). But the earliest stages revealed many features that were difficult to distinguish from those that were common among the invertebrates he also studied. By turning to this simplest verte-brate, Kovalevskii opened up a window on the past that was concealed in the development of higher forms such as the fish. Here was the kind of

8. Haeckel, *Generelle Morphologie*, II, p. cxix. For details of these early ideas, see Beeson, "Bridging the Gap."

9. Alexander Kowalevsky [Kovalevskii], "Entwickelungsgeschichte der einfachen Asci-dien," and "Entwickelungsgeschichte des *Amphioxus lanceolatus.*"

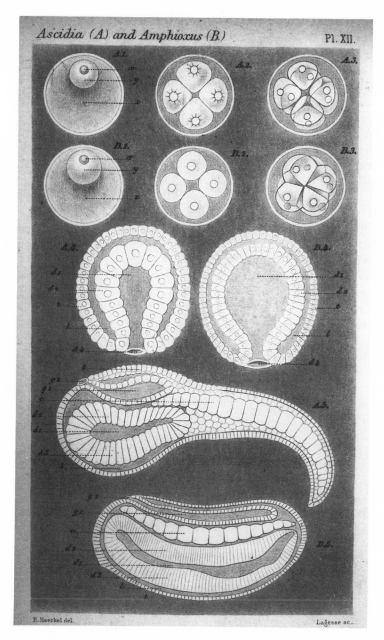

FIGURE 4.2 Development of A: an ascidian; and B: *Amphioxus* (now called *Branchiostoma*). From Ernst Haeckel, *The History of Creation* (New York, 1883), II, plate XII.

material upon which Haeckel's search for the most fundamental links in the tree of life could be based—although Kovalevskii himself did not at first treat his studies as the basis for a new approach to the reconstruction of phylogenies.

Kovalevskii's study of the simple ascidian, or tunicate, suggested that the early stages of its developmental history were so close to those of *Amphioxus* that many structures must be seen as homologous, that is, as having a common evolutionary origin (fig. 4.2). Many different kinds of ascidians, or sea squirts, exist; some are free-swimming, but most enter a sessile adulthood when their mobile, tadpole-like larvae attach themselves to rocks and transform themselves into sacklike structures through which water is pumped to sustain life (figs. 4.3 and 4.4). Because of certain superficial similarities, the ascidians had been classified as molluscs, but Kovalevskii's work showed that this was no longer tenable. The ascidian larva showed many features that unmistakably linked it to the vertebrate type, including a neural canal very similar to that of *Amphioxus*. By tracing these structures back to their earliest origins in the embryo, Kovalevskii showed that they were genuine homologies, not mere analogies that might be explained in terms of similar adaptations in unrelated types. Beeson notes that Kovalevskii's own intepretration of these facts was that the larval ascidian might be the actual ancestor of the vertebrates. As Albert von Kölliker had suggested, a larval form might achieve sexual maturity, thus sloughing off the original adult form and pioneering a new step in evolution.[10] Haeckel and others would, in fact, develop the ascidian theory in a rather different way.

Kovalevskii's observations attracted a good deal of attention. A detailed summary by Michael Foster was published in the *Quarterly Journal of Microscopical Science* in 1870.[11] But long before this, the implications of these homologies had been explored in Haeckel's preliminary version of the ascidian hypothesis of vertebrate origins. The theory was introduced in his popular *Natürliche Schöpfungsgeschichte* of 1868, which used the recapitulation theory as the basis for much speculative reconstruction of phylogenies.

The hypothesis on the origin of the vertebrates went a good deal further than the data justified, although Haeckel was willing to speculate in this way in order to stimulate further research. He argued that the vertebrates and the tunicates, or ascidians, were derived from a common root.

10. Beeson, "Bridging the Gap," p. 128, translating Kovalevskii, "Entwickelungsgeschichte der einfachen Ascidien," p. 12.

11. This account is included in the anonymous review article, "The Kinship of Ascidians and Vertebrates."

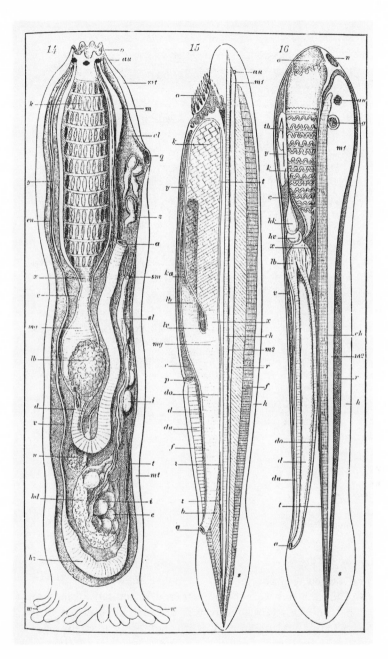

FIGURE 4.3 Internal structure of left: an ascidian; center: *Amphioxus;* right: the larva of the lamprey, *Petromyzon.* From Ernst Haeckel, *The Evolution of Man* (London, 1879), I, plate XI.

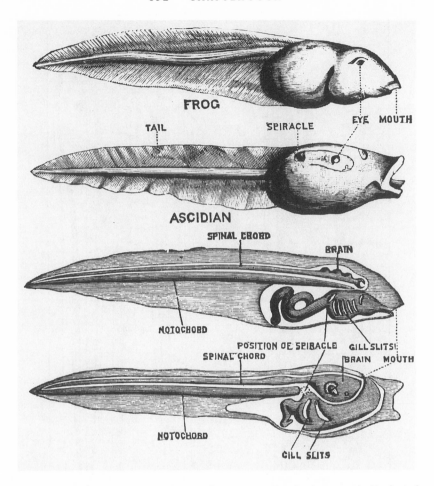

FIGURE 4.4   Tadpole of a frog and an ascidian. From E. Ray Lankester et al., *Zoological Articles* (London, 1891), p. 174.

In his search for a monophyletic interpretation of animal evolution, Haeckel wanted to argue that all the higher animals were derived from worms. The tunicates were, he now argued, a remnant of the branch of the worms most closely related to vertebrates. The skull-less vertebrates represented now by *Amphioxus* were the next stage of evolution, but the modern tunicates were a degenerate offshoot. Their larvae represented the same stage as *Amphioxus,* but the later evolution of this branch had led to degeneration through the adoption of sessile habits. "It is quite evident that genuine Vertebrate animals developed progressively during the primordial period (and the skull-less animals first) out of a group of worms, from which the degenerate Tunicate animals arose in another and retrograde direction." [12]

This suggestion differed from Kovalevskii's hint in that it did not require the vertebrates to be derived from anything like a modern ascidian by what Walter Gastang would later call "neotony" (loss of the original adult stage). Instead, the vertebrate ancestor was a sexually mature form otherwise equivalent to the ascidian larva, with the modern ascidians being developed by degenerate evolution at the same time. Yet by insisting that the tunicates were still to be regarded as a branch of the worms, Haeckel had his link between the lowest vertebrates (the jawless cyclostomes) and the rest of the animal kingdom, as shown in the following diagram:

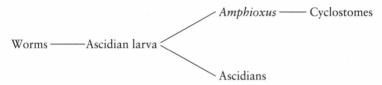

In the introduction to his *Descent of Man,* Darwin noted the extent to which his views paralleled those of Haeckel, and he certainly adopted a more or less identical interpretation of the origin of the vertebrates. Kovalevskii was in touch with Darwin directly—the *Descent of Man* records that Kovalevskii had written to say that his later research confirmed his published conclusions. Darwin could now proclaim that if embryology provided a reliable guide, we at last had an insight into the question of vertebrate origins. "We should then be justified in believing that at an extremely remote period a group of animals existed resembling in many respects the larvae of our present Ascidians, which diverged into two great branches—the one retrograding in development and producing the present class of Ascidians, the other rising to the crown and summit of the animal kingdom by giving birth to the Vertebrata." [13]

Some time later the theme of progress and degeneration was taken up as the topic of Lankester's 1879 B.A.A.S. lecture, published as *Degeneration: A Chapter in Darwinism.* Here Lankester stressed that progress was always the result of adaptation to an active lifestyle. Any species taking up a passive, especially a sessile, form of life would inevitably degenerate. Although inspired by Anton Dohrn, who pioneered a rival theory of vertebrate origins, Lankester chose to use the degeneration of the ascidians

12. Haeckel, *History of Creation,* II, p. 201.
13. Darwin, *Descent of Man* (2d ed.), p. 160.

as a prime example. They were the descendants of much higher and more elaborate ancestors, and "in fact are degenerate Vertebrata, standing in the same relation to fishes, frogs and men, as do the barnacles to shrimps, crabs, and lobsters." [14] The implications of all this for human affairs were spelled out in some detail. Noting that there was no sign of physical or mental progress in the human race since the time of the ancient Greeks, Lankester wondered if we, like the ascidians, might give up our advanced characters and sink back into ignorance and superstition. If we allow ourselves to adopt too soft a lifestyle, we must inevitably slide toward degeneration. The only hope of staving off this prospect, he maintained, was the continued stimulus that science offers to the human intellect.

Lankester went on to admit what was becoming an increasing problem for the ascidian theory: the association between the ascidians and vertebrates was now becoming so close that it no longer threw any real light on the invertebrate origins of the vertebrate type. Lankester insisted that the ascidians must be classified within the vertebrate subkingdom. [15] But if the ascidians were only degenerate vertebrates, where did the vertebrates actually come from? Haeckel still treated the ascidians as worms, but Lankester had recognized that the evidence now suggested that they ought to be transferred into the vertebrate phylum itself. We shall see how increasing recognition of this point began to direct attention to other forms, especially the acorn worm *Balanoglossus* as a better indication of the true origin of the vertebrates.

Endorsement by Haeckel and Darwin did not guarantee the new theory universal support. Karl Ernst von Baer acknowledged the importance of Kovalevskii's embryological techniques but retained his commitment to the type concept and refused to accept the possibility of cross-type homologies. [16] For von Baer, the ascidians were still molluscs. Alfred Giard, who pioneered support for evolution in the late nineteenth-century French scientific community, also adopted a cautious approach. He was sympathetic to the general principle that one might seek cross-type homologies but thought that the assumed link between ascidians and vertebrates went far beyond what the evidence allowed. The similarities detected by Kovalevskii were analogies caused by the ascidian tadpole and *Amphioxus* independently adapting to the same method of locomotion. He did, however, admit that having seen how the nerve cord had evolved in the ascidian tadpole, one could see how another animal with a similar

---

14. Lankester, *Degeneration*, p. 41, and on human degeneration pp. 59–61. On the influence of degenerationism, see Bowler, "Holding Your Head Up High."

15. Lankester, ibid., p. 45.

16. K. E. von Baer, "Entwickelt sich die Larve der einfachen Ascidien in der ersten Zeit nach dem Typus der Wirbelthiere?"

structure might have gone on to progress rather than degenerate, thus giving rise to the vertebrates.[17]

More interesting is the reaction of Carl Gegenbaur. In the second edition of his *Grundzüge der vergleichenden Anatomie* of 1870, Gegenbaur adopted a favorable approach to the ascidian theory, tempered by his preference for anatomical rather than embryological criteria for determining homologies. He emphasized the similarities of overall structure, which gave much weaker evidence than the embryological origins of the nerve cord and notochord. Gegenbaur was also concerned over the problem of metamerism (segmentation), which was not present in the ascidians. We saw in chapter 3 that he was prepared to admit the independent evolution of arthropod characteristics, and he adopted the same approach to the origin of segmentation. He implied that the metamerism of the annelid-arthropod phylum was not homologous with that of the vertebrates: each had independently acquired its segmented character.[18] Gegenbaur also noted another important discovery made by Kovalevskii: the gill slits of the acorn worm *Balanoglossus* (fig. 4.7). This differentiated *Balanoglossus* from all the other worms and brought it closer to the ascidians and vertebrates. *Balanoglossus* might thus offer a better impression of the wormlike ancestor from which the vertebrates had ultimately sprung.[19] Gegenbaur created a new class of worms, the Enteropneusta, to accomodate *Balanoglossus*.

Some of the reasons for caution urged by Gegenbaur had already been addressed by Carl Ritter von Kuppfer in two articles published in 1869 and 1870. Kuppfer was professor of zoology at Kiel and an acknowledged expert on the embryology of fish. He had originally decided to repeat Kovalevskii's observations in the expectation of refuting them. But he was soon convinced, and went on to become a leading exponent of the ascidian theory. His first account was an enthusiastic letter to the editor of the *Archiv für mikroskopische Anatomie*, which was rapidly translated into English.[20] Even this brief account made the implications of the observations clear: describing the ascidian larva, Kuppfer wrote, "It is impossible to think of a more beautiful model of a vertebrate embryo."[21] He extended Kovalevskii's work by tracing the tunicate nerve cord all the way back to the tail. He also gave an extended list of homologies between ascidians and *Amphioxus*, which he explicitly interpreted as evidence of a

17. A. Giard, "Étude critique des travaux d'embryogénie," p. 281.

18. Gegenbaur, *Grundzüge der vergleichenden Anatomie*, pp. 576–7.

19. Ibid., pp. 248–249.

20. Carl von Kuppfer, "Die Stammverwandschaft zwischen Ascidien und Wirbelthieren," (1869), translated in the anonymous "The Kinship of Ascidians and Vertebrates."

21. Translation from "The Kinship of Ascidians and Vertebrates," p. 68.

phylogenetic link.[22] Kuppfer went beyond Haeckel in stressing the arrangement of organs and the significance of other characters in addition to the nerve cord and notochord, including the gill slits and musculature. In addition to satisfying some of the requirements laid down by Gegenbaur, Kuppfer lent his considerable prestige to the theory.

In 1872 Haeckel published his Gastraea theory, which provided a new foundation for the recapitulation theory and for the claim that the whole animal kingdom is monophyletic. The claim that all higher animals develop from a two-layered sac, the gastrula, allowed the phylogenetic equivalent of this form, the Gastraea, to serve as the foundation from which all branches of the animal kingdom were derived. Haeckel's *Anthropogenie* of 1874 incorporated this theory into its survey of the pedigree of the human species and made significant additions to the ascidian theory of vertebrate origins. Germ-layer homologies were now seen as fundamental to the reconstruction of phylogenies, thus providing a firm foundation for the interpretation of Kovalevskii's ascidian-vertebrate homologies.

Haeckel devoted two chapters to exploring the structural similarities and the germ-layer homologies between *Amphioxus* and the ascidians.[23] Drawing on the work of Richard Hertwig and Wilhelm Müller, he noted that the ascidian endostyle was probably homologous with the thyroid gland of *Amphioxus* and the cyclostomes (lampreys, etc.). When he came to define the steps by which the first vertebrates had evolved, Haeckel retained the view that the ascidians were worms. *Amphioxus* was a surviving relic of the original primitive vertebrate, which had possessed a spinal cord, a notochord, and gill slits. He insisted that metamerism had evolved independently in the articulate (annelid-arthropod) line and in the vertebrates.[24] Following Gegenbaur, he now saw *Balanoglossus* as the best illustration of the type of worm from which the form equivalent to the ascidian larva and *Amphioxus* had evolved.[25]

The ascidian theory had now matured and expanded to take in the enteropneusts as the most likely invertebrate ancestors of the vertebrate phylum. Once the majority of biologists agreed to transfer the ascidians themselves into a protochordate category, the way was cleared for *Balanoglossus* to take on the status once accorded to the ascidian larva, now seen as little more than a phylogenetic equivalent of *Amphioxus*. The

22. Kuppfer, "Die Stammverwandschaft zwischen Ascidien und Wirbelthiere" (1870); a brief account in English is given in the anonymous review article "The Genetic Relationship between Ascidians and Vertebrates."

23. Haeckel, *The Evolution of Man*, chaps. 13 and 14.

24. Ibid., II, 94.

25. Ibid., II, 86–87.

adult form of the modern ascidian was merely a degenerate side branch from an early point on the chordate stem. In the 1880s Bateson would modify this position significantly, transferring *Balanoglossus* itself to the chordates and hinting at an even more remote link to the larvae of echinoderms. By the mid-1870s, however, a signficant rival to this whole approach had emerged in the form of Anton Dohrn's annelid theory.

THE ANNELID THEORY

If the empirical inspiration for the ascidian theory was Kovalevskii's embryological observations, the annelid alternative drew its main support from comparative anatomy. Indeed, as Russell observed, it can be seen as an evolutionary update of Geoffroy Saint Hilaire's effort to extend the scope of transcendental morphology by portraying a vertebrate as an inverted arthropod.[26] In both the arthropods and the annelid worms, the nerve chord runs along the body in a position which is ventral in relation to (i.e., below) the esophagus or alimentary canal. Since in vertebrates the position is reversed—the esophagus is ventral—a vertebrate can be seen as an annelid walking on its back. What had been an attempt to see a structural homology between two types could be turned into an account of vertebrate origins if a convincing explanation could be given for the inversion of the ancestral annelid. For zoologists such as Dohrn, who had begun to distrust the recapitulation theory, this was a possibility worth investigating.

In both the annelid theory and its later modification, the arthropod theory (discussed below), it is taken for granted that the metamerism, or segmentation, of the vertebrates has been inherited from the ancestral form. The need to invoke convergence or parallel evolution to explain the independent invention of segmentation is avoided. Since both annelids and arthropods have a distinct head and brain, headless chordates such as *Amphioxus* cannot be primitive: the earliest vertebrates must already have had a head, and *Amphioxus* must—like the ascidians—be a degenerate form. Dohrn's approach thus had major implications for theories of the evolutionary process. It followed the strictest logic of the theory of common descent by insisting that a basic phenomemon such as metamerism could not be produced or 'invented' twice, forcing those who followed Haeckel's line of thought to become more self-conscious about the fact that their theories did have to include the possibility of parallel evolution. Dohrn was also forced to explore the extent to which degeneration might lead to a simplification of structure that could mislead the biologist into

26. Russell, *Form and Function*, p. 274.

History of the Human Body (New York, 1909), p. 518.

Wilder, no notochord (*nt*)', and the supposed change is not as simple as it seems. From H. H. Wilder,

necessary to build a new mouth (*ts*) and anus (*pr*) and close the old one; the worm really had

··· and now we have the vertebrate, with nerve cord and blood streams reversed. But it is

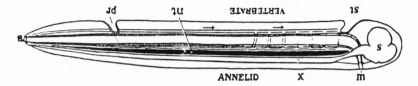

FIGURE 4.5   Diagram to illustrate the supposed transformation of an annelid worm into a vertebrate. In the normal position this diagram represents the annelid with a 'brain' (*s*) at the front end and a nerve cord (*x*) running alongside the upper side of the body. The mouth (*m*) is on the underside of the animal, the anus (*a*) at the end of the tail; the blood stream (indicated by arrows) flows forward on the upper side of the body, back on the underside. Turn the book upside down . . .

thinking that a particular simple form was 'primitive' in the sense of being ancestral to more highly developed types.

The main problem with Dohrn's assumed inversion of the ancestral annelid is that the relationship between the mouth and the brain does not correspond to what the hypothesis predicts. In the vertebrate, the whole nerve cord including the brain is dorsal in relation to (i.e., above) the alimentary canal. In an annelid, the cerebral ganglion (the first and major swelling of the nerve cord equivalent to the brain) is dorsal to the esophagus, while all the rest of the nerve chord is ventral (fig. 4.5). The cerebral ganglion is linked to a lesser swelling (the sub-esophagal ganglion lying at the head of the rest of the nerve cord) by nerves which pass around the esophagus on either side. In effect, the annelid esophagus penetrates through the nerve cord just behind the brain. Since this is not the case in vertebrates, either the vertebrate brain is not homologous with the annelid brain, or the vertebrate mouth is not homologous with the annelid mouth. The supporters of the annelid theory assumed that in the earliest vertebrates the old mouth had closed and that a new one had opened up on the same (i.e., ventral) side of the nerve cord. To any critic who did not believe this massive rearrangement possible, the homologies on which the theory was based were totally false, products of the independent development of metamerism.

An early move toward this type of theory was made by Franz Leydig in his *Vom Bau der thierischen Körpers* of 1864.[27] Leydig had been trained by Haeckel and Gegenbaur and was an expert on insect histology (the

27. This account of Leydig's work is based on Beeson, "Bridging the Gap," pp. 19–63.

study of the microscopic structure of tissues). He was aware of Geoffroy's much earlier efforts to homologize arthropods and vertebrates, which he took as the basis for a highly speculative evolutionary connection. Like Haeckel, he saw Darwin's initiative as opening up a new era for biology in which the rigid distinctions between the animal types would be broken down. For Leydig, the course of evolution should leave visible traces in the structure of living organisms. He was convinced that his histological comparisons between the nervous tissue of insects and vertebrates indicated that the nerve cords of the two types were homologous, whatever the different anatomical relationships.

Carl Semper adopted a sympathetic attitude toward Leydig's idea in his 1876 paper developing the annelid theory.[28] Semper thought that Leydig's evidence was unconvincing because he had compared tissues from the highest form of articulated animal—insects—with tissues from vertebrates. The best comparison would be between the simplest forms of the two types, and Semper himself took the comparison one step further back by relating vertebrates not to arthropods but to annelids. As we saw in chapter 3, it was universally accepted that the arthropods had evolved from annelids: Semper and Dohrn would argue that vertebrates and arthropods are parallel evolutionary developments from the annelid stock.

Leydig's hypothesis was violently condemned in Carl Gegenbaur's 1870 *Grundzüge der vergleichenden Anatomie*.[29] While accepting the value of the histological comparisons, Gegenbaur saw them as merely evidence for similarity of function. Leydig had simply failed to realize what would count as valid evidence for homology. Histological similarity was not enough: the morphological arrangement of the organs must be comparable for a homology to be plausible, and this was not the case. At this point Gegenbaur was still using a pre-evolutionary criterion for defining homology, but his strongly negative attitude did not bode well for the later development of this line of analysis by Semper and Dohrn.

Semper's version of the theory was published in two papers in the *Arbeiten aus dem Zoologische-zootomischen Institut in Würzburg*, 1875 and 1876.[30] The first and shorter paper was only a preliminary study pointing out the similarities between annelids and vertebrates. The fact that both are segmented was seen as a fundamental indication of relationship, backed up by similarities between the nephrida (segmental organs) of annelids and the excretory tubules he had discovered in embryonic sharks.

28. Karl Semper, "Die Verwandschaftsbeziehungen der gegliederten Thiere," pp. 115–116.

29. Gegenbaur, *Grundzüge der vergleichenden Anatomie*, pp. 368–70.

30. Semper, "Die Stammesverwandschaft der Wirbelthiere und Wirbellosen" and "Die Verwandschaftsbeziehungen der gegliederten Thiere."

The second paper included an extensive defense of Geoffroy's principles as the foundation for an evolutionary view in which the annelids, arthropods, and vertebrates were linked into a single unified type. The distinction between dorsal and ventral surfaces was dismissed as essentially arbitrary—what mattered was the basic spatial relationship between the structures. The surfaces should be defined by whether they were on the nerve-cord (neural) or heart/blood-supply (cardial, or haemal) side of the gut. Which of these sides was front (ventral) or back (dorsal) was a matter of how the organism had come to orient itself in the course of its evolutionary history. Semper pointed out the apparent homologies between the annelid and vertebrate structures, and tried to show that the former possessed at least the rudiments of characteristic vertebrate organs such as the gill slits and notochord. He also used evidence from the regeneration of segments in the worm *Nais* to argue that annelids undergo the double-symmetrical type of development which von Baer had depicted as typical of the vertebrates. Embryology was thus brought into the system, although only through indirect evidence.

Semper's enthusiasm for Geoffroy's concept of unity led him to adopt a view of the relationship between brain and mouth that differed from that proposed at the same time by Dohrn. He seems to have believed that each structure in the one type must have a homologue in the other: he preferred a theory which did not require the creation of an entirely new structure. He thus sought a homologue of the vertebrate mouth in the annelids and found it in a pit on the haemal surface of the leech *Clepsine*. Semper believed that the original annelid mouth could no longer be detected in the vertebrates because it was crowded out in individual development by the greater development of the brain. *Amphioxus,* with its lack of brain, could not be a primitive vertebrate—indeed Semper argued that it was not a vertebrate at all. The ascidians were even further removed from the vertebrate stock, since they were not segmented animals.

Anton Dohrn became the better-known exponent of the annelid theory, since his version was proposed in a short but influential book which directly addressed a number of theoretical issues, his *Der Ursprung der Wirbelthiere* of 1875. In chapter 3 we saw how Dohrn began his career with a study of arthropod development which persuaded him that the group was polyphyletic. By 1869 he was convinced of this, and as his conception of the gulf between the crustaceans and the insects grew clearer, so did a conviction that there were links between the insects, the annelids, and the vertebrates.[31] In the same year he gave a lecture on homologies of organ formation between articulates and vertebrates at a conference in

31. See the letter from Dohrn to his father, 29 August 1869 quoted in Heuss, *Anton Dohrn,* p. 81; see also Kühn, *Anton Dohrn und die Zoologie seiner Zeit,* chap. 5.

Innsbruck.[32] One product of the invertebrate work was a growing suspicion of the recapitulation theory, although Dohrn remained convinced of the value of embryology for phylogenetic research. His lack of enthusiasm for simple recapitulationism may have made him less than sympathetic to the ascidian theory that Haeckel was building on Kovalevskii's observations. The rival theory published in 1875 was thus the product of ideas that Dohrn had been developing for some time.

Dohrn believed that his views on insect development harked back to those of von Baer, and the *Ursprung der Wirbelthiere* was dedicated to von Baer despite the latter's strenuous opposition to the search for cross-type homologies. By appealing to von Baer, he went above the heads of Haeckel and Gegenbaur to an older, and by implication wiser, authority whose reputation was founded on the demolition of the original recapitulation theory. In addition to the new hypothesis of vertebrate origins, Dohrn defended a series of theoretical principles which were needed to sustain various parts of the hypothesis. He was convinced that morphology had become too abstract. It was a search for purely formal relationships between structures, which paid little attention to the functions the structures were meant to perform. For Dohrn, morphology and the evolutionary history of structures "are only the content and the process of functions projected as form, and cannot even be conceived of without functions."[33] The loss of the gut in the rear part of the vertebrate body, for instance, was caused by the tail taking on an important locomotory function. Dohrn thus pioneered the search for functional explanations of evolutionary changes. He also advanced his theory of function-change in the course of an organ's history, which avoided the need to postulate the creation of new structures from nothing. In addition, Dorhn stressed the role of degeneration in evolution, a move that was necessary to explain the origin of *Amphioxus,* since according to his theory this could not be a truly primitive (i.e., ancestral) vertebrate.

Dohrn opened his study with a strong appeal to embryological evidence. His theory of vertebrate origins would explain otherwise meaningless facts such as the very late appearance of the mouth in the vertebrate embryo. Like Semper, Dohrn derived the vertebrates from an inverted annelid ancestor. The original mouth had thus been dorsal; it had closed, and a new mouth opened on the ventral side of the body.[34] The two may have coexisted for a while, but the old mouth has now disappeared—

---

32. Heuss, *Anton Dohrn,* p. 83.

33. Anton Dohrn, *Der Ursprung der Wirbelthiere,* p. 70, as translated by Michael Ghiselin; see Dohrn, "The Origin of Vertebrates," p. 74. I am grateful to Michael Ghiselin for making his edition available to me on disk prior to publication.

34. Dohrn, *Ursprung,* pp. 3–5.

although a trace of it may be seen in the sucker of the degenerate tunicates.[35] The new mouth was derived from the coalescence of a pair of gill slits. Dohrn assumed that the ancestors of the vertebrates had been annelids of a higher level of organization than any now living. The ancestors themselves had been exterminated by their more successful descendants long before the first fossils were laid down.

Dohrn insisted that the gill slits were derived from the segmental organs of annelids, and had originally been present in every segment of the body. The gills were at first external, as in many annelids, and became supported by gill arches in the body wall. The annelid ancestors of the vertebrates possessed gills with cartilaginous skeletons and gill arches on every segment.[36] Dohrn argued that the gill arches formed a separate part of the skeleton from the notochord and vertebral column. Eventually the posterior gills disappeared, their arches becoming ribs. Two pairs of ribs subsequently became the foundations for the paired fins of fish.[37] Dohrn thus lent his support to Gegenbaur's new theory on the origin of paired fins, although he would later become one of its harshest opponents.

For Dohrn, one of the advantages of his theory was that it allowed every part of the primitive vertebrate to be derived from a corresponding part of the annelid ancestor. There was no need to suppose that evolution created any totally new structure, even for such a fundamental step. Dohrn rejected in principle the idea that new organs could arise from nothing: it was a mere *deus ex machina* equivalent to the old idea of spontaneous generation (*generatio aequivoca*).[38] He made this point in the context of his own, rather bizarre, theory that the vertebrate penis is derived from the gills of annelids. The reluctance of morphologists to postulate the appearance of new characters in the course of evolution is, perhaps, an inheritance from the old transcendentalism. It would be correspondingly more difficult to invoke cross-type homologies if the production of a new type did require the creation of something new. Obviously, such a principle could not be extended indefinitely—even for Dohrn, the complex structures of the annelids must have appeared at some stage in their evolution. But by minimizing the need to invoke the 'invention' of new structures, he was defending the simplest possible interpretation of a theory of evolutionary progress.

But Dohrn had another problem in mind when he invoked his principle of function-change to explain the transformation of organs. In his *Genesis of Species,* St. George Mivart had attacked Darwin's theory by arguing that it would be impossible for adaptive evolution to explain the appearance of a new organ, because in its early stages of formation, the

35. Ibid., p. 58,

36. Ibid., p. 14.

37. Ibid., pp. 17–18.

38. Ibid., p. 21.

structure would be too incomplete to serve a useful function. Dohrn realized that by invoking only the transformation of existing organs, he could nullify Mivart's objection. The structure was already in existence, and merely changed its function to become something new. Indeed, it was not so much a change of function as a change of priorities among existing functions:

> The transformation of an organ occurs through the succession of functions, the bearer of which remains one and the same organ. Each function is a resultant of many components, of which one forms the main, or primary function, while the others represent subsidiary, or secondary functions. The sinking of the main function and the rising of a subsidiary function changes the total function. The subsidiary function gradually becomes the main function, the total function becomes another one, and the result of the entire process is the transformation of the organ.[39]

Dohrn thus linked his theory of vertebrate origins to what he regarded as an important contribution to the theory of the evolutionary mechanism. In the same vein, he was led to modify the prevailing interpretation of evolutionary progress by stressing the role of degeneration. Not that he was opposed to the basic theory of progress in animal evolution. Indeed, he went out of his way to stress that there was indeed a main line of progress which served as a the trunk for the tree of evolution. Following his friend Karl Snell, he argued that the animal kingdom consists of a perfectible basic phylum, which strives toward humankind.[40] The highest products of this single phylum had gone on to achieve self-consciousness. But the lower reaches of the animal kingdom comprised a range of less successful forms which had degenerated from the main line of ascent. Some animals had more evolutionary potential than others, just as some individual humans have more potential for self-development. To explain why such a collection of apparently primitive forms had come into existence, Dohrn appealed to the environment as the chief stimulus of progress— and the chief cause of degeneration. In his concluding remarks he referred to Herbert Spencer on the role of the environment in stimulating the organism to activity.

Haeckel had highlighted the progressive stages in vertebrate evolution, but his version of the ascidian theory required one to believe that the modern ascidians were degenerate descendants of the more active creature now represented by their tadpole stage. For Dohrn there were many chordates that also had to be degenerate. If the earliest vertebrates came from annelids, then they already had a distinct head region, yet the sup-

---

39. Ibid., p. 60, as translated by Ghiselin in Dohrn, "The Origin of Vertebrates," p. 67.
40. Ibid., p. xi; see also p. 74.

posedly primitive *Amphioxus* lacked a head and showed no sign of other annelid features that could be expected if it were a remnant of the earliest vertebrate form. *Amphioxus* thus had to lose its ancestral status and be treated as a degenerate vertebrate that had lost some features possessed by the earliest members of the type. Dohrn argued that the cyclostomes (lampreys, etc.) are degenerate fish: their jawless state is not an ancestral feature but is the product of their adaptation to a parasitic lifestyle. If a cyclostome were to degenerate further, it would become something resembling *Amphioxus*. Further degeneration would lead to a form like the ascidian larva, although this transformation would also require the nasal aperture to take over the function of the mouth.[41]

To defend his interpretation of vertebrate ancestry, Dohrn was thus forced to extend greatly the role of degeneration as opposed to progress. The attempt to discover the reasons why one animal progressed while another degenerated was the great task of biology, indeed of human understanding. The secret lay in the interaction of the organism and its environment. Degeneration was always possible when organisms adopted a less challenging mode of life such as parasitism. In Lankester's hands, the principle of degeneration would become an object lesson that teaches us the dangers of relaxation. Lankester expressed major reservations about Dohrn's view that some early animals had been predestined to rise to the pinnacle of development. He preferred to invoke the power of the enviroment to stimulate progress or degeneration at any point in the history of life. Just as the ascidian has degenerated, so did the Roman empire, and so might Western civilization if its conquest of nature leaves it with no challenges to face.[42]

The annelid theory was never as popular as its ascidian-based rival, although its opponents differed among themselves as to how great a threat it represented. Patten regarded it as having been "generally and enthusiastically advocated," while Hubrecht thought that the theory was "rapidly gaining ground" in the early 1880s, listing Dohrn, Semper, Hatschek, Leydig, Kleinenberg, and Eisig as its foremost supporters.[43] An exchange of letters between Darwin and Dohrn in 1882 suggests that Richard Owen had borrowed some of Dohrn's ideas without acknowledgement in a talk to the British Association, and Owen published several papers speculating about homologous structures in the brains of vertebrates and inver-

---

41. Ibid., p. 56.
42. Lankester, *Degeneration*, esp. p. 33. For his negative views on Dohrn's image of purposeful evolution, see Lankester, "Dohrn on the Origin of the Vertebrata."
43. Patten, "Gaskell's Theory," p. 360; A. A. W. Hubrecht, "On the Ancestral Form of the Chordata," pp. 349–350.

tebrates.[44] Although these papers are essentially morphological in character, Owen criticized Dohrn for postulating an idealized annelid as the ancestral vertebrate and proclaimed that the search for phylogenetic links between known types should not be given up. He defended Geoffroy Saint Hilaire's position against Cuvier's attacks and argued that the dorsal/ventral distinction was meaningless in this area. Owen was even prepared to homologize the brains of cephalopod molluscs with those of the articulated animals and the vertebrates.

Dohrn's imaginative linkage of morphology to wider evolutionary issues caught the attention of other biologists who could not swallow the whole argument. Lankester offers one obvious example through his work on degeneration. As late as the 1890s, Brooks records that the graduate students in biology at Johns Hopkins were required to study the *Ursprung der Wirbelthiere* and had produced an English translation of it.[45] Brooks himself developed the ascidian theory in important ways (discussed below), but he thought that Dohrn's work was one of the most important contributions to evolutionary morphology.

Others were less impressed. Dohrn's open challenge to the ascidian theory inevitably put him at odds with Haeckel and his followers and contributed to a steady breakdown in relations between them. It was Gegenbaur, however, who took the greatest exception to Dohrn's new idea. In an 1875 article discussing the significance of morphology, Gegenbaur criticized the annelid theory as a "striking example of . . . unscientific procedure" without mentioning Dorhn by name.[46] For him, this kind of cross-type homology was still unacceptable: the different layout of the annelid and vertebrate bodies meant that any similarities between the nerve cord and the gut must be mere analogies. The comparison of the brain and the annelid nerve ganglia was the product of "a lack of critical attention" and "the grossest unawareness of the facts."

Not surprisingly, Dorhn reacted strongly against this kind of criticism. He wrote to Huxley showing how hurt he was by Gegenbaur's attack and went on to become actively hostile to the Haeckel-Gegenbaur

<hr />

44. Darwin to Dohrn, 13 February 1882, in C. Groeben, ed., *Charles Darwin, 1809–1882, Anton Dohrn, 1840–1909. Correspondence*, p.80 and the letter from Dohrn of 9 February on the preceding page. See Owen, "On the Homology of the Conario-hypophysial Tract," esp. pp. 146–148; "On Cerebral Homologies in Vertebrates and Invertebrates"; and "On the Answerable Divisions of the Brain in Vertebrates and Invertebrates," esp. p. 304.

45. W. K. Brooks, "Salpa in its Relation to the Evolution of Life," p. 183.

46. Gegenbaur, "Die Stellung und Bedeutung der Morphologie," translated as "The Condition and Significance of Morphology" in W. Coleman, *The Interpretation of Animal Form*, pp. 39–56, see p. 42.

axis.[47] In 1880 he compared Darwin's theory with the shell of an immense building, the interior of which was to be finished off by his followers. Haeckel and Gegenbaur had "built the backstairs instead of the stately front staircases, and their corridors are nothing but blind alleys. They denounce my staircases for being fantastic products which will break down; up to now the scientific henchmen run after them, with a few talented exceptions, but I have arrived already at the hall, with some secret constructions. And in one or two years it will become apparent who will kick whom downstairs."[48] By the mid-1880s relations had become so bad that Gegenbaur forced their common publisher, Engelmann of Leipzig, to chose between them, and Dohrn had to look elsewhere.

As part of his campaign, Dohrn undertook a long series of studies published in the *Mitheilungen* of his Naples Zoological Station under the common title "Studien zur Urgeschichte des Wirbelthierkörpers." Here he attempted to reconstruct the form of the original vertebrate head with the aim of showing that his theory accounted for the transformations better than its rivals. He looked at the structure of the gills, the cranial nerves, and the thyroid gland, and made important contributions to the segmental theory of the vertebrate head.[49] He also participated in the critique of Gegenbaur's theory of the origin of the vertebate fins and limbs (see chapter 4). For Dohrn, Gegenbaur's reliance on comparative anatomy at the expense of embryology had precipitated a *Competenzkonflikt,* a conflict over which discipline had the right to determine homologies and evolutionary relationships. Instead of cooperating, the two disciplines had become rivals.[50] The bitterness of the controversy with Gegenbaur helped to bring the study of evolutionary morphology into disrepute in Germany, fueling the kind of cynicism that characterized Bateson's attacks in the English-speaking world.

We should be careful, however, not to exaggerate the speed with which this dissatisfaction spread. The debate over vertebrate origins was still very active in the 1890s and only declined seriously in intensity in the early years of the new century. Modified versions of Dohrn's thesis re-

47. See Heuss, *Anton Dohrn,* pp. 178–185. As his relations with Haeckel and Gegenbaur soured, Dohrn corresponded increasingly with British biologists; see Groeben, ed., *Charles Darwin, 1809–1882, Anton Dohrn, 1840–1909. Correspondence.*

48. Dohrn to Robert von Keudell, 3 May 1880, quoted in Heuss, *Anton Dohrn,* pp. 179–180.

49. For a list of the more significant papers in the "Studien zur Urgeschichte der Wirbelthierkörpers," see the bibliography. For a description of Dohrn's work, see Kühn, *Anton Dohrn und die Zoologie seiner Zeit,* chap. 5.

50. Dohrn, "Studien . . . ," XXI, "Theoretisches über Occipitalsomite und Vagus: Competenzkonflikt zwischen Ontogenie und vergleichender Anatomie." See Lynn Nyhart, *Biology takes Form,* chap. 8.

ceived some support. Balfour conceded that *Amphioxus* might have degenerated through adopting burrowing habits, and thus might provide only indirect evidence of the original vertebrate form. The part of the head with the forebrain might be the equivalent of the praeoral lobe, or ganglion, of invertebrates.[51] He accepted Dohrn's view that the vertebrate anus might be a secondary formation in the vertebrates, although he could see no reason why evolution should have made this change.[52] But Balfour could not accept the idea that the vertebrate mouth was a new formation, and thus he could not go along with the whole of Dohrn's thesis. He argued that the nearest ancestors of the vertebrates were not to be found in the Chaetopoda (annelids) themselves, but that both vertebrates and annelids were developed from a common unsegmented ancestor in which the lateral nerve chords coalesced dorsally, not ventrally.[53] This group had left no descendants in the modern world. Balfour noted, however, that its mode of nerve development was preserved in the nemertine worms, a view exploited in Hubrecht's theory (discussed below).

A new twist on the annelid theory was offered by the American morphologist Ethan Allen Andrews in 1885.[54] Andrews dismissed Dohrn's suggestions as purely hypothetical and was more sympathetic to Semper's approach. The presence of segmentation and bilaterial symmetry in annelids and vertebrates was the basis for comparison, but it was possible that both were descended from a common ancestor, rather than one from the other. The ancestor would have been a segmented form like a planarian worm; from it the annelids had developed by degeneration, and they, not the vertebrates, had developed the new mouth. The vertebrates had developed through significant changes of function in the segmental organs of the ancestor, which had become the mouth, the gill clefts, and the genital and urinary ducts. Andrews thus endorsed at least the principles of degeneration and function-change underlying Dohrn's thesis.

In 1888–89 John Beard, a British morphologist working in the Anatomical Institute at Freiburg, published commentaries on the Dohrn-Semper thesis in *Nature*. Dohrn had originally hoped to find traces of the original annelid mouth in vertebrates and had suggested that these might be traced in a structure known as the hypophysis cerebri. Noting that Dohrn had now abandoned this search, Beard commented, "We seem to

---

51. F. M. Balfour, *Treatise on Embryology,* in *The Works of Francis Maitland Balfour,* III; see p. 314. Balfour wrote anxiously to Dohrn assuring him that his dissent from the annelid theory did not imply any lessening of his respect for Dohrn's work; see letters dated 24 October 1880 and 24 November 1880, Archives of the Stazione Zoologica, Naples.

52. Balfour, *Treatise,* III, p. 323.

53. Ibid., p. 314. See also Balfour's "Monograph on the Development of the Elasmobranch Fishes," in *Works,* II, p. 311.

54. E. A. Andrews, "Affinities of Annelids to Vertebrates."

be gradually getting out of the idea that ontogeny is even a fair repetition, much less a perfect one, of phylogeny."[55] Beard accepted the idea that a new mouth had developed in the vertebrates, and wanted to explain why all trace of this massive transformation had vanished from ontogeny. He drew attention to the work of Nicolaus Kleinenberg on the development of annelids, which had shown that both the mouth and the nervous system were replaced in the transition from the larval to the adult form. The implication of this work was that it would be difficult to detect traces of adult annelid structures in vertebrate development, making the theory almost impossible to test. Beard thought that this problem could be overcome, however, and argued that the development of the gill clefts in the early vertebrates had transformed their structure almost beyond recognition. As the eyes and sense organs on the gills had developed, the nerves leading from them had boosted the structure that became the vertebrate brain, deposing the supra-esophagal ganglion of the annleids from its preeminant position in the nervous system, so that it eventually disappeared.

Beard's second contribution dealt with the affinities of the annelid and vertebrate nervous systems. He claimed that one could identify a "combination of Vertebrate characters for which parallel is to be found in the Annelida, and in no other group."[56] The nervous system was the most conservative of all organ systems, the most likely to preserve ancestral traditions. Yet all efforts to homologize the vertebrate nervous system with that of invertebrate groups had failed. Hubrecht's links with the nemertines were very strained, while Bateson ignored the nervous system in his comparison of the chordates with *Balanoglossus*. The gill slits of the latter could easily have been acquired independently of those of the vertebrates. The Dohrn-Semper thesis was the only remaining possibility, yet no vertebrate equivalent of the annelid supra-esophagal ganglion could be found. If all traces of this structure were suppressed in the ontogeny of vertebrates, the spinal ganglia of the vertebrates could be homologized with a ventral chain of nerves in the original annelid, perhaps the parapodial ganglia studied by Kleinenberg. Beard also noted work by Hugo Eisig at the Naples station on the lateral sense organs of the Capitellidae, which seemed equivalent to the similarly placed organs in lower vertebrates.[57] Many regarded Eisig's monograph as one of the best defenses of the annelid theory.

In 1897 Charles Sedgwick Minot, professor at the Harvard Medical School, introduced yet another twist on the theory. By this time, Patten

55. J. Beard, "The Old Mouth and the New," p. 224.

56. J. Beard, "Some Annelidan Affinities in the Ontogeny of the Vertebrate Nervous System," p. 260.

57. Hugo Eisig, *Monographie der Capitelliden der Golfes von Neapel.*

and Gaskell had launched their rival versions of the arthropod theory, but Minot had no time for these. He also rejected Bateson's *Balanoglossus* theory on the grounds that the gill slits of this form could be accounted for as mere resemblances to the corresponding vertebrate structures, not true homologies. He conceded that W. K. Brooks' modification of the ascidian theory was a serious contender but thought that morphologists had not given the annelid theory a fair trial. The proponents of other theories had never dared to confute Dohrn and Semper's arguments and had merely spoken slurringly of the theory in the hope of dismissing it. "I take this to signify that the writers in question have attacked the problem of the ancestry of Vertebrates not in a judicial but in a somewhat partisan spirit."[58]

Minot built his thesis around a reinterpretation of the position of *Amphioxus* in relation to the other disputed groups. He argued that all the evidence suggested that the alleged affinities between *Amphioxus* and the rest of the chordates had been exaggerated. The absence of lateral ears and eyes separates *Amphioxus* from all the true vertebrates, and links it to the ascidians. Embryology too suggests that the link is much closer to the ascidians. Minot suggested that the term "Atriozoa" should be introduced to denote the grouping of *Amphioxus* and the ascidians. The segmentation of the body of *Amphioxus* was much more complete than that of the true vertebrates, a character linking it more with the annelids. *Amphioxus* was thus a chordate with closer affinities to the ascidians than to the true vertebrates, and its differences from the latter indicated a close link to the annelids. Since no one would suggest that the annelids had evolved from *Amphioxus,* it was reasonable to suppose that the annelids were the ancestors of the cephalochorda (*Amphioxus*) and the vertebrates.

Minot offered a new theory to account for the vertebrate head. He believed that the eyes of vertebrates and articulates were homologous. Since the supra-esophagal ganglion of articulates is mainly linked to the visual organs, it was possible that this had been absorbed completely into the retina of the vertebrates. The supra-esophagal ganglion became circular by the fusion of nerves developing from both sides; if this fusion were prevented by the growth of the vertebrate brain, the nerves would form the Y-shape diverging to the two eyes. The vertebrate brain would then be homologous with the sub-esophagal ganglion, which had undergone a development that was "enormous and precocious"[59] The invertebrate mouth had, as Dohrn earlier supposed, become the olefactory pits and hypophysis of the vertebrates, and the new mouth formed as in the ealier

58. C. S. Minot, "Cephalic Homologies," p. 934.
59. Ibid., p. 940.

versions of the theory. Minot felt that he had provided a modified theory of the origin of the vertebrate from the annelid head, obviating the difficulties of the Dohrn-Semper thesis which had led the majority of biologists to dismiss it. He concluded his article with the following phylogenetic tree:

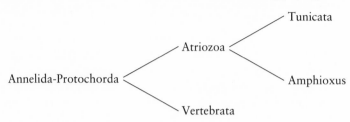

The surveys by Beard and Minot were intended to convey the complex state of the debate over vertebrate origins to a nonspecialist readership. Their work shows that the Dohrn-Semper thesis was not without support, although the majority of morphologists continued to work along lines that derived more from the ascidian theory. Some support for Dohrn's position continued in the early twentieth century. The Dutch morphologist Hendricus Delsman argued that the neural tube was once the stomeodoeal part of the alimentary canal. He too required the formation of a new mouth, the old one surviving as the neuropore. Delsman's submitted a book outlining his theory to the Royal Academy of Sciences in Amsterdam in 1918, only to have one referee report that it should be confiscated and consigned to the flames.[60] The book was published in 1922, although by this time Delsman had moved to the Marine Laboratory in Batavia (Jakarta) in the Dutch East Indies. The referee's reaction indicates the growing frustration of many biologists at the multiplication of incompatible hypotheses on the origin of the vertebrates.

Support for a modified version of the annelid theory also came from the Swiss morphologist Adolf Naef, whose book on idealist morphlogy was mentioned in chapter 2. On the strength of this book, Naef has been treated by some modern Darwinians as a throwback to a pre-evolutionary way of thought, yet he published a whole series of articles in the 1920s on the relationship between vertebrate morphology and evolution. The seventh in the series addressed the question of vertebrate origins in a way that allowed some aspects of Dohrn's thesis to be retained.[61] Naef thought that

60. Hendricus Delsman, *The Ancestry of Vertebrates,* preface, p. iii.

61. Adolf Naef, "Notizen zur Morphologie und Stammesgeschichte der Wirbeltiere, 7: Der Verhältnis der Chordaten zu niederen Tierformen und der typiche Verlauf ihre frühen Entwicklung," and on *Amphioxus,* see Naef, "Notizen zur Morphologie und Stammesgeschichte der Wirbeltiere." Naef's contribution to the debate was brought to my attention by K. Nübler-Jung and D. Arendt, "Is Ventral in Insects Dorsal in Vertebrates?" which gives a

Amphioxus and the ascidians were genuinely primitive, but he neverthless tried to reconstruct an idealized picture of the ontogenetic development of a segmented organism that could be modified to give both the modern annelids and chordates. The axis of development was reversed in the two groups, the original creature having lived in a way that made the distinction between dorsal and ventral irrelevant. Naef's approach differed significantly from that of other evolutionists in that his hypothetical ancestral mode of development was offered as a pattern that could be modified in two different ways: he did not postulate the conversion of one existing type into another. In this respect, his search for relationships still reflected the idealist viewpoint rather than the more functional approach pioneered by Dohrn.

Neither Delsman nor Naef seem to have aroused any wide interest by their hypotheses: the attention of most biologists had shifted elsewhere. Like Delsman, Naef ended up having to take a post outside Europe, at the university of Cairo—although his views are held to have been influential in shaping the subsequent development of German paleontology (see chapter 7). Some time before these last efforts to defend Dohrn's thesis, yet another twist was added to the story by the attempt to link the vertebrates not to the annelids, but directly to the more advanced arthropods. To many this seemed a grasping at straws that illustrated the desperation of those who sought to link the segmentation of the vertebrates with that of the articulated animals.

THE ARTHROPOD THEORIES

In 1890 the *Quarterly Journal of Microcopical Science* carried two consecutive articles advocating the descent of the vertebrates from an arthropod ancestor. The theories were advocated by the American morphologist William Patten and the British physiologist Walter Gaskell. Patten postulated an arachnid ancestor, while Gaskell opted for the crustaceans, but this disagreement was comparatively minor given the profound differences in the kind of transformation required by the two theories. Minot "wondered whether the juxtaposition [in the *QJMS*] was due to editorial design."[62] As editor, Lankester may well have highlighted the unorthodox nature of the new initiatives in this way. Of the two, Patten's was the more acceptable to the community of morphologists; his ideas were treated as worthy of consideration even if few were prepared to accept them.

---

fuller account of his views. Nübler-Jung and Arendt suggest that recent molecular evidence may yet favor Dorhn's thesis.

62. Minot, "Cephalic Homologies," p. 931.

Gaskell's theory suffered a very different fate. It was seen as a challenge from outside the field by someone who was prepared to violate all the accepted canons of morphological research. Lankester refused to accept any more articles from Gaskell, and he was forced to publish in journals that would not normally be used by evolutionary morphologists. The whole episode thus tended to highlight the growing dissatisfaction of those biologists who were turning to experimental biology—a tide which Patten was desperately anxious to stem.

Patten had studied at Leipzig, Vienna, and the Naples zoological station, and then served as an assistant to Charles Otis Whitman.[63] After a period as professor of zoology at the University of North Dakota, he moved to Dartmouth College in 1893. His early work was on the structure of the eyes of arthropods, and it was on the basis of these studies that he was soon led to explore the possibility that the arthropods, rather than the annelids, were the most plausible source of vertebrate origins.[64] The theory he announced in 1890 was presented as an initiative designed to stimulate research which had become bogged down as the annelid theory failed to live up to the expectations of its founders. It was also intended to allow fossils to be brought into the study of vertebrate origins for the first time, a move which fitted the growing role played by paleontologists in the reconstruction of life's ancestry. Patten linked his theory to a definite philosophy of the evolutionary mechanism, a philosophy which emphasized internally programmed trends rather than the effect of the environment, and which had clear overtones of progressionism. In his later writings, Patten chided the new generation of experimental biologists for giving up too easily on serious problems in order to chase after quick success in more superficial areas of research.

Patten's 1890 article, "On the Origin of the Vertebrates from Arachnids," was based on work he had done at the Lake Laboratory in Milwaukee, but it advanced only his theoretical conclusions. He argued that the annelid theory was, after fifteen years, still as far as ever from providing a satisfactory explanation of vertebrate origins.[65] The similarities between the two types were common to nearly all segmented animals. The unspecialized, segmented character of the annelids could not explain the profound specialization of the vertebrate head, and it was thus necessary to find a hypothetical ancestor in which this specialization was foreshadowed. On a priori grounds, the arachnids thus offered the most likely candidate. Patten then summarized what he regarded as the chief factual evidence for a link. The cephalothoracic nerves of the scorpion have the same

63. On Patten's career, see Greg Mitman, "Evolution as Gospel."
64. Patten, "Studies on the Eyes of Arthropods."
65. Patten, "On the Origin of Vertebrates from Arachnids," p. 317.

specialization and arrangement as the corresponding parts of the vertebrate head. Patten believed that the work of Beard and others had destroyed the credibility of the assumed similarities between the annelid and vertebrate nervous systems.[66] The annelid theory could not explain the cranial flexure of vertebrates, but the arachnids had such a flexure and thus fulfilled the conditions demanded by any natural theory of vertebate origins. The cartilaginous sternum of the arachnids represented the primordial cranium of the vertebrates.

Patten used both anatomical and embryological evidence to argue for the similarity between arachnids and vertebrates. He also launched an attack on the gastrula theory and the study of germ layers, claiming that these had misled morphologists for decades.[67] But the most innovative aspect of his theory was its appeal to the fossil ostracoderms as a 'missing link' between the two phyla. All previous theories had assumed that the vertebrates originated long before the known fossil record began. But by the 1890s paleontology was playing a much more significant role in attempts to reconstruct the later episodes in the history of life on earth, while the lack of fossil evidence for the phylogenetic links proposed by the evolutionary morphologists was becoming something of a scandal. Patten saw his theory as an opportunity to address these problems by providing a synthesis resting on both morphological and paleontological evidence.

The heavily armored ostracoderms of the Silurian and Devonian periods (fig. 5.2) were normally treated as a specialized offshoot of the hypothetical jawless fish which, according to the ascidian theory, had evolved from an *Amphioxus*-like ancestor. As Patten noted, some early nineteenth-century paleontologists had recognized that their armored exterior gave them a resemblance to crustaceans. He now suggested that the ostracoderms were the direct link between the Merostomata (the sea scorpions of the Palaeozoic) and the true vertebrates.[68] They were a distinct class forming a half-way house between the arachnid and vertebrate phyla (fig. 4.6).

Over the next two decades or more, Pattern devoted himself to the support of his hypothesis. He worked extensively on the fossil ostracoderms in order to bolster his interpretation of their position in the evolutionary sequence. He also performed extensive studies of *Limulus*, the horseshoe crab, which was the closest living relative of the ancient Merostomata (he became known as "Limulus" Patten among his colleagues).[69] In 1912 he summed up his efforts in a book, *The Evolution of the Verte-*

66. Ibid., p. 326 and p. 346.
67. Ibid., p. 366.
68. Ibid., pp. 359–360.
69. See Mitman, "Evolution as Gospel," p. 449.

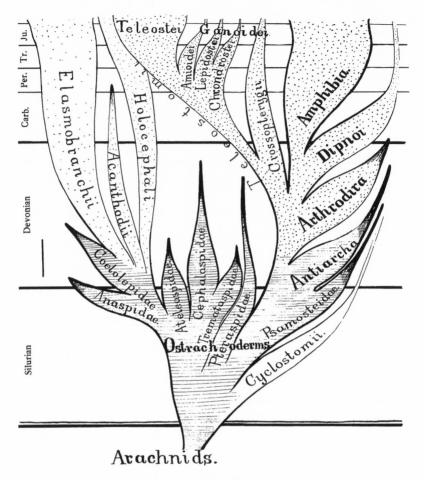

Arachnids.

FIGURE 4.6   Diagram to illustrate the phylogeny of the arachnid, ostracoderm, and vertebrate stock according to Patten's theory. From William Patten, *The Evolution of the Vertebrates and Their Kin* (Philadelphia, 1912), p. 382.

brates and their Kin, in which he also expressed his distrust of the latest trends in biology and an overarching progressionist philosophy of evolution.

Patten began by acknowledging the frustrations that had by now turned many biologists away from the search for vertebrate origins. The first vertebrates sprang into view apparently fully formed in the Silurian period, and the attempt to trace their ancestry seemed to have foundered on a bewildering multiplicity of hypothetical trails. The resulting disappointment had created "an attitude of indifference, the feeling that the

solution to this great problem was forever beyond our reach."[70] The annelid theory, he claimed, had become a dogma which, when it proved incapable of further development, had led many biologists to "turn . . . their
attention toward the Eldorado of cytology, heredity, and experimental
evolution, where 'results' were easy and promised to carry far." The new
biologist was contemptuous of morphology: "'Paleontology,' he cried, 'is
mute, Comparative Anatomy meaningless, and Embryology lies.' "[71] Patten believed that, given the complexity of the problem, it was too soon to
give up after only fifty years of research. His own theory had the advantage of bringing the fossil record firmly into the picture: "Is not the paleontological record more precise and complete than we have supposed?"[72]
It united all opposing views and would "go a long way toward restoring
morphology to its former dominant position as the expounder and
prophet of the biological sciences."[73]

The arachnid theory incorporated all the valid features of the annelid
theory and threw light on the disputes between the supporters of the other
hypothetical ancestors. There was a common arthropod-vertebrate stock
from which the echinoderms, tunicates and *Balanoglossus* were but degenerate offshoots. Patten expanded on all the alleged homologies reported in his first paper, but he now also addressed the question of why
the transformation of the arachnid ancestor into a vertebrate had taken
place. The arachnids were subject to a trend that had led to a steady increase in the size of their nervous system. But since their gut penetrated
through the central nervous system, it had been steadily constricted until
most living arachnids had been forced to adopt parasitic or blood-sucking
habits as the only means of obtaining nutrition. These animals were "making the best of a desperate situation, adapting their lives with great precision to meet the inevitable march of events within."[74] The vertebrates
were the descendants of arachnids which had met this crisis by undergoing
a dramatic transformation, allowing the old mouth to close up and opening a new one directly into the gut from the cephalic navel, a structure
which had been prepared to take over this function when required.[75]

Patten's views on the pressures that led to the transformation derived
from a vision of evolution as a process directed by trends programmed
from within the organism itself. The development of the animal kingdom
had a main line moving in a predictable direction: "the great vertebrate-
ostracoderm-arthropod phylum forms the main trunk of the genealogical
tree of the animal kingdom; that, emerging from unsegmented, coelenter-

70. Patten, *The Evolution of the Vertebrates and their Kin,* p. vi.
71. Ibid., p. vii.   74. Ibid., p. 232.
72. Ibid.   75. Ibid., p. 250.
73. Ibid., p. 2.

ate-like animals, as though driven by some mysterious internal power, moves with astonishing precision, through broad, predetermined channels—from which neither habit, nor environment, nor heredity, can cause it to diverge—toward its goal."[76] The crisis in arthropod evolution that had forced the transformation into vertebrates was an inevitable consequence of the "steady, underlying trend toward a more compact, voluminous nervous system."[77] With hindsight we can "trace with our mind's eye, the slow, inevitable approach and consummation of the most momentous event in organic evolution."[78]

The Merostomata were the highest arachnids in the sense that they had made the greatest progress toward the structure later to be elaborated in the vertebrates.[79] The ostracoderms were the key to this transformation, and Patten now reinforced his early claim that these were not merely degenerate fish, but an entirely distinct type which had flourished temporarily as the intermediate stage on the way to the true vertebrates. He gave a history of the differing interpretations offered of this puzzling group, criticizing those such as Huxley and Lankester who had treated them as merely specialized offshoots of the early fish.[80] Some authorities had actually treated them as arthropods, but the solution to the problem was to recognize their unique intermediate status. Before they disappeared, they had given rise to several distinct, but similar phyla. Patten regarded the elasmobranch fish (sharks, rays, etc.) as having been evolved independently of the line leading to the higher fish and the tetrapods. The ostracoderms had thus been programmed with certain tendencies which were expressed in all their descendants.[81]

In the line leading through to the Amphibia, evolution is "steady, comparatively rapid, and in every respect leads consistently upward to the first air-breathing land vertebrates."[82] This teleological language occurs repeatedly throughout Patten's account: "In this vigorous phylum, evolution follows a logical and consistent course, each important event being a direct or indirect result of the preceding ones, and they themselves creating the conditions that bring about those that follow."[83] This confirmed that the transforming factors arose from within the organism—neither natural selection nor Lamarckism offered a satisfactory explanation of the steady upward trend. These points were repeated in the book's conclusion, where Patten also expressed the view that the human form was the goal toward which the whole process had been moving. "In so far as we can judge from

76. Ibid., p. 2.
77. Ibid., p. 252.
78. Ibid., p. 253.
79. Ibid., p. 338.

80. Ibid., pp. 342–347.
81. Ibid., pp. 381–382.
82. Ibid., p. 383.
83. Ibid., p. 392.

the teachings of morphology, the perfection of physical organization is reached in man."[84]

Patten's enthusiasm for internally directed evolution was shared by many contemporary biologists, yet in his case it did not translate into a theory of parallelism. Where William Bateson, for instance, used the idea of internally generated variation to argue for the independent appearance of metamerism in several different phyla, Patten insisted that metamerism was a unifying feature of the annelid-arthropod-vertebrate stock. For him, the internal pressure to advance the level of organization was confined to this one stock, thus supporting a very different and ultimately more simple view of the pattern of phylogenetic development.

As Gregg Mitman has argued, Patten's vision of inevitable progress was typical of the message that some early twentieth-century evolutionists wanted to extract from their science.[85] But his actual theory did not become very popular, although it was sometimes given a sympathetic hearing. The leading British expert on fossil fish, Arthur Smith Woodward, dismissed the alleged arachnid affinities of the ostracoderms: "They are to be regarded merely as an interesting example of mimetic resemblance between organisms of two different grades adapted to live in the same way and under the precisely similar conditions."[86] An anonymous review in *Nature* professed more enthusiasm for Patten's views on directed evolution than for the arachnid theory itself.[87] The most positive support came from geologists interested in the history of life of earth. In the second volume of their *Geology: Earth History* (1905), Chamberlin and Salisbury accepted that the ostracoderms are "clearly proven to be an entirely distinct class lying between the arthropods and vertebrates."[88] If this intermediate position were accepted, they claimed, Patten's theory of vertebrate origins seemed plausible. Lull's popular textbook, *Organic Evolution*, devoted a couple of pages to the theory and, while conceding that it had not been widely accepted, thought that it was based on "an admirably executed piece of research."[89]

Gaskell's rival version of the arthropod theory received no such sympathetic reception, even Patten joining in the chorus of complaints. Walter Holbrook Gaskell was a physiologist—he became a lecturer in this field at the University of Cambridge—and his interest in the problem of verte-

84. Ibid., p. 472.

85. See Mitman, "Evolution as Gospel."

86. A. Smith Woodward, *Outlines of Vertebrate Palaeontology*, p. xxiv.

87. Anonymous review of Patten, *Evolution of the Vertebrates and their Kin.*

88. T. C. Chamberlin and R. D. Salisbury, *Geology: Earth History*, II, p. 482.

89. R. S. Lull, *Organic Evolution*, p. 469.

brate origins was aroused by studies of the functioning of the nervous system. He saw that the central nervous system of the vertebrates consisted of nervous material surrounding a central tube, and conceived the idea that this tube might be the remnant of the gut of an invertebrate ancestor, probably a crustacean. Gaskell's approach thus differed profoundly from that of both the annelid theory and Patten's version of the arthropod theory. Instead of looking for a new mouth into the old gut, he maintained that the old gut was incorporated into the nervous system, and a whole new alimentary system created in the early vertebrates. As he himself later wrote: "Not having been educated in a morphological laboratory and taught that the one organ which is homologous throughout the animal kingdom is the gut, and that therefore the gut of the invertebrate ancestor must continue as the gut of the vertebrate, the conception that the central nervous system has grown round and enclosed the original ancestral gut, and that the vertebrate has formed a new gut, did not seem to me so impossible as to prevent my taking it as a working hypothesis, and seeing to what it would lead." [90]

This theory was first announced in an article in the *Journal of Physiology* in 1889 and then in the 1890 *Quarterly Journal of Microscopical Science* article that Lankester printed immediately after Patten's.[91] Gaskell undertook an extensive program of research on Ammocoetes, the larva of the lamprey, on the grounds that this most simple of the true vertebrates might best elucidate the homologies he sought. He insisted that the lamprey was not a degenerate organism; on the contrary, its development revealed the actual course of evolution from the arthropod ancestor to the lowest vertebrate.[92] It was the recapitulation theory which offered the best guide to phylogeny, not the new-fangled germ-layer theory, which imposed unrealistic limitations on the search for homologies.[93] Like Patten, Gaskell saw the development of the arthropod nervous system as having provoked a crisis in evolution when it threatened to strangle the old gut. "Truly, at the time when vertebrates first appeared, the direction and progress of evolution in the Arthropoda was leading, owing to the manner in which the brain was pierced by the oesophagus, to a terrible dilemma— either the capacity for taking in food without sufficient intelligence to capture it, or intelligence sufficient to capture food and no power to consume

90. W. H. Gaskell, *The Origin of Vertebrates*, p. 4.

91. Gaskell, "On the Relation between the Structure, Function, Distribution, and Origin of the Cranial Nerves," pp. 190–208, and Gaskell, "On the Origin of Vertebrates from a Crustacean-Like Ancestor." I am in possession of Adam Sedgwick's offprint of the former paper, with a number of impatient comments scribbled in the margin (PJB).

92. Gaskell, *The Origin of Vertbrates*, pp. 60–61.

93. Ibid., p. 456.

it."[94] Unlike Patten, he thought it possible for the early vertebrates to have developed an entirely new gut, arguing that cells from other parts of the body in both the crustaceans and Ammocoetes had slight digestive capacity.[95]

In the later version of his theory Gaskell followed Patten in seeing the Palaeostraca or Merostomata of the Palaeozoic as the hypothetical ancestors of the vertebrates. Thus paleontology took its place in the search for evolutionary links along with evidence derived from embryological recapitulation. It was a mistake, he argued, to assume that a higher class must always arise from an unspecialized member of a lower class. In fact, there was a main line of evolution linking the dominant forms of each geological epoch.[96] Without following Patten's insistence on evolution as an internally directed process, he nevertheless argued for a steady trend in the development of the brain. And, like Patten, he drew a moral message from this: "As for the individual, so for the nation, as for the nation, so for the race; the law of evolution teaches that in all cases brain-power wins."[97]

Gaskell seems to have persuaded his fellow physiologists to take his theory seriously, but the morphologists treated it as something from beyond the pale. He himself complained that the separation of morphology and physiology was detrimental to both, but the morphologists thought that his theory violated the most basic principles of their science. In 1896 Gaskell was president of the physiology section of the British Association and gave his presidential address on the topic of his theory.[98] But there are many signs of the morphologists' displeasure. His later articles were published in the *Journal of Anatomy and Physiology,* and in one of these he records the negative response of the *Quarterly Journal of Microscopical Science* to his attempts to follow up his 1890 article: "Owing to a difference of opinion between the editors of the *Quarterly Journal* and myself upon the value of my further contributions on the subject, I determined to seek another channel of publication . . . "[99] In his book *The Origin of Vertebrates* he recorded the following backhanded encouragement from Huxley: "Go on and prosper; there is nothing so useful in science as one of those earthquake hypotheses, which oblige one to face the possibility that the solidest-looking structures may collapse."[100]

Gaskell also records the almost universal condemnation of morphol-

94. Ibid., p. 57.   96. Ibid., p. 498.
95. Ibid., p. 58.   97. Ibid., p. 499.
98. Gaskell, "President's Address."
99. Gaskell, "On the Origin of Vertebrates, Deduced from the Study of Ammocoetes," p. 516.
100. Letter from Huxley to Gaskell, 2 June 1889, quoted in Gaskell, *The Origin of Vertebrates,* facing p. 1.

ogists at a meeting of the Cambridge Philosophical Society in 1895, where he was accused of violating the central principles of the germ-layer theory by ignoring the ultimate origin of the various structures he studied. Hans Gadow, the lecturer in morphology at Cambridge, was the only speaker who urged that Gaskell be given a serious hearing, a position he defended again at a meeting of the Linnean Society fourteen years later.[101] MacBride spoke strongly against Gaskell at both meetings, while at the later meeting Lankester compared Gaskell's enthusiasm for his position with that of the cranks who try to prove that Bacon wrote Shakespeare's plays. Even Patten wrote a review condemning Gaskell's version of the theory, and the conclusion of his comments may perhaps serve as an epitaph for this whole episode. "Papers like the ones we have just reviewed are unfortunate. They are not a credit to the science of comparative morphology, and the interest in the whole subject of the origin of the vertebrates suffers from the reaction induced by such efforts."[102]

NEMERTINES AND THE ACTINOZOA

Both the annelid and the arthropod theories depended on the assumption that the metamerism of the vertebrate body was homologous with and hence derived from that of the invertebrate ancestor. Its supporters were prepared to accept a variety of major and highly speculative rearrangements of the body-structure in order to preserve this link. But the earlier and rival ascidian theory had been based on the view that—since the ascidian larva was not segmented—this characteristic had appeared later in the chordates and had thus developed independently of the same phenomenon in the articulated animals. Metamerism had evolved in several different groups and was thus, in effect, a product of parallelism that could not be homologized between the types. Among those who did not take up the annelid theory, efforts to demonstrate the independent invention of segmentation at several points in the history of life on earth increased in the last decades of the century. In the case of Bateson, at least, these efforts led to a loss of interest in the original morphological problem and a switch to the study of the processes of variation that might lead to the duplication of segments.

The opponents of the annelid theory also had to face another problem: what exactly was the original invertebrate ancestor? If the ascidians

---

101. The proceedings of the later meeting are recorded in Gaskell et al., "Discussion 'Origin of the Vertebrates'"; for Gadow see, pp. 27–30; MacBride, pp. 9–15; Lankester, pp. 38–40. For Gaskell's account of the earlier meeting, see *The Origin of Vertebrates*, pp. 485–486.

102. Patten, "Gaskell's Theory," p. 369.

were dismissed as degenerate offshoots of the chordate stock, what had the original chordate (equivalent to the ascidian larva) evolved from? Bateson's work on the enteropneust *Balanoglossus* merely added another lowly organism to the list of early offshoots from the chordate stem. Before abandoning the problem altogether, Bateson himself hinted at the possibility of a link between *Balanoglossus* and the larvae of echinoderms, and in the later years of the century this possibility was taken up by a number of morphologists. A rival view expressed by the Dutch biologist A. A. W. Hubrecht traced the vertebrates back to the unsegmented nemertine worms. Hubrecht also contributed to the discussion of the ultimate origin of all the higher phyla, drawing on Adam Sedgwick's suggestion that these could be derived from the Actinozoa, or sea anemones.

Hubrecht was a disciple of Gegenbaur who became professor of zoology at the University of Utrecht in Holland. He began work on the nemertine worms around 1880 and introduced them into the debate in an 1883 paper "On the Ancestral Form of the Chordata." Hubrecht argued that the now universal recognition that the ascidians were degenerate vertebrates meant that they offered no real clue to the nature of the invertebrate ancestor. The annelid theory had gained much support, but had been resisted by "the school of Gegenbaur and Haeckel." [103] Gegenbaur himself had noted that the presence of two lateral nerve cords in the nemertines was a primitive arrangement, and these might have coalesced to give the ventral nerve cord of the annelids. In his "Monograph on the Development of the Elasmobranch Fishes," Balfour had hinted that a similar coalescence on the dorsal side of the body might have given rise to the nerve cord of the chordates. [104] He had then come out more firmly on the side of this hypothesis in his *Treatise on Comparative Embryology.* Balfour firmly rejected Dohrn's theory that the ancestral vertebrate could have formed a new mouth and anus. The ultimate ancestor of the vertebrates had been a now extinct, unsegmented form in which lateral nerve cords similar to those possessed by the living nemertines had condensed dorsally. [105] Hubrecht's work on the nemertines was designed to confirm this suggestion, and he believed that Balfour's comments had forced many younger biologists to reconsider their support for the annelid theory.

Hubrecht pointed out that the great difficulty facing any theory of vertebrate origins had been to find an invertebrate form possessing any trace of the notochord, the stiffening rod which provides the axis for the nerve cord in the most primitive chordates. He believed that he had found

---

103. Hubrecht, "On the Ancestral Form of the Chordata," p. 350.

104. Balfour, "Monograph on the Development of the Elasmobranch Fishes," in *Works,* I, pp. 393–394.

105. Balfour, *Treatise on Embryology,* in *Works,* III, p. 311.

a homologue for the notochord in the sheath that surrounds the proboscis, which is a characteristic feature of the nemertines. The proboscis itself is homologous with the hypophysis cerebri, a rudimentary structure present in all vertebrates for which there was otherwise no explanation.[106] The nemertines thus provided a bridge between the lower flatworms and the vertebrates. Hubrecht argued that what little was known about nemertine ontogeny suggested that the sheath arose in the same manner as the notochord. As the ancestral vertebrate had gradually abandoned the use of a proboscis, the sheath had been adapted to a different function in providing support for the newly coalesced nerve cord. Hubrecht insisted that he was not proclaiming a direct link between living nemertines and vertebrates, merely noting that the nemertines threw light on the character of the ancestral form.

Hubrecht believed that the nemertines did not possess a true coelom, or body cavity, distinct from the gut. He argued that the recent work of Hatschek on the development of *Amphioxus* showed that in this primitive chordate the coelom formed in a manner that suggested an origin in the platyhelminths (flatworms) via something that must have resembled the nemertines.[107] In a later paper, however, Hubrecht argued that Hatschek's work indicated a different interpretation of *Amphioxus:* Dohrn had been right to treat it as a highly degenerate chordate that did not deserve the attention paid to it as a clue to the ancestral form of the phylum.[108]

Hubrecht continued to work on the nemertines and was given the task of describing the new members of this group brought back by the *Challenger* expedition. In a supplement to the descriptive side of this work published in 1887, he again emphasized the significance of the links to the vertebrates. The nemertines threw valuable light on the transitional stage between a diploblastic, or coelenterate, stage of development (jellyfish, sea anemones, etc.) and the triploblastic ancestors of the vertebrates. He accepted the significance of Bateson's work on *Balanoglossus,* but insisted that in recogizing the primitive status of this form in the chordate phylum, we were not necessarily obliged to rule out the nemertines as clues to a far more distant ancestor. Lankester, one of the few biologists to offer temporary support to Hubrecht's theory, saw *Balanoglossus* as a link between the nemertines and the lowest vertebrates. Although a close friend of Dohrn, Lankester criticized the annelid theory on the grounds that segmentation had developed in many different phyla.[109]

106. Hubrecht, "On the Ancestral Form," pp. 351–352.

107. Ibid., pp. 363–364.

108. Hubrecht, "The Gastrulation of the Vertebrates," pp. 416–417.

109. Hubrecht, "The Relation of the Nemertea to the Vertebrata," p. 606. See Lankester's article, "Vertebrata," in Lankester, ed., *Zoological Articles Contributed to the Encyclopae-*

Continuing his attack on the annelid theory, Hubrecht himself addressed the question of metamerism, arguing that it was quite unreasonable to suppose that this characteristic of the vertebrates must have been derived from the invertebrate ancestor. The tendency to duplicate body segments had arisen as a survival mechanism among long-bodied forms which faced a high risk of being subdivided. In effect, metamerism was analogous to the ability to regenerate parts, and was thus something which could have developed independently in any form facing similar risks. The nemertines themselves had a kind of incipient metamerism that indicated how the characteristic could have evolved.[110] Metamerism was thus due not to some inner developmental tendency, but to adaptive pressures that could have been maintained by natural selection. Once this point was accepted, the logic of the annelid theory was undermined, and the way was clear for recognition of the many homologies which Hubrecht believed indicated an origin in something resembling the unsegmented nemertines.

In speculating about the nemertines' position as a link back to even earlier stages of evolution, Hubrecht was drawn into the debate over the origins of the metazoan phyla. He noted that Adam Sedgwick had dismissed the unsegmented worms as of no significance in the attempt to trace the chordates back to a diploblastic ancestor.[111] (The early embryos of diploblastic animals, including coelenterates such as the sea anemones, have two germ layers from which the organs develop. The higher animals are triploblastic, having a third germ layer, the mesoderm.) Yet Sedgwick had pioneered a suggestion as to the course of evolution from the diploblastic to the triploblastic stage which Hubrecht himself now tried to develop. As noted in chapter 2, Sedgwick's 1880 paper, "The Origin of Metameric Segmentation," had sketched in hypothetical lines of descent leading from the Actinozoa (sea anemones) to the higher invertebrates and to the vertebrates. Implicit throughout his argument was the claim that metamerism had developed independently in the two lines.

Sedgwick argued that all the Triploblastica were derived from an ancestral form which could be reconstructed by careful study of the modern Coelenterata. In particular he referred to Oscar and Richard Hertwig's work on the Actinozoa (sea anemones), which showed that the slitlike mouth of these coelenterates could be homologized with the blastopore of

---

dia Britannica, p. 181. Hubrecht's theory did receive enthusiastic, if belated, support in a general text on phylogeny by the botanist John Macfarlane; see his The Causes and Course of Organic Evolution, pp. 421–425.

110. Hubrecht, "The Relation of the Nemertea to the Vertebrata," pp. 610–614.

111. Ibid., p. 618.

higher animals such as *Peripatus*.[112] The blastopore is the opening in the wall of the gastrula, the two-layered sphere from which all higher animals develop. Haeckel had turned the gastrula into the hypothetical Gastraea, the diploblastic ancestor from which all higher animals have evolved. The Hertwigs had now shown that in the Actinozoa (which remain diploblastic in the adult form), the slitlike mouth closes in the middle to leave two openings at either end. Sedgwick argued that this process was repeated in the development of the higher animals: the blastopore closes in the middle to become the mouth and the anus. According to the principle of recapitulation, this suggested that all the higher animals had evolved from an ancestor resembling the Actinozoa. One implication of this was that the mouth and the anus are homologous throughout all the higher phyla (a point rejected in Dorhn's annelid theory).

Sedgwick believed that the vertebrates were a very ancient stock: the separation of the vertebrate and invertebrate lines of evolution was one of the most fundamental zoological divisions. In the line leading from the Actinozoa toward the vertebrates, the mouth and the anus were separated at opposite ends of the body, and the primitive nerve-ring around the mouth became the foundation for the vertebrate nervous system.[113] Sedgwick was unwilling to see the unsegmented worms as a stage in vertebrate evolution (they belonged to the other branch). But Hubrecht now argued that the nemertines could form one stage in the development from an Actinozoan ancestor toward the vertebrates. In his 1887 paper he tried to explain how an originally radiate form could develop an elongated, bilaterally symmetrical body if it acquired the habit of moving preferentially with one part of its body foremost.[114] The part that took the lead would naturally acquire more sense organs and become the head. In a 1906 paper he traced a sequence from the radially symmetrical ancestor of the Actinozoan type, through an elongated, wormlike modification to a protochordate with some differentiation of the head, a notochord, and incipient metamerism.[115] *Amphioxus* was now rejected as an ancestral vertebrate because it had lost the differentiation of the head region, which must already have appeared in the protochordate.

Hubrecht's nemertine theory joined the ideas of Patten and Gaskell as yet another alternative to the two main rivals, the ascidian and the annelid theories. His attempt to trace the vertebrate phylum back to a radiate ancestor, while equally idiosyncratic, matched a growing recognition of

112. Adam Sedgwick, "The Origin of Metameric Segmentation," pp. 54, 68; see Richard and Oscar Hertwig, "Die Actinien."

113. Sedgwick, "Origin of Metameric Segmentation," p. 70.

114. Hubrecht, "Relation of Nemertea to Vertebrata," p. 607.

115. Hubrecht, "Gastrulation of the Vertebrates."

the need to link theories of vertebrate origins with more general ideas on the origin of the higher animals. Haeckel's Gastraea theory more or less required the postulation of a radiate ancestry for all the bilaterally symmetrical phyla. By deriving the vertebrates from other highly developed animals, the annelid and arthropod theories could be treated as self-contained units—the origin of the annelid or arthropod ancestor was a separate issue. But those who saw vertebrate metamerism as an independent invention were forced to enquire more carefully into the nature of the ultimate ancestor. Sedgwick and Hubrecht both chose to trace the vertebrates back to the Actinozoa, and the supporters of the ascidian theory were similarly inclined to link the hypothetical protochordate with earlier phases of invertebrate evolution.

BALANOGLOSSUS AND THE ECHINODERMS

We have seen that several exponents of the ascidian theory noted the possibility that the acorn worm *Balanoglossus* might also be a primitive chordate, the relic of a yet earlier stage in the phylum's history. Exploration of this possibility constituted William Bateson's first major project, and although Bateson soon turned away from evolutionary morphology completely, his work became generally accepted as an important contribution by those who were seeking to extend the original insights of Kovalevskii. As Bateson turned away to study the kind of variation that might generate features such as metamerism, others were building on his observations to strengthen the role played by *Amphioxus* and to seek a link back to the Echinoderms as the ultimate source of the chordate phylum.

Bateson was trained by Sedgwick and became a fellow of St. John's College, Cambridge.[116] His work on *Balanoglossus kowalevskii* was performed in 1883 and 1884 at the Chesapeake Zoological Laboratory under the inspiration of Brooks. He published a series of technical papers on the development of *Balanoglossus* in the *Quarterly Journal of Microscopical Science* in 1884–86—which, significantly in view of his rapid change of interest, were not reproduced in his collected *Scientific Papers*. The first paper included in that collection was his general discussion of the question of chordate ancestry, also from the *Quarterly Journal,* in which his loss of patience with evolutionary morphology was already becoming apparent.

Bateson rejected both the annelid and nemertine theories. The annelid theory required that the notochord must have arisen after segmentation, since the latter was already a feature of the annelid ancestor. Yet the notochord arose very early in vertebrate ontogeny, suggesting that it was also

116. On Bateson's work see, Beatrice Bateson, *William Bateson,* and surveys of the history of genetics such as Robert C. Olby, *Origins of Mendelism.*

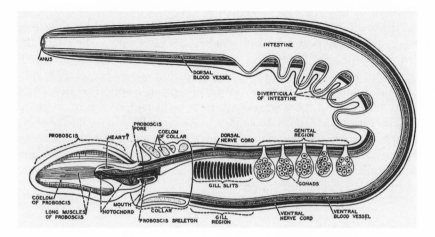

FIGURE 4.7   Internal structure of *Balanoglossus*. From H. V. Neal and H. W. Rand, *Comparative Anatomy* (London, 1936), p. 659. Note the dorsal nerve cord and the gill slits, which show vertebrate affinities.

an early feature of the phylum's evolution.[117] Balfour and Hubrecht's link to the nemertines was rendered implausible by the lack of any evidence for the dual origin of the vertebrate nerve cord.[118] Bateson criticized the general assumption that lower chordates such as *Amphioxus* were degenerate. This assumption was based on their lack of sense organs and parasitic way of life, but their development showed no evidence of the major changes in structure that would result from degeneration.

Bateson suggested that the notochord must have arisen in an unsegmented form, and argued that *Balanoglossus* offered the best evidence for the nature of this form. He admitted that at first sight there seemed few links between the wormlike *Balanoglossus* and the chordates. But certain key features did suggest a connection, especially an outgrowth of the gut which resembled the notochord, and the presence of gill slits similar to those of *Amphioxus* (fig. 4.7). Although the living Enteropneusta were sand-burrowers, the ancestral chordate was probably either pelagic or a bottom-creeper. It would have differed from the living entereopneusts in a number of respects, including the possession of serial gill slits. The ascidians, the enteropneusts, and the cephalochordates (*Amphioxus*) were thus all offshoots from the main stem of chordate evolution. As for where this stem began, Bateson noted that the Tornaria larva of some species of *Bal-*

117. Bateson, "The Ancestry of the Chordata," in Bateson, *Scientific Papers*, I, p. 14.
118. Ibid., p. 17.

anoglossus (not *B. kowalevskii*) suggested a link between the chordates and the echinoderms (sea urchins, etc.).[119] Like Hubrecht, Bateson saw that to reject the annelid theory he had to provide an alternative to the assumption that vertebrate metamerism had been inherited from an already segmented ancestor. He argued that metamerism was simply a product of the general tendency of all living forms to vary by the sudden repetition of existing parts.[120] It was precisely to confirm the existence of this mode of variation, and its potential for the production of new species by saltation, that Bateson began the research that culminated in his *Materials for the Study of Variation* of 1894. The preface to this work offered a vitriolic critique of evolutionary morphology as a sterile discipline and a call for more experimental work on the actual nature of variation.[121] From this, Bateson graduated to the study of heredity and his career as one of the founders of genetics. The preface to his 1886 paper on the ancestry of the chordata had, in fact, already complained that "the attempt to arrange genealogical trees involving hypothetical groups has come to be the subject of some ridicule, perhaps deserved."[122]

Those who followed Bateson in the rejection of morphology must have shared his frustration at the apparent lack of hard evidence (e.g., from the fossil record) to confirm or refute the various mutually inconsistent hypotheses of vertebrate origins. These feelings were becoming apparent even before Patten and Gaskell introduced their rival versions of the arthropod theory. Yet—as the latter initiatives suggest—the 1890s was by no means a slack period in the study of this problem. Only in the early years of the twentieth century did it become necessary for Patten and others to lament the growing lack of interest in their discipline. There was some hostility to Bateson's theory, especially from the professor of zoology at Giessen, J. W. Sprengel, in his monograph on the Enteropneusta published by Dohrn's zoological station at Naples.[123] But in general the interpretation of *Balanoglossus* as a primitive chordate was widely endorsed, and the enteropneusts became known as the subphylum Hemichorda of the phylum Chordata. Furthermore, Bateson's hint at a link with the echinoderms was taken up by a number of biologists seeking to complete the link back from *Amphioxus* to the invertebrates.

A good illustration of how this link could be made is provided by

119. Ibid., p. 26.
120. Ibid., pp. 3–10.
121. Bateson, *Materials for the Study of Variation*, especially p. v.
122. Bateson, "Ancestry of the Chordata," p. 1.
123. J. W. Spengel, *Die Enteropneusten des Golfes von Neapel*, pp. 721–736.

another biologist who would soon follow Bateson in rejecting evolution-ary morphology altogether: Thomas Hunt Morgan. In an 1891 article on the Tornaria larva of some species of *Balanoglossus,* Morgan traced changing ideas about the interpretation of this form and concluded by supporting both Bateson's interpretation of *Balanoglossus* and his sugges-tion of a link to the echinoderms. In 1866 Alexander Agassiz had actually thought that the Tornaria was a larval echinoderm, but its true status as the larva of *Balanoglossus* was discovered by Elie Metchnikoff three years later. Since Bateson's *B. kowalevskii* had no Tornaria stage, he had not stressed the echinoderm affinities (although we have seen that he noted them briefly), and Morgan was now prepared to make this link explicit. He listed a number of features in which the Tornaria larva resembled the larvae of echinoderms. Bateson's reasons for including *Balanoglossus* in the chordates were valid, but the living species did not represent the real character of the ancestral vertebrate, since their extreme length was a secondary feature acquired after they adopted burrowing habits.[124] Any links with the nemertines were purely superficial. After rejecting the idea that the chordates might have acquired a new mouth, Morgan concluded, "If Balanoglossus be related through its larva with the Echinoderms, as I have attempted to show in the preceding pages, we see how old a phylum that of the vertebrates must be, and hence the futility of attempting to derive them from any such highly specialized animals as the Annelids of to-day."[125]

In a subsequent article on *Balanoglossus,* Morgan confined himself to pure description and declared his unwillingness to engage in "extreme phylogenetic speculation."[126] Like Bateson, he was becoming increasingly dissatisfied with the speculative nature of evolutionary morphology. Other biologists were not yet willing to abandon the problem of vertebrate origins, and Bateson's work on *Balanoglossus* continued to attract atten-tion. It was certainly incorporated into one of the last major efforts to use *Amphioxus* as a key to the origin of the vertebrates. Arthur Willey studied at Cambridge with a studentship founded in memory of Balfour and also under Lankester. He then went to America to become a tutor in biology at Columbia University, after which he was professor of zoology at McGill University in Montreal. His *Amphioxus and the Ancestry of the Verte-brates* of 1894 was based both on his research papers and on a series of lectures, and attempted to provide an overview of the question that would link *Amphioxus* to the wider search for vertebrate origins.

124. T. H. Morgan, "The Growth and Metamorphosis of Tornaria," p. 445.

125. Ibid., p. 447.

126. Morgan, "The Development of Balanoglossus," p. 74. On Morgan's later work, see Garland Allen, *Thomas Hunt Morgan.*

Henry Fairfield Osborn wrote a preface for Willey's book emphasizing the extent to which adaptation to similar lifestyles had produced resemblances between vertebrates and invertebrates which misled biologists into postulating false relationships.[127] But "By the side of [these] parallelisms are real invertebrate and vertebrate affinities; so that the problem of resolving the various cases of real and acquired likeness in their bearing upon descent has become one of the most fascinating which modern Zoology affords." The links to *Balanoglossus*, the ascidians, and *Amphioxus* were genuine homologies, with the latter lying well toward the vertebrate end of the spectrum, while the former two still had predominatly invertebrate characters. Osborn concluded by stressing the importance of *Amphioxus:* "Its interest and value as an object of biological education has steadily increased with the knowledge that in contrast with all the related forms, it stands as a persistant specialized but not degenerate type, perhaps not far from the true ancestral line of the Vertebrates."[128]

Willey introduced his study with an attack on the annelid theory. The supporters of this theory had all adopted an *a priori* assumption that the metamerism of the vertebrates was derived from that of the annelids. This seemed natural enough at first sight, but the massive rearrangement of the body required by the theory counted heavily against it.[129] Everyone accepted that the ascidians were degenerate, but "that they have phylogenetically undergone the immeasurable degeneration which was postulated by Dohrn, is a view which is entirely unjustified by the facts."[130]

Turning to *Amphioxus* itself, Willey insisted that it had specializations defining "its own particular line of evolution," but could still be regarded as a close relative of or ancient offshoot from the actual ancestor of the vertebrates.[131] There were indeed affinities with the annelids, including the excretory system, but "It is eminently probable that, in respect to this and the other systems of organs, as well as the segmentation of the body, the Annelids and Vertebrates present an instance of *parallel evolution*."[132] The idea that the annelid ancestor of the vertebrates had begun to swim on its back was dismissed as "a phylogenetic acrobatic feat" which it was difficult to take seriously.[133] Noting that the larva of *Amphioxus* was asymmetrical, Willey insisted that this had no adaptive value and explained it as a relic of the time when the mouth was removed from its primitive mid-dorsal position as a consequence of the notochord's

127. H. F. Osborn, preface to A. Willey, *Amphioxus and the Ancestry of the Vertebrates,* pp. vii–viii.

128. Ibid., p. ix.

129. Ibid., pp. 2–6.

130. Ibid., p. 6.

131. Ibid., p. 42.

132. Ibid., p. 80.

133. Ibid., p. 92.

extension.[134] *Amphioxus* had undergone a secondary multiplication of the gill slits, the vertebrate ancestor probably having between nine and fourteen pairs.[135]

After presenting a fairly orthodox account of the ascidians as degenerate offshoots of the vertebrate stem, Willey turned to *Balanoglossus*. He noted Bateson's claim to have discovered structures homologous with the notochord and the nerve cord of chordates. *Balanoglossus* showed no muscular metamerism, but the gill slits and gonads were arranged metamerically, and the generally more segmented form of the chordate body could be explained as the result of an inherent tendency for the duplication of parts. Willey thus shared Bateson's view that metamerism was a process likely to occur sporadically in many different phyla: "Nothing is more patent than the fact that the metameric repetition of parts has arisen independently over and over again in different groups of animals." [136] The metamerism displayed by the annelids, arthropods, and vertebrates was thus no clue to their evolutionary relationship—it was a product of parallel evolution.

Having stressed the links between *Balanoglossus* and *Amphioxus*, Willey noted the resemblance between the Tornaria larva of the former and larval echinoderms. It was now generally admitted that the radial symmetry of echinoderms was not a primitive characteristic: they had begun as bilaterally symmetrical animals but had become adapted to a fixed, or sessile, mode of life and had thus reverted to a radial form.[137] All except the crinoids had returned to a mobile lifestyle, but had retained the secondarily acquired radial structure. The embryological work of MacBride and others had suggested that the distant ancestors of the echinoderms had structures which resembled those to be expected among the progenitors of the chordates.[138] The echinoderms could thus, in effect, be included as yet another degenerate offshoot from the line leading toward the vertebrates, having branched off at an even earlier stage than *Balanoglossus*. The subsequent development of the true chordate character had involved a concentration of the central nervous system along the dorsal side of the body and its conversion to a hollow tube.[139]

In his conclusion Willey again attacked the annelid theory, claiming that its supporters ignored the prevalence of parallel evolution. "The closer the superficial resemblance between an Annelid and a Vertebrate . . . is shown to be, the more perfect appears the parallelism in their evolution and the more remote their genetic affinity." The proximate ancestor of the vertebrates was "a free-swimming animal intermediate in or-

134. Ibid., p. 161.

135. Ibid., pp. 173–174.

136. Ibid., p. 246.

137. Ibid., pp. 267–268.

138. Ibid., pp. 270–271.

139. Ibid., p. 289.

ganisation between an Ascidian tadpole and Amphioxus," while the more distant ancestor would have been "a worm-like animal whose organisation was approximately on a level with that of the bilateral ancestors of the Echinoderms."[140]

With respect to the echinoderms, at least, Willey thus endorsed MacBride's view (discussed in chapter 2) that the invertebrates were degenerate offshoots from the vertebrate stem. MacBride himself supported Bateson's interpretation of *Balanoglossus* and the link to the echinoderms in his 1896 study of the embryology of *Asterina gibbosa*. Here he proposed that both the echinoderms and the vertebrates had evolved from the Dipleurula, of which *Balanoglossus* was a less modified descendant.[141] Two years later MacBride published a new study of the early stages in the development of *Amphioxus* in which he again endorsed these conclusions. He stressed that the animals comprising the line leading upward to the vertebrates had always been active animals, which is why the echinoderms, *Balanoglossus,* and *Amphioxus* had degenerated after adopting sessile, burrowing, or bottom-dwelling habits.[142] The Tornaria larva gave the best idea of the remote ancestry of the vertebrates and the echinoderms. MacBride believed that *Amphioxus* was actually a more primitive offshoot from the main stem than the ascidians, since the larvae of the latter possessed a structure equivalent to the brain and a muscular tail.

MacBride and Willey thus endorsed a new interpretation of vertebrate origins in which the echinoderms became the closest invertebrate relations of the vertebrate stock, as shown in the following diagram:

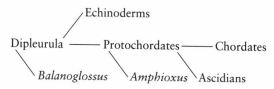

Their position resembled that being developed by German morphologists such as Karl Grobben, who proposed a division of the animal kingdom into two great stems determined by the fate of the blastopore in the earliest stages of development (see chapter 2). These two stems were the Protostomia (annelids, arthropods, and molluscs) and the Deuterostomia (chordates, hemichordates, and echinoderms).[143] According to this classification, the two stems were so far apart that any hypothesis deriving the

---

140. Ibid., pp. 290–291.

141. E. W. MacBride, "The Development of Asterina gibbosa," p. 396; see also MacBride's *Textbook of Embryology,* pp. 560–562 and p. 662.

142. MacBride, "The Early Development of Amphioxus," p. 607.

143. Karl Grobben, "Die systematische Einteilung des Tierreiches."

vertebrates from annelids or arthropods was rendered untenable. Sedgwick's claim that the vertebrates were a very ancient stock, long separated from the main invertebrate stem, was vindicated—provided the echinoderms were seen as an offshoot from the vertebrate stem.

MacBride's claim that the vertebrates represent the end product of the main, progressive trunk of the tree of evolution may sound like a throwback to earlier, developmental models of evolution. Yet his ideas were in line with the latest morphological analysis on the question of vertebrate relationships. Nor was his progressionism all that much different from the views expressed by Patten and other evolutionists in the early twentieth century. MacBride's emphasis on the importance of an active, free-swimming lifestyle as the basis for future progress shows at least some interest in the problem of changing habits as a key to important steps in evolution, but his approach was still essentially that of a morphologist. Species adopted new habits within a fixed range of environments offered by the ocean. But at the turn of the century some biologists began to think more carefully about the possibility that new habits and new environments might act as critical triggers for key advances in evolution. Their work was linked with the theories of geologists who were trying to identify episodes of drastic environmental change that could have provided the pressures needed to stimulate relatively rapid bursts of evolution.

THE ENVIRONMENTAL TRIGGER

Like Dorhn, whom he greatly admired, William Keith Brooks saw morphology as only the starting point for investigations of other biological phenomena. It was while studying with Brooks at Johns Hopkins that Bateson was stimulated to take an interest in the phenomena of variation and heredity. But Brooks himself made no clean break with the tradition of evolutionary morphology, and his 1890s work on the genus *Salpa,* a pelagic ascidian, led him to develop a bold hypothesis on the origins of the higher animal types. While the vast majority of biologists shared Charles D. Walcott's view that a long period of pre-Cambrian time was missing from the fossil record, Brooks sought to explain how a relatively sudden burst of evolution could account for the appearance of the varied animal types found in the Cambrian rocks.

The typical ascidians studied by Kovalevskii and others undergo metamorphosis into a sessile adult stage, spending the rest of their lives attached to rocks and filtering water through their sacklike bodies. *Salpa* belongs to an order of ascidians which remain mobile as adults—they are completely pelagic, spending their entire lives in the open sea, moving by squirting water out of their body cavities. Brooks produced a monograph on the genus and then commented on its evolutionary implications in a long article published by the Biological Laboratory of Johns Hopkins in 1893. He believed that in trying to understand how it became adapted to its pelagic lifestyle, he was illuminating some of the most fundamental problems of evolutionary morphology. Although now completely pelagic, *Salpa* had evolved from a sessile form similar to the more familiar ascidian type. Yet the ancestors of the ascidians were also, like their modern larvae, active animals. So why did evolution not proceed directly from the active ancestor to the active modern form? Was it possible that a bottom-dwelling stage was essential in the evolution of higher animals? [144]

Brooks argued that the earliest animal fauna lived at the surface of the ocean, because only here was there an adequate supply of oxygen. But all the higher animals had evolved from bottom-dwellers, and the burrowing habits of *Amphioxus* suggested that this had been true for the vertebrates. The pelagic lifestyle did not seem to offer the stimulus necessary for evolutionary advance—the earliest fauna was adapted to a simple environment with no competition and little pressure to diversify.[145] We have seen in chapter 3 how Brooks was led to comment first on the origin of the Crustacea. But his ideas were applied more forcefully to the problem of vertebrate ancestry.

Despite his admiration for Dohrn, he began by launching a fierce attack on theories which derived the vertebrates from an already segmented ancestor. The belief that homologies between structures are always derived from common ancestry was a hypothesis which was useful when carefully controlled, but it had now become a dogma which was distorting the work of evolutionary morphologists. This was especially true for the insistence that "the vegetative duplication of parts in animals has a phylogenetic significance, and implies descent from a duplicated ancestor."[146] This had led to "fantastic and grotesque unscientific speculation," the best evidence of which was the simulataneous publication of Patten and Gaskell's rival versions of the arthropod theory. Like Bateson and many others, Brooks believed that metamerism had evolved separately on many occasions. He went on to develop a detailed critique of Dohrn's annelid theory.[147]

This critique allowed Brooks to challenge Dohrn's exaggerated claims that the ascidians were a highly degenerate offshoot from the complex early vertebrates. One modern order of ascidians, the Appendicularia, do not undergo degeneration from the tadpole stage, and most biologists agreed that these were the most primitive (i.e., earliest) members of the

144. Brooks, "Salpa and its Relation to the Evolution of Life," pp. 139–140.
145. Ibid., p. 145.
146. Ibid., p. 153.
147. Ibid., pp. 182–192.

group. All ascidians were thus descended from pelagic ancestors, whose more active descendants would have gone on to become the ancestors of the vertebrates. Brooks agreed with Dohrn that it was vital for the morphologist to go beyond the study of pure form to look at the animals' mode of life and conditions of existence.[148] But it was precisely by speculating on the lifestyles of the various forms that he was led to propose his version of the ascidian theory. The ancestor of both the chordates and the ascidians was a pelagic creature like the ascidian larva. Brooks believed that the Appendicularia were undegenerate remnants of the earliest ascidian type, the better-known sessile forms being a later development. But this later episode was of little interest to him: the real problem was the origin of the highly developed ancestor, and in order to explain this initial development, he postulated a dramatic shift in the environmental locus of evolution.

Brooks' hypothesis of the "discovery" of the seabed was advanced in his *Salpa* paper, developed in an article in the *Journal of Geology* in the following year, and then made the basis of a chapter in his classic *Foundations of Zoology* of 1899.[149] The thesis was a bold attempt to link morphology with the evidence coming in from paleontology concerning the origins of life and what is now known as the Cambrian explosion. Walcott's work on the fossils of the Cambrian rocks had highlighted the diversity of the earliest known metazoan fauna long before his discovery of the Burgess shale fossils.[150] As Darwin himself was well aware, and as Walcott's further discoveries proved, the ancestors of all the major animal phyla appeared suddenly at the beginning of the Cambrian period. Darwin had supposed that there must be a long process of evolution in the pre-Cambrian period hidden by a massive gap in the record, a gap to which Walcott gave the name the "Lipalian interval." Brooks now disputed the assumption that the relative complexity of the first known fossils must imply a long period of slow evolution. His intention was to show that the record might in fact be a fairly accurate representation of a quite sudden leap in the advance of life caused by the opening up of a major new environment.

Brooks believed that the original fauna had been composed of simple pelagic creatures, which since they lived in a relatively uniform and uncompetitive environment, had been under no pressure to evolve further.

148. Ibid., p. 171.

149. Ibid., pp. 160–170; see also Brooks, "The Origin of the Oldest Fossils and the Discovery of the Bottom of the Ocean," and his *Foundations of Zoology,* chap. 9.

150. See Walcott's "Cambrian Geology and Paleontology" articles; on the Lipalian interval, see I, no. 1, "Abrupt Appearance of the Cambrian Fauna." On Walcott's life and work, see Gould, *Wonderful Life.*

At first, the waters of the bottom of the ocean had probably not contained enough oxygen to support life, but eventually this deficiency had been made up. He did not believe, however, that the exploitation of the new environment on the seabed had been a simple consequence of its physical availability. His choice of the term *discovery* was a deliberate attempt to invoke the image of life trying to colonize a new environment. The process was exactly parallel to the colonization of oceanic islands.[151] Life would have made repeated efforts to gain a foothold in the new environment, perhaps unsuccessfully on many early occasions. But once established, the bottom-dwellers soon created a new situation for themselves: in this much more restricted space, they became overcrowded and subject to a much more intense competition than had ever been present in the ocean itself. The result was a rapid diversification and a drive toward increased complexity, which generated the ancestors of all the major animal phyla in a comparatively short period of time. Many animals moved back into the ocean waters to escape the competition on the bottom, taking their advances with them. Among these were forms resembling the modern Appendicularia, which had gone on to to evolve into both the chordates and the other ascidians.

Brooks' hypothesis was certainly noticed, but more by the geologists than by the biologists. The popular British journal *Natural Science* published a substantial and positive review.[152] Walcott himself responded in a later paper in which he defended his own thesis on the sudden appearance of the Cambrian fauna but conceded that this did not rule out Brooks' theory that the earliest life was pelagic.[153] It is significant that Brooks had chosen to publish an account of his idea in a geological journal, perhaps recognizing that here was an audience more interested in the posssibility of an environmental trigger to evolution. A few year later in 1900, the *Journal of Geology* carried another attempt to explain the rapid origin of the vertebrates based on a transition to a new environment, this time by the geologist Thomas C. Chamberlin.

Like Brooks, Chamberlin wanted to explain the sudden appearance of the Cambrian fauna without invoking a gap in the record, and his own geological theories predisposed him to accept the idea of sudden bursts of evolution. He noted that the earliest fossils of fish seemed to indicate a habitat in coastal or inland waters. There was a predisposition to assume that the course of migration was always from the sea to the land, but Chamberlin argued that it was by no means impossible for the reverse to

151. Brooks, "Salpa and its Relation," p. 160.
152. Anon., "The Evolution of Life."
153. Walcott, "Cambrian Geology and Paleontology," I, no. 1, "Abrupt Appearance."

have taken place. He believed that "the chordate phylum is . . . essentially from first to last a terrestrial race, whose main habitat was the land waters and the land itself, though still a race that sends its offshoots down to the sea from time to time from the mid-Paleozoic onwards."[154] The essential feature of inland waters was that they flowed rather than remaining static, and the vertebrate body was an adaptation to swimming in moving water. The unknown ancestors had moved from the sea into inland waters and had there developed the true fishlike form. Only when they reinvaded the ocean did their fossils appear suddenly in the record.

As Chamberlin admitted, his was very much a geologist's attempt to throw light on what had been a morphologist's problem. It is not clear how seriously he took his own suggestion, given that a few years later he seems to have endorsed Patten's arachnid theory.[155] Chamberlin's idea was, however, taken up a few years later by Lull, who, more reasonably, linked it to the echinoderm-larva/*Amphioxus* theory.[156] Neither Brooks nor Chamberlin can be said to have had a major influence on the general debate—their contributions came when serious discussion of the origin of the vertebrates was beginning to die down. Indeed, their efforts suggest that, as the role of pure morphology began to seem more problematic, other lines of evidence were being sought in a somewhat desperate attempt to shift the focus of debate. Just as Patten tried to bring in the fossil record, so Brooks and Chamberlin were using the apparently sudden appearance of the earliest fauna to broaden the scope of the debate. However speculative their efforts, they were in tune with the growing move towards the use of geological and fossil evidence to reconstruct the history of life on earth. The only problem was that this kind of evidence was far more suited to the resolution of problems arising within the more recent phases of the process.

LATER DEVELOPMENTS

The 1890s certainly saw no slackening in the pace of debate on the question of vertebrate origins, but the level of interest dropped sharply in the years following 1900. Hopes that geology and paleontology would solve the problem remained unfulfilled, thus strengthening the suspicions of those who claimed that the whole enterprise was futile. The pace of specialized research certainly slackened, although new initiatives were on the

154. Chamberlin, "On the Habitat of the Early Vertebrates," p. 412. A more general attempt to argue that fresh water was the site of all major evolutionary innovations was made by John Macfarlane; see his *The Causes and Course of Organic Evolution,* e.g., p. 383.

155. See the reference to Chamberlin and Salisbury, note 88 above.

156. Lull, *Organic Evolution,* pp. 462–465 and 472–476.

horizon in the form of biochemical techniques that would test the relationships between phyla. These new techniques helped to reinforce what was emerging as a consensus in favor of the echinoderm-*Amphioxus* theory. This consensus can be seen in a number of popular works on evolution, even though little new work was now being done on the problem. The debate over vertebrate origins, once a centerpiece of evolutionary biology, thus petered out through lack of confidence in the ability of pure morphology to solve the problem, paradoxically just as the debate was fairly successfully resolved in favor of one of the main alternatives.

The books by Patten and Gaskell (1912 and 1908) provided continued support for the arthropod theory, although both authors were aware that by now they were battling a growing trend of indifference. We have also noted the extremely negative reaction to Delsman's attempt to revive the annelid theory. Several popular surveys endorsed the view that the vertebrates were distantly related to the echinoderms via the larvae of the latter and the *Balanoglossus*-ascidian tadpole-Amphioxus sequence. These include Harris H. Wilder's popular anatomy text, *History of the Human Body* (1909). Wilder included an introductory chapter on phylogeny in which he endorsed the recapitulation theory and warned against the misleading effects of convergence. In a concluding chapter surveying the various theories of vertebrate origins, Wilder noted the chordate affinities of *Balanoglossus* and then turned to its embryological development. This led to the "surprising and not very satisfactory result" that the chordates must be related to the echinoderms.[157] Wilder admitted that this theory was consistent with approved lines of biological thought, but ended by quoting Korschelt and Heider's text on invertebrate embryology to the effect that the problem remained essentially unresolved.[158]

Somewhat more positive endorsement came in Lull's textbook *Organic Evolution* of 1917 and Osborn's *Origin and Evolution of Life* of the same year.[159] The latter devoted remarkably little space to the topic, however, contrasting with Osborn's enthusiastic endorsement of Willey's book in the 1890s. Alfred S. Romer's *Vertbrate Paleontology* of 1933 accepted the same view.[160] Yet these surveys were written by paleontologists for whom the problem of vertebrate origins lay, one might almost say, in the prehistory of life. They briefly outlined the consensus offered by the morphologists, but only as an introduction to the study of the fossil evidence for later developments. The text of H. G. Wells and Julian Huxley's

157. H. H. Wilder, *History of the Human Body*, p. 536.

158. Ibid., p. 538, quoting Korscheldt and Heider's *Entwickelungsgeschichte*, p. 1465.

159. Lull, *Organic Evolution*, chap. 28; Osborn, *Origin and Evolution of Life*, p. 162.

160. A. S. Romer, *Vertebrate Paleontology*, p. 19.

Science of Life of 1931 evades the whole issue of vertebrate origins.[161] The issue was no longer thought to be especially important.

Even so, some technical work continued to be done. James F. Gemmill, a lecturer in zoology at Glasgow, published a study of the starfish *Asterias rubens* in 1914 endorsing the link between the echinoderms and *Balanoglossus*.[162] The Swedish zoologist Torsten Gislen produced a lengthy study of the affinities between the echinoderms, Enteropneusta, and chordates in 1930.[163] By far the most imaginative extension of the echinoderm-origin theory by a twentieth-century morphologist came from Walter Garstang, who began his career on the staff of the Marine Biological Assocation and later became professor of zoology at Leeds. Garstang introduced important modifications to the theory and built his arguments into a critique of MacBride's original version. This was part of Garstang's attempt to overthrow the recapitulation theory and replace it with his own theory of "paedomorphosis" or "neotony," the retention of larval stages of development in sexually mature forms. As Stephen Jay Gould has noted, Garstang's theory was acceptable within the framework of twentieth-century evolution theory, where the teachings of Mendelian genetics had rendered the old recapitulation theory untenable.[164] Garstang subsequently produced a book of popular verses encapsulating the philosophy behind his new vision of evolution.[165]

At the beginning of his career Garstang had supported a more or less conventional version of the theory of echinoderm ancestry for the vertebrates. He argued that the modern echinoderms, *Balanoglossus,* and the Chordata had evolved from a bilaterally symmetrical form resembling the larva of the modern *Auricularia.*[166] The echinoderms had secondarily assumed a radial symmetry, while the ancestors of the chordates had continued to build on the original bilaterally symmetrical layout. In the early decades of the new century Garstang emerged as a leading critic of the recapitulation theory, which MacBride, at least, was still trying to defend. His theory that larval forms could become sexually mature and dispense with the original adult form exactly reversed the logic of recapitulation-

161. See Wells et al., *The Science of Life,* p. 422, on the earliest fish. The phylogeny given in the diagram on p. 411 shows no close relationship between the vertebrates and any other phylum.

162. J. F. Gemmill, "The Development of Certain Points in the Adult Structure of the Starfish *Asterias rubens.*"

163. Torsten Gislen, "Affinities between the Echinodermata, Enteropneusta, and Chordonia."

164. Gould, *Ontogeny and Phylogeny,* pp. 177–184.

165. W. Garstang, *Larval Forms and other Zoological Verses.* Extracts from these verses are quoted in Gould's account of Garstang's work.

166. Garstang, "Preliminary Note on a New Theory of the Phylogeny of the Chordata."

ism, allowing him to embody many factual counterinstances to the older theory in a new synthesis that was more in line with modern evolutionism.

This critique of MacBride was clearly evident in Garstang's 1928 article on the morphology of the tunicates (ascidians), in which he advanced his new theory of vertebrate origins. He accepted the logic of the basic sequence: echinoderm-*Balanoglossus*-*Amphioxus*-chordate, but now insisted that it was no longer possible to derive the whole sequence from something resembling the active echinoderm larva. The theory of paedomorphosis was called in to suggest that the ancestors of the vertebrates were fixed animals in their adult state—the active larval phase had become sexually mature in this line of development, and the old fixed adult form had simply been sloughed off. The ascidians were not degenerate offshoots from the vertebrate stem: theirs was the truly ancestral condition from which the chordates had developed. Their active tadpole stage was a specialization for purely larval purposes, and their degeneration to the fixed adult stage was equivalent to that undergone by the larval echinoderm.[167]

Garstang thus accepted the traditional links drawn between echinoderms and protochordates, but held that all earlier interpretations, including his own 1894 theory, had been thrown off course by the assumption that the active larval stage was the adult ancestor—as demanded by the recapitulation theory. In fact, the sessile adult stage of many echinoderms represented the truly ancestral condition. Progressive evolution had taken place by the loss of the old adult stage as the larvae became sexually mature. They would increase in size as this happened—Garstang pointed out that the traditional idea of a minute "larva" representing the original adult condition was untenable simply because so small a creature could not have floated if loaded down with gonads.[168] The various protochordates had branched off at successive stages of this process, undergoing secondary modifications, some of which, such as the multiplication of the gill slits, were parallel developments in their separate stocks. But they were not degenerate modifications of higher stages in the process. The ascidians were not descended from *Amphioxus*-like ancestors: they showed incipient, not vestigial, metamerism. Theirs was the first step along the path which led to the old echinoderm larva becoming the new vertebrate adult. *Amphioxus* was the penultimate stage in the process of paedomorphosis. The possibility that a Dipleurula larva swimming by cilia could transform itself into a muscular chordate larva was illustrated by a similar tendency visible in some living starfish larvae.

---

167. Garstang, "The Morphology of the Tunicata." For a summary of the theory, see pp. 51–63.

168. Ibid., p. 61.

Garstang's theory of paedomorphosis allowed a reinterpretation of the theory of echinoderm origins, but sustained the basic thesis that the two phyla were closely related as compared with other invertebrate types. The relationship itself was confirmed during the 1930s through the emergence of a wholly new line of evidence offered by biochemistry. The first, rather abortive, effort in this area had been made at the turn of the century by George Nuttall, lecturer in bacteriology and preventative medicine at Cambridge. Nuttall developed a technique for estimating the degree of relationship between two organisms based on the immune reaction of the blood.[169] His results were of some interest in confirming the close relationship between humans and apes, but outside this they were wildly erratic and were largely ignored by evolutionary biologists. Nevertheless, Nuttall established the basic principle that biochemical properties might be used to supplement morphological resemblances when assessing degrees of relationship.

A more practical application of Nuttall's insight emerged in the late 1920s, when it was found that the chemical reactions responsible for muscle contraction were not uniform across the whole animal kingdom. P. and G. P. Eggleton found that labile phosphorus played an important role in muscle contraction, and that creatine phosphate was produced in the muscles of many vertebrates. Otto Meyerhof and others discovered that crustacean muscles contained arginine phosphate instead, and in his *Die chemischen Vorgänge im Muskel* of 1930, Meyerhof argued that the production of creatine phosphate was a characteristic vertebrate reaction.[170] The morphological divergence separating vertebrates from invertebrates had apparently been accompanied by the emergence of two different kinds of muscle biochemistry.

In 1932 Dorothy and Joseph Needham, writing along with Ernest Baldwin and John Yudkin, noted that not all invertebrate types showed the creation of arginine phosphate in the muscles. They found creatine phosphate in the muscles of both echinoderms and *Balanoglossus* and suggested that this supported the theory of the development of the vertebrates from echinoderms via the enteropneusts, as proposed by Bateson, MacBride, and Garstang.[171] Thus a primitive form of molecular biology could be brought to bear on the question of the relationships between the phyla and could provide an entirely new source of evidence favoring one

---

169. George H. F. Nuttall, *Blood Immunity and Blood Relationships*. The only reaction which gave positive results outside the vertebrates served to confirm the link between *Limulus* and the arachnids, see p. 362.

170. Otto Meyerhof, *Die chemischen Vorgänge im Muskel*, p. 93.

171. D. Needham et al., "A Comparative Study of the Phosphagens," p. 293.

of the major theories. The issue of phylogeny was only incidental to the Needhams' work, but they were clearly aware that their chemical studies could be interpreted as evidence on one side of a long-standing, if now rather neglected, debate. Only in the modern world of DNA analysis, however, has it proved possible to extend this line of research to give a substantial body of evidence bearing on degrees of relationship throughout the animal kingdom.

By the time the new evidence of the 1930s emerged, interest in the origin of the vertebrates had largely evaporated. The massive development of experimental disciplines such as genetics and biochemistry had forced evolutionary morphology onto the sidelines. Only in those areas where fossil evidence could be brought to bear was there still active work on the reconstruction of phylogenies. The shift of emphasis from morphology to paleontology was particularly obvious in areas such as the debate over vertebrate origins, where it seemed clear that no fossil evidence would ever be forthcoming. Patten and Brooks's efforts to reinterpret the existing fossil record were not convincing. The emergence of the new biochemical techniques offered a new way of confirming the relationship between the main branches of the tree of life, but still threw no light on the actual process of evolution. All efforts to provide an adaptive scenario for the origin of the vertebrates would thus remain untestable. Herbert V. Neal and Herbert W. Rand's *Comparative Anatomy* of 1936, a textbook for medical students, offered a chapter-long account of the problem, but this was atypical. Their account, with its long list of unconfirmed theories, served only to highlight why many biologists had lost interest.[172]

In 1955, the professor of zoology at McGill University in Montreal, N. J. Berrill, published a book on *The Origin of the Vertebrates* which offered the first comprehensive survey of the question for several decades. Berrill had been working on tunicate morphology since the 1930s, and it is worth noting that a certain amount of purely desciptive work of this kind went on even when phylogenetic speculation had been pushed into the background. He argued that all the efforts devoted to the annelid and arthropod theories had succeeded in showing no more than that this whole approach was a dead end. His own interpretation followed in the tradition of Garstang, linking the vertebrates back to the echinoderm larvae via the assumption that the larval mobility developed by both echinoderms and ascidians was an important adaptive feature allowing the animal to choose its site for the fixed adult stage. He was interested in Brooks' idea that the discovery of the seabed was the trigger for the relatively sudden diversification of the higher animals in the late pre-Cam-

172. Herbert V. Neal and Herbert W. Rand, *Comparative Anatomy*, chap. 16.

brian, and favored Chamberlin's view that the transfer to shallow and perhaps inland waters was the stimulus for the development of the original vertebrate form.[173]

The most telling aspect of Berrill's book, however, is its apologetic tone. In his preface he noted that interest was now centered on the "anthropoid past," where plentiful fossils were available. The study of vertebrate origins, once so fashionable, had now been forgotten, leaving him as a voice crying in the wilderness. He admitted that his own theory was purely speculative, based on largely circumstantial evidence. It lacked any hope of being verified by fossil evidence, the only thing which—to quote the zoologist and paleontologist D. M. S. Watson—would give "an intellectual respectability to our procedure."[174] In this defensive tone we see the plight of a biologist conscious that he is advancing a theory quite out of touch with contemporary priorities. Interest had now firmly switched to those areas of the history of life that could be observed in the fossil record. Even so, it is worth remembering, in conclusion, that the rejection of the annelid-arthropod theories constituted a resolution of at least one long-standing debate—the claim that no consensus would ever be reached had been disproved just as the majority of biologists were losing interest.

173. N. J. Berrill, *The Origin of Vertebrates*, pp. 94, 156.
174. Ibid., p. 248.

5

FROM FISH TO AMPHIBIAN

The debate over the origin of the chordates had obvious implications for biologists' views on the nature of the first true vertebrates, the fish. A great deal of attention was also given to the major steps in the subsequent evolution of the vertebrates, the origins of the amphibians, reptiles, birds, and mammals. Certain issues regarding the origin of the vertebrate classes became subjects of intensive research, and sometimes of acrimonious debate. The origin of the paired fins of fishes and their transformation into the limbs of the earliest amphibians was a particularly active area. Much attention was also focused on the transformation of different groups of reptiles into birds and mammals. These issues were addressed both by evolutionary morphologists and by paleontologists, and their discussions provide valuable information not only about changing views on the pattern of life's history, but also about the relationship between biologists using different kinds of evidence to address the same problems.

Some of the key steps in vertebrate phylogeny took on special interest because they defined the 'main line' of evolution leading to the human species. By the end of the nineteenth century it was less fashionable to treat the human form as the self-evident goal of creation, but everyone still accepted that, with hindsight, one could identify the critical phases in the evolution of life toward consciousness. The origin of the land vertebrates, and of the mammals, were two such steps. It was clear, of course, that much vertebrate evolution lay off the main line on side branches which constituted specializations unrelated to the advances culminating in humans. Indeed, it was often held that precisely because they had become too specialized for a particular way of life, later fish and reptiles had 'cut themselves off' from further progress.

Some of these key steps could be seen as purely anatomical transformations, for example, the origin of limbs, or the complex skeletal changes leading to the mammalian jaw and ear. Here morphology and paleontology could hope to illustrate the various stages in the process. A growing interest in functional anatomy led to increasingly more realistic questions about the causes of the evolutionary transformations: what exactly were the mechanical causes reponsible for such far-reaching modifications of structure? Other changes were largely physiological, and here it was much more difficult to bring direct evidence to bear. The origin of the amniote egg, which allowed the reptiles to free themselves completely from the water, seems to have attracted little attention, possibly because it involved physiological changes which lay outside the investigative realm of the morphological sciences.

Popular modern accounts of the history of life on earth tend to regard these questions as lying purely within the province of paleontology. Yet in the late nineteenth century, comparative anatomy and embryology were thought to have an equal right to speak on these topics. The fossil record of the 1870s was deficient in clues concerning all of the major steps in vertebrate history. Even the famous *Archaeopteryx*, the best candidate for a 'missing link' between reptiles and birds, was thought to appear far too late in the record to be the true ancestor of the modern birds. Morphologists thus took it upon themselves, here as in the origin of the chordate phylum itself, to identify the key transitions and the most likely ancestral forms. In some cases they disagreed profoundly among themselves. Certain lines of evolution were postulated which subsequently turned out to be unacceptable. Carl Gegenbaur's theory explaining the origin of paired fins and limbs was eventually rejected, as was T. H. Huxley's claim that the mammals could not be derived from any line which passed through the known reptiles, living or fossil.

In each of these cases, expanding knowledge of the fossil record helped to determine the outcome of a debate which had been opened on purely morphological grounds. Referring to Gegenbaur's archipterygium theory of the origin of the vertebrate limbs, Henry Fairfield Osborn subsequently wrote: "The hypothesis of Gegenbaur, which has been warmly supported by a talented group of his students, is memorable as the last of the great hypotheses regarding vertebrate descent to be founded exclusively upon comparative anatomy and embryology as opposed to the triple evidence afforded by these sciences when reinforced by paleontology."[1] Note that Osborn did not rule out the use of anatomy and embryology, but he believed that where the fossils did provide evidence, it could be decisive. Field studies could also be crucial, as when Richard Semon went to Australia in the 1890s to collect specimens revealing the development of the lungfish *Ceratodus*, widely regarded as a 'living fos-

1. H. F. Osborn, *The Origin and Evolution of Life*, pp. 172–173.

sil' representing a stage in development toward the amphibians. The relationship between morphology, paleontology, and fieldwork thus needs to be evaluated with care: paleontology did not completely displace morphology even in areas where new fossil evidence became most plentiful.

A number of problems continued to plague the reconstruction of phylogenies, whatever the source of evidence. Major disputes erupted over the determination of the most primitive members of each class. The jawless cyclostomes (including the lamprey) were widely regarded as surviving relics of the most primitive vertebrate form, but Anton Dorhn dismissed them as degenerate offshoots of the true fish which threw no light on the main line of descent. The relationship between the cyclostomes and the earliest fossil vertebrates, the ostracoderms of the Silurian and Devonian rocks, was also disputed. The status of crucial fossils was particularly open to challenge when, as in the case of *Archaeopteryx*, they were clearly too late to be the actual link between the two groups whose characters they seemed to share. Both morphologists and paleontologists had also to confront the problem of parallel evolution. All the vertebrate classes were, at one time or another, claimed to be polyphyletic, that is, to be grades of organization reached independently by more than one line of development arising from the previous class. Paleontologists eventually dismissed lungfish such as *Ceratodus* from their role as ancestors of the amphibians, claiming that they had independently evolved the ability to breathe air. But a few zoologists then tried to argue that the Amphibia were diphyletic, some having their origin in the lungfish, others in the crossopterygian fish favored by the paleontologists.

The fossil record rarely provided enough information to eliminate the possibility of convergence or parallelism, and indeed many paleontologists were anti-Darwinians who were predisposed to accept evidence favoring the idea of predictable trends in evolution. Later paleontologists such as William King Gregory, Robert Broom, and D. M. S. Watson were, however, far more willing to look for the functional causes of change than the pure morphologists had been. They wanted to know exactly how the fin of a fish had been transformed into the limb of a tetrapod—what were the mechanical problems involved, and how had they been overcome? This type of investigation had been pioneered by Lamarckians, who assumed that evolution was shaped directly by the mechanical transformations of the individual organism. But in a transition remarkable for the lack of debate over the actual mechanism of evolution, a later generation of paleontologists came to assume that natural selection could produce equivalent transformations in the species. Functional changes in the limbs could be studied in considerable detail, however, without asking about the environmental stresses that might have forced the animals to adopt a new means of locomotion. Functional morphology was still morphology, and did not necessarily trigger an interest in the role played by external factors in determining the organism's behavior.

Those paleontologists who worked closely with geologists were more aware of the evidence for past climates and environments. The late nineteenth century saw a growing interest in the possibility that crucial breakthroughs in evolution might have been triggered by climatic stress. American paleontolgists were especially active in this area, perhaps because they worked more closely with the geologists who were providing evidence of past climatic changes. Strict geological uniformitarianism was now breaking down, to be replaced by the view that some episodes of mountain-building had been rather abrupt, if not actually catastrophic. Attempts were made to explain the sudden appearance of new classes as a response to the climatic stress induced by such events. Even so, few efforts were made to depict what would now be called an adaptive scenario to explain the precise circumstances which had forced the modification of a species' structure in a particular direction. Alfred S. Romer's suggestion that the amphibians might have developed legs as a means of crawling to other pools in a world subject to increasing drought was one of the earliest suggestions of such a scenario, and it was not proposed until the 1930s.

THE ORIGIN OF FISH

A number of preconceptions influenced biologists' beliefs about the process by which the first true fish, with jaws and paired fins, originated. Most obvious, in the context of our study so far, were ideas on the origin of the chordate phylum itself. As we have seen, most authorities accepted that the lancelet, *Amphioxus,* was the best living representative of what the ancestral chordate had been like. Once this point was accepted, it was necessary to reconstruct the steps by which the *Amphioxus*-like ancestor had acquired a head, jaws, and two sets of paired fins with the girdles to which they are attached. Since neither *Amphioxus* nor the most 'primitive' fishes had a bony skeleton, it was natural to assume that the vertebrate skeleton had been originally cartilaginous, with true bone being a later evolutionary development in the higher fish. This was certainly the scheme outlined in Haeckel's pioneering phylogenetic schemes.

But the assumptions upon which such schemes rested were by no means uncontroversial. Dohrn's claim that the vertebrates came from annelid worms led him to dismiss both *Amphioxus* and the cyclostomes (lampreys and hags) as degenerate offshoots from the true vertebrate stem which offered no clue as to the nature of the primitive chordates. After all, *Amphioxus* has no distinct head structure, while the ancestral annelid would already have had a head, even if its structure had to be significantly reorganized. The cyclostomes were merely the best indication we have of the route by which a true vertebrate degenerated to the level of *Amphioxus*. Patten's theory that the chordates had evolved from arthropods had even more radical implications. For Patten, the first fossil fishlike creatures, the ostracoderms of Silurian and Devonian times, had inherited their heavy armor from the arthropod exoskeleton. Again, *Amphioxus* and the cyclostomes had to be degenerate side branches.

Even before detailed theories of vertebrate origins became available, systematists had preconceived ideas about which were the most primitive members of the chordate phylum. The fact that the cyclostomes had no jaws and only a cartilaginous skeleton made it easy to dismiss them as the 'lowest' true vertebrates. As the American ichthyologist Theodore Gill noted, no one disagreed about which were the lowest fish—but there was wide debate about which were the highest. Gill commented, "Perhaps there are no words in science that have been productive of more mischief and more retarded the progress of biological taxonomy that those words, pregnant with confusion, HIGH and LOW, and it were to be wished that they might be erased from scientific terminology."[2] Gill himself provided a genealogy which made it clear why this confusion had arisen: the teleosts, or bony fish, were the most specialized and hence well-developed forms, as fish, but their line of evolution had diverged from that leading to the lungfish and the higher vertebrates.[3] The 'highest' form as defined by the internal development of the class was not the one that gave rise to the next major step in the evolution of life.

The American paleontologist and neo-Lamarckian Edward Drinker Cope proposed his "Law of the unspecialized" to express exactly this characteristic of the evolutionary process. The law was outlined in Cope's 1879 lecture "A Review of the Modern Doctrine of Evolution."[4] It was also included as a conclusion to the chapter on phylogeny in his *Primary Factors of Organic Evolution* of 1896.[5] Here he emphasized that "the point of departure of the progressive lines of one period of time has not been from the terminal types of the lines of preceding ages, but from points further back in the series." The relationship between the fish and amphibians illustrated this point: "It is not the higher fishes (Actinopterygia) which offer the closest points of affinity to the succeeding batrachian class, but that more generalized type of the Devonian period, the Rhypidopterygia."[6] The terms *high* and *low* were relics from a period when naturalists still thought in terms of an unambiguous hierarchy of organi-

2. Theodore Gill, "Arrangement of the Families of Fishes," p. xxxvi.
3. Ibid., p. xliii.
4. Reprinted in Cope, *The Origin of the Fittest*, pp. 215–240; see pp. 232–233.
5. Cope, *The Primary Factors of Organic Evolution*, pp. 172–174.
6. Ibid., p. 172.

zation; they made no sense once it was realized that evolution was a divergent process. The most advanced members of a class were the most specialized, but this very characteristic rendered them incapable of significant change: "The specialized types of all periods have been generally incapable of adaptation to the changed conditions which characterized the advent of new periods."[7] It was thus from those which had preserved a relatively generalized structure, resisting the temptation to specialize, that major new developments arose.

No one doubted that the cyclostomes were the most primitive vertebrates, but unfortunately there were no cyclostomes in the fossil record, and it was open to debate how close the earliest fossil fish were to this primitive state. The ostracoderms, although heavily armored, seemed to have no bony internal skeleton (see fig. 5.2). But the absence of bone was, by itself, no reason to dismiss them as primitive. After all, the elasmobranches (sharks and rays) are cartilaginous too, yet they are sophisticated fish that are still successful in the modern seas. Early students of fossil fish such as Louis Agassiz and Hugh Miller were anti-evolutionists who were anxious to stress that the first fish were not lowly organized.[8] Most late nineteenth-century authorities rejected this interpretation and linked the ostracoderms to the cyclostomes as evidence of the first stage in vertebrate evolution. But later studies of the fossils suggested that the ostracoderms might, after all, have a significant component of bone in their skeletons. The assumption that there had been a gradual ossification of the vertebrate skeleton in the course of time was challenged. The possibility that the early vertebrates had a bony skeleton threw doubts on the status of the elasmobranches: Were the sharks really the stem from which all other fish (and by implication all other vertebrates) had branched out, or were they a side branch which had degenerated, at least in the condition of the skeleton?

The most natural starting point for a more detailed analysis of these questions is Haeckel's genealogy of the vertebrates. In his *History of Creation* he introduced the "round-mouthed animals" (Cyclostoma) as a class of vertebrates standing far below the fish. The mouth of the lamprey is adapted to sucking flesh. Its structure might have been dismissed as a degenerate character produced in response to its parasitic habits, yet Haeckel argued that the absence of jaws was an indication that the animal was truly primitive. He noted that a better characterization was that it had only a single nostril, unlike the paired nostrils of all higher vertebrates. These and other differences warranted separating the cyclostomes from the true fish, with which they had previously been associated. The class

7. Ibid., p. 173.
8. See, for instance, Peter J. Bowler, *Fossils and Progress*, chaps. 3 and 4.

Cyclostoma "represents a very important intermediate stage between the Skull-less animals [*Amphioxus*, etc.] and Fishes, and . . . its few still existing members are only the last surviving remains of a probably very highly developed animal group which existed towards the end of the primordial period."[9]

Haeckel argued that all double-nostriled animals had diverged from a single, common primary form which developed directly or indirectly out of the single-nostriled forms. This ancestor of all the higher vertebrates would already have possessed jaws, a swimming bladder, and two pairs of limbs or fins. The "Primaeval fish," or selachians (sharks and rays), were "a primary group, not only of the Fish class, but of the whole main-class of double-nostrilled animals."[10] The hypothetical "proselachii" probably differed only slightly from the living sharks. From them had radiated two main lines of evolution, one leading to the ganoids and on to the bony fish, the other to the Dipneusta, or lungfish, and the amphibians, shown in the following diagram.[11]

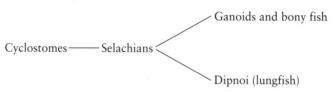

In his *Evolution of Man*, Haeckel noted embryological evidence suggesting that the jaws of the first true fish had evolved from gill arches, and this point was accepted more or less uncontroversially by all later morphologists.[12] Oscar Hertwig subsequently showed that the teeth were derived from the scales of the first fish.[13]

Most morphologists accepted that the cyclostomes were a relic of the transitional stage leading to the first true fish. Opinions differed on whether they should be counted as a distinct class intermediate between the chordate ancestors and fish. Haeckel's position was supported by Cope, Bashford Dean, Smith Woodward, and others, the class being variously known as the Cyclostomata, Marsipobranchii, or Agnatha.[14] Most authorities accepted a distinct class of jawless vertebrates, but a few

9. Ernst Haeckel, *The History of Creation*, II, p. 202.

10. Ibid., p. 206.

11. Ibid., p. 210.

12. Haeckel, *The Evolution of Man*, I, p. 357.

13. On Hertwig's work in this field, see Paul Weindling, *Darwinism and Social Darwinism in Imperial Germany*, p. 50.

14. See Cope, *Primary Factors*, p. 93; Bashford Dean, *Fishes, Living and Fossil*, p. 8; and A. Smith Woodward, *Outlines of Vertebrate Palaeontology*, pp. 1–16.

preferred to treat the cyclostomes as a subclass included within the class Pisces (fish).[15] The most serious challenge to Haeckel's phylogeny of the early vertebrates came from Anton Dohrn, for whom the cyclostomes were merely degenerate offshoots of the true fish.[16] Since they did not fit into his hypothetical genealogy leading from the annelid worms to the first chordates, Dohrn invoked the principle of degeneration to explain how such apparently primitive forms could have evolved. The modern lamprey is parasitic, and Dohrn could thus argue that its loss of the higher verte-brate characters was a consequence of its less active way of life. Further degeneration led to *Amphioxus* and the ascidians. As we saw in chapter 4, few were prepared to accept that degeneration could have produced a series so perfectly matching what the orthodox theory regarded as the natural line of ascent. We should note, however, that Carl Gegenbaur also doubted that the cyclostomes could lie on the main line of evolution lead-ing toward the higher vertebrates.[17]

The absence of fossil cyclostomes made it impossible to substantiate Haeckel's claim that the modern forms were relics of a once numerous class. It was assumed that their cartilaginous skeletons were unlikely to leave fossil remains. Great interest was thus aroused by the discovery of a strange fossil in the Devonian rocks of Scotland in the early 1890s which seemed to offer the first positive indication that the cyclostomes did indeed have a long history behind them: *Palaeospondylus gunni,* named and de-scribed by Ramsay Traquair (fig. 5.1).[18] The American ichthyologist Bash-ford Dean accepted that this was a "fossil lamprey," and Arthur Smith Woodward of the Natural History Museum in London created a new or-der of the subclass Cyclostomi to accomodate it.[19] In the first (1933) edi-tion of his *Vertebrate Paleontology,* Alfred S. Romer presented *Palaeos-pondylus* as a possible intermediate between the armored ostracoderms and the modern hagfish.[20] In his second edition (1945), however, Romer noted traces of jaws and limb girdles in the fossil, as pointed out by J. A. Moy-Thomas, and now preferred to treat it as a descendant of the placo-derms, true fish which had already come into existence in the Devonian.[21]

Haeckel noted the occurrence of fossil teeth and fin spikes in the

15. For example, Adam Sedgwick, *Student's Textbook of Zoology,* II, p. 95.

16. Anton Dohrn, *Ursprung der Wirbelthiere,* p. 56.

17. Carl Gegenbaur, *Grundzüge der vergleichenden Anatomie* (1870), p. 577.

18. Ramsay Traquair, "A Further Description of *Palaeospondylus gunni.*"

19. Dean, *Fishes, Living and Fossil,* p. 65; Woodward, *Outlines of Vertebrate Palaeontol-ogy,* pp. 2–3.

20. A. S. Romer, *Vertebrate Paleontology* (1933), pp. 28–29.

21. Romer, *Vertebrate Paleontology* (1945), pp. 58–59. See J. A. Moy-Thomas, "The De-vonian Fish *Palaeospondylus gunni.*"

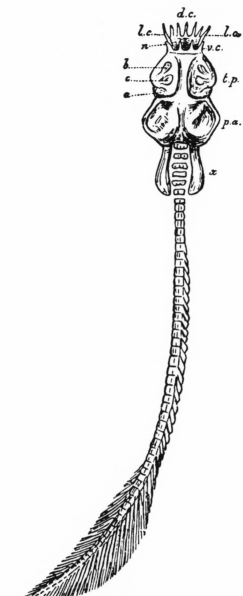

FIGURE 5.1 *Palaeospondylus
gunni.* From A. Smith Woodward,
Vertebrate Palaeontology (Cambridge, 1898), p. 2.

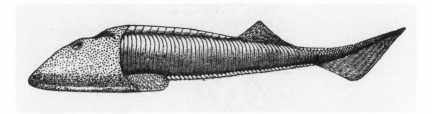

FIGURE 5.2   An ostracoderm, *Cephalaspis murchisoni*. From A. Smith Woodward, *Vertebrate Palaeontology* (Cambridge, 1898), p. 11.

Silurian rocks as evidence that true fish had appeared by that period, but he did not discuss the most controversial fossil evidence of early vertebrates. The heavily armored ostracoderms of the Devonian had been known since the work of Hugh Miller on the Old Red Sandstone of Scotland in the 1840s (see fig. 5.2). The fossils had, from time to time, been identified as crustaceans, but most authorities now accepted them as vertebrates, and the question was, How were they related to the cyclostomes and the ancestors of the whole vertebrate phylum? Patten referred to the earlier misidentifications to support his own view that the ostracoderms were an intermediate class linking the true vertebrates back to an arthropod origin (see chapter 4).[22] Patten's claim that the ostracoderms were a "missing link" was, of course, rejected by the vast majority of morphologists who expected something more like the cyclostomes as an intermediate. They dismissed the similarities between the external armor and the arthropod exoskeleton as a purely superficial resemblance. But to sustain this view, it was necessary to explain how an offshoot of the early cyclostome stem had developed such heavy armor before going extinct.

In the 1860s Huxley and Lankester had treated the ostracoderms as an order of ganoid fishes.[23] By the end of the century, the view had gained ground that they were, instead, a highly specialized form of cyclostome, showing no sign of jaws or limb girdles. Arthur Smith Woodward, who had originally seen the ostracoderms as true fish, converted to this position under the influence of Cope.[24] In 1885 Cope argued that a new species of the ostracoderm *Pterichthys* discovered by the Geological Survey

22. William Patten, "On the Origin of Vertebrates from Arachnids," pp. 359–360, and Patten, *The Evolution of the Vertebrates and their Kin*, pp. 342–347.

23. See, for example, E. Ray Lankester, *A Monograph: The Fishes of the Old Red Sandstone of Britain*, p. 13, acknowledging Huxley's work; see also Alleyne Nicholson, *Ancient Life History of the Earth*, pp. 152–154.

24. Woodward, *Outlines of Vertebrate Palaeontology*, pp. 1–16, and Woodward's obituary of Edward Drinker Cope.

of Canada showed clear evidence of a single nostril, thus associating it with the cyclostomes.[25] He later suggested that the cyclostomes and ostracoderms formed a group intermediate between the ascidians and jawed vertebrates.[26] In the early twentieth century the American paleontologists Lull and Romer both saw the ostracoderms as related to the primitive chordate stem.[27] One genus, *Cephalaspis*, showed elementary pectoral fins, which at first led Romer to suggest that paired appendages had developed at a very early stage in vertebrate evolution.[28] Romer later changed his mind, calling these structures an "independent evolutionary development" unrelated to the paired fins developed later by the true fish. He now thought that there was a general tendency to develop this kind of structure among all the early vertebrates: "Similar essays in the establishment of paired-fin systems, often of curious types, will be seen among the placoderms."[29] The placoderms were the earliest group of true fish with jaws and paired fins, although some still had heavy armor (the genus *Pterichthys* was transferred from the ostracoderms to the placoderms on the strength of later evidence that it did have rudimentary jaws).

According to this, the conventional model, the line of vertebrate evolution led from the hypothetical Palaeozoic cyclostomes through the placoderms to the higher fish. But were the ostracoderms really relatives of the early cyclostomes, thus providing the only real evidence for the existence of a large group of jawless vertebrates preceding the true fish? In 1897 Lankester launched a vicious attack on Smith Woodward (who would shortly become his subordinate at the Natural History Museum in London), arguing that the assumed link between cyclostomes and ostracoderms was unjustified.[30] Lankester did not believe that the ostracoderms formed a natural group in the evolutionary sense—they had been linked for convenience on the basis of purely superficial resemblances. He claimed there was no good evidence that they were, in fact, single-nostriled animals, and argued that they should not be associated with any other vertebrate group. Some aspects of Lankester's position were supported by Ramsay Traquair, who wanted to know what evidence there was that the ostracoderms were jawless.[31] If their internal skeleton was cartilaginous, all

25. Cope, "Position of Pterichthys in the System."

26. Cope, "On the Phylogeny of the Vertebrata," p. 281.

27. R. S. Lull, *Organic Evolution*, p. 120; Romer, *Vertebrate Paleontology* (1933), chap. 2.

28. Romer, ibid., p. 33.

29. Romer, *Vertebrate Paleontology* (1945), p. 35.

30. Lankester, "The Taxonomic Postion of the Pteraspidae, Cephalaspidae, and Asterolepidae."

31. Traquair, "Report on the Fossil Fishes Collected by the Geological Survey of Scotland," pp. 854–855.

traces of a jaw would be missing in the fossils. Unlike Lankester, however, Traquair thought that the fossil record proved the ostracoderms to be degenerate descendants of the elasmobranch fish.

These doubts about the status of the ostracoderms were largely eliminated by the work of Erik Andersson Stensiö on ostracoderm fossils discovered by the Norwegian expedition to Spitzbergen of 1917–1920. In 1927 Stensiö published a monograph in which he showed that the ostracoderms were a group of jawless vertebrates closely linked to the cyclostomes. He argued that the lampreys and hags should be included within the Ostracodermi. One group, the Cephalaspidomorphae, had given rise to the modern lampreys by a process of specialization and degeneration; the other, the Pteraspidomorphae, had similarly evolved into the modern hags. Far from being a specialized, armored offshoot of the cyclostome line, the primtive ostracoderms had actually evolved into the modern members of the group.[32] Stensiö's work became widely accepted in the 1930s as evidence that the ostracoderms were the first jawless vertebrates. Although the fossil record gave no information on the transitions, the line of vertebrate evolution led from them to the first true fish with jaws, the acanthodians. These were the most primitive members of the order of placoderms, which had flourished in the Devonian and had presumably given rise to the higher fish. Their primitive status was established by Watson in the 1930s.[33] Between 1933 and 1945 Romer added a new chapter to his textbook on vertebrate paleontology stressing the status of the placoderms as the ancestral group for the jawed fish.[34]

The studies by Stensiö and Watson challenged a basic assumption which had been built into most early phylogenies of the vertebrates. The modern cyclostomes have no bony skeleton, and the elasmobranch fish (sharks and rays) also have a cartilaginous skeleton. The modern teleosts, or bony fish, began to flourish only later in the fossil record. Since the vertebrates were widely supposed to have evolved from the ascidians via something like *Amphioxus,* it was natural to assume that the first vertebrates had only a cartilaginous skeleton and that bone was a later development in vertebrate evolution. The sharks were regarded as primitive, in the sense that they were the closest living relatives of the first true fish. Such a view was not contradicted by the bony external skeleton of the ostracoderms, which seemed to be an independent specialization unrelated to the development of an internal bony skeleton. But Stensiö's inves-

32. E. A. Stensiö, *The Downtonian and Devonian Vertebrates of Spitzbergen,* I, *Cephalaspidae,* pp. 378–379. On Stensiö's work see Stephen Jay Gould, *Eight Little Piggies,* pp. 422–424.

33. D. M. S. Watson, "The Acanthodian Fishes."

34. Romer, *Vertebrate Paleontology* (1945), chap. 3.

tigation of the well-preserved Spitzbergen ostracoderms showed that some members of the class had in fact possessed internal structures of bone. The lack of bone in the modern cyclostomes was thus an evolutionary degeneration, and this in turn raised the prospect that the sharks might also be degenerate in this respect, and were thus not truly primitive.

The view that the selachians or elasmobranch fish (sharks, etc.) were the stem form for the later fish and all higher vertebates had gained wide acceptance in the late nineteenth century. It was endorsed on morphological grounds by Balfour, whose first major study was devoted to the embryology of the elasmobranch fish. The phylogeny of the Chordata given in Balfour's *Treatise on Comparative Embryology* presents the cyclostomes as a "degenerate offshoot" of the earliest vertebrates and the elasmobranches as "the nearest living relatives" of the Proto-Gnasthomata, the founders of the jawed animals. Balfour did admit that the living elasmobranches were "a lateral offshoot from the main line of descent," but they were close enough to give a good idea of the common ancestor's structure.[35]

The paleontologists of the late nineteenth century were happy with this arrangement. The leading British and American specialists on fossil fish were, respectively, Smith Woodward and Bashford Dean. Both thought that the earliest fish must have been sharks, although both warned that the skeleton of living sharks was very specialized when compared with that of the earliest members of the group.[36] In 1895 and 1896 both published articles on the most ancient fossil sharks in the popular British journal *Natural Science*. Woodward accepted that the earliest fish, if its remains could be found, would be classified as an Elasmobranch, while Dean's article was entitled "Sharks as Ancestral Fishes."[37] For Dean, the sharks in general represented the persistent ancestral condition of the fishes and other jawed vertebrates.[38] Cope, a leading exponent of the recapitulation theory, thought that cartilage was primitive on embryological grounds. He saw the sequence from cyclostomes through true fish to amphibians and reptiles as an ascending one on the basis of increasing ossification of the skeleton.[39] The elasmobranches thus came before the other orders of fish in the evolutionary series.

35. Francis Balfour, *Treatise on Comparative Embryology,* in *The Works of Francis Maitland Balfour,* III; see pp. 328–329.

36. Woodward, *Outlines of Vertebrate Palaeontology,* p. 39; Dean, "Sharks as Ancestral Fishes."

37. Woodward, "The Problem of the Primaeval Sharks," and Dean, "Sharks as Ancestral Fishes."

38. Dean, *Fishes, Living and Fossil,* pp. 78–79.

39. Cope, *Primary Factors,* p. 95.

New fossil discoveries in the 1890s added to knowledge of the earliest fish. On the basis of the structure of its paired fins, the most primitive shark was *Cladoselache* from the late Devonian of Ohio, described by Dean as the most archaic and generalized form of the sharks from which all other fish had descended.[40] Osborn also wrote of the "Primordial sharks, ancestral to higher vertebrates."[41] Gregory's popular book *Our Face from Fish to Man* of 1929 used the face of the shark as the prototype upon which the human face was based.[42] The sharks had become a classic example of what was popularly known as the 'living fossil.'

The image of the shark as the closest living relative of the primeval vertebrates gained wide currency. But in the early decades of the twentieth century the paleontologists realized that this was to some extent an over-simplification of the situation revealed by the fossil record. The extensive fish fauna of the Devonian contained many different types, including some that were more plausible than the sharks as ancestors of the bony fish and amphibians. Both Osborn and Gregory conceded that their section titles linking the sharks to the main line of vertebrate evolution were not to be taken literally. While accepting *Cladoselache* as the prototype shark, Osborn actually preferred the upper Silurian acanthodian *Diplacanthus* as the more primitive stage in vertebrate evolution.[43] Gregory also pointed out that the ancient lobe-finned ganoid fishes stood closer to the main line than did the sharks.[44] Even in the 1890s, Woodward had offered *Pleuracanthus* from the Carboniferous of France and Germany as a form that "might with very little modification become either a Selachian, Dipnoan, or Crossopterygian."[45] On these models of fish phylogeny, the sharks were already being moved onto a specialized branch that diverged from the line leading to the bony fish and the higher vertebates. If they were 'living fossils' at all, they were not representative of a stage in the main line of vertebrate evolution.

Stensiö's demonstration that the ostracoderms had bony internal structures reinforced this challenge to the shark's central status by suggesting that their cartilaginous skeleton was not in itself a primitive characteristic. Bony skeletons might have developed at an early stage in vertebrate evolution, with some branches reverting back to a cartilaginous state through later degeneration. Both the cyclostomes and the sharks exhibited this de-

---

40. Dean, *Fishes, Living and Fossil,* pp. 78–79, and Dean, "Sharks as Ancestral Fishes."
41. Osborn, *The Origin and Evolution of Life,* p. 167.
42. W. K. Gregory, *Our Face from Fish to Man,* p. 12.
43. Osborn, *The Origin and Evolution of Life,* pp. 167–168.
44. Gregory, *Our Face from Fish to Man,* p. 23.
45. Woodward, *Outlines of Vertebrate Palaeontology,* p. 32.

generation, although they had branched off the main line of evolution at two quite different times. Thus Romer used the fossil record to stress that bone had appeared at a very early stage in vertebrate evolution. The 'classical theory' in which cartilage was the older material, simply did not fit the facts of paleontology: "We find it almost impossible to escape the conclusion that armor, probably bony in nature, and some ossification of the internal skeleton were developed at a very early stage in vertebrate history, and that the cartilaginous skeleton of the lower existing types represents a degenerate, rather than a primitive, condition."[46] True bony fish had appeared by the mid-Devonian, some time before the first recognizable sharks. A new scheme of fish evolution, shown in the following diagram, had thus emerged.

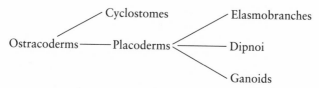

The most extreme version of this position was developed by the Swedish paleontologist Gunnar Säve-Soderbergh, who dismissed the traditional class Pisces as an assemblage of three distinct phyla. The cartilaginous elasmobranches and the modern bony fish, or teleosts, were both independent of a third group which Säve-Soderbergh named the Choanata. This comprised the Crossopterygian fish, the lungfish, and their descendants, the tetrapods, or land vertebrates.[47] In this scheme, the sharks were the ancestors of neither the modern bony fish nor the parallel group which had given rise to the land vertebrates. The old class Pisces was, in effect, merely a grade of organization reached independently by three phyla of jawed vertebrates.

Paleontology certainly did not have all the answers, and fish phylogeny remained a scene of constant debate. Nevertheless, as evolutionary morphology declined in influence, one can see the paleontologists gaining a new level of confidence in their ability to challenge hypotheses that had been erected on purely morphological grounds. In a 1900 address to the zoological section of the British Association for the Advancement of Science, Traquair sought to challenge what he perceived as a widespread opinion that the results of paleontology were too uncertain to be of any value in the reconstruction of phylogenies.[48] A work such as Osborn's

46. Romer, *Vertebrate Paleontology* (1933), p. 32, and 1945 ed. pp. 34–35.

47. G. Säve-Soderbergh, "Some Points of View Concerning the Evolution of the Vertebrates," esp. p. 8 and the phylogenetic tree, p. 16.

48. Traquair, "The Bearing of Fossil Ichthyology on the Problem of Evolution," p. 463.

popular *Origin and Evolution of Life* of 1917 demonstrates the power of the new paleontology to project an image of the science's ability to solve phylogenetic problems. Romer wrote his *Vertebrate Paleontology* of 1933 to provide an authoritative survey of a field that had expanded enormously since the late nineteenth-century overviews by Smith Woodward and Karl von Zittel. He declared that "vertebrate paleontology is essentially a biological science" and insisted on the artificiality of separating the study of fossil and living forms.[49] The main interest of the science lay in its application to the reconstruction of the evolutionary story, and—as we have just seen—Romer was keen to point out where the new fossil evidence overturned assumptions based on pure morphology.

The works of Osborn and Romer do, however, display new avenues of investigation that were being opened up through the use of fossil evidence. As we shall see in the following chapters, Osborn was particularly active in exploiting the link between paleontology and stratigraphy as a means of working out the timing of evolutionary events and the geographical migrations of groups spreading out from their place of origin. Both Osborn and Romer stressed the ability of the paleontologist to reconstruct the overall fauna of a particular period and deduce ecological relationships that might explain evolutionary novelties. Thus Romer addressed the question of what might have forced the ostracoderms to develop their heavy armor plating, and noted that the presence of the predatory eurypterids or, sea scorpions, would have provided just such a challenge.[50] He linked this to a discussion of the geologist Thomas C. Chamberlin's hypothesis that the early vertebrates had developed their undulatory mode of swimming in fresh rather than salt water.[51] Romer agreed that the traditional story of life spreading from sea to land was probably reversed in this case. The vertebrates had originated from ancestors which had migrated to inland waters where they had to swim in running streams. The vertebrate kidney was probably 'invented' to cope with a freshwater habitat and only later adapted to a marine environment. The eurypterids were the only Palaeozoic carnivorous type to be present in the same inland deposits, and were thus the most likely predators against which the armor of the early vertebrates had been developed. Romer thus fleshed out the purely morphological account of fish origins to provide a basic adaptive scenario explaining the main steps revealed by both morphology and the fossil record.

49. Romer, *Vertebrate Paleontology* (1933), p. 1.

50. Ibid., pp. 33–34, and 1945 ed. pp. 36–37.

51. T. C. Chamberlin, "On the Habitat of the Early Vertebrates."

THE FIN PROBLEM

One of the most controversial issues which emerged from the study of fish evolution was the origin of the paired fins. If the cyclostomes offered any clue, the most primitive vertebrates did not have paired fins, and unless they had arisen *de novo* it would be necessary to identify a preexisting structure which could have been transformed to produce them. The topic was an important one, not least because the paired fins would subsequently be transformed into the limbs of tetrapods; they were an essential prelude to one of the most far-reaching revolutions in the history of the vertebrate phylum. Before tackling the problem, morphologists had to decide which was the most primitive form of the paired fins, since this would to some extent determine which was the more likely source for these peculiar structures. Other limb forms would then have to be identified as specialized developments from the primitive original. It would also be important to determine which line of limb evolution had produced a form that made a plausible candidate for the transition to the legs of amphibians.

Two rival theories emerged very rapidly in the post-Darwinian era and were debated fiercely into the twentieth century. Working with evidence from comparative anatomy, Carl Gegenbaur identified the limb of the lungfish *Ceratodus* as the primitive "archipterygium" and argued that this structure had evolved from the gill arches of the early, limbless vertebrates. Almost immediately this was challenged by the American James K. Thatcher and the British anatomist (and strong opponent of Darwinian selectionism) St. George Jackson Mivart. They proposed that the paired fins had evolved from a continuous lateral fin that had once run down either side of the body in the earliest vertebrates. This interpretation was supported by embryological as well as anatomical evidence. By the end of the nineteenth century, the debate seemed to be going in favor of the finfold theory, although Gegenbaur's disciples continued to defend their master's interpretation. It was this issue which exploded into the *Competenzkonflikt* between Gegenbaur and Dohrn, the vicious debate over the relative standing of anatomical and embryological evidence which did much to discredit evolutionary morphology in Germany. Meanwhile the paleontologists were accumulating an ever-expanding wealth of fossil evidence which at last seemed to offer some hope of determining the structure of the most primitive paired fins. As we have already noted, by 1917 Osborn thought that paleontology had finally resolved a debate which could not be settled on purely morphological grounds.

Gegenbaur's conversion to evolutionism was based on his ability to translate the idealist concept of an archetype into the Darwinian model of a hypothetical common ancestor for each group. His work in comparative anatomy led him immediately to the idea of defining the most primitive form of the paired limbs, from which he sought to indentify the most likely origin of these structures. In the second volume of his *Untersuchungen zur vergleichenden Anatomie der Wirbelthiere,* published in 1865, he showed how the shoulder girdle from which the forelimbs are suspended could be traced through the evolution of the higher vertebrates. He also dealt with the pectoral fins of fish, and at this point he took the Elasmobranch (shark) form as the most primitive.[52] By 1870 Gegenbaur had changed his views significantly; he now held that the forelimb of the African lungfish *Protopterus* illustrated the most primitive form: a whiplike rod with traces of rays on one side. Gegenbaur believed that the shark fin, which has strongly developed rays on one side, had been formed from this original archipterygium. Note that in the eyes of Gegenbaur and most of his contemporaries, the lungfish, or Dipnoi, were the most likely ancestors of the amphibians. It was thus possible to trace a direct line from the archipterygium of the Dipnoi to the amphibians, with the sharks and other fishes representing side branches leading to a purely finlike specialization.

An important and final transition in Gegenbaur's position was now prompted by evidence provided by the first detailed studies of the Australian lungfish, *Ceratodus.* This was later known as *Neoceratodus;* the original *Ceratodus* was a fossil dipnoan from the Carboniferous known at the time only from very incomplete specimens. Discovery of the living Australian species allowed ichthyologists such as Albert Gunther of the British Museum to bring to life a veritable 'living fossil'—indeed, for Gunther, who had no time for Darwinian speculations, its persistence over the ages was evidence against the possibility of evolution. The views of those who saw the lungfish as the last step in the advancement of a "struggling ichthyic type toward the higher class, that of the Amphibians," were refuted by the evidence that these fish had persisted unchanged for so long.[53]

The evolutionists, of course, were quite happy to suppose that early stages in the ascent of life had been preserved in 'living fossils.' Of the surviving lungfish, *Ceratodus* has by far the best-developed limbs, consisting of a central rod with rays spreading out from it on both sides. Gegenbaur now decided that this limb structure was the true archipterygium; in effect he projected the *Ceratodus* limb back into the past to make it the stem form from which all other vertebrate limbs had diverged (figs. 5.3, 5.4). Far from being primitive, the assymetrical fins of the sharks had been produced by the disappearance of the rays on one side of the rod.[54]

---

52. On Gegenbaur's early views, see Lynn Nyhart, *Biology Takes Form,* pp. 251–262, and chap. 7; also James K. Thatcher, "Median and Paired Fins," pp. 294–296.

53. Albert Gunther, "Description of *Ceratodus,*" p. 560.

54. Gegenbaur, "Ueber das Archipterygium."

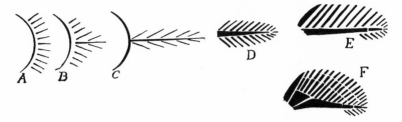

FIGURE 5.3 Diagram illustrating Gegenbaur's theory of origin of fin skeleton from gill-arch and gill rays. A: original structure of gill-arch; C: biserial archipterygium; D, E, F: origin of Selachian pectoral fin-skeleton from archipterygium (after Kingsley). From E. S. Goodrich, *Structure and Development of Vertebrates* (London, 1930), p. 124.

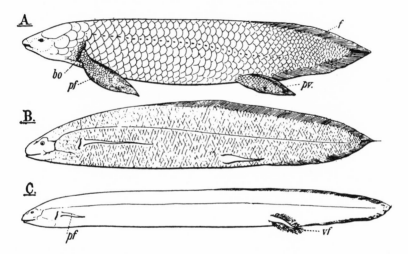

FIGURE 5.4 The living lungfishes. A: *Ceratodus foresteri*; B: *Protopterus annectens*; C: *Lepidosiren paradoxa*. From E. S. Goodrich, *Structure and Development of Vertebrates* (London, 1930), p.146

In the same year, 1872, he made the first suggestion of how this structure had evolved: the archipterygium was formed from gill arches which had moved backwards and assumed a new function. The rays were developed from rudimentary rays already present on the gill arches.[55] This view was then elaborated in Gegenbaur's later works.[56]

55. Gegenbaur, *Untersuchungen zur vergleichenden Anatomie der Wirbelthiere*, III, p. 181 note.

56. E.g. Gegenbaur, "Zur Morphologie der Gliedmaassen der Wirbelthiere," and Gegenbaur, *Elements of Comparative Anatomy*, pp. 472–474.

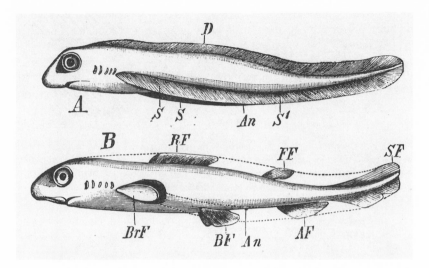

FIGURE 5.5   Formation of the paired and unpaired fins from a hypothetical ancestor with continuous dorsal and lateral fin folds. From E. Ray Lankester et al., *Zoological Articles* (London, 1891), p. 176.

The alternative fin-fold theory was first proposed by James K. Thatcher in a paper published by the Connecticut Academy of Arts and Sciences in 1877. After giving an overview of the development of Gegenbaur's thinking, Thatcher noted that Gegenbaur and Huxley disagreed over the relationship between the archipterygium and what Huxley had called the "cheiropterygium," the basic limb form of the tetrapods.[57] His alternative theory was based on the assumption that the limb form of *Ceratodus* and the Dipnoi (lungfish) in general was a specialization peculiar to that group. Thatcher himself worked on the structure of the sturgeon. The similarities between the fins of the Dipnoi and the elasmobranches did not indicate an evolutionary connection. Thatcher recognized that to be plausible, his rival theory had to show that this resemblance was "a merely superficial one" which "might have arisen in two entirely distinct and different series of developments."[58] Noting that the dorsal and anal fins of fish were widely regarded as derived from the median fin folds of *Amphioxus*, he proposed that the two sets of paired fins were specializations of lateral folds that had run along the body of the primitive vertebrate (fig. 5.5). The fin folds had originally been supported by parallel rays, but when concentrated at two spots along the length of the fish, these had

57. Thatcher, "Median and Paired Fins."
58. Ibid., p. 298.

developed a stronger base which had ultimately joined across the width of the body to form the girdles with which the fins now articulate. Thatcher argued that the ways in which the fins were supplied by nerves was completely consistent with his theory.[59]

Thatcher made some appeal to embryological evidence, arguing that the lateral fin folds were homologous with the Wolffian ridges in the embryos of higher vertebrates.[60] At the conclusion of his article he was able to add references to the work of Balfour, who had independently criticized Gegenbaur's theory on embryological grounds in an article published in 1876.[61] Balfour confirmed his position in his "Monograph on the Development of the Elasmobranch Fishes" of 1878, where he declared, "The facts can only bear one interpretation, *viz.: that the limbs are the remnants of continuous lateral fins."*[62] The shark embryo showed lateral lines running along the length of the body, and it was natural to assume that this had originally been an adult structure in the earliest fish, providing the ideal starting point for the formation of the paired fins.

The following year Thatcher's views were supported on independent grounds by Mivart. Like Thatcher, Mivart commented on how often Gegenbaur had changed his mind, and thought that he would soon have to change it again.[63] Mivart's own position arose from consideration of a point made by Huxley in opposition to Gegenbaur: the fin girdles were not part of the main axial skeleton, which they would have to be if derived from gill arches. They were part of the exoskeleton, bony material which was formed independently of the main vertebral column. (These structures are integrated into the axial skeleton in higher animals, as in the human pelvis, but are separate in the fish.) This point had led him to conclude that the paired fins were formed from a continuous fin fold, the theory already published by Thatcher. Mivart gave a detailed account of Thatcher's views, observing that he had not been able to consult the original journal in the British Museum and had had to get an offprint from the author![64] He insisted that the *Ceratodus* limb form was highly specialized, not primitive.

Mivart's *Genesis of Species,* a vitriolic attack on Darwinism, had appeared in 1871, and by this time he was virtually ostracized by the Darwinian community. At first sight, it might seem significant that Huxley

59. Ibid., pp. 304–306.

60. Ibid., p. 298.

61. Ibid., pp. 306–307.

62. Balfour, "Monograph on the Development of the Elasmobranch Fishes," in *The Works of Francis Maitland Balfour,* I, p. 320.

63. St. George Jackson Mivart, "Notes on the Fins of Elasmobranches," p. 459.

64. Ibid., p. 467.

had supported Gegenbaur's position on the nature of the archipterygium—surely a good excuse for Mivart to take up a rival position.[65] But Huxley had also disagreed with Gegenbaur on the transformation of the archipterygium into the cheiropterygium (the tetrapod limb), and Mivart drew heavily and openly on this aspect of Huxley's views. His anti-Darwinian stance was, however, apparent in one respect: he believed that there were several parallel lines of fin evolution in the various groups of fishes, some of which had produced remarkably similar structures. He argued that naturalists frequently underestimate the plasticity of the organism under evolutionary pressure, and, like Thatcher, he argued that there are "many instances of the independent origin of similar structures."[66]

The Gegenbaur school responded rapidly to the Thatcher-Balfour-Mivart theory. Gegenbaur's assistant at Heidelberg, M. von Davidoff, published an extensive study of the pelvic fins in 1879 which concluded with a critique of the rival theory.[67] He argued that the structure of the nerves to the fins could be explained more readily in terms of the gill-arch theory. Gegenbaur himself provided a backup note supporting Davidoff's position.[68] He now argued that the achipterygium could be seen as the basic form underlying the structure of all the different fin forms, thus confirming that it was the primitive structure from which the other had evolved. The Wolffian ridges which had been seen as an embryological equivalent of the original fin fold could be interpreted as the track left by the fin structure as it migrated backwards from its original position as a gill arch. Gegenbaur also appealed to a point that Dohrn had stressed in his *Ursprung der Wirbelthiere* of 1875. Structures never arose from nothing in the course of evolution: therefore it was necessary to find a homologous structure which had been transformed into the fins. The gill arches provided such a starting point, whereas the fin-fold theory suggested no structure that would be homologous with the fins.[69]

Dohrn was not, of course, a supporter of Gegenbaur's anatomical approach. In his study of vertebrate origins he suggested that the paired fins had developed from gills, but only in the context of his own theory, in which the original vertebrate was a transformed annelid with gills in every segment.[70] His emphasis on the priority of embryological over purely anatomical evidence led him to become a close friend of Balfour, and in 1884

65. T. H. Huxley, "Contributions to Morphology. Ichthyopsida: On Ceratodus fosteri."

66. Mivart, "Notes on the Fins of Elasmobranches," p. 481.

67. Max von Davidoff, "Beiträge zur vergleichenden Anatomie der hinteren Gliedmassen der Fische," esp. pp. 510–515.

68. Gegenbaur, "Zur Gliedmassen Frage."

69. Ibid., p. 522.

70. Dohrn, *Ursprung der Wirbelthiere*, p. 14.

he published a study of the paired and unpaired fins of sharks which came out strongly in support of the fin-fold theory.[71] While not accepting all of Balfour's interpretations, he insisted that the embryological evidence was quite incompatible with the gill-arch theory. The mode of formation of the fins showed no sign of a backwards migration, so unless ontogeny was totally useless as a guide to phylogeny, the gill-arch theory had to be abandoned. Furthermore, embryology confirmed that the fins and girdles arose independently of the axial skeleton, just as predicted by the fin-fold theory. The fin muscles were not homologous with those of the head region, but consisted of metameric segments, suggesting that they were derived from ancient metameric fin-folds running the whole length of the body.

Over the next decade the fin-fold theory gained the support of many German morphologists, including Robert Wiedersheim and Carl Rabl.[72] But Gegenbaur refused to back down and returned to the defense of his theory in 1895. He was particularly concerned to diminish the effectiveness of embryological evidence, insisting that ontogeny was not a good guide to phylogeny because cenogenetic factors obscured those aspects of development that might be a record of the past. The three-volume *Festschrift* for Gegenbaur published in 1896–97 contained a long article by Hermann Klaatsch defending the master's views on the development of fins and limbs.[73] It also included a whole monograph by Max Fürbringer on the structure of the spino-occipital nerve, which dismissed embryology as a source of self-contradictory evidence and held up comparative anatomy as the only true guide to relationships.[74]

Tension already existed between the anatomists and embryologists. Dohrn had fallen out with Gegenbaur completely, and in the twenty-first installment of his "Studien zur Urgeschichte des Wirbelthierkörpers" published in 1902, he proclaimed the *Competenzkonflikt* between ontogeny and comparative anatomy. Dohrn attacked Gegenbaur and Fürbringer for their assumption that anatomy was invariably superior. They accepted embryological evidence when it fitted their own conclusions, but dismissed it as the result of cenogenesis whenever it did not.[75] Gegenbaur was at least a founding father of evolutionary morphology and might

71. Dohrn, "Die paarigen und unpaaren Flossen der Selachier."

72. Robert Wiedersheim, *Das Gliedmassen Skelet der Wirbelthiere;* Carl Rabl, "Theorie des Mesoderms" and "Gedanken und Studien über den Ursprung der Extremitäten"; see Nyhart, *Biology Takes Form,* pp. 251–262.

73. Hermann Klaatsch, "Die Brutflosse der Crossopterygier."

74. Max Fürbringer, "Ueber die spino-occipitalen Nerven der Selachier und Holocephalen," p. 712.

75. Dohrn, "Theoretisches über Occipitalsomite und Vagus," p. 262.

be forgiven for pontificating, but it was intolerable when disciples such as Fürbringer took on the role of Minos, Aeakos, and Rhadamanthos (the judges of Hades in Greek mythology) and dismissed embryology from on high.[76]

Dohrn was supported by Carl Rabl, who also found the self-confidence of the anatomists unacceptable. A letter written by Rabl to Haeckel in 1893 gives some impression of the tensions that had already emerged in German morphology:

> The antagonism between my standpoint and Gegenbaur's is very sharply expressed in Klaatsch's words: "Without comparative anatomy, ontogeny cannot make the simplest process comprehensible." I turn this sentence (which characterizes the Gegenbaur school splendidly) on its head and say: without comparative embryology, anatomy cannot make the simplest fact comprehensible. Gegenbaur proceeds from the finished form; I proceed from the growing form.[77]

In a paper published in 1898 Rabl openly proclaimed the view that comparative anatomy was impoverished, while embryology was still a vigorous science.[78] He did not want to defend the recapitulation theory but nevertheless insisted that it was only by studying the early stages of development that evolutionary relationships could be confirmed. By distancing himself from recapitulationism, Rabl undermined the effectiveness of Gegenbaur's frequent appeals to cenogenesis as a process that would obscure the parallel with phylogeny. Embryology revealed relationships between living species more effectively than anatomy, even if it did not allow one to reconstruct ancestral forms.

In a 1901 paper Rabl used a study of fish ontogeny to defend the finfold theory of the origin of the paired fins.[79] A long appendix to his paper surveyed the development of Gegenbaur's archipterygium theory and attacked its author's scientific competence. Rabl compared Gegenbaur's speculative, yet dogmatic, approach to that of the old *Naturphilosophen,* and insisted that embryology would triumph over a method which encouraged such dishonesty. This was a forthright attack, and it called forth an equally pugnacious response from Fürbringer in which Rabl himself was damned for his dishonesty and bad scientific method.[80] The details

76. Ibid., p. 263.
77. Rabl to Haeckel, 5 November 1893, quoted in G. Uschmann, *Geschichte der Zoologie,* p. 131; translated in Nyhart, *Biology Takes Form,* pp. 262–263.
78. Rabl, "Ueber den Bau und die Entwickelung der Linse," p. 80.
79. Rabl, "Gedanken und Studien über den Ursprung der Extremitäten."
80. Fürbringer, "Morphologisches Streitfragen."

of the resulting controversy have been outlined by Lynn Nyhart, who concludes that it was instrumental in discrediting phylogenetic research within German morphology.[81] Gegenbaur received some support on personal grounds, but his death in 1903 removed the chief foundation supporting the anatomical school of thought, and Fürbringer found himself increasingly isolated over the following years. Nyhart argues that the embryologists also began to turn away from phylogenetic research to *Entwickelungsmechanik* and the study of the processes that actually cause ontogenetic differentiation. The conclusion that evolutionary morphology collapsed must, however, be treated with some caution. As we shall see in chapter 6, W. K. Gregory thought that German morphology was still in a flourishing state in 1913, citing as evidence the definitive work of Ernst Gaupp on the homologies between the bones of the reptile jaw and the mammalian ear.[82]

The English-speaking world saw no corresponding split between comparative anatomy and embryology, although interest in the phylogenetic implications of embryological research certainly waned in the early twentieth century (Gregory thought this contrasted with the situation in Germany). No equivalent of the Gegenbaur school existed, and the fin-fold theory became widely accepted. The English translation of Gegenbaur's *Elements of Comparative Anatomy* appeared with an introduction by Lankester reminding the reader that the fin-fold theory provided an alternative to that proposed in the main text and supporting Huxley's view that the achipterygium was derived from the elasmobranch fin.[83] In a paper published in 1881, the year before his untimely death, Balfour reiterated his support for the theory and provided a detailed summary of Thatcher's and Mivart's contributions. He rebutted the objections raised against the theory by von Davidoff and Gegenbaur.[84]

The Oxford comparative anatomist Edwin Goodrich published an extensive survey of the German debates in a 1906 paper which came out decisively in favor of the fin-fold theory. According to Goodrich, one of the most telling arguments against Gegenbaur's gill-arch theory was that it did not explain the similarities between the paired and unpaired fins.[85] Like Dohrn and his supporters, he stressed the metameric origins of the fins and was particularly concerned about the problem of how the fins (or

81. Nyhart, *Biology Takes Form*, chap. 8.

82. Gregory, "Critique of Recent Work on the Morphology of the Vertebrate Skull," p. 2.

83. Lankester, preface to Carl Gegenbaur, *Elements of Comparative Anatomy*, pp. xi–xii.

84. Balfour, "On the Development of the Skeleton of the Paired Fins of Elasmobranchii."

85. E. S. Goodrich, "Notes on the Development, Structure, and Origin of the Median and Paired Fins of Fishes," p. 343.

limbs) could migrate along the body.[86] A later paper explored the implications of this process for the concept of homology. Goodrich noted that the limbs arise from different body segments in different animals and insisted that this did not undermine the traditional assumption that they were homologous. He argued that this migration took place through the transposition of "formative substances" from one segment to another in the early embryo.[87] Support for the fin-fold theory was repeated in Goodrich's classic *Studies in the Structure and Development of Vertebrates* of 1930, where he declared that the gill-arch theory could only have been proposed at a time when the ontogeny of the fins was scarcely known at all.[88]

Meanwhile the paleontologists were also convinced that their evidence favored the fin-fold theory. In 1892 Woodward published a popular survey article on the evolution of fins in which he linked the morphological and paleontological evidence.[89] Embryology, he maintained, supported the fin-fold theory because it revealed a transitory ridge linking the pelvic and pectoral fins in the early embryos of some sharks and skates, resembling the structure of the median fins. The elasmobranch fossils of the Carboniferous and Permian revealed the primitive structure of the fins. In his later *Outlines of Vertebrate Palaeontology,* Woodward argued that a theory based on the gradual crowding together of fin supports to give the paired fins was confirmed by both embryology and paleontology.[90] The oldest known shark, *Cladoselache,* demonstrated the primitive arrangement in which the cartilaginous bars of the fins are parallel, showing no fusion of the base.[91] At the same time Dean endorsed Wiedersheim's version of the fin-fold theory, in which a lateral fold originally used for balancing became concentrated into paired fins that could be used also for propulsion. He too believed that the new evidence provided by *Cladoselache* confirmed the theory.[92] We have already noted Osborn's confident assertion that paleontology had essentially settled the morphologists' debate in favor of the fin-fold theory. In 1915 Gregory wrote of "the farfetched and mystifying character of Gegenbaur's theory" before launching into a discussion of the evidence which had convinced most paleontolo-

86. Ibid., pp. 343–348.
87. Goodrich, "Metameric Segmentation and Homology," p. 239.
88. Goodrich, *Studies on the Structure and Development of Vertebrates,* p. 126.
89. Woodward, "The Evolution of Fins."
90. Woodward, *Outlines of Vertebrate Palaeontology,* p. 19.
91. Ibid., pp. 24–25.
92. Dean, *Fishes, Living and Fossil,* pp. 40–45.

gists that Gegenbaur was also wrong in his derivation of the tetrapod limb from the archipterygium of *Ceradotus*.[93]

THE ORIGIN OF THE AMPHIBIANS

The debate over the origin of the paired fins gained some significance because it served as a foundation for the equally controversial topic of the transformation of those fins into limbs. But the question of the origin of the amphibians raised wider issues. In order to move out of the water onto dry land, the fish ancestor of the amphibians had not only to develop limbs capable of supporting its body but also to acquire a lung capable of breathing air. No fossils had been found to demonstrate the course of the transformation, although some early amphibian remains provided indirect evidence. Morphologists and paleontologists had a field day arguing about the precise relationship between the paired fins and the tetrapod limbs. Much of the early discussion was of a purely morphological character, based on an analysis of the mechanical transformations required by the conversion of a fin into a limb. Only in the twentieth century was there any serious discusion of the adaptive pressures that might have actually encouraged one group of fish to take the risk of venturing into a hostile environment on the land.

There was also a major debate about which group of fish would have been ancestral to the amphibians. Haeckel made the natural assumption that the Dipnoi, the lungfish, were the most likely candidates. In addition to their obvious ability to survive out of the water, detailed anatomical studies revealed a number of resemblances to amphibians. By the end of the century, however, the paleontolgists' attention had increasingly switched to a group of fossil fish which seemed to provide a more plausible ancestry. The fringe-finned ganoids, or crossopterygians, were Palaeozoic fish which also had swim bladders (equivalent to lungs) and bony fins that might serve as the starting point for legs. The Dipnoi had specializations that suggested they were a side branch which had independently acquired characters resembling those of amphibians. By the early decades of the twentieth century, the crossopterygian ancestry of the tetrapods was taken for granted by most paleontologists. The morphologists who studied living species were not so sure, however, and there were occasional warnings that the lungfish might turn out to be the closest living relatives of the amphibians after all. A group of Scandinavian biologists even argued that the amphibians might be diphyletic, the modern salamanders having

93. Gregory, "Present Status of the Problem of the Origin of the Tetrapoda," p. 338.

evolved from the lungfish, while the frogs were descended from the crossopterygians. In a climate where parallel evolution and convergence were taken for granted, anything was possible.

An account of the history of this problem by a modern taxonomist has compared these evolutionary speculations unfavorably to the cladistic approach now preferred. Colin Patterson points out that in a cases where the fossil record offers no time sequence by which ancestor-descendant relationships can be inferred, it makes sense only to enquire about the degrees of relationship between known forms.[94] Since the earliest known amphibians (from the late Devonian) appear almost simultaneously with the various orders of fish from which their descent has been inferred, the search for ancestry is futile. Patterson suggests that the pre-evolutionary debates over how the lungfish should be classified in relationship to the true fish and to amphibians were more closely in tune with the techniques of modern taxonomy. By allowing biologists to slur over the distinction between fish and amphibians (on the grounds that intermediate and hence indeterminate forms were to be expected), evolutionism actually retarded efforts to work out the relationships involved. The cladists need to define groups clearly in order to determine how closely they are related but regard it as illegitimate to infer ancestry on the basis of this information. Even so, the majority of modern authorities still treat the crossopterygians as most closely related to the amphibians, and many still regard this as at least a strong indication that the amphibians evolved from this group of fish.

It is not my purpose here to assess the scientific validity of the attempt to reconstruct phylogenies. My goal is to describe rather than defend the techniques of the evolutionists: in the context of the time, it was inevitable that genealogical relationships would be inferred. Patterson does, however, have a point when he argues that the studies of the late nineteenth century tended to focus attention on paleontology as the science that would ultimately prove or disprove hypothetical phylogenies. Morphologists dealing with living organisms could work out degrees of relationship in a way that would be quite compatible with the rigors of modern cladistic analysis—but it was thought that only the fossil record might demonstrate the actual course of evolution. The paleontologists of the late nineteenth and early twentieth centuries certainly hoped that the flood of new fossil discoveries would eventually resolve the issues. The fact that the record has remained blank in many key areas is the source of much of the twentieth century's disillusionment with the reconstruction of phylogenies. Nevertheless it is worth recording that morphologists continued to work on the structure of living amphibians and lungfish in the twentieth

94. Colin Patterson, "The Origin of the Tetrapods: Historical Introduction to the Problem."

century, and still made some active contributions to the evolutionary debate. In fact it was the morphologists who formed the main residue of support for the original hypothesis that the lungfish were the ancestors of the amphibians.

When specimens of the South American and African species of lungfish were brought to Europe in the late 1830s, they immediately posed a problem for naturalists used to making a clear distinction between fishes and amphibians. Since their air bladders functioned as lungs enabling them to survive out of water, they seemed to serve as a bridge between the two classes. As in the case of the duck-billed platypus, pre-Darwinian evolutionists could exploit them as evidence that there was some form of continuity between the apparent divisions in nature. The South American lungfish, *Lepidosiren,* was actually described by T. L. W. von Bischoff as an amphibian. He thought that the lungs, internal nostrils, and the structure of the heart were amphibian features outweighing the scales and other fishlike characters.[95] Specimens of the African lungfish, *Protopterus,* were brought to London, and the species was described by Richard Owen as a fish. Owen based this decision on his inability to find an internal nostril. He later admitted that he was mistaken on this point but never wavered from his belief that the lungfish were true fish which resembled the amphibians in a few characters. Owen was supported by Louis Agassiz and other experts, so that by the middle of the century it was taken for granted that the lungfish were indeed an order of fish, the Dipneusta, or Dipnoi.

When he came to the origin of the tetrapods in his *History of Creation,* Haeckel appealed to the Dipneusta as a transitional class between true fish and amphibians.[96] He justified this claim on the grounds that morphologists disagreed over which class they should be included in. The surviving lungfish were the relics of a once numerous group, fossil evidence of which was provided by the teeth of *Ceratodus* in the Triassic rocks. The early Dipneusta were, in fact, the primary form from which the Amphibia had sprung. The oldest true members of the higher class were the "mailed batrachians," or Ganocephala, of the coal beds, including *Archegosaurus,* which had been discovered in Germany in the 1840s and had provided clear evidence of the existence of early amphibians.[97] Later in the record came the labyrinth-toothed Amphibia, or Labyrinthodonts, of the Permian. Haeckel argued that the possession of a pentadactyle, or five-digit, limb by all tetrapods confirmed that they were a monophyletic group arising from the primitive amphibians. This latter point was taken

95. For details of these early debates, see Patterson, ibid.

96. Haeckel, *History of Creation,* II, p. 213.

97. Ibid., p. 216. For details of the earlier fossil discoveries and their interpretation, see Bowler, *Fossils and Progress,* p. 101.

for granted by all morphologists into the early twentieth century.[98] Most authorities believed that the number five had been arrived at by the reduction of an originally larger number of digits, although Georg Baur argued on embryological grounds that the ancestors of the known fossil amphibians must have had a smaller number of digits.[99] Modern discoveries have confirmed that some early amphibians had more than five digits, although we still have no satisfactory explanation of why five eventually became standard.

The belief that the Dipneusta, or Dipnoi, were the ancestral form of the Amphibia became widely accepted in the 1870s and 1880s. In 1872 the American ichthyologist Theodore Gill accepted the Dipnoi as the closest fish to the amphibians. He saw them as among the most primitive fish, from which radiated two great branches, one leading to the more specialized bony fish, the other to the amphibians and the higher vertebrate classes.[100] We have already seen how the same point was built into Gegenbaur's theory of the origin of the vertebrate limbs. The discovery of the Australian lungfish, *Neoceratodus*, in 1870 suggested that the fossil Dipnoi, including *Ceratodus* itself, had well-developed bony fins. Gegenbaur insisted that this symmetrical bony structure was a model for the archipterygium or basic limb form from which the fins of ordinary fish and the amphibian limb were developed by divergent evolution. Balfour's *Treatise on Comparative Embryology* also placed the Dipnoi immediately preceding the hypothetical Protopentadactyloidei from which the Amphibia and the higher vertebrate classes had sprung.[101] When Richard Semon, a disciple of Haeckel, went to Australia in the 1890s, one of his chief objects was to study the embryology of *Neoceratodus* because it served as a link between fish and amphibians.[102]

According to this model the amphibianlike characters of the lungfish were genuine anticipations of the the characters which became fully manifested in the true Amphibia. As we shall see, some morphologists retained the view that the similarities could not be coincidental. But by the end of the century a powerful opposing movement had grown up based on the assumption that the dipnoans' resemblance to amphibians was superficial,

98. For example, Goodrich, *Studies on the Structure and Development of Vertebrates,* pp. 150–158.

99. Georg Baur, "The Stegocephali," p. 669. For a discussion of the effects of recent discoveries on evolutionists' evaluation of this problem, see Stephen Jay Gould, "Eight Little Piggies," reprinted in Gould, *Eight Little Piggies,* pp. 63–78.

100. Gill, "Arrangement of the Families of Fishes," pp. xxxv–xxxvi.

101. Balfour, *Treatise on Comparative Embryology,* in *The Works of Francis Maitland Balfour,* III, p. 327.

102. See Richard Semon, *In the Australian Bush,* p. 2, and for the formal reports on his specimens, Semon, ed., *Zoologische Forschungsreisen in Australien,* Bd. 1.

a product of convergent evolution, and was not an indication of true genealogical relationship. The Dipnoi, it was claimed, could not be ancestral to the amphibians because they had already developed specialized characters such as the crushing-plates of the jaw by which the fossil *Ceratodus* was known. This structure was unlike anything possessed by amphibians, and indicated that the dipnoans must lie on a side branch that did not lead toward the higher class. The alternative hypothetical ancestor of the amphibians was another group of fish prominent in the Palaeozoic, the crossopterygian or lobe-finned ganoids. These also had well-developed bony fins which, Gegenbaur's opponents claimed, offered a better starting point for the evolution of the tetrapod limb. In the most extreme version of this theory, the Dipnoi were actually derived from the crossopterygians, as shown in the following diagram:

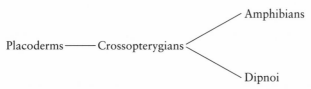

T. H. Huxley had created the ganoid suborder Crossopterygidae in his 1861 memoir on the fish of the Devonian epoch.[103] Originally it included the Dipnoi, although they were soon separated as a distinct suborder by most other systematists. In addition to the fossils which were later identified as potential ancestors of the amphibians, now known as rhipidistians, it also included the coelacanths, which at the time were known only from the fossil record. The only living fish included—which thus became 'living fossils'—were *Polypterus* of the river Nile and its more specialized relative *Calamoichthys*. The presumed existence of living representatives of the crossopterygians became particularly significant later in the century when earlier members of the suborder were postulated as ancestors of the amphibians. Morphologists expended a great deal of effort on *Polypterus* in the hope that it would throw light on this crucial transition. But even Huxley admitted that it exhibited significant differences from the other crossopterygians,[104] and its identification as a member of the suborder would be questioned in the twentieth century.

The claim that the crossopterygians offered a more plausible ancestry than the dipnoans for the amphibians was first suggested by H. B. Pollard and J. S. Kingsley in the early 1890s and soon gained wide support from

103. Huxley, "Preliminary Essay upon the Systematic Arrangement of the Fishes of the Devonian Epoch." On the significance of this memoir, see Mario Di Gregorio, *T. H. Huxley's Place in Natural Science*, pp. 69–75.

104. Huxley, "Preliminary Essay," pp. 441–442.

influential figures such as Cope. Henry Bargman Pollard studied the anatomy of *Polypterus* while working under Robert Wiedersheim at Freiburg (he later returned to England, where he taught anatomy at Owens College, Manchester, and then at Charing Cross Hospital). He identified many characters in which the fish resembled salamanders, arguing that, for instance, "the affinity of the skull of Polypterus to that of the Urodeles is unmistakable."[105] After listing the resemblances, he insisted, "The conclusion to be drawn from the above facts is that the ancestry of the Urodela [salamanders] must be sought among the Crossopterygian forms now represented only by Polypterus and Calamoichthys."[106] The transformation of the crossopterygian fin into a limb would follow naturally from the transition to a more terrestrial way of life. "One has every condition necessary for the 'picking out' of a humerus from the shoulder girdle. When the Crossopterygian Ganoid became a marsh animal and a double-levered arm became of advantage, this portion of the shoulder girdle through Natural Selection would become a perfect humerus."[107] If this genealogical connection were accepted, the Dipnoi would have to be moved to a position quite different from that which they had occupied hitherto. Pollard argued that the skull structure of the Dipnoi differed from that of the Amphibia, and there was no evidence of a phase resembling the Dipnoi in the ontogeny of living Amphibia or in the fossil members of the group. His phylogenetic tree showed the Dipnoi as descendants of the crossopterygians, branching off in a different direction to that taken by the amphibians.[108]

A study of the development of the amphibian skull published by the American anatomist J. S. Kingsley in 1892 made essentially the same point. Kingsley, who had also worked under Wiedersheim, argued that the structure of the skull separated the Dipnoi from the salamanders. He suggested that the lungfish and the amphibians were two divergent branches leading from a common crossopterygian ancestry.[109] In the same year a study of the anatomy of the African lungfish *Protopterus* by W. N. Parker, professor of biology in the University College of South Wales, supported the argument that the Dipnoi were a specialized group. Parker was no enthusiast for genealogical trees, and his warnings strike a note that some modern cladists would welcome:

105. H. B. Pollard, "On the Anatomy and Phylogenetic Position of Polypterus," pp. 339–340.
106. Ibid., pp. 341–342.
107. Ibid., p. 342.
108. Ibid., p. 344.
109. J. S. Kingsley, "The Head of an Embryo Amphiuma."

As it is impossible to decide how far the various resemblances in structure which exist between Dipnoans on the one hand, and certain Fishes and Amphibians on the other, indicate any real relationship, it is perhaps inadvisable to construct hypothetical genealogical trees indicating the possible phylogeny of the group, for, after all, these do not materially add to our knowledge, as they only represent the views of the observers who make them; and, from the nature of the case, these views are founded on insufficient data.[110]

Even so, Parker had no doubts about the degrees of relationship involved: the lungfish could not be linked to the amphibians. He accepted that the lungfish differed significantly from other fish and argued that they deserved to be treated as a separate class. But the structure of the paired fins suggested that they were degenerate and gave no support to the view that they could have evolved into the limbs of tetrapods. Similarities to the amphibians could be explained in terms of the two groups independently moving into the same habitat. This would explain the parallel development of lungs, since there was no reason why these structures "should not have become independently developed in two groups standing widely apart from one another, if the necessity for such respiratory apparatus should occur in both."[111]

Already in 1892 Edward Drinker Cope, originally a supporter of the lungfish-amphibian link, took note of Pollard and Kingsley's work and opted for the new theory, thereby extending it into the realm of paleontology. Cope argued that the structure of the paired fins in Dipnoi did not anticipate that of the tetrapod limb, while that of fossil rhipidistians offered a better model on which the derivation of the limb could be based. In particular the fins of *Eusthenopteron* from the Devonian of New Brunswick well-nigh realized Gegenbaur's ambition of demonstrating the derivation of the tetrapod limb from the fin of a fish.[112] Cope repeated these views in his *Primary Factors of Organic Evolution.*[113]

By throwing his weight behind the new theory, Cope ensured that other paleontologists would take it seriously. In his *Catalogue of Fossil Fishes in the British Museum*, Smith Woodward argued that the Palaeozoic Dipnoi were as specialized as any modern fish, implying that they could not have served as the ancestors for a new class.[114] His textbook of

---

110. W. N. Parker, "On the Anatomy and Physiology of Protopterus annectans," p. 221.
111. Ibid., p. 222.
112. Cope, "On the Phylogeny of the Vertebrata," pp. 279–280.
113. Cope, *Primary Factors*, pp. 88–89.
114. Woodward, *Catalogue of Fossil Fishes in the British Museum*, II, p. xx.

vertebrate paleontology repeated this point and also described the earliest fossil Amphibia as having far closer resemblances to the crossopterygians than to the Dipnoi.[115]

Perhaps the most decisive intervention in the debate came from the respected Belgian paleontolgist Louis Dollo. His 1895 reappraisal of lungfish phylogeny transformed ideas about the group's evolution in a way that seemed to confirm their status as a specialized offshoot from the stem leading to the amphibians. Dollo was already an internationally known figure, in part because of his role in reconstructing the spectacular dinosaur fossils discovered at Bernissart. He was also the author of the much misunderstood 'law' of the irreversibility of evolution. It was in part because of this law that Dollo became interested in lungfish evolution. If the conventional interpretation set up by Smith Woodward and others were accepted, it looked as though evolution did indeed run backwards in this case: the earliest lungfish were the most specialized, and they had evolved into apparently 'primitive' descendants. As Stephen Gould has pointed out, Dollo did not see his law as a consequence of rigid orthogenetic trends, but it did imply that a primitive structure, once lost, could not be re-evolved.[116]

Smith Woodward had inferred that the Devonian lungfish were more specialized than the living members of the group because he took the diphycercal tail of the modern *Ceratodus* to be primitive. This tail consists of more or less symmetrical dorsal and ventral fins which meet at the tip of the body. It was assumed that the earliest lungfish would have had heterocercal, or unevenly lobed tails like the majority of their contemporaries, and (on the fin-fold theory) they would have evolved from an earlier diphycercal form. If this were so, the primitive tail would have been re-evolved, thus overturning the law of irreversibility. Dollo drew upon work by Balfour and Parker which suggested that the tail of the modern lungfish was not truly primitive.[117] It was a new structure formed to replace the old tail, which had been lost in the course of evolution. The dorsal and anal fins had extended backwards to give a new tail which only superficially resembled the old. Dollo was able to proclaim that the law of irreversibility remained intact: the new tail of *Ceratodus* was a sign of specialization from the original heterocercal type.[118]

Dollo now interpreted lungfish evolution in ecological terms, as a specialization for living in impure water. Devonian lungfish such as *Dipterus*

---

115. Woodward, *Outlines of Vertebrate Palaeontology*, p. 60 and pp. 123–124.

116. Louis Dollo, "Les lois d'évolution"; see S. J. Gould, "Dollo on Dollo's Law."

117. F. M. Balfour and W. N. Parker, "On the Structure and Development of Lepidosteus"; see Dollo, "Sur la phylogenie des Dipneusts," pp. 79–80.

118. Dollo, ibid., p. 96.

had moved into this environment, and the living members of the group illustrated the stages of further specialization. The Australian *Ceratodus* still had working fins and could not live out of water, while *Protopterus* and *Lepidosiren* had better-developed lungs and almost totally degenerate paired fins. These later forms were adapted to living actually in the mud, and other fish adapted to the same environment shared a similar eel-like structure, acquired by convergent evolution.[119] Dollo then went on to look for the most likely ancestry of the earliest dipnoans and found it in the crossopterygians. The latter were already adapting in the same direction: they were bottom-dwellers rather than swimmers in the open water, and their lobed fins had been developed to enable them to 'walk' over the bottom surface.[120] In effect, then, the lungfish were the end-product of a specializing trend that had been started by the Devonian crossopterygians.

Having denied that the Devonian lungfish were as specialized as their modern descendants, Dollo was by no means prepared to see them as ancestors of the amphibians. *Dipterus* was less specialized than *Ceratodus,* but it was already too far along the path to have evolved into the new class. Dollo threw his weight behind the views of Cope, Pollard, and Kingsley, arguing that even the earliest lungfish had too specialized a dentition to have played this role. The crossopterygians themselves were the ancestors of the amphibians: some may have already had lungs, and their lobed fins were already preadapted to walking on the bottom.[121] Dollo concluded with an extended diagram of fish phylogeny in which the dipneusts and the amphibians were represented as two divergent stems arising from a crossopterygian ancestry.[122] From a life on the bottom of shallow waters, one group had moved deeper into the mud, while the other had ventured out onto dry land.

Ramsay Traquair thought that Dollo's essay was a brilliant analysis of the problem and acknowledged his conversion to the new theory of lungfish origins.[123] The Dipnoi were the children of the crossopterygians, not—as some had once claimed—their parents. Traquair noted that one implication of this theory was the need to concede a good deal of parallelism in fish evolution. The archipterygium of *Ceratodus,* for instance, had its equivalent in certain selachians (sharks), and according to the new theory there could be no direct link between them, so the structure must have arisen independently as a parallel development.[124] Traquair seemed

119. Ibid., pp. 99–100.
120. Ibid., p. 107.
121. Ibid., pp. 111–112.
122. Ibid., p. 113.
123. Traquair, "The Bearing of Fossil Ichthyology," pp. 519–522.
124. Ibid., p. 521.

unperturbed by this, since there were many other examples of the independent appearance of similar structures, and the rival theory would also require parallel evolution in the crossopterygians and sharks.

Some morphologists supported the theory of crossopterygian ancestry. In the early twentieth century the work of Edward Phelps Allis on the structure of the snout seemed to show that the nasal apertures of the lungfish were not homologous with the internal nostrils of amphibians.[125] But attention was increasingly switching to the fossil record as the preferred source of information on the relationship between fish and amphibians. In 1896 a study of the early armored amphibians, the Stegocephalia, by Georg Baur lent support to the new theory. Baur was born and trained in Germany but came to America in 1884 as assistant to the eminent and controversial paleontologist O. C. Marsh. He subsequently became assistant professor of comparative osteology at Yale. His paper was based on a lecture given to the Biological Club at the University of Chicago in April 1895, before the appearance of Dollo's analysis of lungfish phylogeny, which Baur endorsed completely in the published version.[126] Like Cope, Baur had once supported the lungfish theory, but he now insisted that the structure of the earliest amphibians could best be explained by supposing that they had evolved from crossopterygians. The condition of the ribs alone was sufficient "to refuse completely the opinion of the origin of the Batrachia (Stegocephalia) from the Dipnoi."[127] In Baur's view, "All the resemblances between the Dipnoi and Batrachia are not genetic resemblances, but parallelisms."[128] The lungfish were specialized descendants of the earliest crossopterygians, from which the first amphibians had also evolved.

The same point was taken up in the early decades of the new century by D. M. S. Watson, the professor of zoology at University College, London. Watson spent much of his career trying to identify trends in the evolution of the fossil Amphibia.[129] In 1926 he gave the Royal Society's Croonian Lecture on the topic of the origin of the Amphibia, coming out firmly against any direct link to the Dipnoi. Watson saw paleontology as the main source of support for the rival theory. Early studies based on the morphology of living forms had centered on the Dipnoi, but had been led astray by similarities to amphibians which paleontology had shown to be

---

125. E. P. Allis, "The Lips and the Nasal Apertures in the Gnathostome Fishes," and "Concerning the Nasal Apertures."

126. Baur, "The Stegocephali," p. 671.

127. Ibid., p. 664.

128. Ibid., p. 670, note.

129. See, for example, D. M. S. Watson, "Structure, Evolution, and Origin of the Amphibia." On Watson and orthogenesis, see Bowler, *Eclipse of Darwinism,* p. 171.

merely superficial, due to "the parallel evolution of not distantly allied stocks."[130] Watson was a strong supporter of orthogenesis, and it was thus quite easy for him to maintain such a position. There was no direct relationship between the earliest amphibians and the Dipnoi, but Dollo and others had shown that the Devonian *Dipterus* was the most primitive member of the group, resembling the contemporary osteolepids (the order Crossopterygii was now called the Osteolepidoti). Watson believed that the amphibians evolved from a hypothetical ancestor which had also given rise to the known osteolepids and lungfish. It was the osteolepids, however, that were closest in structure to this hypothetical ancestor.[131]

We shall see in the following section how Watson, Gregory, and others tried to explain the actual transformations that gave rise to the amphibians. But all of these studies were based on the assumption of a starting point in the osteolepid crossopterygians. Popular studies by paleontologists like Gregory and Osborn brought the new theory to the general public. Osborn's *Origin and Evolution of Life* rejected the theory of lungfish ancestry along with Gegenbaur's archipterygium theory of limb evolution and pointed to the lobed fin of the crossopterygians as the most likely starting point for the tetrapod limb.[132] Gregory's *Our Face from Fish to Man* argued that the dipnoans had "definitely and hopelessly removed themselves from the main line of ascent" by losing the marginal bones of the lower jaw.[133] Like Watson, he favored a hypothetical ancestor which had given rise to the amphibians, crossopterygians, and lungfish. Romer's 1933 text on vertebrate paleontology dismissed the lungfish as "not the parents but the uncles of the tetrapods" and sought the origins of the tetrapod limb in the crossopterygian fin.[134] An overview published in 1943 by T. S. Westoll, a paleontologist at the University of Aberdeen, proclaimed the crossopterygians as the only possible source of the amphibians.[135]

Some morphologists challenged the paleontologists' conclusions because they found it difficult to believe that the similarities between lungfish and amphibians were of no phylogenetic significance. To the embarrassment of the British who had colonized Australia, the first really detailed study of the development of *Ceratodus* was published by a German team led by Richard Semon, who—as we have already noted—had gone to

---

130. Watson, Croonian lecture: "The Evolution and Origin of the Amphibia," p. 190.
131. Ibid., pp. 195–199.
132. Osborn, *The Origin and Evolution of Life*, pp. 172–174.
133. Gregory, *Our Face from Fish to Man*, p. 25.
134. Romer, *Vertebrate Paleontology* (1933), p. 92 and p. 104.
135. T. S. Westoll, "The Origin of the Tetrapods."

Australia to collect specimens of this 'living fossil.'[136] Semon had been inspired by Haeckel, and it is worth noting that in his later career he proposed a neo-Lamarckian theory of heredity based on the analogy between inheritance and memory. He was certainly aware of the similarities between the lungfish and amphibians and published an article stressing their phylogenetic significance in 1901. Having gone to study *Ceratodus* under the conviction that it was the link between fish and amphibians, he now found himself forced to respond directly to the challenges of Pollard, Dollo, and Baur.[137] The relationship between the lungfish and amphibians was confirmed by a long list of similar characters which, in Semon's view, far outweighed the significance of the relatively few specializations that had been used to rule them out from tetrapod ancestry.

In his volume on fishes in the *Treatise on Zoology*, edited by Lankester, Goodrich also insisted that the Dipnoi "present many striking points of resemblance to the Amphibia, which cannot all be put down to convergence."[138] Goodrich was particularly concerned with the structure of the heart, and he repeated his doubts about the possibility of parallel evolution in his classic text *Studies on the Structure and Development of Vertebrates* of 1930.[139] An incidental product of Goodrich's work was a growing suspicion that the one supposedly modern crossopterygian, *Polypterus*, did not belong in the group and hence could not be used to confirm the crossopterygian-amphibian link. These doubts were first voiced in a 1907 paper presented to the British Association.[140] In a paper on the origin of land vertebrates delivered to the same forum in 1924, Goodrich argued that the resemblances studied by Pollard and others were deceptive. *Polypterus* was not linked to the ancestry of the amphibians and was probably not a crossopterygian. The amphibians could have evolved from the Dipnoi before that group had acquired the specialized mouth structure which had been used to disqualify it by the paleontologists.[141] A few years later Goodrich boldly stated the possibility that *Polypterus* might be a palaeoniscid—an entirely different order of bony fish—not a crossopterygian.[142] It was now Goodrich's turn to invoke convergence to explain

136. Semon, *In the Australian Bush*, p. 2 and pp. 68–79, 87–92, 96–99; also Semon, ed., *Zoologische Forschungsreisen in Australien*, Bd. 1. On Semon's theory of heredity see Bowler, *Eclipse of Darwinism*, pp. 84–85.
137. Semon, "Über das Verwandtschaftsverhältnis der Dipnoer und Amphibien," pp. 180–181.
138. Goodrich, *Vertebrata Craniata (Cyclostomes and Fishes)*, p. 230.
139. Goodrich, *Studies on the Structure and Development of Vertebrates*, pp. 552–553.
140. Goodrich, "On the Systematic Position of Polypterus."
141. Goodrich, "The Origin of Land Vertebrates," p. 936.
142. Goodrich, "Polypterus a Palaeoniscid?"

the resemblances that had originally led Huxley to include it in the crossopterygians. A modern systematist has described *Polypterus* as a "well-intentioned impostor" intruding into discussions of tetrapod ancestry.[143]

Even when still regarded as a crossopterygian, *Polypterus* had been placed in a suborder of its own, distinct from the Rhipidistia, which most paleontologists saw as the ancestors of the amphibians. It would thus have given only indirect clues about the nature of the true ancestor. Now its status as a 'living fossil' had been eroded, and it is significant that the 1933 edition of Romer's textbook on vertebrate paleontology treated the Crossopterygii as entirely extinct.[144] It was not until 1938 that a living representative of the third division of the crossopterygians, the coelacanths, was discovered in the waters of the Indian ocean. No one expected this: the coelacanths were a group that had transferred from fresh to salt water and were thought to have become extinct in the Cretaceous, since no later fossils were known. The first specimen of the surviving coelacanth *Latimeria* was incomplete and gave little information about this important relic of the past.[145] Better specimens were not obtained until 1952, and since then the coelacanth has taken over the place once occupied by *Polypterus* as the only surviving member of the crossopterygians.

In addition to Goodrich, several other morphologists continued to defend the idea that the lungfish were closely related to the amphibians. In a paper delivered to the British Association for the Advancement of Science in 1901, J. Graham Kerr rejected all previous theories of limb formation and suggested that they might be modified external gills.[146] In his textbook of vertebrate embryology published in 1919, Kerr—now professor of zoology at Glasgow—extended the theory, arguing that fish fins and amphibian limbs might have evolved independently from the external gills.[147] He regarded the limbs of the lungfish *Lepidosiren* as typical of the early form of limb derived in this manner, which might have evolved in two separate directions, towards the crossopterygian and other fish fins, and towards the amphibian leg. No one took this theory seriously, although Kerr defended it again in an article published in 1932.[148]

In this article Kerr also repeated the claim that lungfish are primitive, and stressed the link between them and amphibians. He insisted that relationships should be evaluated on the basis of a wide range of charac-

143. B. Schaeffer, "The Evolution of Concepts Related to the Origin of the Amphibia."

144. Romer, *Vertebrate Paleontology* (1933), p. 88.

145. See, for instance, the brief comment in Romer, *Vertebrate Paleontology* (1945), pp. 121–122. For the story of the Coelacanth, see K. S. Thompson, *Living Fossil*.

146. J. G. Kerr, "The Origin of the Paired Limbs of Vertebrates."

147. Kerr, *Textbook of Embryology (Vertebrata)*, pp. 451–452.

148. Kerr, "Archaic Fishes," p. 428.

ters, not just a single system such as the skeleton. For this very reason, paleontology was not the most reliable guide to ancestry, since it had to work almost exclusively from fossilized bones. The embryology of the surviving primitive fish offered far more comprehensive evidence. Kerr argued that the skeleton was particularly unreliable because bone structure was affected by mechanical stresses during the development of the individual organism.

> The intensive study of histogenesis in *Lepidosiren* makes it in fact difficult to resist the conclusion that skeletal formation in the vertebrate is very much a matter of direct reaction to mechanical stress, that the laying down of skeleton in a particular locality is not merely a matter of ancestral dictation but that it is largely influenced by factors of a purely mechanical kind.[149]

Kerr's thesis would imply that the skeleton is widely plastic within the individual organism, and thus could offer no clues to ancestry because it is not controlled by genetic factors. Not surprisingly, this view was ignored along with his external-gill hypothesis. Nevertheless, his feeling that anatomy and embryology did not support the paleontologists' rejection of the lungfish-amphibian link was shared by other morphologists. A study of the skull of *Ceratodus* by the Australian zoologist H. L. Kesteven in 1931 also emphasized the similarities between this lungfish and amphibians.[150] At least one modern zoologist has defended the proposition that the lungfish are, after all, the ancestors of the amphibians.[151]

Disagreement had thus emerged between the paleontologists, almost all of whom had adopted the crossopterygian theory, and the morphologists who dealt with living lungfish and amphibians, many of whom saw the similarities as being too close to be explained away by convergence. It is a sign of paleontology's increasing dominance that we seldom hear of the rival theory, especially in popular accounts of the history of life on earth. One of the strangest products of this tension between the professionals was the suggestion developed by several Scandinavian biologists that the Amphibia might be diphyletic, having two separate origins within different groups of fish. In 1933 Nils Holmgren published a study of amphibian limbs which stressed the differences between the urodeles (salamanders) and anurans (frogs), no doubt a product of the latter's adaptation to jumping. He seized upon this difference as a means of arguing that the Amphibia are an artifical group composed of two separate phyla. Existing theories of limb evolution were unsatisfactory because no one had

149. Ibid., p. 420.
150. H. L. Kesteven, "Contribution ot the Cranial Osteology of the Fishes."
151. Brian G. Gardiner, "Tetrapod Ancestry: A Reappraisal."

admitted the possibility that the 'amphibian' limb had been formed by two different routes.[152] Holmgren argued that the stegocephalians had evolved from crossopterygian fish and had in turn given rise to the anurans and the reptiles. The urodeles had evolved separately, either from another crossopterygian source, or more likely from the dipnoans.[153] The credibility of the morphological evidence for a relationship between the lungfish and at least one type of amphibian was thus salvaged at the price of splitting the old class Amphibia into two fundamentally different types.[154]

A modified version of this theory was proposed by G. Säve-Soderbergh, who was doing important work on newly discovered, very primitive amphibian fossils from the Devonian of Greenland. Säve-Soderbergh argued that many previous classifications were wrong because they ignored the existence of orthogenetic trends that could produce similar results in several evolutionary lines.[155] He accepted the idea that the urodeles might be derived from dipnoans, while the anurans were linked back to the fossil labyrinthodonts of the Carboniferous and then to the crossopterygian fish.[156] The early amphibians (other than the urodeles) had in turn split into two lines, one leading directly to the reptiles, the other to the known fossil amphibians and the anurans, as shown in the following diagram:

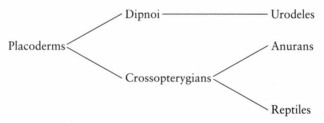

In a study of the snout of crossopterygians published in 1942, Erik Jarvik toned down the more extreme claims of the Scandinavian school, accepting that there could be no direct link between any amphibians and the lungfish. But instead he stressed the possibility that several different groups of crossopterygians might have been preadapted for terrestrial life, so that the amphibians might have diverse origins within the crossop-

152. Nils Holmgren, "On the Origin of the Tetrapod Limb."

153. Ibid., p. 288.

154. See the critical discussion of Holmgren's thesis in I. I. Schmalhausen, *The Origin of Terrestrial Vertebrates*, chap. 19.

155. G. Säve-Soderbergh, "Some Points of View Concerning the Evolution of the Vertebrates," pp. 10–11.

156. See the phylogenetic tree, ibid., p. 16.

terygians themselves.[157] As the Russian paleontolgist Ivan Schmalhausen later observed, Jarvik turned the family tree of the Amphibia into a lawn.[158]

The diphyletic theory gained little support, and Säve-Soderburgh's work on the early Amphibia from Greenland, the Ichthyostegids, only helped to convince most paleontologists that the link back to the crossopterygians offered the best way of explaining amphibian origins. Romer's textbook of 1933 figured the skull of *Ichthyostega* and argued that this Devonian type was probably ancestral to the modern amphibians.[159] Fossils to fill in the actual transition back to the crossopterygians remained elusive, however, and much theoretical effort was expended on trying to show how the crossopterygian lobe-fin could have been transformed into the tetrapod limb. It is, perhaps, a sign of the self-confidence of the early twentieth-century paleontolgists that they were prepared to expend considerable effort on this theoretical reconstruction.

FROM WATER TO LAND

The problem of explaining the transition to a new habitat on the land was a complex one. The physiological transformation was obvious enough: the lungs had to replace the gills as a means of respiration. Darwin himself had argued that this was not as great a problem as it might seem. He pointed out that most fish have swim bladders which contain air and are used to regulate buoyancy. It was relatively easy to imagine how this bladder could be transformed into a lung in an animal that needed to breathe air.[160] Most evolutionists agreed that the lungfish—whether or not they are directly related to the amphibians—show the transitional phase in which the bladder has become modified to absorb air in circumstances where the fish has to exist at certain times out of water. In fact Darwin had here fallen into the trap of assuming that the structure typical of the fish must be more primitive than that of the higher vertebrates. The American biologist Charles Morris seems to have been the first to suggest the modern view that the original function of the swim bladder was respiratory—only in the later bony fish had it degenerated into a mere regulator of buoyancy as the gills took over the whole function of respiration.[161] He

157. Erik Jarvik, "On the Structure of the Snout in Crossopterygians."

158. Schmalhausen, *The Origin of the Terrestrial Vertebrates*, p. 275. See also K. S. Thompson, "A Critical Review of the Diphyletic Theory of Rhipidistian-Amphibian Relationships."

159. Romer, *Vertebrate Paleontology* (1933), p. 106 and p. 115.

160. Darwin, *Origin of Species* (1859 ed.), pp. 190–191. For a critique of Darwin's assumption in the light of modern knowledge, see Stephen Jay Gould, "Full of Hot Air," reprinted in Gould, *Eight Little Piggies*, pp. 109–120.

161. Charles Morris, "The Origin of Lungs."

pointed out that the sharks did very well without a swim bladder, which certainly suggested that it was not a necessary fish structure.

In this respect, Morris's view did not anticipate the modern idea that the earliest vertebrates had possessed two means of respiration, gills and an air bladder. (The latter structure has been lost in the sharks, but transformed in the bony fish.) He imagined that the bladder had actually been evolved by fish which needed to breath air. Morris believed that ganoids and lungfish had been driven into shallow waters by the early sharks, which had been the "monarchs of the seas." Both types of fish had developed the bladder as a lung and may even have adapted to spending some time on land. They had retreated back into the muddy waters only after the true amphibians had begun to dominate the land. Most early twentieth-century evolutionists rejected Morris's claim that fish with lungs had invaded the land, but there was certainly a strong presumption that the crossopterygians had bladders preadapted to breathing air, which would have prepared them to move into the new environment.

More crucial was the problem of locomotion. The necessary transformations involved far more than the fins acquiring the ability to move the body over the ground. As Dollo argued, the crossopterygian fin was preadapted to pushing the fish along the bottom in shallow water, and it was relatively easy to suppose how the same structure could be used to propel a primitive amphibian over a muddy surface. But to provide efficient movement on the land, the limbs had to become far more powerful and had to be anchored into the body in a way that would transmit the force efficiently. To function out of water, the whole body had to be supported in such a way as to allow breathing to take place against the pressure created by gravity. A complex series of morphological transformations must thus have taken place to give rise to the first amphibians.

Despite the lack of fossils illustrating the actual transformation, paleontologists became increasingly willing to use their studies of crossopterygians and primitive amphibians to explore the details of how it might have taken place. In part the problem would be solved by identifying homologies: Which bones in the ancestral fin support have been transformed into the bones of the tetrapod limb? This was not as straightforward as it might seem. The fish fin is an essentially rigid structure which articulates with the body only at the 'shoulder.' The tetrapod limb articulates at the 'elbow' and 'wrist' as well, and the upper and lower parts of the limb have evidently been twisted with respect to the body in a way that confused many early morphologists who tried to work out the homologies involved. In addition it was necessary to study the transformation of the shoulder and pelvic girdles, which are not integrated with the axial skeleton in fish. The fish shoulder girdle is attached to the rear of the skull; to avoid transmitting the shock of each step to the head it must have been moved backwards and become connected more closely with the spine. The pectoral girdle of the fish, which floats freely in the muscles, had to be enlarged and also become connected to the spine.

Even when they came to an agreement over the basic transformations by which the lobe-fin of a crossopterygian had been transformed into an amphibian leg, paleontologists were no longer satisfied. Increasingly, they saw themselves as functional morphologists, trying to understand the pattern of stresses and strains that would have shaped the transformation as the ancestral fish began to move out of the water. Finally, they tackled the question of ultimate causation: How had the transitional forms coped with a way of life that was partly aquatic and partly terrestrial, and—perhaps more important—why would a fish have taken the risk of first venturing out into a new and hostile environment? Evolutionists were no longer satisfied with the construction of phylogenetic trees based on morphological relationships. They were now beginning to construct adaptive scenarios to explain particular transformations, exploiting information about changing environments derived from geology.

In the final version of Gegenbaur's theory (discussed above) the archipterygium modeled on the fin of *Ceratodus* was seen as the most primitive form which had been converted both into the fins of other fishes and into the amphibian limb. But few, apart from Gegenbaur's own disciples, were entirely happy with the theory. The archipterygium consisted of a central rod of bones with rays branching out symmetrically on either side. Yet the tetrapod lower limb consists of *two* bones, the radius and ulna in the anterior limb or arm, the tibia and fibula in the posterior or leg. These in turn must articulate in a particular way. In the human arm, the wrist is a simple hinge, while the elbow also permits the lower arm to rotate as a unit with respect to the upper. In the leg it is the opposite way around: the *lower* joint, the ankle, permits both bending and rotation, while the upper, the knee, is a simple hinge. Gegenbaur certainly tried to identify the bones of the leg and arm with elements of the symmetrical archipterygium. He believed that the homologues of the main axis in the achipterygium were (for the forelimb) the humerus, the radius, and the first digit.[162] The pentadactyle limb was thus derived from only one side of the archipterygium.

In 1876 T. H. Huxley published a study of *Ceratodus* in which he evaluated Gegenbaur's theory. He noted the problem that in fish and tetrapods the limbs rotated in different directions with respect to the trunk.[163] While accepting that the archipterygium of *Ceratodus* was the fundamental form of the limb, he was forced to dissent from the rest of Gegen-

162. Gegenbaur, *Grundriss der vergleichenden Anatomie*, p. 497.
163. Huxley, "Contributions to Morphology: Ichthyopsida, I: On Ceratodus fosteri," pp. 109–110.

baur's theory. As he understood the homologies of the limb bones in fish and tetrapods, the rotations required by the theory would create a torsion of the humerus which he found quite implausible.[164] Gegenbaur thought that the tetrapod limb was produced by a continuation of the same process as that which generated the asymmetrical fins of other fish, but Huxley argued that by abandoning this assumption, a simpler explanation became possible. The tetrapod limb, or cheiropterygium, and the fish fin were developed by different kinds of specialization, starting from the archipterygium. Huxley gave a diagram to illustrate the comparable bones in a shark fin and an amphibian limb.[165] Further investigation of Palaeozoic amphibians was needed to test the different suggestions which had been put forward to explain the relationship.

Gegenbaur accepted Huxley's criticism, later editions of his work showing the main axis running through to the fifth digit.[166] Little further progress was made while the majority of biologists continued to believe that the lungfish were the starting point for amphibian origins. But when it was recognized in the 1890s that the crossopterygians offered a more plausible ancestral form, new developments became possible. It was immediately obvious that the fins of rhipidistian crossopterygians could much more easily have been transformed into tetrapod limbs than could the archipterygium of *Ceratodus*. In 1896 one of Gegenbaur's own students, Herman Klaatsch, proposed a scheme in which the archipterygium was transformed into the tetrapod limb via the limb form of *Polypterus* (then still considered a surviving crossopterygian).[167] His idea did not gain much support: the paleontologists were losing patience with the anatomists' use of living organisms to study a problem that, from their point of view, involved relationships between the extinct forms to which they alone had access. Klaatsch subsequently went on to propose a polytypic theory of human origins which was also dismissed by most of his contemporaries. His biographer blamed his lack of recognition on enmity created in part by his boldness in challenging accepted views on the evolution of limbs.[168]

Morphologists continued to study the problem of the origin of limbs in the twentieth century, confirming that this branch of biology remained active long after it had been dismissed as worthless by its critics. Richard

164. Ibid., p. 118.

165. Ibid., p. 114 and p. 120.

166. Gegenbaur, *Elements of Comparative Anatomy*, p. 480 (this is a translation of the 1877 edition of the *Grundriss der vergleichenden Anatomie*). See also Gegenbaur, *Vergleichenden Anatomie* (1898), II, p. 520.

167. Klaatsch, "Der Brutflosse der Crossopterygier."

168. See the introduction by Adolph Heilborn in Klaatsch, *The Evolution and Progress of Mankind*, p. 15. On Klaatsch's work on human origins, see Bowler, *Theories of Human Evolution*, pp. 135–140.

Semon, Carl Rabl, Max Fürbringer, Ivan Schmalhausen, and others continued to publish on the topic.[169] Much of their work continued to be based on the study of lungfish anatomy and embryology, exploring a relationship that the majority of paleontologists had already rejected. The results were certainly not conclusive. Thus in 1930 Goodrich, who still saw his work as an attempt to understand evolutionary relationships, claimed that none of the efforts made to reconstruct the evolution of the tetrapod limb were convincing and concluded that "as yet nothing for certain is known about the origin of the cheiropterygium."[170] In the detailed study of amphibian limb anatomy that led him to propose that the class was diphyletic, Holmgren also noted that "it is fairly clear that the problem of the origin of the tetrapod limb is today nearly as far from solution as it was in Gegenbaur's time."[171] In this case, at least, the morphologists themselves were forced to admit that their work was getting nowhere.

The confusion expressed by some morphologists only served to convince the paleontologists of the futility of using the structure of living organisms to study a problem that was really centered on the relationships between the more primitive forms revealed by the fossil record. The early twentieth century saw a rush of work by paleontologists seeking to exploit the new theory that the amphibians had evolved from crossopterygians. In an article written in 1915, Gregory recorded a remarkable coincidence of scientists independently moving toward the hypothesis that the fins of certain fossil crossopterygians could be used as a model for the origin of the early amphibian limb.[172] In his *Evolution of the Vertebrates and their Kin* of 1912, Patten cited the fin of the Carboniferous rhipidistian *Eusthenopteron* as the form that would become characteristic of the land vertebrates.[173] This came in a brief survey of vertebrate evolution offered as an extension of his theory of vertebrate origins, but it matched the views of the paleontologists themselves. In the following year both Watson, an expert on fossil amphibians, and Robert Broom, better known for his work on the mammal-like reptiles, independently identified *Eusthenopteron* or the late Devonian *Sauripterus* as the best models from which to derive the tetrapod limb (fig. 5.6).[174] Gregory records that he became

169. For a survey of this work, see Holmgren, "On the Origin of the Tetrapod Limb," pp. 203–208.

170. Goodrich, *Studies on the Structure and Development of Vertebrates*, pp. 159–160.

171. Holmgren, "On the Origin of the Tetrapod Limb," p. 208.

172. Gregory, "Present Status of the Problem of the Origin of the Tetrapoda," p. 358.

173. W. Patten, *The Evolution of the Vertebrates and their Kin*, p. 390.

174. Watson, "On the Primitive Tetrapod Limb"; and Robert Broom, "On the Origin of the Cheiropterygium."

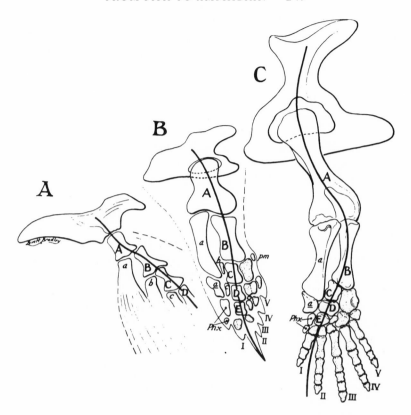

FIGURE 5.6   Pelvis and pelvic appendages of A: *Eusthenopteron;* B: graphic intermediate; C: the amphibian *Trematops.* From W. K. Gregory and H. C. Raven, *Studies on the Origin and Early Evolution of Paired Fins and Limbs* (New York, 1941), p. 333.

aware of these publications while he was himself investigating the fin of *Sauripterus,* having been alerted to its amphibianlike structure by the publication of a photograph in a museum catalogue.[175] This fin has a single proximal element equivalent to the humerus, two distal elements equivalent to the radius and ulna, and a number of radials from which the digits might be derived. Lull reported these studies in his textbook on evolution of 1917 and added an illustration of a fossil footprint from the upper Devonian which seemed to indicate that the earliest amphibian foot had not yet developed the full complement of five digits.[176]

Over the next couple of decades, a number of paleontologists tried to reconstruct the details of a process by which the crossopterygian fin could

175. Gregory, "Present Status," p. 358.
176. Lull, *Organic Evolution,* pp. 488–489.

be transformed into the tetrapod limb. The best available fossil amphibians were studied in an attempt to understand the structure of the early amphibian limb and the way in which it was used. As Watson wrote in his Croonian Lecture of 1926, the mere search for homologies was no longer satisfying: "the centre of interest has passed from structure to function, and it is in the attempt to realise the conditions under which the transformation took place, and to understand the process by which the animals' mechanism was so profoundly modified whilst remaining a working whole throughout, that the attraction of the problem lies."[177]

It was William King Gregory and his students, including Alfred Sherwood Romer and Roy Waldo Minor, who were most active in carrying forward the program sketched out in Watson's words.[178] They created a paleontology based on functional morphology, using living examples to reconstruct not only the skeleton but also the musculature of fossil species. Both fish and amphibian fossils were studied in an effort to bridge the gap. Gregory continued to work on the fins of both *Sauripterus* and *Eusthenopteron.* The relatively unspecialized Permian amphibian *Eryops* was a favorite source of comparisons. Minor undertook a detailed reconstruction of the musculature of *Eryops,* using the primitive New Zealand reptile *Sphenodon*—"practically a 'living fossil' "—as a model.[179]

In 1931 Romer and F. Byrne drew attention to the problem posed by the fact that in the forelimb it is the wrist that is a simple hinge, while in the hindlimb it is the knee which is the simpler joint.[180] They argued that this difference could be explained by the fact that in the crossopterygian ancestors of tetrapods, the leading edge of the pectoral fin was pointed downwards towards the belly, not towards the back as in most modern fish. This orientation influenced the way in which the fin could twist as it acquired joints for walking, thus explaining why the pectoral and pelvic fins had adapted to the same function in different ways. Unfortunately they did not appreciate the possibility of reorientations within the limb, and reversed the homologies of the radius and ulna. They were also forced to postulate unnatural twistings of the limbs and a reversal of function for some of the muscles.

In a further study of *Sauripterus,* Gregory rejected several aspects of the theory proposed by Romer and Byrne, while applauding their basic

---

177. Watson, Croonian Lecture: "The Evolution and Origin of the Amphibia," p. 189.

178. For details of the work done by Gregory's team, see Ronald Rainger, *An Agenda for Antiquity,* chap. 9.

179. R. W. Miner, "The Pectoral Limb of *Eryops.*"

180. A. S. Romer and F. Byrne, "The Pes of Diadectes." For details of their work and its influence, see J. S. Rackoff, "The Origin of the Tetrapod Limb."

FIGURE 5.7   Protetrapod, illustrating the transitional phase between fish and amphibian. From W. K. Gregory and H. C. Raven, *Studies on the Origin and Early Evolution of Paired Fins and Limbs* (New York, 1941), p. 343.

approach to explaining the difference between the fore and hind limbs.[181] In 1941 Gregory and Henry C. Raven published an extensive study of the evolution of the limbs, using *Eusthenopteron* as a starting point (fig. 5.7). They used a large flexible model to demonstrate the different positions taken up by the limb as it became bent and twisted to form a functioning leg. They were particularly insistent that the transformation should be explained as far as possible by seeking transitions between forms already known from the fossil record. Some authorities were all too ready to dismiss known forms as too specialized, so that they could postulate hypothetical ancestors.

> The habit of inventing new hypothetical orders rather than of making full use of those which we have has arisen, we maintain, through a misreading of the doctrine of the irreversibility of evolution and through a tendency . . . to expect remote ancestors to display only those specializations which are found in greater degree in their descendants. We, on the other hand, maintain that the wholly new habit of using the paired appendages to support the weight of the body on land became possible only through a drastic revolution in construction, with the loss of many ancient uniformities and the rise of such new and unexpected entities as the digitiferous zone on the margin of the rhipidist 'fleshy lobe.' [182]

Even when the known fossils occurred too late in the record to be the actual ancestor (this was certainly the case with *Eusthenopteron*) the later form could be used as a model on the assumption that close relatives of the true ancestor might have survived unchanged into later epochs. Gregory adopted a similar approach in the study of human origins, where he

181. Gregory, "Further Observations on the Pectoral Girdle and Fin of Sauripterus."
182. Gregory and H. C. Raven, "Studies on the Origin and Early Evolution of Paired Fins and Limbs," pp. 343–344.

also rejected the common tendency to dismiss known fossils as the products of side branches away from the main line of evolution.[183]

Gregory was evidently willing to allow major and unexpected transformations in response to the adoption of new habits. His technique of imagining how the body could be molded by mechanical forces had been pioneered in the late nineteenth century by neo-Lamarckians such as Cope and John A. Ryder.[184] They had imagined the forces imposed by new habits acting directly on the body via the inheritance of acquired characteristics. Gregory was no Lamarckian—indeed as Ronald Rainger has shown, he was strongly opposed to the anti-Darwinian theories of his mentor, Osborn.[185] Nevertheless he was aware of Ryder's work and quoted it approvingly.[186] He was able to assume that natural selection could perform the same function indirectly: if the stresses and strains were imposed by a new habit, selection would favor the appropriate bone and muscle structures. Romer too favored natural selection over Lamarckism to achieve the same goals.[187]

The development of a functional approach to the study of fossils seems to have been possible among a wide group of paleontologists, some of whom retained an enthusiasm for older, non-Darwinian ideas, while others were aware that new areas of biology were increasingly tending to promote the synthesis of natural selection and genetics. Of those mentioned in the preceding section, Watson certainly advocated orthogenetic trends in the later phases of amphibian evolution, while Broom was an ardent supporter of an almost mystical view of evolution as the unfolding of God's design.[188] It would seem that serious work on the major transitions between classes could be done by paleontologists with very different views about the actual mechanism of evolution. In this area, it was obvious that living organisms had moved into new environments requiring new habits. Whatever the mechanism of change, the functional problems would be the same.

The kind of study done by Gregory and his school represented an important development from the mere reconstruction of relationships toward an attempt to understand evolution as a process which worked by

183. On Gregory's work in human origins, see Bowler, *Theories of Human Evolution*, pp. 70–71, and Rainger, *An Agenda for Antiquity*, pp. 230–236.

184. For details of Cope and Ryder's work, see Bowler, *Eclipse of Darwinism*, p. 125 and pp. 137–138.

185. Rainger, *An Agenda for Antiquity*, chap. 9.

186. Gregory, "Present Status," p. 327.

187. Romer, *Vertebrate Paleontology* (1933), p. 4.

188. On Watson's views, see above, note 129; on Broom, see his *The Evolution of Man*. For more details of Broom's views, see the discussion of his work on the mammal-like reptiles in chapter 6 of this text.

modifying a functioning animal body. Yet to some extent it took for granted the fact that the earliest amphibians, for instance, had adopted the new habit of walking on land. Some paleontologists were now willing to go further and ask a far more 'Darwinian' question: What had the earliest amphibians actually gained from the transition to the new habitat that made it sufficiently attractive to be worth the major transformations required? It was no longer possible simply to assume that life would mount inevitably up the ladder toward humankind. Yet most evolutionists would still have subscribed to the notion that the move onto dry land was an essential prelude to the development of higher forms, and ultimately of the human mind. In the 1920s Charles Schuchert, for instance, could still maintain that the transition to land was vital for the evolution of higher levels of mentality; there could have been no further advance in the sea where unintelligent prey was so abundant, because constant vigilance was the driving force of progress.[189] But Schuchert, like most of his generation, now felt the need to ask why the ancestors of the amphibians had ventured into this new environment. However important the consequences of the step, its immediate causation had to be sought in some real advantage to the organisms involved.

Given the significance of the move from water to land, there were surprisingly few attempts to treat the problem in a unified way. Heinrich Simroth's *Die Entstehung der Landthiere* of 1891 was one of the few efforts to make a systematic overview, bringing together a study of how animal life in general, both vertebrate and invertebrate, had made the transition.[190] It was known that plants had already begun to colonize the dry land by the Devonian, soon to be followed by some invertebrates. Presumably this created an environment within which carnivorous vertebrates could function, provided they could cope with the problems of respiration and locomotion. Shallow waters and swamps offered a natural intermediate environment which would serve as a stepping-stone on the way to complete terrestrial adaptation. In a sense, the clothing of the land with vegetation created an opening which the vertebrates might exploit— but was it enough to suppose that this opportunity, merely by its very existence, had been enough to precipitate the revolution? Or did one need to ask more detailed questions about the adaptive pressures that could have forced so major a transformation of the vertebrate body?

The morphologists could simply take it for granted that lungfish or crossopterygians had acquired the habit of moving around outside the wa-

189. Charles Schuchert, *Textbook of Geology*, II, pp. 304–305.

190. Heinrich Simroth, *Die Entstehung der Landthiere;* see chapter 22 on the amphibians. There is also a short chapter on the transition to the land in W. D. Matthew, *Outline and General Principles of the History of Life*, chap. 16.

ter and could investigate the functional changes of the body which made this possible. In the late nineteenth century all morphologists, and most paleontologists, limited their enquiries to this level. They were not interested in postulating what a modern evolutionist would call an adaptive scenario to explain the transition. The first steps towards what we might call a more Darwinian approach to the question (in the modern sense of that term) were prompted by the interaction between paleontologists and geologists, especially in America. Here new theoretical developments in geology encouraged the search for evidence of past climatic changes and were linked to an active use of vertebrate paleontology in stratigraphy. By the end of the nineteenth century geology was no longer dominated by a philosophy of complete, steady-state uniformitarianism. Geologists such as Thomas C. Chamberlin were now convinced that there were episodes of intense (but not actually catastrophic) change in the earth's physical conditions. From this source came the inspiration to enquire whether some of the more dramatic steps in the history of life might have been triggered by the resulting environmental stresses.

The American geologist Joseph Barrell was the first to apply the new philosophy of earth history to the question of the origin of land vertebrates. In 1906 he began a series of studies on sedimentation which provided information on the climates of the successive geological periods. Over the following ten years he became convinced that climatic stress was the trigger for major evolutionary changes, and in 1916 he published a paper on the "Influence of Silurian-Devonian Climates on the Rise of Air-Breathing Vertebrates." The main driving force of evolution, he maintained, was the pressure of the environment on the organism. Periods of climatic stress imposed a more intense struggle for existence which eliminated the less hardy and adaptable types and favored the survival of advanced mutations.[191] It is worth noting that Barrell was not a convinced Darwinist. Like many of his contemporaries, he thought that natural selection was not the sole driving force of evolution—but he did insist that it is "nevertheless a broad controlling force which compels development within certain limits of efficiency."[192] He also thought that it coordinated changes in different parts of the organism.[193] It was within the context of this rather vague sense of an environmental pressure upon the organism that he began to ask exactly what kind of incentive would have been enough to drive the ancestors of the amphibians out of the water.

Barrell rejected several possible explanations which were independent

191. Joseph Barrell, "Influence of Silurian and Devonian Climates on the Rise of Air-Breathing Vertebrates," p. 414.

192. Ibid., p. 390.

193. Ibid., p. 422.

of a change in the physical environment. Fish would not have been driven out of the water to escape predators, since the numbers of predators would be regulated by the numbers of the prey species itself—and in any case the amphibian tadpole has to spend its life in the water. They were unlikely to have been tempted out by the search for food; a potential supply of food had been available on land since the early Palaeozoic, and most fish species continued to ignore it anyway.[194] It was more efficient to absorb oxygen directly rather than via solution in water, but again most fish made no use of the air for respiration, even those which lived near the surface. The ability to breath air was "an adaptation which has been forced repeatedly to a greater or less degree on fishes by the recurrence of an unfavorable environment, rather than assumed within a constant environment because of inherent advantages."[195] Barrell referred to a suggestion by Charles Morris that the Devonian ganoids had been driven out of the water by sharks, but argued that this inverted the logic of causation.[196] These fish (the crossopterygians) had probably gulped air to supplement their respiration when the oxygen level of the water was diminished. The direct cause of their move into the air was the pressure of the Silurian-Devonian climate, which had created large bodies of warm and stagnant water in which the need to breath air had become essential. The physical environment was the trigger of change—although it is significant that Barrell's theory did not explain why the fish eventually became so modified that they could live permanently on the land.

The particular adaptations the fish had developed to make air-breathing possible were fortuitous in the sense that there were other conceivable ways of making an equivalent change: "This blind choice of the ganoid fishes, directed by minor and unessential factors, fixed for all time the lines of evolution of vertebrates in regard to the fundamental life activities of respiration and circulation."[197] Barrell even suggested that alternative systems of respiration might have been better suited to the future evolution of warm-bloodedness.[198] He concluded with a hymn of praise to the role of varied conditions in promoting divergent and progressive evolution, and with a suggestion that the particular geological events which had triggered the development of air-breathing were an example of "the rythmic pulses of climatic and diastrophic changes which have remorselessly urged forward the troop of living creatures."[199]

194. Ibid., pp. 414–416.
195. Ibid., p. 416.
196. Ibid., pp. 417–419; see also Charles Morris, "The Origin of Lungs."
197. Barrell, ibid., p. 428.
198. Ibid., pp. 432–434.
199. Ibid., p. 435.

Barrell's ideas were taken up almost immediately in Richard Swann Lull's 1917 textbook *Organic Evolution*. Lull introduced his chapter on the "Emergence of Terrestrial Vertebrates" with a three-page discussion of Barrell's thesis.[200] The topic reemerged in the epilogue, "The Pulse of Life"—a classic expression of the theory that the advance of life came in sudden spurts triggered by climatic revolutions. Here Lull again noted that the appearance of the amphibians had been triggered by the drying climate produced by the uplift of land in Silurian times. The air-breathing fishes were confined to a series of shallow pools. Then, "when the final dwindling of their habitat left them stranded, such as could become exclusively air-breathing survived, giving rise to the amphibia, but those which could not, perished, except that in some remote asylums where a vestige of their habitat persisted the lung-fishes also survived, for their descendants, few as to kinds, are still extant."[201] Some years later Lull waxed even more poetic in an article on the same theme. In the crowded waters, "a premium would be placed upon ability to crawl ashore and maintain an active life, while the less fit would sleep the sleep that knows no waking, to their racial extinction."[202]

The same theme was taken up in the discussion of the history of life in Schuchert's *Textbook of Geology*. As noted above, much of Schuchert's account was suffused with the progressionism of an earlier generation of evolutionists. He wrote of an "invasion of the land" by the air-breathing fish, "prophetic of vertebrate ascendancy, hereafter never to be questioned in its onward sweep to its culmination in man."[203] Yet a few pages before this, Schuchert had depicted a far less triumphant scenario. Legs were developed from fins "through the enforced hobbling about of the fringe-finned ganoids in their search for water holes in desert regions."[204] Here was a new twist to the story, carrying the implication that the invasion of the land was really only a by-product of the fish's desperate attempts to find the few remaining sources of water in a drying landscape.

Barrell's thesis was also mentioned in Watson's 1926 Croonian Lecture on the origin of the amphibians.[205] Watson believed that the later phases of amphibian evolution were governed by rigid orthogenetic trends, but he contrasted this with a different type of evolutionary change "with an obvious adaptive significance" that was involved in the actual origins

200. Lull, *Organic Evolution*, pp. 477–479.
201. Ibid., pp. 688–689.
202. Lull, "The Pulse of Life," p. 124.
203. Schuchert, *Textbook of Geology*, II, p. 307.
204. Ibid., p. 302.
205. Watson, Croonian Lecture: "The Evolution and Origin of the Amphibia," p. 200.

of the class.[206] The transformation of the limbs and limb girdles was an example of such a change, and "the first tetrapods in fact swam upon land." Watson saw the first amphibians as still living most of the time in the water "and only passing over the land from one pool to another either to find deeper water or to escape the attacks of the large fish."[207] The first of these possibilities again echoes the idea that the move onto land was in part an attempt to find water in an ever more dessicated world.

It was Romer who became most closely associated with the theory that the first amphibians had acquired the ability to walk on land as a means of moving to new pools. Although Barrell is not mentioned by name in Romer's 1933 textbook of vertebrate paleontology, the discussion of "the 'why' of tetrapod origins" followed very much along the same lines. Romer asked why the amphibians left the water. It could not have been to find food, because they were carnivores for whom there was as yet little to eat on the land, nor to escape enemies, because they were the largest vertebrates in the waters from which they came. Instead, "Their appearance on land seems to have resulted as an adaptation for remaining in the water."[208] While pools remained in good supply, air-breathing fish had no incentive to go ashore. But when the pools began to dry up, "the amphibian, with his short and clumsy but effective limbs, could crawl out of the pool and walk overland (probably very slowly and painfully at first) and reach the next pool where water still remained."[209] Eventually his descendants would have stayed longer on land to feed off stranded fish or the newly evolved insects, and thus a true terrestrial fauna might be established.

The theory had already been described for nonspecialist readers in Julian Huxley's popular survey *The Science of Life* of 1931 (written in conjunction with H. G. Wells). Huxley invoked the image of the Devonian as a period when numerous lagoons and lakes disappeared in periodic cycles of drought. The abilities to breath air and to walk across dry land were both advantageous in these conditions. The lungfish developed only the first of these abilities and were condemned to a marginal existence. The early amphibians acquired both—yet Watson's work suggested that many of them remained most of the time in the water. Huxley postulated that this was because there was little food for carnivorous animals on the land, apart from a few early scorpions and insects. The legs were thus used

206. Ibid., p. 207.
207. Ibid., p. 203.
208. Romer, *Vertebrate Paleontology* (1933), p. 105.
209. Ibid.

only for the "occasional necessity of moving from one pool to another."[210]
Only a few amphibians had adapted to eating food on the way, thus becoming able to spend significant amounts of time on land, and their descendants had inherited the earth. Huxley noted that this episode was
"not the only time in the history of life that some device, evolved only to
meet emergencies, has been turned into an everyday necessity of fuller
living."

It is not my purpose here to evaluate the plausibility of this scenario.
A recent popular account by Michael Benton challenges it on the grounds
that insufficient evidence exists for a drying climate at the time and that
only the lure of a new food supply could explain the development of true
terrestrial locomotion.[211] Curiously, Benton attributes the thesis to Romer
in the 1950s—although we have seen that he was not the only person to
suggest it, even as early as the 1920s. A gap of three decades occurs between the theory's origins and our perception of where it 'belongs' in the
history of evolutionism. The misperception probably arises from the assumption that the theory fits more naturally into the era of the modern
synthetic version of Darwinism—it is no accident that Julian Huxley was
one of its early supporters. The claim that the amphibians moved onto
land in order to remain in the water is counterintuitive in a very Darwinian way. It portrays an important step in the history of life as a more or
less accidental by-product of an almost trivial process of adaptation. The
triumphant progressionism of the early twentieth century, still visible in
parts of Schuchert's account, contrasts very strongly with the more realistic viewpoint of modern Darwinism. Yet the fact that this theory of amphibian origins was actively discussed in the 1920s and 1930s suggests
that American and British paleontologists were starting to move toward a
more Darwinian way of thinking some time before the emergence of the
modern synthesis of the 1940s.

210. H. G. Wells, Julian Huxley, and G. P. Wells, *The Science of Life*, p. 441.
211. M. Benton, "Four Feet on the Ground," chapter 3, in S. J. Gould, ed., *The Book of
Life*, p. 79.

6

THE ORIGIN OF BIRDS AND MAMMALS

The amphibians had moved out onto the land but were still tied to the water by their mode of reproduction. The vertebrates could consolidate their hold on the land surface of the earth only by developing a method of reproducing that did not require the eggs to develop in water. The evolution of the amniote egg by the first reptiles could thus be represented as a major step in vertebrate evolution. Most evolutionists hailed the significance of this step, although they found comparatively little to say about the issue. In this case the transition was essentially physiological and left no trace in the fossil record. The topic thus offered much less scope for research by the paleontologists who were becoming increasingly active in the reconstruction of the history of life. It was simply assumed that the amphibians would have evolved a way of allowing their eggs to develop out of the water, perhaps in response to a drying up of their environment.

This absence of debate did not hold true for the next major step in vertebrate evolution: the emergence of warm-bloodedness. Here a physiological development was associated with morphological changes of a kind that provoked controversy among and between morphologists and paleontologists. It was clear from the start that the birds and mammals had evolved the ability to control their internal temperature independently. No one doubted that their origins would lie in very different reptile groups, each of which had evolved in parallel to a level where temperature control was possible. T. H. Huxley even suggested that the mammals had not descended from reptiles at all but had arisen directly from amphibians which had evolved the key mammalian advantages without ever passing through a reptilian phase. By any standard, the birds were a side issue in the story of vertebrate evolution as told from an anthropocentric perspective. But birds were familiar and attractive creatures, the subject of much interest by naturalists both professional and amateur, and their origin was thus debated widely in the evolutionary literature.

In the case of both the birds and the mammals, evolutionists came to disagree over which were the most 'primitive' members of the class. These debates were in part a product of the tension between morphology and paleontology. Huxley argued that the birds had evolved from running dinosaurs, which implied that the modern flightless birds are relics of the class's earliest stage of development. Others treated the flightless birds as degenerate and pointed out that the famous fossil intermediate *Archaeopteryx* did not support the idea of a running ancestry. In the case of the mammals, Huxley's dismissal of a reptilian stage in their ancestry was based on morphology and led to the fossil mammal-like reptiles of South Africa being ignored for several decades. Debates also erupted over whether the living monotremes and marsupials were relics of earlier stages in the evolution of the placental mammals.

Conflict between morphology and paleontology was not inevitable, however. When the significance of the mammal-like reptiles was eventually recognized, it was seen that they confirmed theories being worked out independently by morphologists such as Ernst Gaupp to explain the homologies between the bones of the reptile jaw and the mammalian ear. Morphologists and paleontologists then united in an effort to understand the functional changes that could have promoted such a strange transformation. Field studies also played a role in the debates. The discovery that the duck-billed platypus lays a reptilian type of egg was the first major challenge to Huxley's theory of mammalian origins, and played a significant role in removing the prejudices against the study of the mammal-like reptiles.

As in the case of amphibian origins, the possibility that the birds and mammals might be diphyletic or polyphyletic soon emerged. After all, if warm-bloodedness had been developed independently by birds and mammals, why not by several different kinds of birds and mammals? Some ornithologists tried to save Huxley's theory by supposing that the flightless birds had a separate reptilian origin from the carinate, or flying, birds. Morphologists and paleontologists eventually united around the idea that several different lines of reptile evolution had independently acquired mammalian characters. These theories posed the question of what kind of evolutionary pressures could have forced several groups to move in more or less the same direction. Some, at least, saw parallelism as evidence against Darwin's theory of natural selection.

In the early decades of the twentieth century a more functional approach to morphology and the reconstruction of fossil animals promoted stronger interest in the adaptive pressures that might have led to the acquisition of flight. It was no longer enough to look for mere morphological resemblances between reptiles and birds: paleontologists wanted to know why a particular kind of reptile had found it advantageous to develop wings. Adaptive scenarios of the kind we have already encountered in the study of amphibian origins became popular. The complex morphological and physiological transformations that created the mammals were less easily tackled by this approach. Even so, paleontologists became increasingly willing to look for adaptive pressures that might have triggered the evolutionary developments. In this case, perhaps a dramatic change in the physical environment had increased the pressure to evolve what were universally regarded as the more advanced characters of mammals.

FROM REPTILE TO BIRD

The transition from amphibians to reptiles confirmed the vertebrates' hold on the land. The development of the amniote egg, with a membrane to retain water, meant that reproduction could take place on dry land without the need for a periodic return to the aquatic environment. Presumably some amphibians had eventually acquired the ability to lay eggs of this kind. Haeckel noted the presence of fossils that were possibly reptilian in the Permian and of undoubted reptiles in the Triassic. The emergence of the reptiles was the vertebrates' contribution to the great turning point that marked the beginning of the Mesozoic.[1] The hypothetical 'protamnion' probably resembled the *Proterosaurus* of the Permian, but the stock had rapidly split into two, one branch leading to the dinosaurs, modern reptiles, and birds, and the other leading to the mammals.[2]

Most later authorities agreed that the reptiles had evolved from amphibian ancestors, although the date was soon pushed back to the Carboniferous. By the end of the Permian most of the various branches of the reptiles had already become differentiated. Following the argument pioneered by Barrell in the context of amphibian origins, it was assumed that the process had been driven by the increasing aridity of the climate.[3] Romer's 1933 textbook of vertebrate paleontology highlighted the significance of the reptilian process of reproduction. The amphibians were "nothing but a rather aberrant fish type," still chained to the water, and "the initiation of this new mode of development [the amniote egg] was an epochal event in vertebrate history."[4] The reptiles were more highly or-

1. Ernst Haeckel, *The History of Creation*, II, p. 220.
2. Ibid., II, p. 222.
3. For example, R. S. Lull, *Organic Evolution*, pp. 492–493.
4. A. S. Romer, *Vertebrate Paleontology* (1933), p. 120.

ganized in all respects, including their brain structure, with small cerebral hemispheres which indicated "rudiments of the higher mental centers."[5] Even so, the transition was a hard one for the paleontologist to trace. The new mode of reproduction was manifested only in the soft parts and thus could have left no fossil evidence. In modern reptiles and amphibians there are skeletal characteristics which are uniquely associated with the two modes of reproduction. The reptile has one occipital condyle in its skull, the amphibian two. But as Romer noted, primitive reptiles and amphibians were so similar that it was "almost impossible to tell when we have crossed the boundary between the two classes."[6] He classified the fossil *Seymouria* from the lower Permian of Texas as a primitive reptile but accepted that it still had a number of amphibian features.

For all its significance, then, the amphibian-reptile transition attracted comparatively little attention from paleontologists. Far more controversial were the next dramatic steps in the history of vertebrate life, the emergence of the birds and the mammals. Both of these classes had developed the property of warm-bloodedness, although the differences between them were so great that there was never any doubt that the transition had been achieved independently in two very different lines of evolution originating in the reptiles. The basic similarity between birds and reptiles had long been accepted, but in this case the morphological and physiological transformations were associated with the conquest of yet another medium, the air. Evolutionists began by looking for the best morphological and paleontological evidence linking the two clases, but in the end they were forced to confront another kind of question. As with the amphibians' move onto the land, it was necessary to ask what kind of adaptive pressures could have driven the birds to fly.

The two questions were not unrelated. Huxley became the leading advocate of the view that the Ratitae, or nonflying birds (ostriches, etc.), were the primitive form from which the Carinatae, or flying birds, had evolved. The Ratitae were an offshoot of the running dinosaurs, whose hindquarters were indeed remarkably birdlike. Huxley, as a typical morphologist, said little about the adaptive pressures that would have encouraged running birds to fly—it was left for a later generation to create the imaginary picture of them flapping primitive wings to increase their running speed as a prelude to true flight. Other evolutionists took the more obvious view that the flightless birds are degenerate descendants of flying ancestors. If the birds evolved their wings and feathers directly as a consequence of flying, it was natural to link them back to a more primitive reptile which might have taken to climbing in trees. But in this case, the

5. Ibid.
6. Ibid., p. 121.

similarities between flightless birds and dinosaurs would have to be explained as convergences or parallelisms. The only alternative would be to make the birds diphyletic, the Ratitae evolving from the dinosaurs, while the flying birds had emerged quite separatly from a different reptile stock. If so, then the other bird-defining characters (beak, feathers, etc.) would have to be the results of parallel evolution.

The whole question was bound up with the growing influence of paleontology. In this area, at least, the evolutionists had a few good fossils from the very start of the debate. The story of the fossil *Archaeopteryx*— the reptile-bird intermediate from the Jurassic limestone of Solnhofen— has been told over and over gain in the context of the debates between Huxley and his arch-rival Richard Owen (fig. 6.1). But studies by Mario Di Gregorio and Adrian Desmond have stressed that the classic image of Huxley, the triumphant evolutionist, seizing upon the fossil's intermediate character to checkmate the anti-evolutionist Owen are a vast oversimplification.[7] Huxley's views on bird origins were skewed by his own strange ideas about the timing of major evolutionary events, and *Archaeopteryx* was by no means crucial to his case. Owen, far from being an anti-evolutionist, proposed a rival theory in which birds were derived from pterosaurs. As Stephen Jay Gould has shown, the classic oversimplification of the Owen-Huxley confrontation has even misled that modern anti-Darwinian Fred Hoyle into claiming that the feathers on the fossils of *Archaeopteryx* are a forgery, engraved on the stone as part of a plot by Owen to discredit Huxley.[8]

The story of these early debates need only be sketched in briefly here. The first skeleton of *Archaeopteryx*, showing feathers but minus the head, was found in 1861 and acquired by Owen for the Natural History Museum in London. Owen described it in 1864, dismissing its reptile affinities and proclaiming it a bird, albeit one with an "embryonal and transitory" structure that bore a "closer adhesion to the general vertebrate type" than modern birds.[9] The discovery was widely reported by the periodical press in the context of the Darwinian debate.[10] Huxley was already a bitter foe of Owen, although, as Desmond has stressed, he was remarkably slow to begin using fossils as ammunition in the evolutionary debate. Huxley had earlier committed himself to the idea of the "persistence of type"—the

7. See M. Di Gregorio, *T. H. Huxley's Place in Natural Science*, pp. 76–96; A. Desmond, *Archetypes and Ancestors*, pp. 124–146; also Desmond, *The Hot-Blooded Dinosaurs*, chap. 6.

8. S. J. Gould, "The Fossil Fraud That Never Was;" see Fred Hoyle, *Archaeopteryx: The Primordial Bird.*

9. Richard Owen, "On the Archaeopteryx of von Meyer," p. 46.

10. See A. Ellegard, *Darwin and the General Reader*, pp. 230–231.

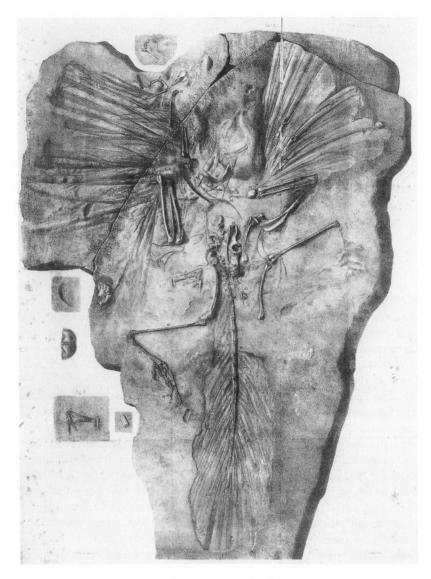

FIGURE 6.1   The first specimen of *Archaeopteryx*. From Richard Owen, "On the *Archae-opteryx* of von Meyer," plate 1.

lack of fundamental change in organic classes over vast periods of geo-logical time—and when translated into evolutionary terms, this implied that classes had probably originated long before their first fossil remains were found. He was thus unwilling to admit that *Archaeopteryx* could be the genuine transitional form leading to the birds, even if it shared the characters of the two classes.

It was his reading of Haeckel's *Generelle Morphologie* that inspired Huxley to begin the search for fossil intermediates. The reptile-bird link was certainly one of the first he addressed, but his interest in the apparently birdlike character of some dinosaurs led him immediately to a theory in which *Archaeopteryx* was sidelined. In January 1868 he wrote as follows to Haeckel:

> In scientific work the main thing just now about which I am engaged is a revision of the Dinosauria, with an eye to the 'Descendenz Theorie.' The road from the Reptiles to Birds is by way of Dinosauria to the Ratitae. The bird 'phylum' was struthious [ostrichlike], and wings grew out of rudimentary forelimbs.
>
> You see that among other things I have been reading Haeckel's *Morphologie.*[11]

In the same year Huxley published his own account of *Archaeopteryx,* ridiculing Owen for describing the fossil upside-down, but insisting that it was a bird—and would still be a bird even if a head with teeth were found (a prediction that would be fulfilled in 1877). Huxley was disposed to think that *Archaeopteryx* was "more remote from the boundary-line between birds and reptiles than some living *Ratitae* are."[12] In the same year Huxley published his classic paper, "On the Animals Which Are Most Nearly Intermediate between Birds and Reptiles," based on a popular lecture at the Royal Institution. Here he certainly used the fact that *Archaeopteryx* was more reptilian than any living bird to confound those who claimed that the fossil record was devoid of intermediates.[13] But he passed straight on to a different topic: the birdlike feet and limbs of some dinosaurs. Huxley claimed that many large dinosaurs, including *Iguanodon,* walked at least part of the time on their hind legs. (Owen had originally reconstructed *Iguanodon* as a quadruped, and it was not until better specimens were unearthed, especially Louis Dollo's Bernissart fossils, that Huxley's prediction was confirmed.) He drew attention to a small dinosaur, *Compsognathus longipes,* whose hind limbs were so birdlike that it could be counted as a "missing link" between reptiles and birds (fig. 6.2).[14] Huxley concluded with a warning that the known fossils were merely il-

---

11. T. H. Huxley to Haeckel, 21 January 1868; see L. Huxley, ed., *The Life and Letters of Thomas Henry Huxley,* I, p. 437.

12. Huxley, "Remarks upon Archaeopteryx lithographica," p. 345.

13. Huxley, "On the Animals Which Are Most Nearly Intermediate between Birds and Reptiles," p. 305.

14. Ibid., p. 311.

FIGURE 6.2   Restoration of *Compsognathus longipes*. From Gerhard Heilmann, *The Origin of Birds* (London, 1926), p. 167.

lustrative of what the much earlier transitional forms would have looked like, but insisted that "surely there is nothing very wild or illegitimate in the hypothesis that the *phylum* of the class *Aves* has its roots in the Dinosaurian reptiles; that these, passing through a series of such modifications as are exhibited in one of their phases by *Compsognathus*, have given rise to the *Ratitae*; while the *Carinatae* are still further modifications and differentiations of these last." [15]

Huxley explicitly ruled out a link between birds and pterosaurs despite their many similarities. The flying reptiles were equivalent to bats in the mammalia: their wings were built on entirely different principles than those of birds, and they had acquired the power of flight independently.[16] In a paper on the classification of dinosaurs published in 1870, Huxley insisted that those characters in which the birds did resemble pterosaurs had relation "to physiological action and not to affinity." [17] In other words, they were convergences. The same year saw the appearance of another paper reinforcing the dinosaur-bird link. Here Huxley conceded that Cope had advocated the same connection in a short note to the

15. Ibid., p. 312.

16. Ibid., p. 311.

17. Huxley, "On the Classification of the Dinosauria," p. 494.

Academy of Natural Sciences in Philadephia in 1867.[18] He brought forward further evidence from John Phillips based on dinosaur fossils in the Oxford museum. In the discussion following the presentation of this paper at the Geological Society of London, Harry Govier Seeley challenged several of Huxley's points. Seeley was making a study of pterosaur fossils, and was already moving toward a view which saw this group as distinct from the other reptiles. He also dismissed the similarities between the hind limbs of dinosaurs and birds as due to the similar functions they performed in running, not to genetic affinity—a point that would become central for later critics of Huxley's theory.[19]

In 1876 Huxley gave a series of lectures on evolution in America. Here he introduced—in addition to *Archaeopteryx*—the toothed birds described from the Cretaceous rocks of the American midwest by Othniel Charles Marsh.[20] Huxley insisted that these were 'intercalary types" rather than truly transitional forms: they blurred the distinction between classes without necessarily marking the actual course of evolution. The true line of descent was from the birdlike dinosaurs, and Huxley suggested that the fossil tracks from the Triassic of Massachusetts, long attributed to birds, may have been made by dinosaurs.[21] The legacy of the concept of the persistence of type was still apparent, however: Huxley thought that the actual transition to birds had probably taken place in the Palaeozoic.[22]

The toothed birds, or "Odontornithes" as Marsh called them, had been discovered in Kansas in 1872. Like *Archaeopteryx,* they seemed to confirm that in the earlier part of their existence, the birds had still retained reptilian characteristics such as teeth. There were two very different types, suggesting that the birds were already well established by the Cretaceous. *Hesperornis regalis* was a waterbird something like a grebe; the other, *Ichthyornis dispar* was a more normal carinate bird (its toothed head has since turned out to belong to a marine reptile). Marsh and Huxley got on well together during the latter's visit to America, and Marsh endorsed Huxley's theory of bird origins in his monograph *Odontornithes* of 1880. The birds were an ancient class, he claimed, dating back at least to the Triassic, although the birds of that epoch would have been consid-

18. Huxley, "Further Evidence of the Affinity between the Dinosaurian Reptiles and Birds," p. 480; see E. D. Cope, "An Account of the Extinct Reptiles Which Approached the Birds."

19. H. G. Seeley, discussion following Huxley, "Further Evidence," p. 486.

20. Huxley, "Lectures on Evolution," in Huxley, *Collected Essays,* IV, *Science and Hebrew Tradition,* pp. 46–138; see also pp. 93–113. The lectures were published separately as Huxley, *American Addresses.*

21. Ibid., p. 109.

22. Ibid., p. 110.

erably more reptilian than any so far discovered.[23] Marsh doubted that the reptilian ancestor would have been a dinosaur—it was more likley a generalized sauropsid from which both birds and dinosaurs had diverged. But the early birds had certainly been flightless, even when they had developed feathers and warm blood. The modern Ratitae were the surviving remnants of this early form.[24] Marsh thought that the flying birds had evolved when the smaller of these early types had taken to an arboreal life and had begun jumping from branch to branch.

Marsh's assistant at the Yale Museum, Georg Baur, supported his view that the birds had evolved from a predinosaurian reptile via the Ratitae.[25] A study of the *Archaeopteryx* fossils by Wilhelm Dames in 1884 linked this fossil to the theory as an early example of the flying birds branching off from the original ratite stock.[26] Much later on, Lankester's popular survey *Extinct Animals* expounded Huxley's views on the bird-like character of the hind limbs in *Iguanodon* and proposed a somewhat more fanciful explanation of how flight had evolved: "It is now certain that reptiles similar to the Iguanodon were the stock from which birds have been derived, the front limb having become probably first a swimming flipper or paddle, and then later an organ for beating the air and raising the creature out of the water for a brief flight."[27] A more orthodox version of Huxley's theory was defended in the 1920s by Percy Lowe, who had made detailed studies of the ostrich and its allies. Lowe thought that the Struthiones (Ratitae) were a natural group, the descendants of a stage in bird evolution before the acquisition of flight. Their forelimb was a degeneration of the primitive, nonvolant sauropsid forelimb, not of the carinate wing.[28] Lowe believed that *Archaeopteryx* was an evolutionary "blind alley failure" which had developed only gliding flight.[29] He observed that the feathers of an adult ostrich resemble those of the chick in normal birds. In a later paper on the same theme, this led him to a literary expression of the recapitulation theory: The ostriches "are the Peter Pans of the avian world. They have never grown up."[30]

Huxley's theory thus survived into the twentieth century, yet from the

---

23. O. C. Marsh, *Odontornithes,* p. 188.

24. Ibid., p. 189.

25. G. Baur, "Notes on the Pelvis in Birds and Dinosaurs."

26. W. Dames, "Ueber Archaeopteryx," pp. 78–79 (of monograph), pp. 194–195 (of volume).

27. E. Ray Lankester, *Extinct Animals,* p. 202.

28. P. Lowe, "Studies and Observations on the Phylogeny of the Ostrich and Its Allies," p. 244.

29. Ibid., p. 245.

30. Lowe, "On the Relationship of the Struthiones to the Dinosaurs," p. 420.

start there were some who thought that the Ratitae were degenerate rather than primitive birds. Thus even Haeckel, while accepting that *Compsognathus* probably indicated the kind of reptile stock from which the birds arose, had doubts about the position of the ostrichlike birds in Huxley's theory. He argued that they were probably derived from the carinate birds by degeneration of the wings, while merely noting as a possibility Huxley's alternative view that they were the nearest relatives of the dinosaurs.[31] In a study of the second *Archaeopteryx* fossil, discovered in 1877, Carl Vogt of Geneva suggested that the similarities between the flightless birds and the dinosaurs were independent adaptations to an upright position. He conceded that the ratites might have evolved from the dinosaurs, while the carinate birds had descended directly from *Archaeopteryx*.[32] Having been exiled from Germany for his radical political views, Vogt also criticized Kaiser Wilhelm I for his lack of interest in the fossil (it was subsequently bought for the Berlin museum by a German industrialist). Vogt was a strong advocate of parallelism, having also suggested that the human races had independently evolved from separate great ape ancestors.

A more serious critique of Huxley's position came in the massive survey of bird morphology and systematics by Max Fürbringer. He insisted that the Ratitae were not a distinct subclass, but a collection of degenerate species which had originated in several different groups of flying birds.[33] This undermined the whole logic of the claim that the flightless birds were the relic of a primitive stage in the class's evolution. Fürbringer was supported by Hans Gadow and Alfred Newton.[34] Gadow agreed that the resemblances between the flightless birds were analogies produced in several groups which had independently given up flying. Newton—who regretted that Fürbringer's work was not better known by English-speaking scientists—accepted that the Ratitae were not a primitive group, although he was unwilling to endorse the claim that they were a collection of independently evolved flightless forms.

Huxley's theory of bird origins was also attacked by his great rival, Richard Owen, who had gained part of his reputation for a study of the extinct flightless Moa of New Zealand.[35] But in addition to dismissing the

31. Haeckel, *History of Creation*, II, pp. 229–230.

32. Carl Vogt, "L'Archaeopteryx macroura." On Vogt's theory of human origins, see Bowler, *Theories of Human Evolution*, pp. 132–133.

33. Max Fürbringer, *Untersuchungen zur Morphologie und Systematik der Vögel*, p. 1624 and p. 1630.

34. Hans Gadow, "The Morphology of Birds;" Alfred Newton, *A Dictionary of Birds*, pp. 100–101.

35. On Owen's work on the Moa, see Jacob Gruber, "From Myth to Reality," and Nicolaas Rupke, *Richard Owen*, pp. 123–129.

Ratitae as an evolutionary sideline, Owen proposed an alternative theory to explain the origin of birds. Huxley had argued that the pterosaurs were unrelated to the birds, but Owen was inclined to see a closer link. In the first volume of his *On the Anatomy of Vertebrates* of 1866, he wrote, "An important link, the *Pterosauria*, or flying reptiles, with wings and air sacs . . . , more closely connecting birds with the actual remnant of the reptilian class, has passed away."[36] It is not clear at this point that Owen saw the pterosaurs as a phylogenetic link between reptiles and birds (as opposed to what Huxley called an intercalated type). His detailed descriptions of British pterosaurs stress their strong crocodilian affinities and explicitly rule out a connection with the birds.[37] Yet in a monograph on the dinosaur *Osmosaurus*, Owen went out of his way to establish an evolutionary link between pterosaurs and birds. He dismissed the allegations of W. H. Flower, one of Huxley's supporters, that he still thought the origin of species to be supernatural. To say that we do not know the actual mechanism of change is not to deny that the change has taken place.[38]

Having made it clear that he accepted the basic idea of evolution, Owen nevertheless challenged Huxley's account of the origin of birds. He found it difficult to accept the amount of modification required "to evolve an ostrich out of an Iguanodon."[39] To explain the flightless birds, he said, "I own to finding more help from the Lamarckian hypothesis than the Darwinian one."[40] If we imagine an island where food was plentiful and predators absent, birds would find it unnecessary to fly and their wings would degenerate, just as the Dodo had degenerated from something like a dove. "But whence the Dove?" asked Owen: if the birds did not evolve from dinosaurs via the ratites, where did they come from? The answer lay with the pterosaurs: "Every bone in the Bird was antecedently present in the framework of the Pterodactyle; the resemblance of that portion directly subservient to flight is closer in the naked one to that in the feathered flyer than it is to the fore-limb of the terrestrial or aquatic reptile."[41]

The detailed structural similarities between birds and pterosaurs could be used to embarrass the Darwinians even when it was recognized that their different wing structures made a direct relationship unlikely. In his *Genesis of Species* of 1870, Mivart brought forward many cases of

36. Owen, *On the Anatomy of Vertebrates*, I, p. 6.

37. Owen, *A Monograph of the Fossil Reptilia of the Liassic Formations*, p. 73 and p. 76; also Owen, *A Monograph on the Fossil Reptilia of the Mesozoic Formations*, p. 14.

38. Owen, "A Monograph of a Fossil Dinosaur (Osmosaurus armatus)," in Owen, *A Monograph on the Fossil Reptilia of the Mesozoic Formations*, pp. 45–93; see pp. 87–88.

39. Ibid., p. 88.

40. Ibid., p. 89.

41. Ibid., p. 91.

parallelism to argue that variation was predetermined rather than random, as the selection theory assumed. He seized upon the bird-pterosaur relationship as a classic illustration of the point he wished to make. Some bones were so similar in structure that "it was hardly possible for the Darwinian not to regard the resemblance as due to community of origin"— and Seeley had now shown that the brains too were almost identical in structure.[42] Yet Huxley and his supporters had denied that the characters were shared as a result of common descent and had claimed that they were developed independently in response to similar physiological needs. This kind of independent invention was a "coincidence of the highest improbability" that was incompatible with the selection theory.[43] The claim that the birds were diphyletic, the ratites evolving from the dinosaurs and the carinates from the pterosaurs, was equally damaging because it still implied a degree of parallel evolution that could only be explained by the assumption of predetermined trends in variation. Ten years later Mivart explicitly supported the diphyletic theory of bird origins, noting Vogt's endorsement of the idea on the basis of the resemblances he had found between *Archaeopteryx* and the pterosaurs.[44]

The strongest exponent of pterosaur-bird affinities was Harry Govier Seeley, who began studying the British pterosaurs under the aged geologist Adam Sedgwick at Cambridge. Sedgwick had been one of Darwin's earliest scientific tutors, but his opposition to evolutionism was legendary. Seeley's own opposition to Darwinism was so pronounced that—unlike Owen and Mivart—he was indifferent to the very idea of evolution. As Adrian Desmond has noted, Seeley's world view was Platonic rather than materialistic, and he delighted in giving almost geometrical representations of organic affinities that defied translation into any naturalistic genealogy.[45] Seeley developed strong links with Owen later on when he began to work on the mammal-like reptiles (see below). But initially he was violently critical of Owen over the latter's early efforts to depict pterosaurs as purely reptilian creatures with no avian affinities.[46] As Mivart noted, it was Seeley who discovered that the brains of pterosaurs were very similar to those of birds. He also insisted that the hollow bones of pterosaurs probably helped with the animals' respiration (as in the case of birds) and were thus a sign of warm-bloodedness.[47] In an 1870 monograph on what

42. St. G. Mivart, *On the Genesis of Species*, pp. 83–84.

43. Ibid., pp. 85–86.

44. Mivart, "A Popular Account of Chamaeleons," p. 338.

45. Desmond, *Archetypes and Ancestors*, pp. 186–193; for Seeley's geometric diagrams, see, for example, his *Dragons of the Air*, pp. 190–191.

46. Seeley, "Remarks on Prof. Owen's Monograph on Dimorphodon."

47. Ibid., pp. 149–150.

he called the "Ornithosauria," Seeley tried to establish the pterosaurs as a new vertebrate subclass allied to birds.[48]

Ten years later Seeley commented on Vogt's account of *Archaeopteryx* and the idea that the birds might be diphyletic. The whole network of relationships raised the question of whether or not identical functions could lead to the development of identical structures in two distinct but allied groups.[49] Seeley himself saw *Archaeopteryx* as a true bird and again stressed the avian characteristics of the pterosaurs. *Archaeopteryx* was not a modified pterosaur, but the two forms had enough in common to make the pterosaurs a better clue to *Archaeopteryx*'s affinities than the dinosaurs.[50] In a popular book, *Dragons of the Air,* published as late as 1901, Seeley still insisted on the complexity of the relationship between reptiles, pterosaurs, birds, and mammals. He gave a diagram implying that pterosaurs had affinities with all of the others.[51] Pterosaurs were not derived from birds, but the two formed parallel groups that might be regarded as "ancient divergent forks of the same branch of animal life."[52] Neither could be descended directly from reptiles, however.[53] In the end, the reader would have been quite confused as to the exact nature of the the evolutionary relationships involved, which was probably Seeley's intention.

The possibility of a genealogical relationship between pterosaurs and birds was thus maintained by those biologists with strong anti-Darwinian feelings. Most others were prepared to accept Huxley's suggestion that the resemblances were merely convergences produced in two quite different groups that were adapting to flight. But it was still possible to argue that the resemblances upon which Huxley based his own case were susceptible to the same nonevolutionary explanation. A popular and well-illustrated account of *Extinct Monsters* by the Rev. H. N. Hutchinson stressed Owen's scepticism about the dinosaur-bird link and noted that the resemblances between bipedal dinosaurs and running birds might be convergences: "Similar habits, if continued for age after age, must tend to produce somewhat similar appearances, at least externally."[54] This line of thought led a number of early twentieth-century evolutonists to suggest that the ancestor of the birds might have been an earlier, less specialized reptile. Without denying that the dinosaurs were closely related to the

48. Seeley, *Ornithosauria.*

49. Seeley, "Professor Carl Vogt on the Archaeopteryx," pp. 306–307.

50. Ibid., p. 308.

51. Seeley, *Dragons of the Air,* p. 191.

52. Ibid., p. 223.

53. Ibid., p. 226.

54. H. N. Hutchinson, *Extinct Monsters,* p. 127.

birds, they argued that they possessed specializations that would rule them out as the direct ancestors.

TAKING TO THE AIR

Doubts about the ancestral status of the dinosaurs were linked to another problem with Huxley's theory of bird origins. It had been founded upon purely morphological resemblances, paying little attention to the adaptive pressures that might have led a running dinosaur to develop feathers and wings. Of what use to an ostrichlike bird were its rudimentary wings, and what mode of behavior could have led the wings to become sufficiently developed to let the animal fly? The paleontologists of the next generation were far more concerned about such questions, and the new century saw a flurry of activity centered on the problem of the origin of flight. No one took Lankester's suggestion of a preliminary aquatic phase seriously, and the search was on for a more realistic adaptive scenario.

The eccentric Hungarian paleontologist Baron Francis Nopsca made an attempt to put adaptive flesh on the bare bones of Huxley's morphological relationships in his article "Ideas on the Origin of Flight" of 1907. Nopsca argued that birds, bats, and pterosaurs had all evolved flight in different ways. Only the birds had evolved from a bipedal as opposed to a quadrupedal ancestor, and the long tail of *Archaeopteryx* indicated that it was a running rather than a perching creature.[55] It had also been a poor flyer, indicative of the earliest phase in the evolution of flight by a feathered, running dinosaur. In Nopsca's theory, flight evolved as a means of running more quickly over the ground: "Birds originated from bipedal, long-tailed cursorial reptiles which during running oared along in the air by flapping their free anterior extremities."[56] He appealed to the fossil record to show that the Ratitae had been very common in the Eocene; they were descendants of the Mesozoic ancestors of the whole bird class, now very much in decline. The carinate birds had evolved when the descendants of *Archaeopteryx* had moved into the trees to escape predators on the ground, and had increased their originally limited powers of flight.

Few evolutionists found Nopsca's scenario plausible. Huxley's theory deriving birds from running dinosaurs was taken seriously in a period when relationships were judged primarily in terms of morphology, but a later generation more concerned with the 'why' of evolution found it difficult to see why wings should have developed merely to aid running— even assuming that feathers had already evolved for some other purpose,

55. Francis Nopsca, "Ideas on the Origin of Flight," pp. 230–232. On Nopsca's life, see Desmond, *Hot-Blooded Dinosaurs*, pp. 57–58.
56. Nopsca, ibid., p. 234.

such as insulation. Nopsca had to postulate a later move into the trees to perfect the power of flight, and to most biologists, it seemed obvious that the birds had evolved from arboreal rather than ground-dwelling ancestors. This in turn meant that the resemblances between the ratites and the bipedal dinosaurs upon which Huxley's theory was based had to be the result of convergence. As Owen and Fürbringer had argued, the Ratitae were not primitive birds, but were the descendants of flying ancestors which had returned to the ground and, in developing their running abilities, had acquired hind limbs resembling those of the bipedal dinosaurs. Birds were related to dinosaurs, but were not descended from them. Both had evolved, as shown in the following diagram, from smaller, more generalized reptiles, some of which had acquired the ability to climb into the trees.

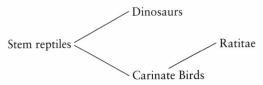

This more distant relationship between birds and dinosaurs was outlined by Osborn in 1900. Osborn described Huxley's theory and the arguments that had been brought forward against it. As a strong exponent of the power of homoplasy (convergence) to produce similar structures in different groups adapting to the same way of life, Osborn was well aware of the strength of the claim that the dinosaur-ratite resemblances might be convergences.[57] He was particularly impressed by Ernst Mehnert's study of the pelvis, which suggested that the resemblances between birds and dinosaurs were not true homologies.[58] On the other hand, he hinted that "where there is so much smoke there may be some fire": there *was* a relationship between birds and dinosaurs in the sense that both were descended from a common stem.[59] The common ancestor was probably bipedal, but the bird branch had separated off when some members of the group had adopted an arboreal lifestyle. Osborn's later survey, *The Origin and Evolution of Life,* made it clear that jumping in trees was an essential starting point for any explanation of the origin of flight.[60]

57. H. F. Osborn, "Reconsideration of the Evidence for a Common Dinosaur-Avian Stem," pp. 781–783. On modern debates concerning the dinosaur-bird link, see Desmond, *The Hot-Blooded Dinosaurs,* and Stephen Jay Gould, "The Telltale Wishbone," in Gould, *The Panda's Thumb,* pp. 267–277.

58. See Ernst Mehnert, "Untersuchungen über die Entwicklung des Os Pelvis der Vögel."

59. Osborn, "Reconsideration," p. 789.

60. Osborn, *The Origin and Evolution of Life,* pp. 227–230.

Similar views were developed by the Austrian paleontologist Othenio Abel in his 1911 study *Die Vorfahren der Vögel und ihre Lebensweise*. Abel considered the structure of the birds' feet to be an adaptation to the arboreal way of life. But he pushed the arboreal phase further back, to include the common ancestors from which both the birds and the dinosaurs had sprung. "The birds and the Theropoda are descended from a common arboreal stem group with climbing feet, from which the Theropoda, for their part, at an early period returned to a terrestrial mode of life, while for the birds, which remained arboreal, the return to terrestrial life did not take place til a long time after the faculty of flight was acquired."[61] The divergence had taken place very early, probably early in the Triassic.

The search for a common ancestor of birds and dinosaurs was taken up by Robert Broom in a paper dealing with a group of fossil reptiles known as the pseudosuchians. He had worked with a South African specimen which he called *Euparkeria*, but he noted its affinities with the birdlike *Ornithosuchus* from the Triassic Elgin sandstones, several species of which had already been described by E. T. Newton and G. A. Boulenger.[62] Despite minor disagreements, all three accepted that these were unspecialized reptiles that were probably ancestral to several later groups, including dinosaurs, pterosaurs, and birds. Broom had already hinted that birds might be descended from a group ancestral to the theropod dinosaurs, and now he saw the pseudosuchians as just such a group. In those characters where the dinosaurs were too specialized to be the ancestors of birds, the pseudosuchians were still primitive enough.[63] Broom was convinced that these early reptiles had acquired a hopping gait similar to that of many modern birds. He thought it was easy to imagine a hopping animal taking to the trees, where its forelimbs would then have become adapted to flight.

While the paleontologists searched for a fossil reptile adaptible enough to take to the trees, ornithologists approached the origins of flight from a different direction. William Pycraft of the zoology department at the Natural History Museum in London included two chapters on phylogeny in his *History of Birds* published in 1910. Some time earlier Pycraft had produced a popular account of *Archaeopteryx* in which he argued that it had been able to fly properly. The fact that the wing still had claws did not count against true flight: the claws were probably used only by the

---

61. O. Abel, "Der Vorfahren der Vögel," as translated by G. Heilmann, *The Origin of Birds*, p. 190. I have not seen the original source of the material translated.

62. R. Broom, "On the South African Pseudosuchian *Euparkeria;*" see E. T. Newton, "Reptiles from the Elgin Sandstone," and G. A. Boulenger, "On Reptilian Remains from the Trias of Elgin."

63. Broom, "On the South African Pseudosuchian," p. 631.

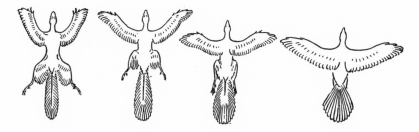

FIGURE 6.3   Evolutionary stages in a hypothetical four-winged bird (after Beebe). From H. F. Osborn, *The Origin and Evolution of Life* (New York, 1923), p. 228.

chick to climb back into the nest and were retained as useless vestiges by the adult.[64] In his later study Pycraft used *Archaeopteryx* to reconstruct a hypothetical "pro-aves" and its way of life. He believed that this ancestral form had taken to the trees, and had then substituted leaping for climbing, from which the development of the wing as a parachute followed naturally. This early bird probably had fimbriated scales (fringed with hairs) from which feathers evolved.[65] Pycraft attacked the view that the Ratitae were a natural group which might correspond to the ancestral birds: they were a diverse collection of forms which had each branched off separately from the main stem of flying birds and taken to the ground.[66]

The most innovative theory to explain the origin of flight came from C. William Beebe, curator of birds for the New York Zoological Society. Beebe saw *Archaeopteryx* as evidence that birds had evolved from small, arboreal, lizardlike creatures which had jumped and eventually parachuted from branch to branch. The amazing thing was how well-developed the wings of *Archaeopteryx* had already become, considering that it probably had only limited powers of flight.[67] The problem was to explain the intermediate stage between parachuting and flying. Beebe's solution was to suggest a "tetrapteryx," or four-winged stage: there was embryological evidence suggesting that the earliest birds had well-developed feathers on their hind as well as their forelimbs (fig. 6.3). A very young dove still shows evidence of this stage in quills sprouting from its hind legs (Beebe supplied photographic evidence).[68] Thus "for some unknown reason, Nature makes each squab pass through the Tetrapteryx stage. . . . No fossil bird of the ages prior to Archaeopteryx may come to light, but the memory of the Tetrapteryx stage lingers in every dove-

64. W. P. Pycraft, "The Wing of Archaeopteryx."
65. Pycraft, *A History of Birds*, pp. 38–39.
66. Ibid., pp. 41–43 and the phylogenetic tree, p. 57.
67. C. W. Beebe, "A Tetrapteryx Stage in the Ancestry of Birds," pp. 40–41.
68. Ibid., pp. 42–43.

cote."[69] Supported by four rather than two wings, the early birds had found it easier to develop the power of flight. *Archaeopteryx* showed a later stage, when the forelimbs were already taking over as the main wings.

In 1916 William King Gregory surveyed "Theories of the Origin of Birds" in a paper which drew particular attention to Beebe's idea. Gregory records that every year, the students in Columbia University's course on vertebrate evolution were required to debate the various theories, each being assigned the task of defending a particular hypothesis.[70] His own suggestions reflected the consensus which had emerged from these discussions. Broom's pseudosuchians represented the active reptiles whose incipient adaptation to bipedalism had paved the way for life in the trees. The appearance of feathers was crucial for flight: they had probably evolved first as insulation, and perhaps as protection from falls.[71] As these almost bipedal reptiles took to the trees, they began to perch on branches and to jump from branch to branch in pursuit of prey. They extended their jumps by making a vigorous flap with their forelimbs on takeoff—from the start flight was a more active mode of locomotion than mere gliding or parachuting. Feathers were naturally adapted to extend the wing area and increase the power of flight. Gregory was convinced that the ratites were the descendants of flying birds which had readjusted to a life on the ground.

Beebe noted important work being done by Gerhard Heilmann on the origin of birds, although this was little known because it was published in Danish. After considerable difficulty, Heilmann eventually gained the support of Arthur Smith Woodward, who encouraged a London publisher to issue his *Origin of Birds* in 1926.[72] The book has since become a classic, owing to its beautiful illustrations and the striking accounts it provides of the lifestyles of extinct animals. Heilmann stressed the reptilian affinities of *Archaeopteryx,* although his frontispiece showed a male with brightly colored plumage very much like a modern bird. He noted Pycraft's argument that the claws of *Archaeopteryx,* like those of the young of the modern South American Hoatzin, were used to climb around in trees even though the adults may have been capable of true flight.[73] But he was adamant that the birds could not have evolved directly from dinosaurs.

69. Ibid., p. 51.

70. W. K. Gregory, "Theories of the Origin of Birds," p. 36.

71. Ibid., p. 37.

72. See Desmond, *Hot-Blooded Dinosaurs,* p. 160, citing a letter from Heilmann to Smith Woodward, 10 October 1925, in the Smith Woodward papers at the D. M. S. Watson Library, University College, London.

73. G. Heilmann, *The Origin of Birds,* pp. 102–107.

Heilmann began from Dollo's law of the irreversibility of evolution. Any reptilian group which had already lost a bone possessed by the birds could not be the ancestor of the class, because the same structure could not be re-evolved. The loss of the clavicle in several reptiles, including the dinosaurs, thus ruled them out as potential ancestors.[74] He was particularly scathing about Lankester's assertion that *Iguanodon* represented the stock from which the birds evolved. It was wrong to call them "bird-like dinosaurs," because their resemblance was to the degenerate ratites, not to true birds.[75] The similarities were all due to convergence: "On going through the various points of resemblance to birds, moreover, we shall find that they denote less affinity to the Jurassic birds than an evolutionary tendency parallel to the development of some birds that have left off flying and in some respects are degenerate."[76] They were "analogous adaptations owing to the same bipedal mode of progression."[77]

Heilmann accepted Broom's pseudosuchians as ancestors of both the birds and the dinosaurs. They were probably partly bipedal, and had scales with ribs that were clearly the starting point for feathers.[78] Like Osborn and Abel, Heilmann pointed to the adoption of an arboreal mode of life as the critical turning point in the evolution of the birds; Nopsca's "running Pro-avis" was totally implausible.[79] Beebe's account of his tetrapteryx stage was quoted at length, but Heilmann professed himself dissatisfied with his own efforts to follow up the theory. There was no sign of a pelvic wing in any of the birds he had studied, and the legs of *Archaeopteryx* did not fit into the the sequence of development predicted by Beebe.[80] Nevertheless, the basic scenario of a transition from arboreal jumping to flying was sound. The hypothetical proavis had jumped further and further from tree to tree, gradually developing feathers so that the wings could serve as parachutes (fig. 6.4).

Heilmann was confident that he had solved the problem of the origin of birds, claiming that "the evidences now at our disposal are so manifold and conclusive that only a few minor particulars are still dubious."[81] Yet several popular surveys published in the early twentieth century suggest that the debate between the rival theories was not considered settled. Lull's *Organic Evolution* cited both the dinosaur-ratite and the arboreal theories without commenting on their relative merits. Lull also pointed to a paper by F. A. Lucas in 1916 reviving the view that the ratite and carinate birds might have separate reptilian origins, the one from a terrestrial,

74. Ibid., pp. 138–142.
75. Ibid., pp. 147–148.
76. Ibid., p. 182.
77. Ibid., p. 183.

78. Ibid., p. 189.
79. Ibid., p. 190.
80. Ibid., p. 190.
81. Ibid., preface.

FIGURE 6.4 Restoration of hypothetical proavian. From Gerhard Heilmann, *The Origin of Birds* (London, 1926), p. 199.

the other from an arboreal ancestor.[82] H. G. Wells and Julian Huxley's *Science of Life* also described both theories, admitting that it was not yet clear how flight had originated.[83] Romer's *Vertebrate Paleontology* was similarly inconclusive on the topic, although the possibility of the ratites being primitive rather than degenerate was firmly ruled out.[84] In 1946 George Gaylord Simpson launched a strong attack on Lowe and those who still defended a dinosaur-bird link, claiming that all the resemblances were parallelisms and convergences.[85] In recent decades the tide has turned once again in favor of the dinosaur link, in part as a result of the controversy surrounding the suggestion that the dinosaurs themselves may already have been warm-blooded.[86] The question of whether flight developed for getting from branch to branch or for leaping up from the ground is still not completely resolved, although the arboreal theory seems still to have the edge. Edwin Colbert's survey of vertebrate evolution still cites both theories, but a recent popular account by Michael Benton—while admitting both possibilities—favors the arboreal origin of flight on the strength of *Archaeopteryx*'s evident ability to use its claws for climbing.[87]

MONOTREMES, MARSUPIALS, AND MAMMALS

No one in the post-Darwinian era doubted that birds and mammals had evolved warm-bloodedness independently. At the very least, birds and mammals had evolved from different branches of the reptile stock. But the emergence of the mammals was seen as a key step forward in the ascent of life toward humankind, while the birds were merely a side branch. Huxley argued on morphological grounds that the mammals were so different from any known reptile, living or fossil, that the class would have to be traced back directly to the amphibians. This view became increasingly untenable as the significance of the so-called mammal-like reptiles of South Africa became apparant. In the end, paleontology provided the evidence that linked the mammals back to the earliest reptiles via a branch that was unrelated to the dinosaurs and birds. We shall return to this episode in the final section of this chapter.

In the meantime, a number of problems faced those who sought to understand the relationship between the living mammals. Whatever the

---

82. Lull, *Organic Evolution,* pp. 532–537; see F. A. Lucas, "The Beginnings of Flight."

83. H. G. Wells, Julian Huxley, and G. P. Wells, *The Science of Life,* p. 455.

84. Romer, *Vertebrate Paleontology* (1933), pp. 211, and p. 215.

85. G. G. Simpson, "Fossil Penguins," pp. 92–95.

86. See Desmond, *The Hot-Blooded Dinosaurs.*

87. E. H. Colbert, *The Evolution of Vertebrates* (1980), pp. 184–185; M. Benton, chap. 4 in S. J. Gould, ed., *The Book of Life,* see pp. 144–147.

actual starting point, the modern placental mammals must have evolved from an animal which had laid eggs like a reptile. The monotremes of Australia, the duck-billed platypus and the echidna, appeared to be far more 'primitive' than any other mammal. Although there was some dispute at first, it was soon confirmed that they even laid eggs. Detailed information about the platypus's reproductive system was an important contribution made by a more systematic approach to field studies in hitherto unknown areas. But did this mean that the monotremes were, in effect, living fossils—only slightly modified descendants of the earliest ancestors of the higher mammals? The same possibility could be raised in the case of the marsupials, whose mode of reproduction involving the birth of very immature offspring also seemed less efficient than that of the placental mammals. Perhaps they too were only modified descendants of the next, intermediate stage in mammalian evolution. Haeckel encouraged the setting up of this almost linear evolutionary scale linking the three grades of mammalian evolution. Such a model made it seem obvious that the few mammalian fossils from the Mesozoic era, the age dominated by reptiles, should belong to the lower levels. The Mesozoic mammals would have been the original monotremes or marsupials.

Some had doubts about this neat linear arrangement, however, suspecting that it was but a hangover from the old idea of a 'chain of being' linking all living forms into a single hierarchy. The earliest mammals might well have passed through a stage when they did actually lay eggs, but it did not follow that the living monotremes must be the direct and only slightly modified descendants of these founding fathers of the mammalian class. A number of naturalists began to argue that the monotremes were so peculiar that they might have to be ranked as a separate class altogether, having evolved from a different reptilian stock than that which gave rise to the modern placentals. A good deal of parallel evolution would be required: the independent acquisition of several key mammalian characters (hair, warm-bloodedness, and the ability to secrete milk) by two independent lines of evolution. It was also argued that the marsupials, far from representing a stage in the evolution of the placentals, were a highly specialized offshoot from an early stage of mammalian evolution, so specialized that they threw no light on what that early stage would have looked like. Debates over the early phases of mammalian evolution had thus emerged even before the significance of the mammal-like reptiles became generally appreciated.

The first generation of evolutionary morphologists saw a sharp contrast between the very obvious reptile-bird affinities and the more elusive question of mammalian origins. The reptiles—living and fossil—and the birds were united as the Sauropsida, but the mammals clearly did not belong to this group. From what animal then, had the mammals originated?

Would the hypothetical, egg-laying promammal be classed as a reptile, perhaps belonging to a branch of that class as yet unrecognized in the fossil record? Or would it be an intermediate between the amphibians and the egg-laying mammals that would not even be counted as a reptile when it was discovered? The mammal-like reptiles were to resolve this problem conclusively by showing that the original stem-reptiles had given rise to two widely differing branches. One led to the 'ruling reptiles' of the Mesozoic, including the dinosaurs, and the birds. The other led to the mammal-like reptiles, which, before their eclipse by the rival Sauropsida, had given rise to the primitive mammals that had survived through the age of the ruling reptiles and had inherited the earth after their extinction.

In the 1860s, however, few had begun to appreciate the significance of the fossil mammal-like reptiles already being found in South Africa. On purely morphological grounds, Haeckel and Huxley both argued that the reptiles did not constitute a stage in the evolution of the mammals. The main line of evolution led straight from the amphibians to the mammals, and it was unlikely that the hypothetical intermediates would be included in the class Reptilia. The whole reptile-bird line was, in effect, a side branch leading away from the main line of evolution—however successful its members had been during the Age of Reptiles. The morphological arguments brought forward to sustain this position were so strong that the significance of the mammal-like reptiles was ignored by many evolutionists even after their potential status as transitional forms had been described in print. The growing status of paleontology in evolutionary studies toward the end of the nineteenth century is mirrored in the sudden explosion of interest in the mammal-like reptiles, despite the arguments of those morphologists who continued to defend Huxley's original position.

The lack of interest in the reptiles as potential ancestors of the mammals is evident from the rather evasive position taken up by Haeckel. In the main account of the origin of the mammals given in his *History of Creation,* Haeckel studiously avoided mentioning the reptiles. He certainly wished to emphasize the significance of the transition to the mammals, insisting that they had a privileged position as the head of the animal kingdom.[88] But the subsequent discussion leads off from the egg-laying monotremes, not from a hypothetical reptilian ancestry. An evolutionary tree shows a thin line leading back from the mammals to the amphibians, parallel to but distinct from that leading to the reptiles and birds (see fig. 9.1).[89] The list of evolutionary stages leading to the human species given later in the book does not include the reptiles. The fifteenth stage is the hypothetical Protamnia, from which both the reptiles and the mam-

88. Haeckel, *History of Creation,* II, p. 231.
89. Ibid., facing p. 222 (see p. 427 below).

mals diverged: the sixteenth stage is already the Promammalia.[90] The implication seems to be that nothing corresponding to a known reptile forms a stage in the evolution of the mammals.

In his *Evolution of Man*, Haeckel again insisted that the line leading to the mammals was quite distinct from that leading to the modern reptiles and birds. He did, however, concede that the two were connected at their root in the earliest reptiles, although they diverged completely soon afterwards.[91] His position resembled that eventually taken up by those advocating two main lines of reptilian evolution, one mammal-like and one birdlike. The famous evolutionary tree given in the *Evolution of Man* does not, however, give the same impression (see fig. 2.5). It shows the reptiles and birds as a side branch, while the main trunk runs directly from the amphibians to the earliest mammals.[92]

The most decisive statement of the view that the reptiles should not be included in the ancestry of the mammals came in Huxley's 1879 paper, "On the Character of the Pelvis in the Mammalia, and the Conclusions Respecting the Origin of Mammals Which May Be Based on Them." Here Huxley took issue with Gegenbaur on the interpretation of the pelvis, insisting that the reptilian modifications were opposite to those found in mammals. The fossil Dicynodontia had a backward extension of the subsacral part of the ilium characteristic of the mammals, but had no obturator fontanelle. Thus, argued Huxley, "it seems to be useless to attempt to seek among any known Sauropsida for the kind of pelvis which analogy leads us to expect among those vertebrated animals which immediately preceded the lowest known mammals."[93] The Promammalia would have had a pelvis intermediate between that of *Orthnithorhynchus* (the platypus) and that of a land tortoise, and this would have been the type from which all modifications of the Sauropsida and Mammalia had diverged. This type, he claimed, could be seen in the pelvis of the salamander. Hence, "these facts appear to me to point to the conclusion that the Mammalia have been connected with the Amphibia by some unknown 'pro-mammalian' group, and not by any of the known forms of Sauropsida."[94] There was other evidence pointing in the same direction, especially the fact that the main conduit of arterial blood from the heart is the right aortic arch in the Sauropsida, but the left in the Mammalia. The implication was that two lines of evolution had diverged from the heart of the Amphibia.

90. Ibid., pp. 289–290.
91. Haeckel, *The Evolution of Man*, II, pp. 138–139.
92. Ibid., plate XV, facing p. 188 (see p. 92 above).
93. Huxley, "On the Character of the Pelvis in the Mammalia," p. 350.
94. Ibid., p. 354.

Huxley concluded his article by suggesting that the successful establishment of the reptile-bird link should encourage the search for the unknown ancestors of the mammals in the fossil record. Yet, as Adrian Desmond has pointed out, the significance of the fossil mammal-like reptiles of South Africa was already being stressed by Richard Owen.[95] Huxley pointedly ignored Owen's suggestions that these fossils provided evidence of the missing line of evolution leading to the mammals. Desmond suggests that Huxley's refusal to follow up this lead reflects his antagonism to Owen's almost idealist philosophy of evolutionism, in which the history of life has been governed by divinely implanted trends. The problem was that the South African fossils were being described as *reptiles,* not as members of a hitherto unknown transitional type. Two lines of evolution leading from the amphibians had independently acquired the characters typical of reptiles, implying a degree of parallelism that was favorable to Owen's anti-Darwinian view of evolution. But Huxley was not a narrow-minded Darwinian and was quite willing to accept the view that evolution could be led in certain directions by predetermined trends affecting variation.[96] As we shall see below, he insisted that there were parallel lines of evolution running through the main stages of mammalian evolution. One is left to wonder if Huxley's refusal to ackowledge the mammal-like reptiles was merely a product of his hostility to Owen: it would simply have been too galling for him to admit that his great rival had solved this major problem in phylogeny. Having committed himself to the claim that the ancestors of the mammals were not reptiles, he was unable to admit that any new fossil being described as a reptile could fit the bill.

The consequence of Huxley's refusal to consider a modification of his position was that the mammal-like reptiles did not receive the attention they deserved until much later in the century. Even when the case for a distinct line of reptile evolution leading toward the mammals had become generally accepted among paleontologists, we shall see that some morphologists continued to support Huxley's original position. Active debates on the origin of the mammals continued in the meantime, centered principally on the question of whether or not the monotremes and marsupials represented stages in the ascent toward the placentals. Haeckel had defended this position, which Simpson would later dismiss as "a recrudescence of the old naive conception of a *scala naturae,* masked by its application to great groups instead of small ones."[97] Huxley himself became

95. Desmond, *Archetypes and Ancestors,* pp. 193–201.

96. See Huxley's essay "Evolution in Biology," reprinted in Huxley, *Collected Essays,* II, *Darwiniana,* pp.187–226; see p. 223. On Huxley as a 'pseudo-Darwinian,' see Bowler, *The Non-Darwinian Revolution,* pp. 76–80.

97. G. G. Simpson, *A Catalogue of the Mesozoic Mammalia,* p. 162.

associated with this position, rather unfairly, as Simpson notes, because his intepretation differed significantly from Haeckel's. Huxley certainly saw a sequence of developmental stages in the evolution of the mammals which he quite explicitly compared to the old chain of being. But for him, the living monotremes and marsupials were only illustrative of the stages; they were not directly related to the ancestors of the placentals, who had passed through the same stages quite independently. Huxley thus set up a system of parallel evolution in which several different lines of evolution had advanced to different levels of the same developmental hierarchy.

This interpretation was outlined in his 1880 paper, "On the Application of the Laws of Evolution to the Arrangement of the Vertebrata and More Particularly of the Mammalia."[98] Huxley called the stage now reached by the monotremes the "Prototheria," while the marsupials were at the stage of the "Metatheria." The placentals were the "Eutheria." Huxley made it clear that the living monotremes and marsupials were specialized in ways that meant they could not represent the actual ancestors of the placentals. He suspected that the primitive marsupials had evolved the technique of preserving their young in pouches because they had been arboreal in habit, a point taken up by several later writers.[99] He hoped for the discovery of fossil representatives of the metatherian stage from the Mesozoic that would not have born their young in pouches. A diagram given to illustrate his idea shows a series of parallel lines of evolution passing through the three stages, which may be simplified as follows:

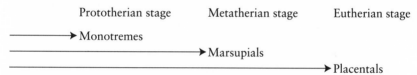

Prototherian stage   Metatherian stage   Eutherian stage

→ Monotremes

→ Marsupials

→ Placentals

Even the main placental groups each had a separate line that had passed through the prototherian and metatherian stages independently.[100] The living monotremes and marsupials were additional lines that had never advanced all the way through the scale. Huxley again noted that the prototherian stage would have had an amphibian, not a reptilian origin.

Huxley's willingness to invoke parallel evolution in the mammals suggests that he was not inherently opposed to the idea of 'grades' of evolution. He did not rule out the possibility of several lines independently

98. Huxley, "On the Application of the Laws of Evolution to the Arrangement of the Vertebrata"; on the chain of being, see p. 461.

99. Ibid., p. 465.

100. Ibid., p. 469. This diagram is reproduced in Di Gregorio, *T. H. Huxley's Place in Natural Science*, pp. 102–103.

evolving through the same grade as a concession to idealism or to the claim that evolution is predetermined by mysterious forces. But his position on this topic does not imply the absence of any real difference between his view of evolution and that promoted by Mivart.[101] Huxley had no doubt that mammals were superior to reptiles, as reptiles were to amphibians—witness his support for the idea of a hierarchy modeled on the chain of being. Any amphibian/reptile which moved toward the mammalian type of organization would thus gain an advantage, and Huxley would have seen no reason to rule out the possibility that several lines might independently have achieved at least part of the transition. The process was no more implausible than the one by which both birds and pterosaurs adapted to flight: the independent invention of a real improvement was no surprise to a Darwinian. It was only anti-Darwinians such as Mivart who seemed to think that any evidence of parallelism automatically implied that evolution was governed by internally programmed directing forces. The difference was that Mivart welcomed any trend, including nonadaptive ones, as grist to his mill. Whatever his ideas about the origin of variations, Huxley refused to admit that any trend could become established unless it conferred some advantage to the organism. In this respect, he could support Darwin and oppose Mivart on a clearly defined principle.

In Huxley's view, the monotremes were equivalent to, but not the superficially modified descendants of, an early stage in mammalian evolution. It was precisely this point which left his wider theory linking the mammals directly back to the amphibians open to challenge. For the zoologists of the 1880s, the strongest line of evidence suggesting that the reptiles were, after all, the parents of the mammals came not from the fossil record but from a more detailed study of monotreme reproduction. Ever since it had become generally accepted in the early decades of the century that the monotremes laid eggs, much research had focused on the precise nature of their reproductive process. In the early 1880s Francis Balfour suggested to one of his students, W. H. Caldwell, that he go to Australia to study these elusive animals. Following Balfour's death in 1882, Caldwell was awarded a studentship in his memory to complete the task. Jacob Gruber has described Caldwell's efforts as typical of a new and more intrusive form of fieldwork.[102] He hired over fifty aborigines to collect specimens, artificially pushing up the price of the food they bought "to keep the lazy blacks hungry." Whatever the morality of the approach,

101. See David Hull, "Darwinism as a Historical Entity."

102. J. Gruber, "Does the Platypus Lay Eggs?" See also Roy MacLeod, "Embryology and Empire." For an account of modern challenges to the view that marsupials are primitive, see Gould, "Sticking up for Marsupials," in Gould, The Panda's Thumb, pp. 289–295.

Caldwell got his specimens and was able to send his famous cable to the president of the British Association for the Advancement of Science to read at the 1884 meeting in Montreal. The cable read: "Monotremes oviparous, ovum meroblastic." [103] The significance of the latter point was that this was a reptilian kind of egg. If the monotremes had anything to do with the early phases of mammalian evolution, the discovery suggested that the class had indeed evolved from a reptilian ancestry.

It was seen as a great blow to the prestige of British science that the next phase in the study of the question was undertaken by a German team led by Richard Semon. We have already noted Semon's journey to Australia in search of the lungfish *Ceratodus*, but the monotremes and marsupials were also high on his list. As he noted in the popular account of his expedition, it had been inspired in part by the view that "the oviparous mammals form a link between reptiles and the higher mammals." [104] From the material collected in the 1890s, three fat volumes of detailed morphological work were eventually published on the monotremes and marsupials. [105]

Many later writers record the effect of Caldwell's annoucement in helping to shift attention toward the possibility of a reptilian ancestry. [106] The paleontologists' efforts to point out the mammalian character of some fossil reptiles now could be seen to have greater significance. Yet, at the same time, the status of the monotremes remained problematic. Were they really the surviving relics of the ancestors from which the whole mammalian class had evolved, or were they so peculiar that they might have to be seen as atypical mammals, perhaps even an independently evolved type that had acquired some characters equivalent to the lowest stage in the evolution of the true mammals? Huxley had, in effect, taken up this position in his 1879 paper: parallel evolution from the amphibians had produced mammals on at least two separate occasions. Earlier suggestions by Huxley that considerable parallelism might have occurred in mammalian evolution had already been greeted with delight by the anti-Darwinian Mivart, on the grounds that such trends gave clear evidence of evolution being predetermined. [107] In 1888 Mivart went on to develop quite explic-

---

103. This incident is described in W. H. Caldwell, "The Embryology of the Monotremata," see p. 466.

104. R. Semon, *In the Australian Bush*, p. 2.

105. Semon, ed., *Zoologisches Foschungsreisen in Australien*, Bds. 2, 3i, and 3ii.

106. E.g. Robert Broom, *The Mammal-Like Reptiles of South Africa*, introduction.

107. Mivart, *On the Genesis of Species*, pp. 81–82. Mivart is here referring to Huxley's unpublished Hunterian Lectures for 1866, in which he suggested that several different orders of marsupials had independently evolved into equivalent placental orders.

itly the claim that the monotremes had evolved independently of the rest of the mammals.

Mivart's paper, "On the Possible Dual Origin of the Mammalia," conceded the growing evidence for reptilian ancestry but evaded it by suggesting that it was limited to the monotremes alone. Mivart first noted new evidence for the existence of rudimentary teeth of a reptilian pattern in the platypus. These rudiments might indicate that the monotremes were degenerate, but he preferred an alternative explanation. The superior mammals had evolved from an amphibianlike stock, as Huxley had suggested, but the monotremes had originated from reptiles that had independently acquired fur and warm-bloodedness. They would thus tell us nothing about the early stages of placental evolution. Mivart speculated that "the Monotremes are an example of hypothetical higher mammals in the making, the future evolution of which may probably be hindered by man's presence, but which, did they appear, would produce mammalian forms more or less parallel to, but of course radically distinct from, the placental and marsupial series of mammals."[108] Mivart concluded by emphasizing the general significance of parallelism as part of his ongoing campaign to argue that much evolution was governed by predetermined trends rather than natural selection.

The separate origin of the monotremes was not a totally unreasonable suggestion. It was supported on paleontological grounds by Harry Govier Seeley in 1896.[109] Seeley described the fossil of an anomodont reptile which, he maintained, showed strong affinities to the monotremes. This suggested that it might be the ancestor of a line leading to the monotremes independently of the other mammal-like reptiles. The same possibility was endorsed by Osborn in 1898 and in Simpson's analysis of the Mesozoic mammals of 1928.[110] Having played an important role in helping to convince many zoologists that the mammals did, indeed, have a reptilian origin, the monotremes were now being moved off to a sideline which had independently acquired some mammalian characters.

Some went even further than this in their suggestion of parallel trends. In an 1892 study of dental evolution, Willy Kükenthal of Jena suggested that the marsupials, the placentals, and the mammal-like reptiles had all evolved their mammalian characters independently from some Palaeozoic form.[111] A notice of Kükenthal's paper in the popular English periodical

108. Mivart, "On the Possible Dual Origin of the Mammalia," p. 376.

109. Seeley, "Researches . . . IX. On the Complete Skeleton of an Anomodont Reptile."

110. Osborn, "The Origin of Mammals," p. 92, and Simpson, *A Catalogue of the Mesozoic Mammalia*, p. 183.

111. W. Kükenthal, "Ueber die Entstehung und Entwicklung des Säugertierstammes."

Natural Science elicited an outraged response from the anonymous pale-ontologist who reported it:

> The 'ancestors' of certain well-marked groups of animals recede further and further into remote antiquity, just in proportion to the amount of research devoted to them. The reputed 'ancestors' of one author become the 'parallel line' of the next; and the geologist, who is supposed to unearth the desired pedigree, begins to despair of following a kind of will-o'-the-wisp. We cannot help thinking that someday a serious misapprehension will be discovered in the prevailing ideas of animal pedigrees. The 'imperfection of the geological record' has its limits, and will not account for everything.[112]

It was precisely this kind of thing, the commentator grumbled, that was leading Bateson and others to reject the whole search for ancestries as a waste of time.

Kükenthal's position was indeed an extreme version of the tendency to use trivial specializations as an excuse to block the possibility that one fossil could illustrate the ancestral form of another. But a far more sensible case could be made for eliminating the marsupials from the position they had occupied as the intermediate phase of mammalian evolution. Huxley had noted that their peculiar reproductive habits might reflect an adaptation to an arboreal mode of existence rather than an intermediate step between egg-laying and the development of the placenta. In fact Darwin himself had queried the intermediate status of the marsupials as early as 1860. In a letter to Charles Lyell he had expressed his preference for a branching scheme in which the marsupials and placentals diverged from a common ancestor that would itself be neither a marsupial nor a placental.[113] At the end of the century a number of zoologists tackled this question along the lines suggested by Huxley. J. P. Hill of the University of Sydney, Australia, argued that the evolution of the placenta need not have passed through a phase resembling the marsupial method of reproduction. He had found evidence of an allantoic placenta in the marsupial *Perameles* (the bandicoot). He suggested that the marsupials had evolved from a stock that had already acquired the placenta; they were, in effect, degenerate placentals rather than primitive mammals.[114]

This point was taken up by Louis Dollo in his essay of 1899 on the arboreal ancestry of the marsupials. Dollo studied the foot structure of the living marsupials and argued that they all exhibited peculiarities that could only be explained on the assumption that they had diverged from a

112. Anon., "The Evolution of the Mammalia."

113. Darwin to Lyell, 23 September 1860, in F. Darwin, ed., *The Life and Letters of Charles Darwin*, II, pp. 341–344.

114. J. P. Hill, "The Placentation of Perameles," p. 436.

stock which had been adapted to living in the trees. The marsupials as a whole represented a highly specialized group, not a primitive one. Dollo quoted Huxley's suggestion that the peculiar marsupial mode of reproduction, expelling the offspring at a very immature phase into a pouch, was itself an adaptation to the arboreal way of life.[115] He argued that once this highly specialized character was confirmed, the marsupials were barred from their presumed position as placental ancestors, because no one could imagine that the mammals as a whole had passed through an arboreal phase. (This last position was, in fact, defended briefly by William Diller Matthew in 1904.[116])

The aberrant character of the marsupials was also proclaimed by B. Arthur Bensley of Columbia University in 1901. After carefully reviewing all the evidence provided by Dollo and others, Bensley suggested that the marsupials might be even more distinct from the placental stock than they had implied. Perhaps Hill's *Perameles* had independently acquired its rudimentary placenta, rather being a degenerate form.[117] This was coming closer to Kükenthal's position, although two years later Bensley backed off to a more moderate claim that the marsupials and placentals had diverged from a common, primitive mammalian stock.[118] He was now particularly interested in the origin of the Australian marsupials, and was arguing on the grounds of geographical distribution that they represented a very late radiation from an opossumlike ancestral form.

T. Thomson Flynn of the University of Tasmania confronted the alternatives squarely in 1922. One possibility was that the marsupials had descended from the placentals with almost complete loss of the placenta (Hill's position), as in the following diagram.

Or the alloplacenta of *Perameles* might be an example of convergence—the independent acquisition of this structure in a line paralleling the true placentals:

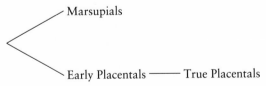

115. L. Dollo, "Les ancêstres des Marsupiaux étaient-ils arboricoles?" p. 191.

116. W. D. Matthew, "The Arboreal Ancestry of the Mammalia."

117. B. A. Bensley, "On the Question of the Arboreal Ancestry of the Marsupalia," p. 129.

118. Bensley, "On the Evolution of the Australian Marsupalia," p. 85.

Flynn himself thought that the structure in *Perameles* was so similar to that of the true placenta that it could not have been produced by convergence. The marsupials were descendants of the Protoplacentalia, the hypothetical first stage in the emergence of the true placentals.[119]

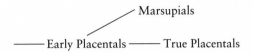

The position of both the monotremes and the marsupials as illustrative of stages in the evolution toward the placentals had thus been seriously undermined by the turn of the century. The image of these types as relics of the past lingered on only in literature intended for popular consumption. Even so careful a paleontologist as Matthew referred to the opossum as a "living fossil" in the synopsis of his lectures published in 1928, while Gregory used the same terminology in his 1929 survey *Our Face from Fish to Man*.[120] But neither of these authorities would have wanted to suggest that the opossum gave a detailed indication of what the first mammals had looked like.

Recognizing that the living marsupials were, in fact, a specialized off-shoot rather than truly primitive mammals inevitably transformed pale-ontolgists' interpretation of the early fossil record of the mammals. As Bensley lamented, arboreal animals leave few fossils, so the record of the events postulated in these theories was bound to be incomplete. But a small number of mammalian fossils had been found in the Mesozoic. Did this fragmentary record throw any light on the stages of development leading through to the placentals that had taken over the earth after the demise of the dinosaurs?

The earliest discoveries of Mesozoic mammals had given rise to great controversy in the 1820s and 1830s because they seemed to violate the whole notion of an 'age of reptiles.'[121] Given the prevailing belief that the marsupials were 'lower' than the placental mammals, it had eventually been decided that these were the fossil remains of marsupials. Significant discoveries in the 1850s extended paleontologists' knowledge of the Mesozoic mammals, while continuing to throw doubts on the idea of a simple progression in the history of life. It was already being recognized that these

119. T. Thomson Flynn, "The Phylogenetic Significance of the Marsupial Allantoplacenta."
120. Matthew, *Outline and General Principles of the History of Life*, p. 88; Gregory, *Our Face from Fish to Man*, pp. 47–48.
121. On the early reception of the Mesozoic mammalian fossils see Bowler, *Fossils and Progress*, p. 21.

fossils might not be marsupials.[122] In the 1860s and 1870s, the influence of Haeckel's linear scheme of mammalian evolution encouraged a temporary reemergence of the belief that the Mesozoic mammals must be lowly developed, perhaps still at the monotreme or marsupial stage. But this situation did not last, and by the 1880s there was an increasing acceptance of the idea that the Mesozoic mammals should not automatically be assigned to categories defined by the living mammals.

In 1877, P. Martin Duncan, president of the Geological Society of London, suggested that the Triassic mammals of Europe might be quite unrelated to the line of evolution which had given rise to the Australian monotremes.[123] In the following year Seeley argued that the mammals of the Stonesfield slate (Jurassic) could be traced back to a generalized order of mammals from which the true marsupials might later have evolved.[124] We shall see below how Seeley—who played a major role in drawing attention to the South African mammal-like reptiles—also endorsed Mivart's view that the monotremes had evolved independently of the other mammals. At about the same time, O. C. Marsh changed his views on the status of the American Mesozoic mammals. In 1878 he described *Dryolectes* as an opossumlike marsupial, but by 1880 he was describing the same genus as "manifestly low generalized forms, without any distinctive marsupial characters.[125] He erected a new order of submarsupials, the Pantotheria, to include most Mesozoic mammals. By 1887 Marsh was insisting that the modern placentals had not evolved from the marsupials; each had evolved separately from a primitive original form, of which the monotremes were the direct but highly specialized descendants.[126] In 1899 Osborn insisted that the placentals had not evolved from marsupials, while the monotremes—far from being the common ancestor—were so distinct that they might have evolved from an entirely separate reptilian stock, as Mivart and others had claimed.[127] Even so, Osborn still accepted that some Mesozoic mammals might have been marsupials. Robert Broom argued in 1914 that the Mesozoic multituberculates were ancestral to the monotremes.[128] The living monotremes were, in effect degenerate multituberculates—although Broom later noted that if this were the case,

122. See ibid., pp. 75–76.

123. P. M. Duncan, "Anniversary Address of the President," p. 87.

124. Seeley, "Note on the Femur and Humerus of a Small Mammal."

125. O. C. Marsh, "Fossil Mammals from the Jurassic of the Rocky Mountains," and "Notice of Jurassic Mammals Representing Two New Orders," p. 239.

126. Marsh, "American Jurassic Mammals," p. 346. Note, however, that Marsh followed Huxley and did not link the Mesozoic mammals back to the reptiles; see below, note 155.

127. Osborn, "The Origin of Mammals," p. 92.

128. Broom, "On the Structure and Affinities of the Multituberculata."

the other Mesozoic mammals were unlikely to be closely related to this group.[129] He did, however, insist that the mammals as a whole were monophyletic. Given the similarity in ear structure between monotremes and the higher mammals, he thought it unlikely that the two could have emerged from different reptilian ancestors.[130]

It was left for the young George Gaylord Simpson to provide the most comprehensive reassessment of the relationships between the Mesozoic and living mammals. Simpson came to London in 1926 on a National Research Council fellowship and was asked to prepare a catalogue of the British Museum's collection of Mesozoic mammals. We have already noted his rejection of Haeckel's linear scheme as a relic of the old chain of being. Simpson seems to have believed that most authorities still regarded these fossils as the remains of marsupials and went to considerable lengths to refute this interpretation.[131] He also refused to accept the significance of the relationships used by Broom to claim a link between the multitu-berculates and the monotremes. It was "impossible to avoid the conclu-sion that they [the multituberculates] represent an independent, special-ized group" with a separate origin in the reptiles from any of the living mammals.[132] Only one group, the Pantotheria, had a structure that would not rule it out as ancestral to the diverging branches that had given rise to the marsupials and the placentals. The monotremes went back to an en-tirely separate reptilian origin via a line which had left no fossil remains.[133] Simpson accepted four parallel lines of evolution from the reptiles, as shown in the following diagram, each independently producing a form of mammal, of which only two had left descendants today.

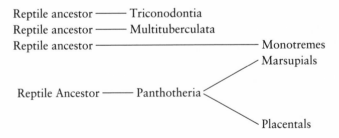

Like virtually every paleontologist from the 1880s onwards, Simpson accepted that the mammals, however, complex their origin, had arisen

129. Broom, *The Mammal-Like Reptiles of South Africa*, p. 331.
130. Ibid.
131. Simpson, *A Catalogue of the Mesozoic Mammalia*, p. 165.
132. Ibid., p. 170.
133. Ibid., pp. 182–183.

from a reptilian rather than an amphibian stock. Huxley's theory excluding the reptiles from mammalian ancestry had been undermined first by
Caldwell's discovery of the reptilian character of the monotreme egg and
then by the growing emphasis placed on the mammal-like reptiles. It is to
the latter that we turn to complete the story, but we should first note that
a significant number of morphologists resisted the trend. They were still
impressed by the morphological affinities between the mammals and the
amphibians (despite the nature of the monotreme egg), and were unwilling
to let the paleontologists get away with what they regarded as an evasion
of real morphological objections to the proposed sequence leading from
the reptiles to the mammals. Whatever the apparent continuity in the development of the jaw and other characters leading from the mammal-like
reptiles to the mammals, they felt that Huxley had been quite right to
insist that the true ancestry of the mammals would have to be sought directly in the amphibians.

In 1887, G. B. Howes—assistant professor of zoology at the Normal
School of Science, where Huxley himself taught—reported the discovery
of an epiglottis in the larynx of certain amphibians. The discovery of such
a peculiarly mammalian character in the amphibians must, he claimed,
support the view that the mammals had originated from animals lower
than the modern reptiles.[134] A number of German morphologists also saw
closer similarities between mammals and amphibians than between mammals and reptiles. Carl Rabl argued this point for the development of the
heart, and Hermann Klaatsch for the the intestinal arteries.[135] Friedrich
Maurer, professor of anatomy at Heidelberg, argued in 1893 that there was
no equivalent of hair on the skin of reptiles, whereas some amphibians had
dermal sense organs that could easily have evolved into mammalian hair.[136]
Probably the most active critics of the new paleontological theory were the
American J. S. Kingsley, and the Dutch embryologist A. A. W. Hubrecht,
both of whom we have already encountered in previous chapters (Kingsley
supported the polyphyletic origin of the arthropods, while Hubrecht proposed the nemertine theory of vertebrate origins).

In an article published in 1901, Kingsley acknowledged the discoveries that had turned many against Huxley's position: the reptilian character
of the platypus egg and the mammalian features of the fossil theromorph
reptiles. Significantly, he thought that the former had been more important, since "embryology was then the ruling force in deciding questions of

134. G. B. Howes, "On a Hitherto Unrecognized Feature in the Larynx of the Anurous
Amphibia."
135. C. Rabl, "Ueber die Bildung des Herzens der Amphibien;" H. Klaatsch, "Zur Morphologie der Mesenterialbildungen am Darmkanal der Wirbelthiere."
136. F. Maurer, "Zur Phylogenie der Säugethierhaare."

phylogeny."[137] But he insisted that there were still insuperable objections to the theory that the mammals had a reptilian ancestry. Morphologists studying the soft parts of the body were coming up with evidence that ran counter to the osteological links uncovered by the paleontologists—so perhaps the latter were susceptible to another explanation.[138] One of the great strengths of the theory based on the mammal-like reptiles was its coherence with new morphological evidence on the evolution of the bones in the mammalian ear from those in the reptilian jaw (discussed below). But Kingsley had already challenged this, arguing that the auditory ossicles could only be derived from a urodele amphibian, not a reptilian form.[139] The resemblances in teeth between the theromorph reptiles and mammals might easily be explained by parallel development from similar conditions.[140]

In his main article, Kingsley launched a direct attack on the new theory of the origin of the mammalian ear bones, insisting that the whole question of the ancestry of the mammals hinged on the decision to be made regarding the fate of the reptilian quadrate bone. The theory that it became transformed into the incus of the mammalian ear had been supported by morphologists working with adult animals, but Kingsley insisted that embryology should still be given the primary role—and embryology suggested that the mammalian ear developed more along the lines laid down in the amphibians.[141] He regarded the transformation of bones in the hinge of the jaw to an auditory function as implausible, considering that the organs were in constant use throughout. The evidence that primitive mammals had a reptilian egg was inconclusive, because again the similarities could have been developed independently.[142]

On this last point, Kingsley appealed to the work of Hubrecht, which he summarized at length on the grounds that it was not available in English. In fact, however, Hubrecht had already provided a substantial account of his ideas in the published version of a lecture delivered at the sesquicentennial celebrations of Princeton University, his *Descent of the Primates* of 1897.[143] The main purpose of this study was to argue against

---

137. J. S. Kingsley, "The Origin of the Mammals," p. 193.

138. Ibid., pp. 195–196.

139. Kingsley and W. H. Ruddick, "The Ossicula Auditus and Mammalian Ancestry," p. 227.

140. Ibid., p. 229.

141. Kingsley, "The Origin of Mammals," p. 199.

142. Ibid., p. 202.

143. A. A. W. Hubrecht, *The Descent of the Primates;* Hubrecht's views had first been published in German in 1895; see his "Die Phylogenie des Amnions und die Bedeutung des Trophoblasts."

the conventional view that humans had evolved from apes. Hubrecht thought that the various primates, humans, apes, and lemurs, had evolved in parallel from a much more primitive mammalian type.[144] But in the concluding section of his little book, Hubrecht turned to his by now equally controversial views on mammalian origins. He openly sought to revive Huxley's views on the amphibian character of the hypothetical mammalian ancestor.[145] The evidence that the monotremes had a reptilian type of egg was irrelevant, because this group might have evolved independently of the placentals. Looking at the reproductive system, Hubrecht declared, "There is really not one cogent reason which would prevent us from deriving arrangements as we find them in placental mammals directly from viviparous amphibian ancestors."[146]

Hubrecht had proposed a new theory to explain the origin of the amnion which, he claimed, eliminated any need for a reptilian phase in the evolution of the placental mammals.[147] He argued that the membrane of the amnion must have been a closed sac from the start in order for it to be effective. Embryologists had been misled into assuming that the amnion of the chick provided a model that could be applied to the placental mammals. But the model was good only for a few placental species: in many others, including the human species, there was a different mode of development more easily explicable in terms of the direct modification of the amphibian system of reproduction. Everyone had simply assumed that the reptile-bird model was a good one on the basis of the by now increasingly popular view that the reptiles were also the ancestors of the mammals.[148] Hubrecht cited the work of those morphologists who remained skeptical of the reptilian ancestry of the mammals, and concluded with a note that all the fossil discoveries of the last decade or so remained mute on the question of how the internal organs had evolved.[149]

Morphology thus came into direct conflict with paleontology on this point, and some morphologists were not yet willing to give up their long-established theories in the face of the fossil evidence. In practice, however, their position became increasingly isolated as paleontology took on an ever more influential role in debates about evolutionary ancestry. As late as 1917, Lull noted the rival position staked out by Huxley before passing

144. On Hubrecht's views on human origins, see Bowler, *Theories of Human Evolution,* p. 118.
145. Hubrecht, *Descent of the Primates,* pp. 27, 29.
146. Ibid., p. 31.
147. Hubrecht, "Die Phylogenie des Amnions."
148. Hubrecht, *Descent of the Primates,* p. 31.
149. Ibid., p. 39.

on to discuss the evidence of the mammal-like reptiles.[150] But Lull, like most paleontologists of his generation, regarded Huxley's position as a hangover from a previous era when morphology had ruled the roost. He was no longer willing to set aside the evidence of transitional forms in the fossil record on the grounds that certain problems in the anatomy of the soft parts indicated a closer link with the amphibians.

THE MAMMAL-LIKE REPTILES

As Desmond has noted, the first fossils of mammal-like reptiles from the Karoo formation of South Africa were sent to England as early as the 1840s by Andrew Bain.[151] They came at first to the Geological Society of London and were described by Owen. After he became superintendant of the natural history collections at the British Museum in 1856, Owen used the museum's resources both to finance the acquisition of more specimens from Bain and to extract the fossils from the very hard matrix within which they were embedded. By the late 1850s Owen had begun to recognize that the fossils were reptiles with mammal-like teeth, and the names he gave to them emphasized this relationship: names such as *Tigrisuchus* (tiger-crocodile) and *Cynosuchus* (dog-crocodile). Owen created the order Theriodontia (beast, i.e., mammal, toothed) to include these reptiles. His papers of the 1870s and his *Descriptive and Illustrated Catalogue of the Fossil Reptilia of South Africa* of 1876 made the evolutionary significance of these resemblances clear. He asked if the transfer of tooth structure from these reptiles to the mammals was a "seeming one, delusive, due to accidental coincidence in animal species independently (thaumatogenously) created" or real, the consequence of "nomogeny or incoming of species by secondary law, the mode of operation of which we have still to learn?"[152] Although Owen denied the adequacy of both the Lamarckian and Darwinian mechanisms, the implication that some kind of evolutionary relationship was involved seems clear.

As we have seen, Huxley refused to acknowledge the potential significance of the South African fossils, and it was only in the late 1880s, following Caldwell's discovery of the platypus egg, that opinions began to

---

150. Lull, *Organic Evolution*, p. 540. In the following year John Macfarlane's *Causes and Course of Organic Evolution* endorsed Huxley's theory (see p. 479)—although, since the author was a botanist, this work had little influence on the debate in zoology and paleontology.

151. Desmond, *Archetypes and Ancestors*, pp. 193–201.

152. Richard Owen, "Evidence of a Carnivorous Reptile (Cynodracus major, Ow.)," p. 101; also Owen, *Descriptive and Illustrated Catalogue of the Fossil Reptilia of South Africa*, p. 76.

change. In the meantime, Edward Drinker Cope deflected attention in a different direction by suggesting a link between the pelycosaurs of the American Permian formations and the origin of mammals. The pelycosaurs of the Texas Permian included *Dimetrodon,* best known for the large 'sail' on its back. Cope created the order Theromorpha to include this and other reptiles in which he recognized mammalian affinities. After listing the various characters, he declared, "The preceding comparison renders it extremely probable that the Mammalia are descended from the Pelycosaurian Reptilia." [153] He noted, however, that these were also the most amphibianlike of reptiles, that is, the least likely to have acquired those characters that Huxley had seen as lying off the line of mammalian descent.

Cope did not maintain this position for long, and in his *Primary Factors of Organic Evolution* of 1896 he gave a phylogentic tree tracing the mammals back to Owen's Theriodontia. [154] The source of his conversion was acknowledged as the work of Georg Baur, which Cope now saw as having been confirmed by his own work on Permian reptiles. We have already noted Baur's important contributions to the debate over amphibian origins. He had come to America in 1884 as assistant to Cope's great rival, O. C. Marsh, at Yale, and in his short career (he died of mental exhaustion produced by overwork in 1898) he made key contributions to several phylogenetic debates. We should note that Marsh himself was not impressed by the new work on mammal-like reptiles, although we have seen that he helped to reinterpret the status of the American Mesozoic mammals. In a debate on the origin of mammals at the 1898 International Congress of Zoology, Marsh recorded that he had been converted to Huxley's position when the latter had visited America in 1876. He had discussed the question with Balfour on a visit to England in 1881 and had hoped that the latter's embryological research would solve the problem. Balfour's death had prevented the fulfillment of this hope, and Marsh now lamented that the debate had come to resemble a medieval disputation. [155] Yet it was his own assistant, Baur, who had helped to put the matter on a firmer foundation as far as many paleontologists were concerned.

In two papers published in 1888, one in Germany and one in America, Baur challenged Cope's classification of the reptiles and his view that the pelycosaurs were the ancestors of the mammals. [156] The structure

153. Cope, "Fifth Contribution to the Knowledge of the Fauna of the Permian Formation of Texas," p. 45.

154. Cope, *Primary Factors of Organic Evolution,* p. 115.

155. Marsh, "Discussion on the Origin of Mammals."

156. Baur, "Über die Kanäle im Humerus der Amnioten," and "On the Phylogenetic Arrangement of the Sauropsida."

of the humerus suggested that the Theromorpha (including the pelycosaurs) were too specialized to be the ancestors of mammals. The pelycosaurs and the mammals were probably derived from unknown Permian
forms, the "Sauro-mammalia." These would, nevertheless have been reptiles, not amphibians. In the 1890s Baur realized that the South African
fossils studied by Owen and later by Harry Govier Seeley were essentially
what he had predicted in this hypothetical ancestry. Adrian Desmond
notes a letter from Baur to Seeley in 1895 admitting that the mammal-like
reptiles were close to the sauro-mammalia.[157] In a note published in 1897
in collaboration with E. C. Case, Baur openly proclaimed that the South
African fossils were the key to the origin of the mammals.[158]

Harry Govier Seeley became interested in Owen's mammal-like reptiles in the late 1880s and published a long series of "Researches on the
Structure, Organization, and Classification of the Fossil Reptilia." In 1888
he noted that Owen's *Pareiasaurus bombidens* seemed to represent the
transitional phase between the labyrinthodont amphibians and true reptiles, yet at the same time showed certain mammalian characters. These
characters were almost certainly transmitted through to the mammals, at
least by collateral derivation from a common ancestor.[159] Another paper
published in the same year actually presented one of the South African
fossils as a mammal, although Seeley was forced to withdraw this interpretation some years later.[160] *Pareiasaurus* was an anomodont reptile, belonging to a group which, along with the pelycosaurs, made up Cope's
order Theromorpha. Seeley assessed the degree of change that would be
required to convert an anomodont to a mammalian skull.[161] He also presented the anomodonts as a group whose affinities ran out radially in several directions to the amphibians, the mammals, and to other reptilian
orders.[162] The diagram he gives to illustrate this point is—like those he
was using to describe the complex affinities of the pterosaurs and birds—
anything but an evolutionary tree, confirming once again Seeley's less than
enthusiastic endorsement of conventional phylogenetic schemes. Nevertheless, his assessment of the intermediate position of the anomodonts

157. Baur to Seeley, 7 May 1895, quoted by Desmond, *Archetypes and Ancestors,*
pp. 245–246.

158. Baur and E. C. Case, "On the Morphology of the Skull of the Pelycosauria and the
Origin of Mammals."

159. Seeley, "Researches . . . II. On Pareiasaurus bombidens," pp. 106–108.

160. Seeley, "Researches . . . III. On Parts of the Skeleton of a Mammal from the Triassic
Rocks of Klipfontein"; and for the withdrawal of this interpretation, see his "Researches . . .
IX, pt. 2. The Reputed Mammals of the Karoo Formation."

161. Seeley, "Researches . . . IV. On the Anomodont Reptilia and Their Allies."

162. Seeley, "Researches . . . VI. Further Observations on Pareiasaurus"; for the diagram of
relationships, see p. 368.

between amphibians and mammals sparked a more general recognition of Baur's point that there might be a distinct line of mammal-like reptiles splitting off from the other reptilian groups at an early stage in the class's evolution.

In 1889 Seeley went to South Africa to see for himself the Karoo deposits from which the mammal-like reptiles were coming.[163] He continued to publish on these reptiles through the 1890s, and also noted the affinities of a European anomodont to the monotremes.[164] The implication was that Mivart had been right to see the monotremes as products of an independent line of reptilian evolution. In the discussion on the origin of mammals at the 1898 International Congress of Zoology, Seeley expressed a more cautious view of the South African fossils. The anomodonts were too specialized to be the parents of the mammals, although they were a "collateral and closely related group."[165] The true ancestors of the mammals would be found in rocks older than the Permian, in the Devonian or even the Silurian. Seeley thus backed away from the position that others would build on the discovery of the mammal-like reptiles. His lukewarm attitude to evolutionism seems to have led him toward a position similar to that proposed by Kükenthal and the more extreme exponents of the independent acquisition of similar characters in multiple lines of development. Osborn, while noting the significance of Seeley's contribution, nevertheless commented on his "indefinite views of evolution."[166] Some time later Robert Broom also referred to Seeley's work as providing confirmation of Baur's ideas, but conceded that it was difficult to get any clear idea of Seeley's views on the relationships involved from his papers. Broom (whose own views on the nature of evolution were anything but conventional) thought the problem lay in Seeley's acceptance of Mivart's idea that the mammals were diphyletic.[167]

Seeley's emphasis on the mammalian characters of the South African Theromorpha shows how a biologist with unconventional ideas about the actual mechanism of evolution can nevertheless play a key role in developing new ideas about phylogeny. His willingness to trace links back to the amphibians as well as forward to the mammals helped others to see a way of evading Huxley's original line of argument that excluded the reptiles altogether from mammalian ancestry. By recognizing a major split in

163. Seeley's letters from South Africa are reproduced in W. E. Swinton, "Harry Govier Seeley and the Karoo Reptiles."

164. Seeley, "Researches . . . IX. On the Complete Skeleton of an Anomodont Reptile (Aristodesmus Rütimeyeri, Wiedersheim)."

165. Seeley, "Discussion on the Origin of the Mammals," p. 70.

166. Osborn, "The Origin of the Mammalia," p. 313.

167. Broom, Croonian Lecture: "On the Origin of Mammals," pp. 4–5.

reptile evolution at a very early stage following their emergence from the amphibians, it could be supposed that Huxley was merely overenthusiastic rather than completely wrong. Flower and Lydekker's *Introduction to the Study of Mammals* of 1891 endorsed this interpretation, stressing the retention of characters derived from the Labyrinthodont amphibians in the South African anomodont reptiles.[168] To drive the point home even further, they suggested that the mammals had actually originated from a line which branched off during the transitional phase from amphibian to reptile. The South African fossils were (as Seeley himself eventually claimed) too specialized to be the true ancestors; they were only a parallel branch of reptile evolution which had independently acquired some mammalian characters.

Osborn claimed that his notes from Huxley's lectures in 1879–80 recorded Huxley himself predicting that older reptiles would be found to serve as a bridge to the mammals, a more successful hypothesis than the one he actually published.[169] In a series of papers in the late 1890s, Osborn proclaimed his support for the Baur-Seeley hypothesis of the origin of the mammals from the Theromorpha. Although himself sympathetic to the view that the monotremes might be the product of a different branch of reptilian evolution, Osborn evaded Seeley's later claim that the known theromorphs were too specialized to be the actual ancestors of the mammals by suggesting that "within the order may well have existed some small insectivorous types, far less specialized in tooth structure than either the carnivorous Cynodonts or herbivorous Gomphodonts, as one of those conservative spurs of adaptive radiation which forms the focus of a new progressive type." [170]

The early years of the twentieth century saw the phylogenetic significance of the mammal-like reptiles being explored in earnest. A key role in this process was played by the eccentric medical doctor and paleontologist Robert Broom. Broom spent the years 1893–97 in Australia, where he made embryological studies of the marsupials in the hope of throwing light on the evolution of the mammalian shoulder girdle.[171] There was a larger structural difference between the monotreme shoulder and that of the higher mammals than there was between the reptile and the monotreme—one of the factors that had led Mivart and others to argue that the

168. W. H. Flower and R. Lydekker, *Introduction to the Study of Mammals*, p. 83.

169. Osborn, "The Origin of the Mammalia," p. 310.

170. Osborn, "The Origin of Mammals," p. 96.

171. These early embryological researches were published in Broom, "On the Development and Morphology of the Marsupial Shoulder Girdle," and "On the Early Condition of the Shoulder Girdle in the Polyprotodont Marsupials *Dasyurus* and *Perameles*." On Broom's life, see his autobiography, *Finding the Missing Link*, and G. H. Findlay, *Dr. Robert Broom*.

monotremes were separately evolved. Broom showed that by looking at the early stages in the development of the marsupial shoulder girdle, this difference could be minimized. His interest in the origin of the mammals having been aroused, he moved to South Africa in 1897 with the aim of studying the mammal-like reptiles. Over the next several decades he produced a series of studies leading to his Royal Society Croonian lecture "On the Origin of Mammals" of 1915 and culminating in his book, *The Mammal-Like Reptiles of South Africa* of 1932. He then became inspired by Raymond Dart's discovery of *Australopithecus* and went on to make a major contribution to the discovery of fossil hominids in South Africa.

Broom noted how Huxley's position had deflected many anatomists from the possibility that any reptilian fossil might serve as a link to the mammals. Caldwell's discovery of the nature of the monotreme egg was the first serious blow to this theory (Broom lamented that it was a German, Semon, who did most to follow up this discovery, a fact which, he felt, reflected badly on British and Australian science).[172] Seeley had then drawn attention to the South African fossils, despite his insistence that they resembled mammals "only by a parallel development." Broom also paid tribute to Baur's work, cut short by his untimely death.[173] His own work had led him to the conclusion that the ancestor of the mammals would have emerged from the South African cynodont reptiles, or at least from other therocephalian reptiles closely allied to the cynodonts (fig. 6.5).[174] The resemblances between these reptiles and the mammals were too strong to be merely the product of convergence. Broom also thought seriously about the causes lying behind the transformation. He was convinced that the peculiarities of the mammalian limbs and shoulder girdle were "distinctly related to the habit of walking with the body off the ground."[175] Earlier reptiles had merely crawled along the ground like their amphibian ancestors, but the South African fossil reptiles had "well developed limbs which enabled the animals to walk with a mammal-like gait with the body well supported off the ground."[176] The key transition must have been a change of habit, possibly in response to a transformation of the environment. The greatly increased level of activity in mammals was a direct consequence of this transformation, as was the increase in the size of their brains.

By the time he collected his work for his 1932 book, Broom had created a new suborder of therapsid reptiles, the Ictidosauria, to accomodate

172. Broom, *The Mammal-Like Reptiles*, pp. 1–2.
173. Broom, Croonian Lecture, "The Origin of Mammals," p. 4.
174. Ibid., pp. 24–25, 31.
175. Ibid., p. 31.
176. Ibid., p. 32.

fossils discovered at Bloemfontein. These were much closer to the mammals than the cynodonts, and were thus the most plausible ancestors.[177] His book again addressed the cause of the transformation, postulating a change of habit produced by a new diet:

> It seems not improbable that some small Ictidosaurians finding difficulties in killing and devouring Eosuchians [small reptiles] came to feed on insects and finding the new diet very satisfactory soon became more adapted to the new requirements. Greater rapidity of movement was acquired resulting in the gradual development of a loose skin and with it changes that resulted in the evolution of hair.[178]

By now, however, Broom had abandoned the idea that a change in the environment might have triggered the adoption of new habits. Indeed, he expressed open support for a kind of theistic evolutionism in which life had been led to progress by a supernatural power. Most branches of evolution were governed by a short-sighted policy of specialization which led to blind alleys incapable of further development. But a few small generalized types always escaped this pressure and remained free to move onto a higher level, and here "the evolutionary force is of a different type and seems to have forseen the future."[179] Broom insisted that further progress in evolution was now impossible: all living animals were too specialized to become the foundation for any transformation to higher types:

> Apart from minor modifications evolution is finished. From which we may perhaps conclude that man is the final product; and that amid all the thousands of apparently useless types of animals that have been formed some intelligent controlling power has specially guided one line to result in man.[180]

Like Seeley, Broom demonstrates how zoologists with bizarre and outdated views on the process of evolution could nevertheless do invaluable work on important phases of the history of life on earth.

In the course of his work Broom confronted one of the great debates generated by nineteenth-century morphology: the relationship between the bones of the reptile jaw and the mammalian ear. The jaw of a mammal consists of a single bone, the dentary, articulating directly with the squamosal bone in the skull. The reptilian jaw consists of a series of bones in

177. Broom, *The Mammal-Like Reptiles*, pp. 299–307.

178. Ibid., pp. 316–317.

179. Ibid., p. 332.

180. Ibid., p. 333; see also Broom's book, *The Coming of Man: Was It Accident or Design?* and Bowler, *Theories of Human Evolution*, pp. 219–222. Julian Huxley later took up Broom's idea that all forms of life except the human species are now too specialized to progress further; see Marc Swetlitz, "Julian Huxley and the End of Evolution."

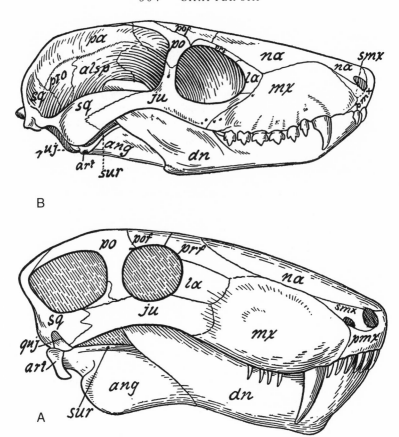

FIGURE 6.5 Transition from mammal-like reptile to mammal. Left, A: earlier; B: later mammal-like reptile. Right, A: advanced mammal-like reptile; B: modern opossum. From W. K. Gregory, *Our Face from Fish to Man* (New York, 1929), pp. 35, 49.

addition to the dentary, the articulation being between the articular bone of the jaw and the quadrate of the skull. In 1837 Carl Reichert had suggested that the reptile quadrate was homologous with the incus of the mammalian ear, while the articular was homologous with the malleus. Translated into evolutionary terms, this would mean that the bones originally forming part of the reptilian jaw had not disappeared, but had somehow been transformed into the sequence of tiny bones which transmit sound in the mammalian ear (fig. 6.6). As Broom himself noted, the evidence for this came mainly from embryology; anatomists were more impressed by the resemblances between the auditory chains in reptiles and mammals, while paleontologists found it difficult to see how there could

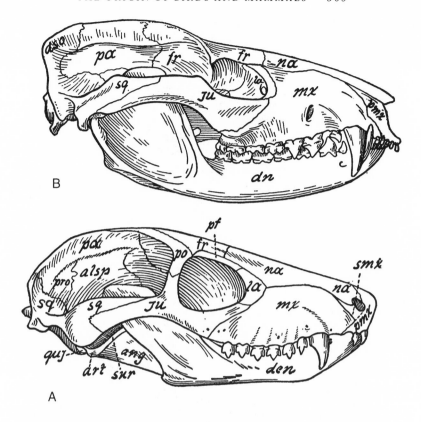

FIGURE 6.5   Continued.

be functional intermediate stages by which the quadrate could become transformed into the incus.[181] Broom himself resisted the quadrate-incus theory until 1912, when more detailed work on the South African anomodonts convinced him that his obections were invalid. He had actually been able to dissolve out the bone from a fossil anomodont skull, leaving a perfect cast of the internal ear. The cranial nerve passages of the fossil were now as well known as those in a living animal.[182] In fossils such as *Cynognathus* the transformation of the jaw toward the mammalian condition was already underway. The posterior part of the dentary was al-

181. Broom, "On the Structure of the Internal Ear," p. 419. For an account of the embryological evidence originally reported in Reichert's *Entwicklungsgeschichte der Gehörknöckelchen,* see Stephen Jay Gould, "An Earful of Jaw," reprinted in Gould, *Eight Little Piggies,* pp. 95–108.
182. Broom, ibid., p. 420.

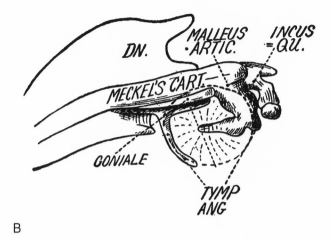

FIGURE 6.6   Origin of auditory ossicles (above) from the articulation of the reptilian jaw (below). From W. K. Gregory, *Our Face from Fish to Man* (New York, 1929), p. 218.

ready taking part in the joint with the skull. The quadrate and articular were degenerate, and it was easy to see how they might eventually have been lost altogether "had not the attachment of the stapes to the quadrate compelled them to take on an auditory function."[183] He was able to give

183. Ibid., p. 415.

a diagram illustrating the successive stages in the transformation, some stages confirmed by fossils, others hypothetical.[184]

It was Gregory who pushed this line of argument further in an effort to understand how the change of function had come about. But as Gregory himself noted, the availability of the fossil mammal-like reptiles to demonstrate the stages in this transition dovetailed neatly with developments taking place quite independently in German morphology. In a critique of recent work on the morphology of the vertebrate skull published in 1913, Gregory lamented the diminishing enthusiasm for comparative anatomy among English-speaking biologists. The rise of statistical and experimental biology had led many to dismiss phylogenetic work as "too much the product of unchecked imagination." But German morphologists had continued to work actively in this field: "There the problems of the vertebrate skull still have a human interest; men still take sides over questions of homology and even get to the point of abusing each other in print."[185]

Gregory singled out the work of Ernst Gaupp of the University of Freiburg im Breisgau on the incus-quadrate theory as a classic example of how this approach was still yielding dividends. In a series of papers published in 1910 and 1911, Gaupp put the theory on a new footing, using morphological analysis of living species as his only evidence—at this point he had not realized that fossil reptiles might actually demonstrate the transformations he proposed.[186] The new biologists, sneered Gregory, would probably dismiss Gaupp's work as "mere flimsy plausibility," but the few who still had patience to follow the detailed argument realized that he has "developed a perfectly consistent body of doctrine, resting upon many independent lines of evidence and offering a highly probable explanation of the two-fold problem of the lower jaw and the auditory ossicles."[187] The only fault of the morphologists was that they, in turn, did not realize that the paleontologists were now coming up with exactly the kind of evidence needed to confirm their theories. Gaupp himself, along with other workers such as Hans Gadow, had excluded the theriodont reptiles from consideration because they had decided on embryological grounds that the mammals must have evolved from a form with a freely moving quadrate.[188] The fact that the quadrate was freely movable in the mammalian embryo

---

184. Ibid., p. 424.

185. Gregory, "Critique of Recent Work on the Morphology of the Vertebrate Skull," p. 2.

186. Ernst Gaupp, "Beiträge zur Kenntnis des Unterkiefers der Wirbeltiere," parts I–III.

187. Gregory, "Critique of Recent Work," p. 2.

188. Ibid., p. 23; see Gadow, "The Evolution of the Auditory Ossicles," p. 407; and Gaupp, "Beiträge . . . III. Das Probleme der Entstehung eines 'sekundären' Kiefergelenkes bei den Säugern," p. 635.

did not mean that it must have been the same in the ancestral reptile: this was carrying the biogenetic law too far.

Gaupp himself claimed that he had been inspired to undertake over twelve year's work in morphology by Dohrn's principle of function-change.[189] He had long believed that the evolution of the amphibians and of the mammals represented two major steps in vertebrate evolution which could only be understood in the light of this theory. The transformation of bones in the jaw hinge to an auditory function would indeed provide a classic illustration of Dohrn's principle.

Gregory now believed that the fossil record showed exactly how and why this transformation had taken place. The application of Reichert's theory to the theriodont reptiles required only that the vestigial quadrate should be freed from the squamosal socket and that the articular should be brought into contact with the stirpes, or the primary auditory rod. But it was necessary to "conceive an adaptive mechanical motive for this extraordinary change."[190] This could be provided by showing that the vestigial jaw elements had come into contact with the auditory mechanism and had thus begun to participate in its function. In the end, they had been incorporated completely into the system for transmitting sound. Broom's work on the mammal-like reptiles showed how this could have come about. D. M. S. Watson saw Gregory's idea of the transfer of function as a "brilliant hypothesis" which followed up Gaupp's work in putting the quadrate-incus theory beyond question. Watson was convinced, however, that once these functional changes were initiated, they established trends which drove many groups of mammal-like reptiles to evolve inexorably toward the mammalian level of structure.[191]

Gregory placed strong emphasis on trying to explain how the bones studied by the paleontologist actually functioned in the once-living animal. He made important studies of the musculature of the reptilian jaw and the mammalian ear to back up the theory of funtion-change in the bones.[192] He also worked out the general musculature of the mammal-like reptiles, showing that *Cyognathus* had limbs and girdles of a reptilian pattern, but with well-marked modifications towards the mammalian, especially the monotreme, type.[193] This supported Broom's conclusions,

189. Gaupp, "Beiträge . . . III," pp. 609–610, quoting his own paper from twelve years earlier.

190. Gregory, "Application of the Quadrate-Incus Theory to the Conditions in the Theriodont Reptiles," p. 600.

191. D. M. S. Watson, "The Deinocephalia," pp. 779, 784–785.

192. Gregory and L. A. Adams, "The Temporal Fossae of Vertebrates in Relation to the Jaw Muscles."

193. Gregory and C. L. Camp, "Studies in Comparative Myology and Osteology, no. III." On this aspect of Gregory's, work see Rainger, *An Agenda for Antiquity*, pp. 220–224.

FIGURE 6.7   Charles Knight's restoration of mammal-like reptiles; on the left, three specimens of *Cynognathus;* on the right, an individual *Kannemeyeria.* Knight's caption explains
that the three carnivores are depicted as attacking in a group, a mammal-like mode of behavior. From Charles R. Knight, *Before the Dawn of History* (New York, 1935).

and allowed Gregory to propose the following characterization of these
reptiles: they would have been sluggish animals with imperfect heat-
producing adaptations, capable of hunting by stealth, but not by running
down their prey.[194] Charles Knight, however, depicted the mammal-like
reptiles as having already developed a mammalian level of intelligence
and socialization, allowing *Cyognathus,* for instance, to hunt in packs
(fig. 6.7).

By 1917 the new interpretation of mammalian origins was sufficiently
well established for E. S. Goodrich to propose a new basis for the classification of the Reptilia to take account of its implications. He declared that
the reptiles were a grade, not a class in the phylogenetic sense.[195] They
were an assemblage containing the ancestors of the birds and mammals,
along with other forms derived from the amphibians. The class would
probably have to be broken up eventually, but for the time being it could
be retained on the understanding that it united the early stages of two
massively diverging lines of evolution.

194. Gregory and Camp., ibid., p. 557.
195. E. S. Goodrich, "On the Classification of the Reptilia."

The solution to the problem of mammalian origins showed that morphology and paleontology could still unite to solve phylogenetic problems, whatever the skepticism of the new generation of experimental
biologists. But at the same time it posed a problem that forced paleontologists to think more carefully about the nature of the evolutionary
process. Accepting Broom's conclusion that the emergence of the mammals was linked to the adoption of new habits, perhaps following a
change of diet, still left the question of what had triggered this step forward. It was also necessary to ask why the mammals had then remained
subordinate to the dinosaurs and other 'ruling reptiles' throughout the
Mesozoic. More than any other event in the history of life on earth, the
long-delayed triumph of the mammals forced evolutionists to confront
the fact that biological progress could count for nothing if the circumstances of the time were unpropitious. Recognition of this point may
have played a major role in the development of mid-twentieth-century
Darwinism.

Lull supported Broom's idea that the evolution of the mammals was
associated with the adoption of a new posture that allowed an increased
level of activity, but suggested that the event had been triggered by an
environmental change. The early Permian had witnessed a massive drying
up of the climate, and one fact that was obvious from modern animals
was that desert conditions promoted the ability to run fast.[196] The origin
of the mammals could thus be fitted into Lull's metaphor of "the pulse of
life"—the claim that the major episodes of progressive evolution were
triggered by dramatic transformations of the environment by geological
forces.[197] The mammals' subsequent stagnation through the Mesozoic
was a consequence of the fact that the dinosaurs evolved large carnivorous
species first. The dinosaur tooth held the mammals in check, suspended
over them like a sword of Damocles until it was removed by the extinction
of the whole class. Although he was aware that the evidence was flimsy,
Lull liked to think that this second event was also triggered by a major
climatic revolution produced by the Laramide episode of mountain-
building.[198]

Matthew also interpreted the relationships between the vertebrate
classes in terms of changing climate. In the climatic extremes of the
Permian, the reptiles had adapted to hot, dry regions, the birds to cold,
arid regions, and the mammals to cold, wet forests. When much of the

196. Lull, *Organic Evolution,* pp. 306–307, 541.
197. Ibid., p. 689.
198. Ibid., p. 690, but note Lull's quotation on p. 548 from Osborn's *The Age of Mammals,*
p. 92, which concedes that the evidence for a climatic change coinciding with the extinction
of the dinosaurs is inconclusive.

world developed a hot, moist climate, the reptiles were able to spread out and marginalize the two warm-blooded classes in the polar regions. The reptiles specialized for these humid conditions, and were unable to respond when more extreme climates reappeared at the end of the Cretaceous. Meanwhile, the mammals' uniform body temperature had encouraged greater activity and the development of a bigger brain: "This . . . was what was going on behind the scenes while the dinosaurs held the stage." [199] The mammals were thus well prepared to dominate the earth when the reptiles were killed off in one of the "cyclic alternations of climate during geologic time."

Like Lull, Romer also depicted the early mammals as having to remain inconspicuous because "the threat of death from the great carnivorous reptiles lay constantly over them." He tried to find a purpose for this long period of "trial and tribulation" by arguing that it was

> a period of training during which mammalian characteristics were being perfected, wits sharpened, developmental processes improved, and the whole organization undergoing a gradual evolutionary change from reptilian to true mammalian character. As a result, when (at the close of the Cretaceous) the great reptiles finally died out and the world was left bare for newer types of life, higher mammals prepared to take the leading place in the evolutionary drama had already evolved.[200]

Romer avoided the question of the ultimate cause of the reptiles' extinction. But even so, it remained clear that the mammals would not have gained the chance to dominate the earth had not the great reptiles died out. The story of the mammals' rise was inevitably bound up with that of the reptiles' eventual demise. Questions relating to the rise and fall of particular groups in the history of life open up another range of issues which we must explore in a separate chapter.

Before turning to these developments, however, it is worth noting that German paleontologists had begun to turn their backs on the search for gradual transitions between major groups. Some German-speaking morphologists and paleontologists retained an interest in functional morphology and were drawn to such problems (Gaupp and Abel are good examples). But in the 1930s some paleontologists began to develop a distinctive approach to the fossil record which accepted the sudden appearance of new types without intermediate stages. This was especially true among invertebrate paleontologists, who had a strong link to the geological community through their work in stratigraphy. Otto Schindewolf and Karl Beurlen developed the outlines of what would later be called the ty-

---

199. Matthew, *Outline and General Principles of the History of Life*, p. 146.
200. Romer, *Vertebrate Paleontology* (1933), p. 261.

postrophe theory.[201] They supposed that new taxa were produced suddenly by massive saltations, and that the subsequent evolution of each group was controlled by internally programmed orthogenetic forces. Beurlen adopted an almost vitalist concept of saltations as the product of a creative life force, while Schindewolf opted for a more materialistic theory based on the geneticist Richard Goldschmidt's concept of macromutations (the "hopeful monster"). It has been argued that their work was strongly influenced by Adolf Naef's revival of idealist morphology.[202] Idealists were certainly reluctant to visualize transitions between the major archetypes, or *Baupläne,* of major groups, although we have seen that Naef himself did not insist on an anti-evolutionary approach (chapter 4). Whatever the source of his views, Schindewolf became immensely influential in German paleontology, effectively limiting the influence of the more Darwinian view of evolution that was emerging in the English-speaking world. In his scheme, there was no room for the investigation of transitions from one class to another, no point in searching for intermediate fossils, and no emphasis on adaptive or functional concerns in the shaping of evolution.

201. See Beurlen, "Funktion und Form in der organischen Entwicklung," and the English translation of Schindewolf's *Basic Questions in Paleontology,* pp. 168–170. On Goldschmidt's macromutational theory, see Bowler, *Eclipse of Darwinism,* pp. 209–210.

202. Wolf-Ernst Reif, "Afterword" in Schindewolf, *Basic Questions in Paleontology,* p. 440.

7

PATTERNS IN THE PAST

The origin of each major phylum could be treated as a distinct event in the history of life on earth. For morphologists, such an event represented a major branching point in the tree of life, the node from which all the diverse members of a group were derived. Paleontologists might hope to pinpoint the exact period at which the founders of the new phylum or class appeared and even to explain the causes lying behind the transformation. But similar problems were confronted by evolutionists dealing with transformations on a smaller scale: developments taking place within the major groups leading to the evolution of particular genera or species.

Systematists who accepted that their classification schemes were a representation of the evolutionary tree had to develop hypotheses about the actual structure of the tree for the group they were studying. Even before the advent of evolutionism, paleontologists had been uncovering patterns of development within particular groups in the course of geological time. They could, to some extent, trace the unfolding lines of specialization radiating from the most primitive founders of the group. Evolutionism gave substance to the existing concepts of divergence and of trends linking sequences of fossils through geological time. Systematics remained an important branch of biology, and at least some systematists began to articulate the evolutionary principles that would become the foundations of modern cladistics. But paleontology could address a much wider range of evolutionary problems and provide concrete evidence of steps taking place in time. As more fossils were discovered, this activity gained in importance, so that by the end of the century it had become one of the most active—and most widely publicized—manifestations of evolutionism. Paleontologists sought not only to reconstruct the relationships between sequences of fossil forms, but also to define the underlying evolutionary trends (sometimes called 'laws') and even to explain the forces which produced these effects.

There have been claims that the actual practice of systematics was scarcely affected by the advent of evolutionism. But the evidence seems to suggest that systematists were becoming increasingly aware of the need to think in phylogenetic terms, especially those dealing with invertebrates. In areas where the fossil record was scanty, the morphology of living forms and the interpretation of structural relationships for the purpose of classification might offer the only window on the past history of a group. Thus entomologists in Britain, America, and Germany sought to interpret the relationships between insect groups in terms of an evolutionary tree. Even in areas where the fossil record was good, systematists dealing with living forms were aware of the need to think in phylogenetic terms. Some of this work anticipated the cladistic principles later articulated by Willi Hennig. In effect, systematics and morphology worked on similar problems and used similar techniques. Many systematists worked with superficial characters, while morphologists used the minute analysis of internal structures, but both disciplines created a demand to know why the relationships among species fall into the kind of 'branching tree' pattern we observe.

Systematists could apply evolutionary principles by attempting to identify the branching-points in the treelike process of diversification which had the living-species as the end-points of its branches. Their work was, however, necessarily limited in its ability to depict the history of life as a process going on against the backdrop of the earth's geological history. It was the paleontologists who were able to exploit the idea of evolution as a natural process in time, as opposed to a rather abstract genealogy. The fossil record became, and has remained, the most visible source of information on the history of life on earth. Of course the paleontologists themselves had to classify the fossil organisms they studied: these relationships, along with the sequence in time based on stratigraphy, were the source of their hypothetical phylogenies.

Systematists and paleontologists had different interests, however, that sometimes led to conflicts of interpretation. For a phylogenetic systematics to be possible, it was necessary to assume that the invention of each new structure or feature had occurred only once—if this were not the case, then what are now called 'shared derived characters' would not be a sign of common ancestry. In modern taxonomy, only genuine homologies (i.e., features derived from a common ancestor) count as characters that may be taken into account when classifying. But before the emergence of the modern synthesis, paleontologists were quite willing to imagine evolutionary trends affecting parallel lines of development, leading to the multiple production of identical structures. The products of parallel or convergent evolution are not homologies and do not count as 'characters' in modern classification. For this reason, taxonomists have been reluctant to admit the possibility of significant amounts of convergent evolution generating illusory homologies. Yet some early twentieth-century paleontologists wanted to restructure the whole system of biological taxonomy around the parallel lines of development they perceived in the fossil record. Systematists and paleontologists could thus disagree significantly over the nature of the evolutionary process and over the basis of taxonomy. We tend to lose sight of the systematists' perspective because the fossil record seems so obviously the primary source of information on evolutionary questions—but previous generations of paleontologists advocated non-Darwinian versions of evolutionism that would be quite unacceptable today.

Given the massive increase in the number of fossils becoming available, especially from later periods in the history of life, most paleontologists developed a sense of confidence that at least some areas were yielding up their secrets. Paleontology had strong links with stratigraphy, of course; this was particularly important for those employed by museums or geological surveys, and the subject was often taught in geology departments of universities, especially in Germany. Some of the earliest evidence for small-scale evolutionary sequences came from the minute examination of the invertebrate remains found in Tertiary strata. Thus Franz Hilgendorf demonstrated a continuous sequence of nineteen varieties of the snail *Planorbis* in the strata at Steinheim, while Melchior Neumayr produced more convincing sequences for the Miocene gasteropod *Paludina* in 1875.[1] On a larger scale, but with significant gaps between the component species, there were many efforts to reconstruct the ancestry of modern mammals. The increasing precision with which the geological strata could be assigned to a sequence valid for many different parts of the world was a major factor in opening up new areas of study. As we shall see in the following chapter, paleontologists could now interact with biogeographers to chart the migration of animal groups from one continent to another in the course of their evolution. It was also possible to study the interaction between different animal groups, including cases where one drove another to extinction. The possibility of environmental changes triggering important episodes in evolution could also be explored. In this area, at least, paleontology was king: it alone could provide hard evidence of the actual course of a group's evolution.

1. See Melchior Neumayr, *Erdgeschichte,* p. 18, and for a good outline of Hilgendorf and Neumayr's work, see O. Abel, *Paläobiologie und Stammesgeschichte,* pp. 25–36. On German paleontology, see W.-E. Reif, "The Search for a Macroevolutionary Theory in German Paleontology." See also M. Rudwick, *The Meaning of Fossils,* p. 255, which also discusses the continuous sequence of *Micraster* fossils in the English chalk obtained by C. J. A. Meyer.

Although much attention focused on fossils that might fill in genealogies for living species, paleontology also helped to generate a more complex image of the history of life by providing evidence of bizarre extinct forms that had to be counted as dead-ends on the tree of life. The dinosaurs were perhaps the classic example of a group that had flourished for a while only to be swept away. It is worth noting that the true complexity of the 'age of reptiles' only became apparent in the late nineteenth century. When Richard Owen created the term *dinosaur* in 1841, there were only three species available, all based on very incomplete specimens. Owen's famous reconstructions of the dinosaurs at Crystal Palace in London showed them as merely gigantic lizards.[2] The much better preserved specimens of *Iguanodon* discovered at Bernissart in Belgium allowed Louis Dollo to reconstruct this particular dinosaur far more effectively, showing that (unlike Owen's reconstruction), it had been bipedal.[3] The 'fossil feud' between Edward Drinker Cope and Othniel C. Marsh turned up more dinosaurs from the American west, revealing an ever-increasing range of bizarre reptilian forms.[4]

In the early twentieth century the American artist Charles Knight gained a reputation for his realistic portrayals of the life of earlier geological periods. Knight described the problems of reconstruction and claimed that after years of experience a team of paleontologists and artists could gain "a certain canniness" in working out what an animal would have looked like when alive (see fig. 6.7; see also Heilmann's reconstructions, figs. 6.2, 6.3).[5] Unfortunately it was not always this simple, and rival reconstructions of the same fossil were produced even where the specimen was reasonably complete. These rival interpretations were based on different views on what the animal's habits had been, even where geology gave some clue about the environment within which it had lived. Adrian Desmond has recorded the debates that took place over reconstructing the various dinosaur species, including one centered on whether or not the gigantic *Diplodocus* had lived mostly in the water.[6]

Late nineteenth-century surveys of the fossil record such as Neumayr's

2. See, for instance, Adrian Desmond, *Archetypes and Ancestors,* chap. 4.

3. Louis Dollo, "Première note sur les Dinosauriens de Bernissart." See David B. Norman, "On the History of the Discovery of Fossils at Bernissart."

4. On this episode, see, for instance, Url Lanham, *The Bone Hunters;* Robert Plate, *The Dinosaur Hunters;* Elizabeth Shor, *The Fossil Feud between E. D. Cope and O. C. Marsh.* For biographies, see H. F. Osborn, *Cope: Master Naturalist,* and C. Schuchert and C. M. Levene, *O. C. Marsh: Pioneer in Paleontology.* More generally, see Desmond, *The Hot-Blooded Dinosaurs,* and F. Buffetaut, *A Short History of Vertebrate Paleontology,* chap. 7.

5. Charles R. Knight, *Before the Dawn of History,* p. 18.

6. Desmond, *Hot-Blooded Dinosaurs,* chap. 5.

Erdgeschichte or Arthur Smith Woodward's *Outline of Vertebrate Palaeontology* make frequent references to evolutionary questions without being dominated by the topic. By the early decades of the new century, it had become fashionable for paleontologists to write popular works more directly aimed at giving an overall view of the history of life. These included evolutionary trees and reconstructions of the appearance of extinct animals, giving the books a much more 'modern' appearance. Among such books were Henry Fairfield Osborn's *Origin and Evolution of Life* of 1917 and Othenio Abel's *Paläobiologie und Stammesgeschichte* of 1929. The Americans were particularly active in this field, perhaps because the rocks of their own continent were yielding many of the best fossils, and even a German survey such as Abel's uses a great deal of American material, including diagrams and photographs.

At the same time, a few paleontologists shared the misgivings of the experimental biologists who rejected the reconstruction of phylogenies as a wild goose chase. Reviewing Samuel Wendell Williston's *Water Reptiles of Past and Present* in 1915, Richard Swann Lull noted that the discovery of so many bizarre fossil reptiles had "weakened our faith in all attempts to trace out the genealogies of the reptilian orders."[7] Lull accepted that "classification is mere genealogy" and insisted that paleontology was absolutely necessary to help the systematist work out the pattern of evolutionary relationships—yet at times he seems to have despaired of putting the project into practice.

Such comments merely reinforced doubts that had been expressed since the 1890s, when Karl von Zittel and Charles Depéret had both warned about the speculative character of the proposed genealogies based on incomplete fossil evidence.[8] William Berryman Scott's *History of Land Mammals in the Western Hemisphere* of 1929 admitted that phylogenetic trees had become a laughingstock in some quarters due to the divergence of paleontologists' opinions. This situation had emerged, he claimed, because biologists adopted different assumptions about the nature of the evolutionary process, disagreeing in particular over the extent of convergence and parallelism.[9] Scott's book is indeed rather evasive about the actual pattern of evolution in the mammalian groups: in effect he was presenting an outline of the history of life that did *not* depend on tracing detailed phylogenies between the inhabitants of successive geological epochs. Others—including Lull himself in his book *Organic Evolution*—were less defensive about the postulation of hypothetical genealogies. But

7. R. S. Lull, review of S. W. Williston, *Water Reptiles of Past and Present.*

8. See C. Depéret, *The Transformations of the Animal World,* chap. 12, discussing Zittel's address to the International Congress of Zoology in Zurich.

9. W. B. Scott, *A History of Land Mammals in the Western Hemisphere,* pp. 645, 649–650.

we need to recognize that paleontologists' efforts to reconstruct the history of life were not tied down by the limitations of the phylogenetic approach. Their increasing appeals to geographical and environmental factors were in part an attempt to show that significant advances could be made even where exact genealogies could not be traced. Zittel included discussions of changing geographical relationships through time in his *Handbuch der Palaeontologie,* despite his misgivings about genealogies.[10] Knight's popular illustrations of the inhabitants of past periods evoked an image of the overall sweep of life's history without the need to reconstruct detailed evolutionary links.

For those groups where the fossil record was good, paleontologists could trace the ancestry of modern forms such as the horse, sometimes recognizing side branches that led off to extinction. In most cases, these phylogenies led back to less specialized ancestors, thus bringing the major branches of the tree closer to their hypothetical junctions. Within each branch, paleontologists could recognize certain regularities. They became aware of the point summed up in Louis Dollo's law of the irreversibility of evolution: A structure once lost could never be recreated. But they were increasingly tempted by the ideas of parallelism and convergence. They thought they could trace developments in which a number of related species advanced in parallel through a fixed pattern of development. Such ideas were linked to the recapitulation theory and to non-Darwinian theories in which evolution was driven by predetermined trends. The recapitulation theory lost popularity in the later decades of the nineteenth century, but it was not eclipsed completely. It was embryological evidence that forced paleontologists to reconsider the trituberculate theory advanced by Cope and others to account for the evolution of mammalian molar teeth.

Some of the trends revealed by the fossil record also dovetailed with the growing sense that the history of life on earth was governed by predictable cycles. Extinct groups such as the ammonites or some of the archaic mammals seemed to appear, flourish, and then decline toward extinction in a sequence that paralleled the life history of an individual organism. The age of reptiles provided an example of a whole class undergoing such a cycle of development, and in this case the demise of the dinosaurs had been directly responsible for the belated rise of the mammals. Anti-Darwinian evolutionists made much of these cycles of rise and decline: they might indicate that each group was somehow endowed with a fixed quantity of evolutionary energy, or that adaptive trends might

10. See the extract from Zittel's *Handbuch der Palaeontologie,* IV, pp. 721–768, translated as "The Geographical Development, Descent, and Distribution of the Mammalia"; see also Zittel, *Textbook of Palaeontology,* III, pp. 290–310.

get out of hand and produce dangerously overdeveloped structures. But predetermined evolutionary forces were not the only way of explaining the cyclic pattern of development. Darwin himself realized that a successful new group would expand its territory, wiping out its less highly evolved predecessors, until it in turn was eclipsed. Struggle occurred between different types of animals, with extinction the penalty for failure. Evolution became a kind of experimental process, with the actual course of progress being determined by trial and error.

Another important change took place in the early twentieth century. In the immediately post-Darwinian era, morphologists such as Haeckel and Huxley had tended to assume that the main branches of a class such as the Mammalia had a hidden ancestry stretching far back into geological periods antecedent to the one in which the first known specimens appeared. By the early years of the new century it was becoming much more fashionable to assume that the bursts of what Osborn called "adaptive radiation" had occurred relatively suddenly at certain points in the history of life. In the case of the mammals, at least, the major radiation did not coincide with the class's origin: it took place much later when the eclipse of another class, the reptiles, created the opportunity for expansion. This new interpretation paved the way for a significant shift away from the old idea of a relatively continuous evolutionary process toward one in which continuous trends were occasionally interrupted by quite dramatic events. Geographical factors such as the breakdown of barriers might trigger major extinctions. More important in the eyes of some geologists was the possibility that the earth's physical history might be punctuated by relatively sudden transformations of the climate that would initiate episodes of mass extinction and evolutionary innovation. Both the origin of new classes and the elimination of previously successful ones might be triggered in this way. The progress of life on earth depended on a series of unpredictable episodes of environmental transformation.

It would be impossible to give a comprehensive account of the efforts made to reconstruct all branches of the tree of animal life in a single book. What follows is a survey of a representative sample of debates in both invertebrate and vertebrate phylogeny, aimed at providing an overview of the issues confronted by the biologists involved. If I have tended to concentrate on vertebrate, and especially mammalian, evolution, it is because some of the most innovative techniques and ideas pioneered in the early twentieth century depended on the availability of a much better fossil record for the recent geological past, where mammalian remains are plentiful. Fortunately, this area of the history of paleontology is well served by existing secondary sources, and parts of this chapter will present an overview of points already established by other scholars.

PUTTING THINGS TOGETHER

In chapter 13 of the *Origin of Species*, Darwin expressed the view that all biological classification was, or should be, genealogical. If the concept of 'relationship' was to have any meaning beyond an imaginary plan of divine creation, it would have to rest on the notion of descent. Two forms were closely related if they shared a recent common ancestor: the large number of characters they shared in common were inherited from that ancestor. Evolution explained why biologists were able to classify species into nested groups in the manner pioneered by Linnaeus. But Darwin also implied that recognition of this point would not necessarily alter the actual practice of classification, since the technique already in use anticipated the evolutionary explanation: "Community of descent is the hidden bond which naturalists have been unconsciously seeking."[11] If this were the case, then it would not matter whether the naturalists constructed a hypothetical evolutionary tree for a group or not; the actual classification scheme would remain the same. In fact, however, considerable disagreement emerged between those systematists who self-consciously sought to base their classifications on trees and those who did not. Working out a hypothetical genealogy did sometimes affect the way species were classified, and some naturalists felt that this approach was being taken too far.

In his *Evolution: The Modern Synthesis*, Julian Huxley echoed the claim that evolution had not had much effect on taxonomy, attributing it to the botanist John Gilmour.[12] Mary P. Winsor has pointed out that although Gilmour had made this point in a 1936 address to the British Association, he had not published it.[13] He later became notorious for claiming that taxonomy should ignore evolution altogether. Biological species should be classified like anything else, by simply assessing the number of taxonomically significant characters in which they resemble each other.[14] Huxley excused the apparent lack of influence postulated by Gilmour with an appeal to Darwin's own argument that a good classification was, in effect, a genealogical one anyway. But to judge from the fact that some biologists went out of their way to *reject* the use of phylogenetic trees in classification, it would seem that the influence of evolutionism was greater than either Gilmour or Huxley were prepared to admit. In some cases systematists had drawn up evolutionary trees and had allowed their assessment of which characters were most significant to be influenced by their preconceived view of which were the 'real' relationships.

11. Darwin, *Origin of Species*, p. 420.
12. Julian Huxley, *Evolution: the Modern Synthesis*, p. 401.
13. Mary P. Winsor, "The Lessons of History."
14. J. Gilmour, "Taxonomy and Philosophy."

Some systematists ignored evolution anyway—the ichthyologist Albert Gunther is one example we have already encountered. Even so convinced an evolutionist as T. H. Huxley could publish an *Introduction to the Classification of Animals* in 1869 which talks only of a "morphological classification" based on underlying structural affinities, with no mention of why these relationships exist.[15] Huxley certainly accepted that the ancestry of particular modern forms might sometimes be traced out in the fossil record—a theme to which we shall return below. But in most areas he thought that the uncertain state of phylogenetic knowledge ruled out its use as a basis for taxonomy.[16] This uncertainty did not necessarily mean that taxonomy was incompatible with an evolutionary world view. Huxley seems to have shared Darwin's opinion that, in the end, a good morphological classification would probably reflect phylogenetic relationships. His classic textbook *The Crayfish* presents a treelike diagram of "demonstrable" morphological relationships but then goes on to insist that evolution is the only reasonable explanation of the facts.[17]

Huxley realized that some groups present a bewildering array of living species that need to be classified. They either have a very poor fossil record or a record that simply adds to the range of diversity without offering direct clues about ancestry. Darwin had faced the problem of classifying in these circumstances when he wrote his monographs on the living and fossil barnacles. He had used this work to throw light on what he believed to be the phylogenetic relationships involved, although the evolutionary implications were concealed in the published texts.[18] In areas where there is such a bewildering diversity, the systematist could either chose to ignore phylogeny—for instance, by expressing relationships in a form that could not be mapped onto an evolutionary tree—or could propose hypothetical trees to be used as guides to assess which morphological relationships were the more significant for classification. The success of Darwin's theory ensured that many systematists of the following generation would explore the possibility of constructing evolutionary trees.

At a fairly early stage in the process, however, some systematists decided that the problems lying in the way of setting up hypothetical genealogies were so great that this whole approach was counterproductive. The earliest expression of such a protest was Alexander Agassiz's 1880 address as vice president to the American Association for the Advance-

---

15. T. H. Huxley, *An Introduction to the Classification of Animals*, p. 2.

16. Huxley, "On the Classification of the Animal Kingdom," pp. 36–37.

17. Huxley, *The Crayfish*, pp. 252, 319.

18. On Darwin and the barnacles, see Michael Ghiselin, *The Triumph of the Darwinian Method*, chap. 5.

ment of Science.[19] Agassiz was a specialist on echinoids (sea urchins), a group with a good fossil record, and he began by sketching in exactly the kinds of parallels between evolutionary and embryological development that many paleontologists were using. But he then dropped a bombshell by insisting that this technique gave only illusory relationships and should be abandoned. A tree constructed by using one character as the basis for classification would differ from that based on another, and given the vast number of characters that could be used, there was no way of knowing which, if any, tree was a valid representation of the group's history. One reason for the lack of an obvious phylogenetic pattern was the possibility that an ancestral character could sometimes reappear later in a group's history. Another was, of course, convergence: the production of the same character independently by two different lines of evolution within the group.[20]

In effect, Agassiz was saying that the evolutionary process is too complex to allow us to reconstruct the phylogeny of any group from a comparison of the characters produced in individual species. We simply do not know which resemblances are indications of community of descent and which are the products of other evolutionary processes. All we can do is classify by degrees of resemblance without reference to how those resemblances are produced. Many later biologists would adopt the same position, although Agassiz was going against the prevailing enthusiasm for evolution at the time. Paleontologists certainly thought that they could trace genealogies, at least in some areas of the fossil record. And—despite Agassiz's skepticism—the recapitulation theory was still thought to provide an independent method of determining which characters were primitive. Perhaps the closest equivalent to Agassiz's skepticism may be found in the views of non-Darwinian anatomists such as St. George Mivart, whose enthusiasm for the idea of parallel evolution also led them to reject the creation of monophyletic trees in favor of multistemmed bushes that would be far harder to reconstruct.

A determined body of zoologists had set out to use the idea of evolutionary descent as a means of understanding the relationships between species. The classic starting point was Fritz Müller's *Für Darwin* of 1864, which we discussed in chapter 3. Müller's appeal to the recapitulation theory was in part an attempt to show that another line of evidence could be brought to bear in determining which characters were truly primitive. He warned against the use of highly variable characters in taxonomy and against the possibility that adaptation might produce similar results in un-

---

19. Alexander Agassiz, "Paleontological and Embryological Development." See Mary P. Winsor, *Reading the Shape of Nature*, pp. 154–163.

20. Agassiz, ibid., p. 396.

related species. Müller dealt with the relationships within a single crustacean genus, *Melita*, showing how the possession of a unique, specialized character was the best evidence for common descent. He produced diagrams resembling modern cladograms to depict the relationships.[21] His was a pioneering effort to show how an understanding of evolutionary principles could be used as a guide to taxonomy.

Robin Craw has argued that a generation or more of German-speaking systematists built upon the principles laid down by Müller to create a phylogenetic systematics.[22] Entomologists were particularly active, their work culminating in Anton Handlirsch's *Die fossilen Insekten und die Phylogenie der rezenten Formen* of 1908. Some defined relationships solely in terms of branches in the evolutionary tree, a position subsequently embodied in modern cladistics. Craw sees a transition from the traditional evolutionary tree, with side branches depicted as somehow more 'primitive' than the 'highly evolved' members of the group, to the modern convention of listing all extant species in a terminal position, which eliminates the distinction between primitive and advanced forms. Robert J. O'Hara has traced similar developments in the classification of birds.[23] Max Fürbringer's 1888 study of bird phylogeny depicted a three-dimensional evolutionary tree with the typical, rather confused attempt to use the vertical axis as a measure of both geological time and evolutionary progress. But R. B. Sharpe in 1891 contrasted this ambiguous phylogeny with a branching diagram in which all living species are equivalent. In the following diagram, the contrast between the phylogeny on the left and the cladogram (to use a modern term) on the right indicates the distinction that Sharpe was beginning to recognize; C may be a 'primitive' species, but it has to be classified in the same way as A and B:

Some taxonomists were thus recognizing that the concept of a multiple-branched evolutionary tree could be divorced from the old image of a scale of development. Discussions of the philosophy of systematics, and the relationship between systematics and evolutionism, continued in German

21. Fritz Müller, *Facts and Arguments for Darwin*, p. 11.
22. Craw, "Margins of Cladistics."
23. Robert J. O'Hara, "Diagrammatic Classification of Birds," and "Representations of the Natural System."

biology through the early decades of the twentieth century. Important contributions by Walter Zimmerman and the ethologist Konrad Lorenz helped to inspire the later work of Willi Hennig, who built the framework of modern cladistics.

The same issues were also explored by English-speaking systematists. Pamela Henson has drawn attention to the work of the American entomologist John Henry Comstock, professor of invertebrate zoology at Cornell University from 1874 to 1914.[24] Comstock created an active research school based on an evolutionary approach to insect taxonomy, and his students were influential throughout America. He recognized that the evolutionist had to determine which were the more primitive characters and which had undergone significant change in the course of a group's evolution. By analyzing the structure of wing veins in the Lepidoptera (butterflies), he was able to determine which structures were the basis from which others evolved. He argued that the loss of a character was not a sign of relationship, since this might happen independently in two different lineages. Relationships were based on the appearance of unique new characters. Paleontological and embryological evidence was used where available to help determine the sequence in which characters had appeared. Comstock was aware of the widespread belief that convergent evolution might produce apparently identical structures that could mislead the taxonomist into thinking they were a homologous character. But he believed that he could distinguish between similar structures produced by two independent evolutionary routes, thus avoiding the problem.

The British entomologist Edward Meyrick also used an evolutionary approach to taxonomy in his influential *Handbook of British Lepidoptera,* published in 1895 and reissued in 1927. Meyrick argued that the metaphor of an evolutionary tree was misleading: it was better to visualize species as bubbles on the surface of a sea, the surface itself rising at a steady rate to represent the passage of time. One bubble can occasionally bud off from another, representing the appearance of a new species, and the relationship between the bubbles can be traced if each trails a filament behind it as it rises. The taxonomist was, in effect, trying to reconstruct the branchings of these filaments.[25] Meyrick proposed three laws to govern the determination of which are primitive characters (the 1927 edition noted with pride that Comstock had subsequently called them "Meyrick's laws"):

1. No new organ can be produced except as a modification of some previously existing structure.

24. Pamela Henson, "The Comstock Research School in Evolutionary Entomology."
25. Edward Meyrick, *Revised Handbook of British Lepidoptera,* p. 12.

2. A lost organ cannot be regained.
3. A rudimentary organ is rarely redeveloped.[26]

Note that item 2 is the law of the irreversibility of evolution proclaimed a few years later by Louis Dollo and usually attributed to him.[27] Meyrick suggested a phylogeny for the Lepidoptera based on principles of trans-formation of wing structure similar to those used by Comstock.

Entomologists were not the only systematists involved in the effort to lay down the foundation for an evolutionary taxonomy. In a 1905 pa-per on the structure of the intestinal tract of mammals, the secretary to the Zoological Society of London, Peter Chalmers Mitchell, articulated one of the basic principles of modern cladistics, sometimes still called 'Mitchell's theorem.'

> Likenesses which are due to the common possession of primitive features cannot be regarded as evidence of near relationship; that certain members of a group have retained what was once the property of all the members of that group can be no reason for placing such creatures close together in a system, if the system is to be based on blood relationship.
>
> Resemblances, on the other hand, that are new acquisitions, that depend on definite anatomical peculiarities, must be the most likely field for the discovery of clues to affinity, simply because it appears to be less probable that the same anatomical device should have been produced independently than that it should have been acquired only once.[28]

In modern terminology: Only shared derived characters can be made the basis for a monophyletic group.

Note that Mitchell explicitly rules out the possibility of the indepen-dent invention of the same structure by convergence; to assume that such a process can occur is to throw the whole principle of phylogenetic tax-onomy into disarray. It seems that systematists working with some groups found it far easier to believe that the effects of convergence could be de-tected and eliminated. The whole point of Agassiz's objection had been that (in his experience, at least) it was impossible to make conclusive de-cisions on which features were homologies and which were homoplasies. Yet many zoologists in the decades around 1900 accepted that evolution could be used as a guide to classification. Arthur Dendy, himself an inver-tebrate systematist, proclaimed that the Linnean taxonomic hierarchy was equivalent to the evolutionary tree in his popular 1908 textbook *Outlines*

26. Ibid., p. 14.
27. See Dollo, "Les lois d'évolution," and S. J. Gould, "Dollo on Dollo's Law."
28. P. Chalmers Mitchell, "On the Intestinal Tract of Mammals," pp. 528–529.

of Evolutionary Biology.[29] Some naturalists, however, felt that the proliferation of rival evolutionary trees confirmed Agassiz's misgivings. We can judge the scope of the problem from the fact that Bashford Dean's 1895 survey of ichthyology depicted seventeen different phylogenies for the class of fish offered by other systematists.[30]

Equally serious were the reservations of those dealing with more limited groups. Mary P. Winsor notes that the monograph on crinoids by Charles Wachsmuth and Frank Springer, published by Agassiz's own Museum of Comparative Zoology in 1897, explicitly refused to offer a phylogenetic tree because neither author felt that he understood evolution well enough to construct one.[31] They were criticized in a review by the British expert Francis Bather, who conceded that phylogenetic speculation was often taken to excess, but insisted that "of all the virtues prudence is the most uninteresting, and an author should at least have the courage of his convictions."[32] In effect, Bather was supporting the claim that phylogenies, even if proved wrong, served as a stimulus to further work (he would, no doubt, have welcomed Karl Popper's view of the scientific method). A purely morphological classification was safer, but offered no further stimulus to understanding the relationships. Those who hoped to set up a phylogenetic scheme were, in fact, sometimes thwarted by the impossibility of dealing with the practical problems. Thus Carl Eigenmann's classification of the Characidae, a group of South American fish— also published by the Museum of Comparative Zoology—noted how the phenomena identified by Agassiz interfered with the attempt to determine phylogenetic links. He gave a diagram of relationships radiating from a center that could not be mapped onto an evolutionary tree.[33]

The problems were also recognized in W. K. Gregory's 1910 monograph, "The Orders of Mammals." The fossil record for the mammals was comparatively good, yet Gregory acknowledged the difficulties faced by a systematist seeking to erect a phylogenetic classification.

> The greatest stumbling blocks of the phylogenist lie: first in the difficulty of distinguishing between primitive and specialized characters, secondly in the tendency to assume relationships between two given forms on the basis of

29. A. Dendy, *Outlines of Evolutionary Biology,* chap. 16.

30. Bashford Dean, *Fishes, Living and Fossil,* pp. 282–283.

31. See Winsor, "The Lessons of History," and C. Wachsmuth and F. Springer, "The North American Crinoidea Camerata," vol. 20, p. 166.

32. F. Bather, "Review of . . . Wachsmuth and Springer," p. 123.

33. See Winsor, *Reading the Shape of Nature,* pp. 228–230, and C. Eigenmann, "The American Characidae," p. 49.

resemblances that may have been brought about by either parallel or convergent evolution.[34]

Gregory noted the generally accepted point that the greatest taxonomic value should be ascribed to so-called nonadaptive characters, since these were less likely to have the ancestral state obscured by later adaptive modifications.[35]

In 1927, Francis Bather, originally a strong proponent of the phylogenetic approach to taxonomy, gave a presidential address to the Geological Society of London in which he too now proposed that it might be time to call a halt. He accepted the theory of parallel evolution, in which several related species might independently transfer from one genus to another in the scale of development. A classification that treated the lineage of each species as more important than the generic characters would require a complete revision of the system that might be totally impractical. But the implications of parallelism were trivial when compared to the possibility of exactly equivalent forms produced from different origins by convergence: "To claim that the same species can blossom twice on the same stem is a simple matter compared with the suggestion that two stems of different history can terminate in the same species."[36] The water snail *Planorbis* acquired a similar shell structure whenever it adapted to fresh water; the species possessing this structure had not inherited it from a common freshwater ancestor—so they could not be placed in the same genus despite their great similarity. The character would have to be treated as a grade into which many species had moved independently.[37] In the end, Bather concluded that the whole history of life was "riddled through and through with polyphyly and convergence."[38] Even if it were possible to reconstruct the tree of life, it might be impossible to use it as the basis for a usable classification.

Bather's call for a halt to phylogenetic theorizing, at least until the implications of the policy were better understood, marks a growing willingness in early twentieth-century systematics to admit that, in effect, Agassiz may have been right to caution against the attempt to use evolution as a guide to classification. When Julian Huxley came to edit *The New Systematics* in 1940, he proclaimed that new developments were putting evolutionism on a firmer footing but conceded that in some areas

34. W. K. Gregory, "The Orders of Mammals," p. 105.

35. Ibid., p. 111.

36. Bather, President's Adress: "Biological Classification," p. xciv.

37. Ibid., p. xcvi.

38. Ibid., p. ci.

parallelism and other forms of nondivergent evolution made it impossible to construct a phylogenetic classification.[39] Several articles in his book illustrated these difficulties, including the discussion of paleontology by W. J. Arkell and W. A. Moy-Thomas,which stressed the role of parallelism.[40] It is significant that those systematists who expressed the strongest doubts about the value of a phylogenetic classification were often those who had most to do with the fossil record. Paleontology had become the strongest source of support for the belief that parallelism and convergence were common products of essentially non-Darwinian evolutionary processes. If confidence in the possibility of an evolutionary classification was renewed in the 1940s, it was because the "modern synthesis" proclaimed by Huxley and others helped to undermine the concerns generated by the paleontologists. Modern taxonomists continue to confront the same problem: the attempt to define groups by the appearance of evolutionary novelties will work only if we assume that the same character cannot normally be generated more than once. The rise and fall of confidence in a phylogenetic taxonomy reflects changes in the prevailing view of the possibility of parallelism and convergence.

ADAPTIVE RADIATION

To explore the role played by paleontology in the reconstruction of the history of individual groups, we turn first to the growing recognition that many classes reveal a pattern of divergence from a common ancestral form. This pattern had already been detected by Richard Owen and others in the 1850s, and would soon be seen as evidence for divergent evolution. Some of the most impressive lines of support for Darwin's theory came from tracing the ancestry of well-known and highly specialized modern forms such as the horse back to more generalized mammalian ancestors. These fossil discoveries influenced systematics by demonstrating that the sharp divisions between modern orders become much less distinct when earlier species are taken into account. The image of lines of adaptive specialization radiating from a generalized ancestral form would soon become a standard component of evolutionary biology.

By the turn of the century, however, paleontologists had begun to change their view of the time scale against which the radiation should be measured. Huxley and many of the first generation of evolutionists were convinced that the actual point of divergence must lie far back in the geo-

39. J. Huxley, *The New Systematics*, pp. 16–18. For a modern statement of the view that convergent evolution occurs more often than most systematists allow, see A. J. Cain, "On Homology and Convergence."

40. W. J. Arkell and J. A. Moy-Thomas, "Palaeontology and the Taxonomic Problem." See also Gilmour, "Taxonomy and Philosophy."

logical record, long before the first known examples of the major groups appeared. It was thought that the sudden flowering of the placental mammals in the Eocene, for instance, must be merely the final act of a long process of mammalian divergence concealed by the inadequacy of the fossil record in the Mesozoic. By the end of the century geologists and paleontologists were beginning to assemble far more complete sequences of strata from different parts of the earth. They became less willing to invoke the imperfection of the record to explain the absence of higher mammals in the later Mesozoic. The concept of a relatively sudden burst of adaptive radiation for the mammals at the beginning of the Cenozoic became the standard model. The possibility that there might have been bursts of rapid evolution had major implications for evolutionists' thinking on related questions such as the geological and biological causes of the rise and fall of particular groups in the history of life.

Although Darwin was forced to devote a whole chapter of the *Origin of Species* to the imperfection of the geological record, his next chapter argued that (allowing for the anticipated gaps) the pattern of development revealed by the fossils is compatible with his theory of branching, adaptive evolution. Highly distinct modern groups could be traced back to generalized forms that blurred the distinctions visible in their descendants: "Cuvier ranked the Ruminants and Pachyderms, as the two most distinct orders of mammals; but Owen has discovered so many fossil links, that he has had to alter the whole classification of these two orders; and has placed certain pachyderms in the same sub-order with ruminants: for example, he dissolves by fine gradations the apparently wide difference between the pig and the camel."[41] Note that it was to Owen, often caricatured as an antievolutionist, that Darwin turned for his evidence. Through the 1850s Owen had become convinced that he could trace out a process of specialization within the mammalian orders in the course of geological time, a process which he saw as parallel to that observed by Karl Ernst von Baer in the development of the embryo.[42] The divergence was now to be interpreted in evolutionary terms, and the most continuous of the fossil sequences would become positive evidence for transformation.

The evidence was not always clear cut, and Darwin referred to problems raised by the British paleontologist Hugh Falconer, who pointed out that in many cases the sequence of fossils does not fit the evolutionist's expectations.[43] Darwin argued that, given the imperfection of the record,

41. Darwin, *Origin of Species,* pp. 329–330.

42. See Dov Ospovat, "The Influence of Karl Ernst von Baer's Embryology," and Bowler, *Fossils and Progress,* chap. 5.

43. See Falconer's *Palaeontological Memoirs,* for example, "On the American Fossil Elephant," II, pp. 212–291.

it was not surprising that sometimes the first recorded example of a species might come from a comparatively late point in its history. A newer species might already have budded off before the first fossils of the parent species were laid down. In this case one might actually find the younger species at an earlier point in the record than the parent form.

The evidence attracted most attention when it seemed possible to trace the ancestry of well-known modern forms such as the horse. An important step in this direction was taken by Albert Gaudry, who accompanied an expedition organized by the Paris Academy of Sciences to the rich fossil beds of Miocene age discovered at Pikermi in Greece. These fossils were geologically intermediate between the hitherto best-known Cenozoic faunas, the Eocene and the Pleistocene. Martin Rudwick has described how Gaudry interpreted the fossils as components of an evolutionary sequence linking the earlier and later faunas.[44] He gave treelike diagrams linking the horses of the Pleistocene back to a less specialized Miocene member of the horse family, *Hipparion*. Gaudry found enough fossils of *Hipparion* to show that there was a wide range of variation within the species. He warned against the all-too-common tendency to create distinct species for individual specimens that could well be only variants of a single species. Gaudry thus provided sound support for Darwin's evolutionism, although he disliked the theory of natural selection and continued to argue that the history of life revealed the harmonious plan of the Creator.

Gaudry's analysis of horse evolution was extended by the Russian paleontologist Vladimir Kovalevskii (Kowalevsky), brother of Alexander Kovalevskii, the zoologist who worked on vertebrate origins. Daniel Todes has shown that Kovalevskii's work on fossils was inspired by a desire to confirm Darwin's theory, to which he had been drawn on ideological grounds.[45] Although he was active for only a few years and was never recognized in his native Russia, Kovalevskii produced a series of memoirs on the evolutionary significance of the fossil record of the ungulates (hoofed animals) which were long regarded as models of analysis. He published in German, French, and English, so that his work reached a wide international audience and was hailed by Darwin, Huxley, and many others. Kovalevskii restudied some of Cuvier's fossils and was able to show that Gaudry's *Hipparion* could be linked back in a convincing evo-

44. M. J. S. Rudwick, *The Meaning of Fossils,* pp. 244–249; see A. Gaudry, "Animaux fossiles et géologie de l'Attique." See also Buffetaut, *Short History of Vertebrate Palaeontology,* chap. 6.
45. Daniel Todes, "V. O. Kovalevskii: The Genesis, Content, and Reception of his Paleontological Work."

lutionary sequence to *Anchitherium* and the Eocene *Palaeotherium*.[46] He was also able to demonstrate the functional purposes of the changes revealed by the fossils (discussed below).

The story of horse evolution was taken up by Huxley and by Marsh, who eventually found the definitive sequence of fossils in the American midwest. Huxley had at first been unwilling to offer paleontological evidence in support of Darwin's theory. He had committed himself to the idea of 'persistence of type,' according to which the major groups of animals have continued largely unchanged through most of the fossil record. His 1862 address to the Geological Society, entitled "Geological Contemporaneity and Persistent Forms of Life," stressed the same point despite his conversion to Darwinism.[47] Huxley still insisted that there was little evidence of progress in the fossil record—that is why Darwin's was the only acceptable theory of evolution, because it did not demand constant progress. But the same principle made it unlikely that we would ever find the fossil evidence of evolutionary transformations. He also insisted that there was no evidence of a gradual specialization within any group in the course of the known fossil record.[48]

By the time he gave his 1870 address entitled "Palaeontology and the Doctrine of Evolution," Huxley was prepared to admit that "there is much ground for softening the somewhat Brutus-like severity with which, in 1862, I dealt with a doctrine, for the truth of which I should have been glad enough to be able to find a good foundation."[49] Better evidence for evolution had become available, he argued, especially for the later evolution of mammalian groups. He made the important point that it was necessary to distinguish between "intercalary types," which merely illustrated intermediate stages between earlier and later forms, and genuine evolutionary transitions. The distinction was necessary because "it is always probable that one may not hit upon the exact line of filiation, and, in dealing with fossils, may mistake uncles and nephews for fathers and sons."[50]

Huxley appealed to the work of Gaudry and Kovalevskii on the horse family. Not surprisingly, they had sought the ancestry of the horse in European fossils—the horse had been extinct in modern North America and

46. V. O. Kovalevskii, "Sur l'*Anchitherium aurelanse* Cuv. et sur l'histoire paléontologique des cheveaux."

47. Huxley's addresses, including "Geological Contemporaneity and Persistent Forms of Life," are reprinted in his *Scientific Memoirs,* but the bibliography lists the more easily accessible reprints in his *Collected Essays.* See Desmond, *Archetypes and Ancestors,* chap. 7, and Mario Di Gregorio, *T. H. Huxley's Place in Natural Science.*

48. Huxley, "Geological Contemporaneity and Persistant Forms of Life," p. 303.

49. Huxley, "Palaeontology and the Doctrine of Evolution," pp. 347–348.

50. Ibid., pp. 349–350.

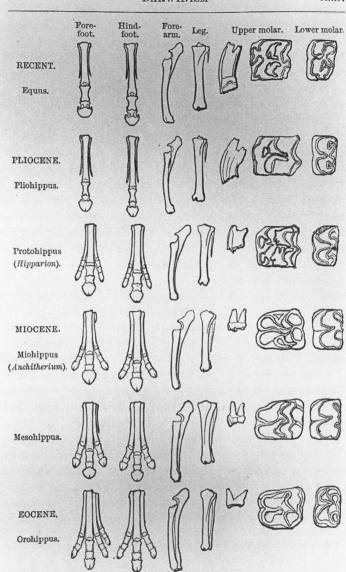

Fore-foot.	Hind-foot.	Fore-arm.	Leg.	Upper molar.	Lower molar.

RECENT.

Equus.

PLIOCENE.

Pliohippus.

Protohippus
(*Hipparion*).

MIOCENE.

Miohippus
(*Anchitherium*).

Mesohippus.

EOCENE.

Orohippus.

FIG. 33.—Geological development of the horse tribe (Eohippus since discovered).

FIGURE 7.1 Modification of the teeth and lower parts of the limbs from the four-toed *Orohippus* of the Eocene to the modern horse. This sequence was constructed from American fossils by O. C. Marsh and was widely regarded as convincing evidence for the evolution of a modern form from a less specialized ancestor. The diagram was frequently reprinted; this example was taken from A. R. Wallace, *Darwinism* (London, 1889), p. 388.

had been introduced by the Europeans. But when Huxley went to America in 1876, he visited Marsh and was shown a far more convincing sequence of fossil horses discovered in the American midwest. In his "Lectures on Evolution" Huxley announced his conversion to Marsh's theory of the American origin of the horse and reproduced Marsh's diagram showing a series of six stages leading from the four-toed *Orohippus* of the Eocene to the modern horse (fig. 7.1).[51] The sequence was, he proclaimed, "demonstrative evidence of evolution."[52] He was also able to announce, in a footnote to the printed text, Marsh's discovery of the even more primitive *Eohippus*, or "dawn horse," which carried the sequence back almost to the hypothetical five-toed ancestor that would link the horse family to the general mammalian type. Marsh's diagram was widely reproduced, and *Eohippus* attracted considerable popular attention over the next few decades, including a story in the *Saturday Evening Post*.[53]

Huxley had originally opposed the view that any evolutionary changes could be traced in the fossil reptiles, but in his 1875 paper "On Stagonolepis Robertsoni and the Evolution of the Crocodilia" he again reversed his original position. He now arranged the crocodiles into three orders—the Parasuchia, Mesosuchia, and Eosuchia—arguing that the sequence represented a transition from a generalized, still lizardlike form to the specialized modern crocodiles. Since this was also the order in which they appeared in the fossil record, the fossil sequence again offered clear evidence of evolution. Desmond notes that Huxley's thesis mapped nicely onto his views on human progress: the crocodiles had undergone an evolutionary spurt and then stagnated, just as social progress is interrupted by occasional "dark ages."[54] Desmond also notes that Huxley made no effort to explain *why* the crocodiles evolved in the way they did. Indeed, he argues, it was Huxley's great rival, Owen, who tried to understand crocodile evolution in terms of the changing ecological balance through geological time. Crocodiles grew larger to prey on the browsing mammals which appeared in the Tertiary. In this case, Owen's search for the 'purpose' of evolution led him to a more modern interpretation of what was going on than was possible with Huxley's purely morphological approach, which merely demonstrated relationships.[55]

The bewildering array of Mesozoic reptiles discovered in the late nineteenth century confirmed that here was a range of diversity fully equiva-

51. Huxley, "Lectures on Evolution," p. 130.

52. Ibid., p. 132.

53. See W. D. Matthew, *Outline and General Principles of the History of Life,* p. 160.

54. Desmond, *Archetypes and Ancestors,* pp. 170–174.

55. Richard Owen, "On the Influence of the Advent of a Higher Form of Life in Modifying the Structure of an Older and Lower Form."

lent to that found in the mammals, although there were few cases like the crocodiles where the fossils were numerous enough to permit the tracing of lineages. Even the dinosaurs were split into two major branches, the Saurischia and the Ornithischia, by Harry Govier Seeley.[56] The latter, the "bird-hipped," showed by their pelvic structure that they might be related to the birds. Since the reptiles were not as well understood as the better-known mammals, it was often difficult to determine exactly what kind of lifestyle these bizarre reptiles had become specialized for, and experts frequently disagreed over how to reconstruct a particular specimen.

On the surface, at least, Huxley and the other paleontologists were creating a 'Darwinian' view of the evolutionary process based on the image of a tree rather than a ladder. The history of life could have no 'main line' of development when each class was seen to diverge into many specialized branches. Cope's "law of the unspecialized" encapsulated this aspect of the new thinking: The most highly developed members of a class were the most specialized, and hence the least likely to give rise to the kind of major transformation that would generate an entirely new class.[57] Robert Broom was so struck by this phenomenon that he saw normal evolution as a process which pushed living things away from the line of progress in search of short-term adaptive gain. It required some kind of divine forethought to preserve a few unspecialized types within each class to serve as the basis for continued progress toward humankind.[58] Specialization, not progress, was the normal pattern of evolutionary development. Beneath the surface, however, progressionist ideas continued to influence everyone's thinking. Even Huxley dismissed the development of the teleostean fish as of little interest because "they appear to me to lie off the main line of evolution—to represent, as it were, side tracks starting from certain points of that line."[59] The image of a main line died hard, despite the new emphasis on specialization.

One characteristic of the early evolutionists' attempts to reconstruct the branches within particular classes was the assumption that a vast timescale was involved, even by geological standards. With his early emphasis on the persistence of type, Huxley was inevitably led to suppose that each class had originated long before the first examples of it appeared in the fossil record. The Eocene mammals were already partly specialized,

56. H. G. Seeley, "On the Classification of the Fossil Animals Commonly Known as Dinosauria."

57. E. D. Cope, The Origin of the Fittest, pp. 232–234, and The Primary Factors of Organic Evolution, pp. 172–174.

58. This is the thesis of Broom's The Coming of Man: Was It Accident or Design?

59. Huxley, "On the Application of the Laws of Evolution to the Arrangement of the Vertebrata," p. 471.

and this led Huxley to argue that the class must have appeared and begun its diversification far back in the Mesozoic: "It is difficult to escape the conclusion that a large portion of time anterior to the Tertiary period must have been expended in converting the common stock of the *Ungulata* into Perissodactyles and Artiodactyles."[60] In chapter 6 we saw that Huxley invoked a good deal of parallelism in the early stages of mammalian evolution. In his 1868 Hunterian Lectures he even suggested that the various marsupial groups had independently evolved into their placental equivalents—placental carnivores from marsupial carnivores, and so on. This was much to the delight of Mivart, although the latter noted that Huxley had soon returned to the orthodox idea of the marsupials as an offshoot of early mammalian evolution.[61] Nevertheless, this episode illustrates just how far Huxley was prepared to go in pushing the separation between the modern orders of mammals as far back in geological time as possible.

Not everyone shared these views. Ernst Haeckel postulated a radiation of the marsupials into four herbivorous and four carnivorous groups in the late Secondary. He noted the possibility of the independent evolution of the placenta in several different lines of marsupial evolution but rejected it in favor of the monophyletic origin of the placentals in the Eocene.[62] Two major radiations had occurred, one for the marsupials and then one for the placentals, with that for the placentals ending just before the known Eocene fossils appeared. Marsh's "Introduction and Succession of Vertebrate Life in America" of 1877, however, supported Huxley's position by insisting that all groups would turn out to have ancestors much earlier than anything seen in the fossil record.[63] Marsh saw the initial branching within the ungulates as having begun in the early Cretaceous, a view shared by Kovalevskii.[64] Charles Depéret insisted in his *Transformations of the Animal World* that the generalized character of the Eocene mammals had been exaggerated, and postulated a divergence beginning far back in the Secondary.[65] Paleontologists were reluctant to admit that the idea of continuous evolution might have to be modified to accept relatively sudden bursts of change at certain points in the history of life. If everything was as slow and gradual as Darwin assumed, then the apparently sudden transition from the age of reptiles to the age of mam-

60. Huxley, "Palaeontology and the Doctrine of Evolution," p. 363.

61. St. G. Mivart, *On the Genesis of Species,* pp. 82–83. Huxley's Hunterian Lectures were not published.

62. Ernst Haeckel, *The History of Creation,* II, pp. 236, 247–248.

63. O. C. Marsh, "Introduction and Succession of Vertebrate Life in America," p. 351.

64. See, for instance, the phylogenetic tree in Marsh's *Dinocerata,* p. 173, and Kovalevskii, "On the Osteology of the Hyopotamidae," p. 150.

65. Depéret, *Transformations of the Animal World,* p. 174; see also p. 107.

mals must, to some extent, be an illusion created by the imperfection of the record. Neither the extinction of the reptiles nor the diversification of the mammals could be as abrupt as it seemed.

By the turn of the century, however, paleontologists began to stress the need to qualify the principle of continuity. As knowledge of the Mesozoic rocks expanded, it became impossible to believe that the mammals could have diversified under the very noses, so to speak, of the dinosaurs. Von Zittel admitted in 1893 that the expected Cretaceous ancestors of the carnivores and ungulates had not been discovered.[66] The mammals had remained insignificant until the great reptiles were removed, and had then radiated very rapidly to fill the range of ecological niches made available. Discoveries in the early Tertiary led to the creation of a new geological period, the Paleocene, preceding the Eocene. These periods marked the earliest stage in the radiation of the placentals, although it also emerged that there had been a radiation of 'archaic' placentals just prior to that which established the more advanced modern groups. The archaic mammals may have begun their evolution in the Cretaceous, but the abrupt flowering of the modern placentals at the beginning of the Tertiary marked the rise of this new class to its dominant position. The evolutionary trees of the early twentieth century thus took on a new appearance. Osborn linked his term *adaptive radiation* to Darwin's principle of divergence, but in fact there was a significant difference in the emphasis of the two expressions.[67] For Darwin, divergence was a tendency at work throughout the history of life, but for Osborn the radiation of the mammals, at least, was a sudden explosion of diversification followed by mainly parallel evolution in the groups thus established (fig. 7.2). The initial branching of the mammals was no longer merely an abstract node in the morphologists' tree of relationships—it was a distinct event in the history of life requiring explanation.

We can see this new approach most clearly in Osborn's monumental *The Age of Mammals* of 1910. Osborn played a major role in coordinating the work of the American and European paleontologists who were establishing a far more detailed model for the sequence of geological formations on a worldwide scale. One product of this was an insistence on the part of vertebrate paleontologists that the disappearance of the dinosaurs marked the Cretaceous-Tertiary boundary. In the later nineteenth century it was still possible to believe that the great reptiles had disappeared gradually. Angelo Heilprin, curator of paleontology at the Academy of Natural Sciences in Philadelphia, argued in 1887 that the

---

66. Zittel, "The Geological Development, Descent, and Distribution of the Mammalia," p. 403.
67. H. F. Osborn, *The Age of Mammals,* pp. 22–23.

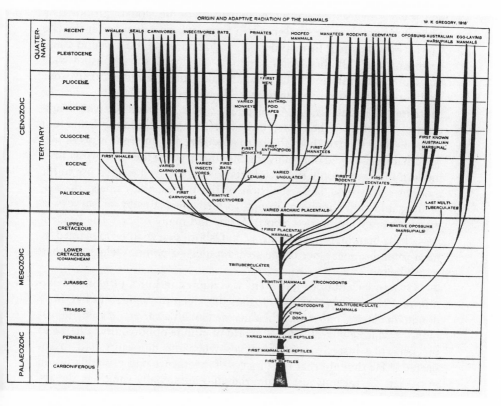

FIGURE 7.2   Ancestral tree of the mammals. From H. F. Osborn, *The Origin and Evolution of Life* (New York, 1923), p. 236.

dinosaurs must have lingered on into the Eocene. But in the early years of the new century this position was challenged by a new generation of paleontologists led by William Diller Matthew, who established that there was a clear break in both the vertebrate and invertebrate fossil sequences.[68] It was also becoming possible to coordinate American and European formations more exactly, which allowed the creation of a far more accurate impression of the biogeography of earlier periods.

In his survey, Osborn went out of his way to stress the suddenness of mammalian radiation. Although the class had appeared much earlier, it had developed very slowly until its rapid expansion in the Eocene.[69] In fact two mammalian radiations had taken place, the first by archaic types,

68. Angelo Heilprin, *The Geographical and Geological Distribution of Animals*, p. 203. On Matthew's involvement in the debate over the Cretaceous-Tertiary boundary, see Ronald Rainger, *An Agenda for Antiquity*, pp. 187–188.

69. Osborn, *The Age of Mammals*, p. 82.

which were then wiped out by the expansion of the modern groups. The archaic mammals were small-brained forms whose origin lay in the late Cretaceous; they had formed the first radiation of the Eocene and could be regarded as "the first grand attempts of nature" to establish a wide range of specializations.[70] The main archaic groups were the carnivorous Creodonta, the primitive ungulates known as Condylarthra, and the short-footed ungulates, or Amblypoda, which included Marsh's Dinocerata.[71] Apart from their small brains, their unspecialized tooth and foot structure had limited their ability to specialize for different ways of life. Most had been wiped out following the emergence of the better-organized modern mammals, only the creodonts having survived to give rise to the modern carnivores. The expansion of the horses, the camels, the elephants, and other modern groups formed the mainstay of surveys of mammalian evolution by Osborn, Scott, and many others. The suddenness of the mammals' takeover was, in Osborn's view, to be explained by the abrupt disappearance of the great reptiles—a point to which we shall return below.[72]

Osborn's account of this event was quoted in Lull's *Organic Evolution* of 1917.[73] Other paleontologists described the sudden radiation of dominant classes, enlivening their accounts with a variety of colorful metaphors. Matthew wrote of the "world of infinite opportunity" offered to the primitive insectivorous mammals when the dinosaurs had disappeared.[74] He thought that the process of divergence had probably begun in the late Cretaceous just before the dinosaurs finally vanished, the apparently abrupt appearance of the placental groups being explained by a migration from some unknown center of evolution.[75] The American Museum of Natural History's Central Asiatic Expedition subsequently found primitive insectivore fossils in the Cretaceous rocks of Mongolia, described by Gregory and Simpson in 1926.[76] Of the Taeniodonts, one obscure group of archaic mammals, Romer wrote, "Obviously the group constitutes an early 'experiment' on the part of the insectivor stock in the creation of a herbivorous form with grinding teeth."[77] H. G. Wells and

70. Ibid., p. 96.

71. For a concise contemporary account of the archaic mammals see also Lull, *Organic Evolution*, chap. 31, esp. pp. 548–559.

72. Osborn, *The Age of Mammals*, p. 97.

73. Lull, *Organic Evolution*, p. 548.

74. Matthew, *Outline and General Principles of the History of Life*, p. 148.

75. Ibid, p. 150; see also Matthew's "The Evolution of the Mammals in the Eocene," p. 960.

76. For brief accounts of these discoveries, see Gregory, *Our Face from Fish to Man*, pp. 51–52, and A. S. Romer, *Vertebrate Paleontology* (1933), p. 270.

77. Romer, *Vertebrate Paleontology* (1933), p. 278.

Julian Huxley wrote of the archaic mammals "blossoming out" to dominate the Eocene.[78] The earlier radiation of the reptiles was described as a series of "striking adventures."[79] The rise of each dominant class was now seen as the expression of a kind of evolutionary creativity, in which life rushed to explore all the avenues opened up to it by a new set of circumstances.

LAWS AND TRENDS

More detailed knowlege of the fossil record allowed paleontologists to study a much wider range of evolutionary problems. They could certainly look for morphological relationships that, if they fell in the right sequence, helped to confirm the basic fact of evolution. But they could also look for generalizations that seemed to apply to many branches of the tree of life and try to understand why those trends occurred. Reconstructing the phylogenies of particular groups thus served as the basis for a more general assault on the question of how and why life had evolved in the manner we observe. Some of the results were codified as 'laws of evolution,' and the status of such laws has plagued the philosophy of biology ever since. Ernst Mayr writes, "Today, the word *law* is used sparingly, if at all, in most writings about evolution."[80]

Some of the general trends in the fossil record were seen as merely very probable, but by no means inevitable, outcomes of the relationships between organisms, or between organisms and their environment. The basic trend toward specialization noted by Owen and Darwin would fall into this category. Any Darwinian or neo-Lamarckian evolutionist would assume that in most cases, the pressure to adapt would cause species to become more specialized for a particular way of life. Functional morphology allowed the paleontologist to study the mechanical causes of such modifications without worrying too much about whether the results were predictable in the sense required by the traditional notion of a law of nature. Dollo (and Meyrick's) law of the irreversibility of evolution falls into the same category: it merely states the extreme unlikelihood of an identical character being recreated in a species that has lost the original version.[81] As Stephen Jay Gould has observed, some later evolutionists incorrectly assumed that Dollo was supporting the idea of orthogenesis, that is, the belief that evolution was driven inexorably in a certain direction by trends

78. Wells et al., *The Science of Life*, p. 473.
79. Ibid., p. 457.
80. Ernst Mayr, *Toward a New Philosophy of Biology*, p. 19.
81. Dollo, "Les lois d'évolution."

somehow built into the germ plasm of the organisms.[82] For those who did believe in orthogenesis, evolution could be governed by 'laws' in a much stricter sense of the term. Thus Osborn—although he appreciated the true purpose of Dollo's law—felt able to write about the "polyphyletic law" and the "law of analogous evolution" because for him these applied to genuinely predictable trends built into the very nature of the organisms themselves.[83] Because the driving force was internal to the organism, the supporters of orthogenesis could imagine a trend being driven to maladaptive extremes, so that it contributed to the species' extinction. Orthogenesis also promoted a belief in evolutionary parallelism, with consequent implications for systematics.

In 1874 O. C. Marsh postulated a "law of brain growth" to denote one of the most general trends observable in the evolution of the mammals, and perhaps in other vertebrate classes.[84] The law figured prominently in his 1877 address, "Introduction and Succession of Vertebrate Life in America." Marsh presented the increase in brain size as a necessary consequence of natural selection: "In the long struggle for existence in Tertiary times, the big brains won, then as now."[85] But he saw the trend as having implications beyond a mere increase in brain size in the course of any evolutionary lineage. In his monograph on the extinct group of gigantic mammals, the Dinocerata, Marsh argued as follows:

1. The brain of a mammal belonging to a vigorous race, fitted for long survival, is larger than the average brain of that period, in the same group.
2. The brain of a mammal of a declining race is smaller than the average of its contemporaries of the same group.[86]

He thus saw evolution as throwing up a succession of new groups, some with bigger brains than others, with selection essentially working to weed out those phyla which were less well endowed (fig. 7.3). Marsh's Dinocerata came to be regarded as typical of the small-brained archaic mammals that were displaced by modern groups. As Osborn wrote in 1893, "Their bodies went on developing while their brains stood still," and the

---

82. Gould, "Dollo on Dollo's Law."

83. Osborn, *The Age of Mammals,* pp. 30–34. Depéret's *Transformations of the Animal World* also lists many laws; see part 2, "The Laws of Palaeontology." On orthogenesis, see Bowler, *The Eclipse of Darwinism,* chaps. 6 and 7.

84. On Marsh's law of brain growth, see Bowler, *Fossils and Progress,* p. 137.

85. Marsh, "Introduction and Succession of Vertebrate Life in North America," p. 377.

86. Marsh, *Dinocerata,* p. 59.

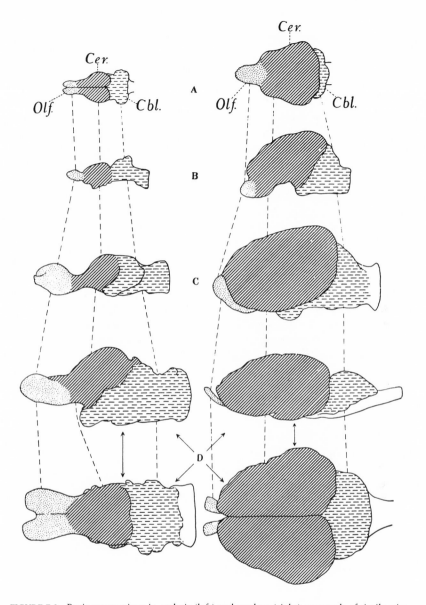

FIGURE 7.3   Brain proportions in archaic (left) and modern (right) mammals of similar size:

Artocyon A Canis
Phenacodus B Sus (domestic)
Coryphodon C Rhinoceros
Uintatherium D Hippopotamus

From H. F. Osborn, *The Age of Mammals* (New York, 1921), p. 173.

"stupid giant fauna" soon "gave way to the small but large-brained modern types."[87]

In an 1880 paper T. H. Huxley invoked the "laws of evolution" to describe the kind of specialization that transformed *Eohippus* into the modern horse. He was "employing the term 'law' simply in the sense of a general statement of facts ascertained by observation," although he was emboldened to use the term by the analogy with the specialization of the developing embryo.[88] He detected three aspects of the trend: the excess of development of some parts, the suppression of others, and the coalescence of originally distinct parts.[89] Although the point is not made clear in this study, we should note that elsewhere Huxley defended the view that evolution would not be absolutely continuous. He seems to have believed that variation would often consist of saltations that would form new species instantaneously. He also thought that variation was normally directed along predetermined channels. If selection had a role, it would be to weed out the maladaptive forms produced in this way.[90]

As Huxley indicated, evolution often proceeded by the elimination of certain parts, compensated by the elaboration of those which remained. One of the most general trends of this nature was the progressive reduction in the number of bony elements in the skull, a phenomenon that became known as 'Williston's law' after the work of the American vertebrate paleontolgist Samuel Wendell Williston.[91]

It was in an attempt to understand the mechanical causes of this kind of transformation that Vladimir Kovalevskii began his study of ungulate evolution. He distinguished two lines of development within what he called the Paradigita (the artiodactyls, or even-toed ungulates), defined by their acquisition of bunodont (tubercular) or selenodont (crescent-shaped) teeth. All too often paleontologists had failed to distinguish between members of the two lines, not realizing that "both lines present a series of parallel modifications, and the parallelism is often carried to the minutest details."[92] Kovalevskii analyzed the adaptive benefits of the re-

---

87. Osborn, "The Rise of the Mammalia in North America," p. 456.

88. Huxley, "On the Application of the Laws of Evolution to the Arrangement of the Vertebrata," p. 458.

89. Ibid., p. 459.

90. See Huxley, "Evolution in Biology," p. 223.

91. See, for instance, Gregory, *Our Face from Fish to Man,* p. 114.

92. Kovalevskii, "On the Osteology of the Hyopotamidae," (abstract), p. 153; see Todes, "V. O. Kovalevskii." Note that the version of this paper designated as an abstract (published in the *Proceedings of the Royal Society*) contains most of Kovalevskii's theoretical conclusions. The longer version published under the same title in the *Philosophical Transactions of*

duction in the number of digits on the feet, founding a tradition of functional analysis in paleontology that was regarded as a model by Osborn and many others.[93] The way in which the bones of the feet coalesced as the number of digits reduced determined the efficiency of the feet in locomotion. Kovalevskii thought that the process could occur in two ways: one, the most straightforward, was inadaptive in the sense that it provided only limited advantages. A more complex kind of coalescence produced the adaptive version of the artiodactyl foot. Although many early lines of artiodactyl evolution had tried the inadaptive method, they had all died out; only those which had adopted the adaptive method had survived to the present.[94] Note that the adoption of the adaptive mode was portrayed as an unusual occurrence, due to "some happy chance" which deflected evolution away from the "more simple and obvious" inadaptive mode.[95] Kovalevskii was thus prepared to see unpredictable changes at certain points in evolution, even though the subsequent changes could be understood as necessary responses to the mechanical constraints of the new structures. He noted that when the most efficient ruminants appeared, they developed "the luxury of all sorts of appendages" in the form of horns.[96]

A functional approach to the study of mechanical specializations became the distinctive feature of the American school of neo-Lamarckism in paleontology. Inspired by the work of John A. Ryder on the modification of tooth structure by mechanical forces, Edward Drinker Cope set out to analyze the causes of specialization in many lines of evolution displayed by the fossil record. Cope introduced the term *kinetogenesis* to denote the direct and inherited effects of motion upon the organism.[97] For the Lamarckian, it was easy to imagine how the bones could be shaped by mechanical forces in the body and the results transmitted to accumulate in future generations. Cope rejected natural selection on the grounds that the apparently linear trends towards specialization seen, for instance, in the horse family, proved that evolution was a directed process—there was no sign of the irregularities one would expect if variation were random.[98]

---

the *Royal Society* (and which alone appears in *The Complete Works of Vladimir Kovalevsky*) does not, in fact, contain the more theoretical material.

93. See, for instance, Osborn, *The Age of Mammals*, p. 6.

94. Kovalevskii, "On the Osteology of the Hyopotamidae," pp. 153–155.

95. Ibid., p. 157.

96. Ibid., p. 164.

97. See Cope, *Primary Factors of Organic Evolution*, chap. 6. On this aspect of neo-Lamarckism, see Bowler, *Eclipse of Darwinism*, pp. 123–126.

98. Cope, *Primary Factors*, pp. 146–150.

Evolution was thus directed by the consciously chosen habits of the organ-isms, a point which Cope emphasized to bring out the moral advantages of Lamarckism.[99] One less-than-obvious extension of the theory was Cope's belief that the process would not be absolutely continuous. He thought that the pressure for change would build up within the organisms over a period of time, and that resistance would give way suddenly at what he called an "expression point," giving a fairly rapid transformation to a new stable type.[100] Cope thus joined the ranks of those paleontologists who believed that evolution would occur through saltations, so that each phylum would still consist of a series of more or less distinct species.

One influential product of Cope's work was his trituberculate theory of the evolution of the mammalian teeth. He postulated a sequence of stages by which additional cusps were added to the original reptilian cone-shaped tooth, the starting point being the tritubercular tooth of the earli-est fossil mammals.[101] The idea was taken up by Osborn and Scott as a means of trying to understand the wealth of mammalian remains, espe-cially teeth, being discovered in the American west. Rival models were proposed, revealing a tension between the paleontologists and the embry-ologists. Cope himself supported the recapitulation theory, but seems to have ignored it in this instance because the development of the teeth in the individual did not fit the model he had proposed. Osborn criticized the alternative theory proposed by W. Kükenthal as the product of "a one-sided morphology which regards only the wonderful though mutilated chapters of embryology when the untorn pages of paleontology are at hand."[102] But the British paleontologist Arthur Smith Woodward warned that embryology threatened the trituberculate theory, while the embryolo-gist Adam Sedgwick rejected the theory as "not worth the effort needed to master its complexities."[103] Goodrich rejected the American theory of the origin of the tritubercular tooth in his 1894 analysis of the British Meso-zoic mammal fossils, although he conceded that the common ancestor of all placentals had teeth of the tritubercular form.[104] American paleontol-ogists themselves abandoned the theory in the early twentieth century. Gregory supported an alternative "wedge theory" in 1910 for the origin

99. For example, ibid., chap. 10.

100. Ibid., p. 24.

101. Cope, "On the Trituberculate Type of Molar Tooth in the Mammalia," reprinted in his *Origin of the Fittest*, pp. 359–362.

102. Osborn, "The Rise of the Mammalia in North America," p. 448.

103. A. Smith Woodward, *Outlines of Vertebrate Palaeontology*, p. 269; A. Sedgwick, *Stu-dent's Textbook of Zoology*, II, p. 504.

104. E. S. Goodrich, "On the Fossil Mammalia from the Stonesfield Slate," pp. 426–429.

of mammalian teeth.[105] In his 1933 textbook, Romer admitted that while the terminology for tooth structure introduced by the theory was useful, it should be stripped of its phylogenetic meaning.[106]

The neo-Lamarckians' emphasis on the organism's direct response to stresses and strains was by no means devoid of influence, even though the inheritance of acquired characteristics came under increasing fire in the early twentieth century. Many of the effects they explored could be explained, if somewhat less directly, by the theory of natural selection. We have noted in preceding chapters that the functional morphology of paleontologists such as Gregory was in part inspired by the work of Cope and Ryder. But there was another aspect of their theory which had a more detrimental influence on developments in paleontology. Even before he became a Lamarckian, Cope had expressed his support for the idea of parallelism in evolution. His 1868 paper "On the Origin of Genera" had postulated evolutionary trends in which several related species could advance independently through a series of developmental stages which defined the genera.[107] The fact that species belonged to the same genus did not mean they were related by common descent—only that their phyla had reached the same stage in the sequence. This idea could be incorporated into the neo-Lamarckian theory because the element of directed evolution allowed the paleontologist to visualize how a number of species could be forced to evolve through the same pattern of changes.

The Lamarckians' emphasis on the power of habit to direct evolution encouraged them to believe that convergence could sometimes allow two long-separated phyla to produce very similar end-products. A good example of this enthusiasm for convergence is provided by the British anatomist Frederick Wood Jones, who pioneered the view that the resemblances between humans and apes are not a sign of close phylogenetic relationship. Beginning in his 1918 article "The Origin of Man," Jones argued for what became known as the tarsioid theory of human origins. Humans and apes had no common ancestor closer than the tarsier; all the resemblances that had led generations of anatomists to link them together were the product of convergent evolution in two stocks which had spent most of their history adapting to an arboreal habitat.[108] A similar position, but without Jones's explicit appeals to Lamarckism, had been suggested ear-

---

105. Gregory, "The Orders of Mammals," pp. 191–194.

106. Romer, *Vertebrate Paleontology* (1933), pp. 249–250.

107. For details of Cope's views, see Bowler, *Eclipse of Darwinism*, pp. 122–123.

108. F. Wood Jones, "The Origin of Man," and *Man's Place among the Mammals;* see Bowler, *Theories of Human Evolution*, esp. pp. 190–193.

lier by A. A. W. Hubrecht.[109] In an era which saw a resurgence of opposition to Darwinian 'materialism,' the advantages of separating humans from apes in this way was not lost on those who had to defend evolutionism. In the 1920s Osborn argued strongly for the claim that the 'dawn man' was not an ape, but the product of a long line of evolution on the plains of Central Asia which had merely paralleled the apes in some respects.[110] Other authorities offered many other examples of similarities between groups that might not be closely related, including the New World and the Old World monkeys and the toothed and whalebone whales.[111]

Osborn's support for parallelism was deeply ingrained in his philosophy of evolution, the product of a tradition which was able to divorce itself from the increasingly discredited Lamarckian theory. For him, parallel evolution was produced by internally programmed trends which pushed variation in a predetermined direction. This was linear evolution, or "orthogenesis," not Lamarckism, and it represents the culmination of a tradition that had been popular among many late nineteenth-century paleontologists. According to this model, once adaptive radiation had divided a class into a multitude of separate branches, each branch continued to evolve in a direction which had been somehow programmed into it at the beginning. Each family consisted of a number of independent phyla which advanced in parallel through a rigid sequence of developmental stages. Since the trend was produced by forces that were not controlled by the interaction between the organism and the environment, the trends might be nonadaptive or even harmful, contributing in their later stages to the eventual extinction of the phyla.

The claim that the fossil record displayed trends in which groups seemed to follow a predetermined course of evolution was made soon after the general theory of evolution became popular in the 1860s. In 1869 the German paleontologist Wilhelm Waagen, who spent much of his career with the Geological Survey of India, introduced his theory of "mutations" to explain trends he found in the evolution of fossil ammonites.[112] The mutations were not sudden changes, but gradual transformations over long periods of geological time in which the phylum seemed to ad-

109. A. A. W. Hubrecht, *The Descent of the Primates;* see Bowler, *Theories of Human Evolution,* p. 118.

110. Osborn, "Recent Discoveries Relating to the Origin and Antiquity of Man"; see Bowler, *Theories of Human Evolution,* pp. 176–179, 195; and Rainger, *An Agenda for Antiquity,* pp. 145–150.

111. The classic account is Arthur Willey's *Convergence in Evolution;* see also Richard Lydekker, *Life and Rock,* chap. 5.

112. W. Waagen, "Die Formenreihe des *Ammonites subradius*"; see Bowler, *Eclipse of Darwinism,* p. 163.

vance steadily in a fixed direction. In a later study Waagen insisted that the trend had nothing to do with the environment; it was produced by "a tendency of organisms to produce an offspring varying in a certain and defined direction."[113]

A similar interpretation of trends in fossil cephalopoda became the hallmark of one of the most prominent supporters of directed evolution, the American paleontologist Alpheus Hyatt. Beginning with an 1866 paper in which he compared the evolution of the whole group Tetrabranchiata to the development of a single individual, Hyatt promoted the view that the history of any group was dominated by trends in which a number of lines advanced in parallel, as though toward a predetermined goal. The ammonite shells became progressively more coiled and elaborate as the group reached maturity, but then gradually simplified as it declined toward extinction.[114] A more fine-grained study of the Tertiary snail *Planorbis* revealed similar apparently predetermined trends.[115] In his later writings, Hyatt accepted a Lamarckian explanation of the progressive phases of evolution, but his main interests continued to center on the idea of parallel lines advancing through a fixed sequence of development. He became best known for his suggestion that a group's decline toward extinction was an inevitable product of built-in evolutionary trends, a point we shall return to below.

Ronald Rainger has shown that Hyatt's approach to the interpretation of fossil sequences was extended by a number of American invertebrate paleontologists, of whom the most influential was Charles E. Beecher of the Peabody Museum at Yale University.[116] We have already noted Beecher's contributions to the study of trilobites (see chapter 3). He also wrote a study of the evolution of spines in various invertebrate groups, a phenomenon he interpreted in orthogenetic terms (fig. 7.4). Like Hyatt he held that each group began with a store of vital energy—this gave an evolutionary plasticity which explained its rapid adaptive radiation.[117] Early forms were smooth, the development of spines being a sign of the onset of decadence after the group had passed its climax: "The first

113. Waagen, "The Jurassic Fauna of Kutch," p. 242.

114. A. Hyatt, "On the Parallelism between the Different Stages of Life in the Individual and Those in the Entire Group of the Molluscous Order Tetrabranchiata," and *Genesis of the Arietidae*. See Bowler, *Eclipse of Darwinism*, 128–130, and Gould, *Ontogeny and Phylogeny*, pp. 91–95.

115. Hyatt, "The Genesis of the Tertiary Species of Planorbis at Steinheim."

116. Rainger, "The Continuation of the Morphological Tradition in American Paleontology."

117. C. E. Beecher, *Studies in Evolution*, section 1, "The Origin and Significance of Spines"; see p. 29 (originally published in the *American Journal of Science*, 1898).

74

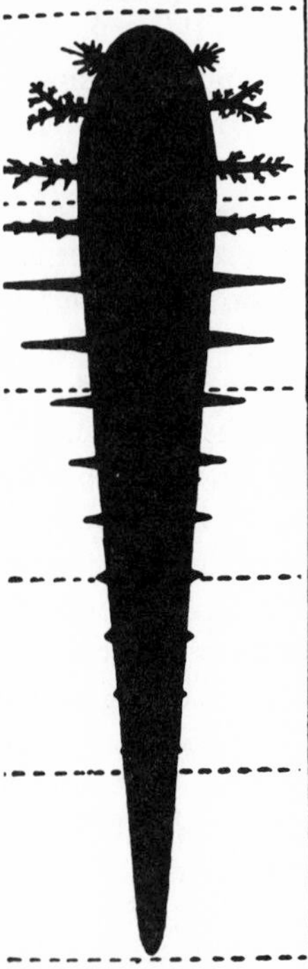

Ontogeny stages.	Ontogeny condition.	Phylogeny stages.	Phylogeny condition.	Chronology.
Old age or gerontic	Paraplasis	Phylogerontic	Paracme	5
Adnlt or ephebic	Metaplasis	Phylephebic	Acme	4
Immature or neanic	Anaplasis	Phyloneanic	Epacme	3
Larval or nepionic	Anaplasis	Phylonepionic	Epacme	2
Embryonic	Anaplasis	Phylembryonic	Epacme	1

FIGURE 7.4   Table showing the development of spinose condition in the course of both individual development and phylogeny of an imaginary invertebrate group. From C. E. Beecher, *Studies in Evolution* (New York, 1901), p. 100.

species are small and unornamented. They increase in size, complexity and diversity, until culmination, when most of the spinose forms begin to appear. During the decline extravagant types are apt to develop, and if the end is not then reached, the group is continued in the small and unspecialized species which did not partake of the general tendency to spinous growth." [118] Beecher argued that the same phenomenon was observed in the horns of vertebrates, citing Richard Lydekker on the parallel between ontogeny and phylogeny in the horns of deer. [119]

Invertebrate paleontology was seen as a major source of evidence for orthogenesis well into the twentieth century. Increasingly, however, the explanations offered for the trends depended not on the exhaustion of evo-

118. Ibid., p. 99.
119. Ibid., p. 24; see Lydekker, *Phases of Animal Life,* p. 230.

lutionary energy, but on the predetermination of the direction in which individual variation occurred. In 1921 W. D. Lang of the British Museum (Natural History) invoked pressures generated by the need to secrete excess calcium carbonate to explain orthogenetic trends in the Bryozoa.[120] Vertebrate paleontologists also became convinced that many groups showed tendencies to evolve in fixed directions even where the results were nonadaptive. D. M. S. Watson studied trends in the evolution of amphibians which seemed to continue even when the habitat changed, and which he attributed to the mechanism "which determines the characters of the adult being so constituted as to be capable of modification only in certain definite ways."[121] F. B. Loomis wrote of variation acquiring a kind of momentum which eventually pushed characters beyond the point of maximum utility, so that they became overdeveloped.[122] Favorite examples were the great size of many dinosaurs, the teeth of saber-toothed tigers, and the antlers of the 'Irish elk.' The German paleontologist Othenio Abel took up the same theme. He adopted a Lamarckian explanation of adaptive trends, but also believed that such trends could aquire an inertia that could drive the structure concerned beyond the bounds of utility.[123] Like many twentieth-century paleontologists, Abel was suspicious of the earlier generations' appeal to vitalistic forces to explain the trends.

The two most active exponents of this theory of predetermined evolution were William Berryman Scott and Henry Fairfield Osborn, both of whom made their reputations in the study of mammalian evolution with special reference to American fossils.[124] Both began with an interest in Lamarckism, but converted in the 1890s to the view that evolutionary trends were produced by factors internal to the organism. Scott's 1891 article on *Poebrotherium* argued that the fossil sequences available for the camel family were so complete that they offered the best possible way of settling vexed questions such as whether evolution was continuous or discontinuous, and the extent of parallelism and convergence.[125] He offered the following metaphor to illustrate the origins of the various camel-like groups: they "arise independently from the abundant and widespread Eocene

120. W. D. Lang, *Catalogue of the Fossil Bryozoa;* see Bowler, *Eclipse of Darwinism,* pp. 165–166.

121. D. M. S. Watson, Croonian Lecture, "The Evolution and Origin of the Amphibia," p. 194; see Bowler, *Eclipse of Darwinism,* p. 171.

122. F. B. Loomis, "Momentum in Variation."

123. See Abel, *Paläobiologie und Stammesgeschichte,* for example, p. 301. On the extent of orthogenetic thinking in German paleontology, see W.-E. Reif, "The Search for a Macroevolutionary Theory in German Paleontology."

124. See Rainger, *An Agenda for Antiquity,* chap. 2, and Bowler, *Eclipse of Darwinism,* pp. 130–133. For Scott's autobiography, see his *Memoirs of a Paleontologist.*

125. Scott, "On the Osteology of *Poebrotherium,*" p. 9.

type, which may be called the Buno-Selenodont; and which forms, as it were, a lake from which several streams, flowing in partly parallel, partly divergent directions, are derived."[126] But each stream, once established, carried on its way as though directed toward a predetermined goal, leading Scott to argue that Waagen's concept of mutation was superior to the new ideas being offered by William Bateson and the experimentalists.[127] Like Cope, Scott believed that the various members of a genus did not share an immediate common ancestry: they were representatives of parallel phyla that had independenty reached the same stage in the development marked out for their group. In his later years Scott admitted that disagreements over the extent of parallelism and convergence had brought the reconstruction of phylogenies into disrepute, although he clearly retained his own belief that parallelism was a major factor.[128]

Osborn developed similar views during the 1890s and incorporated them into his *Age of Mammals* of 1910 and his massive studies on the evolution of the extinct Titanotheres and the living and extinct Proboscidia (the elephants). He believed that the evolutionary development of each individual character could be interpreted in terms of "rectigradations," or linear variation trends that pushed all affected phyla in parallel in a single direction. The horns of the Titanotheres, for instance, began as minute rudiments appearing independently in the adult forms of all members of the group at different points in geological time. The horns then got steadily bigger, and appeared at an increasingly early stage in ontogeny, until eventually they dominated the skull completely.[129] "Within certain phyla a tendency or predetermination [for the horns] to evolve in breadth or length orthogenetically appears to be established, flowing in one direction like a tide, on the surface of which occur individual fluctuations and variations, like waves and ripples."[130] Although adaptive at first, the tendency to grow horns somehow became locked into the germ plasm of the group until it pushed the structure to positively harmful extremes. Osborn went out of his way to insist that the pattern of evolution was incompatible with the natural selection of random variations.

Since this interpretation of fossil sequences invoked a high degree of parallelism, it had major implications for systematics. The conventional evolutionary approach to relationships was based on the assumption that similarities were a sign of common ancestry. The species in a genus were closely related because they had only recently diverged from a single an-

126. Ibid., p. 72.
127. Scott, "On Variations and Mutations."
128. Scott, *A History of Land Mammals in the Western Hemisphere*, pp. 645, 649–650.
129. Osborn, *The Titanotheres*, II, chap. 11.
130. Ibid., II, p. 844.

cestral form. But in the interpretation set up by Cope and Hyatt, and maintained by Scott and Osborn, the generic relationship meant only that species from independent phyla had reached the same point on the developmental scale for their group. Two species in the same genus might belong to lineages which had been separated for a vast period of geological time. Real phylogenetic relationships thus ran vertically through the fossil record, not horizontally between related species at the same point in time. If the genus was to denote evolutionary relationship, then it would have to apply to the "vertical" lineage. Both Scott and Osborn continued to insist on the need to take account of these vertical relationships when classifying fossil species. By the time he wrote his monograph on the Proboscidea in 1936, the ever self-satisfied Osborn was talking of the replacement of the Linnean system of nomenclature by the "Osbornian."[131] There were forty-one independent lines of generic ascent among the elephants.[132]

A related problem concerned the nomenclature applied to the generalized ancestors from which later, more specialized groups were thought to have diverged. Thus in 1892 Osborn complained about the traditional policy of assigning the early, more generalized forms to families that were distinct from those comprising their later descendants. Since he thought the paleontologist could, in fact, trace the phyla back deep into the early forms, he thought that the more generalized ancestors should be included within the modern families.[133] Significantly, Romer argued against this approach in his 1933 textbook, insisting that when generalized forms ancestral to two families were known, it was better to erect a stem group to include the base from which the later families were derived.[134]

Most authorities, even those who accepted that there was a great deal of parallelism, warned against the chaos that would ensue if the taxonomic conventions were modified in the ways suggested by Osborn and others. We have already noted that Bather used this problem to suggest that a halt be called to the attempt to use phylogeny as a guide to classification.[135] But Bather himself was extremely worried by the tendency of some paleontologists to invoke mysterious vitalistic agencies as the cause of orthogenetic trends.[136] The potential for confusion only began to

131. Osborn, *The Proboscidea*, I, p. xiii.

132. Ibid., II, chap. 23.

133. Osborn and J. L. Wortman, "Fossil Mammals of the Wasatch and Wind River Formations," pp. 90–92.

134. Romer, *Vertebrate Paleontology* (1933), p. 7.

135. Bather, "Biological Classification," and Arkell and Moy-Thomas, "Palaeontology and the Taxonomic Problem."

136. Bather, "Fossils and Life"; see Bowler, *Eclipse of Darwinism*, pp. 179–180.

evaporate when paleontologists began to absorb the implications of the new approach to the evolutionary mechanism promoted by what Julian Huxley called the "Modern Synthesis." These implications included recognition of a much wider range of variability in species than Osborn and his supporters had allowed for, and a general suspicion of orthogenetic trends.

The leading figure in this transition was George Gaylord Simpson. Following his early work on the Mesozoic mammals, Simpson began to use the statistical analysis of large fossil populations to show that specimens once split among several different species fell easily within the range of variation that could exist within a single species.[137] Having been taught that a variation of 15 percent in any linear dimension was enough to mark a distinct species, he now insisted that many geographic races, let alone species, exhibited a greater range of variability than this. The assemblages of independent phyla postulated by Osborn and others could thus be seen as artifacts constructed from different points in the range of variability of a single population.

At the same time, other paleontologists were reasserting the case for a more Darwinian view of the evolutionary process. Romer noted that the evidence for orthogenesis may have been created by the fact that branches leading away from the direction of greatest adaptive value were soon nipped in the bud.[138] Others, including Osborn's former subordinate Matthew, noted that in many families there *were* side branches visible in the fossil record.[139] As we shall see in the next section, Matthew also repudiated the evidence for the production by orthogenesis of characters that were positively harmful. Both Romer and Matthew accepted that natural selection was a more plausible theory of the evolutionary mechanism than Lamarckism or orthogenesis. Simpson's *Tempo and Mode in Evolution* of 1944 finally brought together the arguments showing that the evidence for massive parallelism brought about by orthogenetic trends was illusory. The evolutionary pattern displayed by most families was not a series of parallel lines, but the kind of branching tree that Darwin had postulated decades before.

RISE AND FALL

The belief that a whole family of species could be driven inexorably toward extinction by an orthogenetic trend had ramifications for other models used to describe particular episodes in the history of life. The uni-

137. See Leo F. Laporte, "Simpson on Species."
138. Romer, *Vertebrate Paleontology* (1933), p. 5.
139. Matthew, *Outline and General Principles of the History of Life*, chap. 20.

formitarian image of continuous change built into Darwin's theory was always somewhat at variance with conventional ideas about the structure of the fossil record. The very fact that the record could be divided into discrete periods and eras was taken as evidence that the history of life was to some extent a discontinuous process. When John Phillips created the names Palaeozoic, Mesozoic, and Cainozoic (Cenozoic) in 1841, he was expressing his conviction that there were two wholesale transformations of the earth's population imposed upon the lesser—but apparently equally abrupt—changes which distinguished individual geological periods. The popular image of the Mesozoic as the Age of Reptiles (it had actually been founded on changes in the invertebrates) reflected the widespread belief that life had advanced through a series of discrete episodes, not by a gradual progressive trend.

Like his mentor Charles Lyell, Darwin had challenged the discontinuity of the fossil record by arguing that the sudden 'jumps' in the history of life were an illusion created by the imperfection of the record. It was no longer fashionable to argue that the apparently abrupt changes were the result of massive geological catastrophes followed by bursts of miraculous creation. And yet continued research failed to fill in the missing pieces completely; reduced to the simplest possible terms, it seemed that the age of the great reptiles *had* ended suddenly, with a fairly rapid expansion of the mammals to take their place. The belief that the history of life could be understood as a series of partially discrete episodes was retained and even reinforced in the later years of the nineteenth century. Progress was not a continuous trend, but had occurred by a series of abrupt surges forward, like waves advancing up a beach. I have suggested elsewhere that this model of the history of life had its origins in conservative, idealistic philosophies that stood opposed to the liberal image of gradual transformation built upon by evolutionists such as Darwin and Herbert Spencer. Certainly, the closest analogies we can see in interpretations of human history are found in writers who focused in the rise and fall of empires as discrete episodes in the upward surge of the human spirit.[140]

Whatever the origins of this model, evolutionists who took the relative discontinuity of the record at its face value had to explain why certain eras were dominated by particular forms of life, as in the Age of Reptiles. More generally, they had to account for the fact that many classes and families seemed to exhibit a record of sudden expansion, long-continued dominance, and a relatively abrupt decline to extinction. One possibility explored by the supporters of the recapitulation theory was that each group followed a pattern equivalent to that in the life cycle of an individual organism. Groups too had their birth, childhood, maturity, senility,

140. See Bowler, *The Invention of Progress.*

and death. Perhaps this was a consequence of each new form of life being endowed with a limited quantity of evolutionary energy. A more materialistic explanation was that orthogenetic trends established for adaptive benefit during the group's expansion became locked into the germ plasm and eventually produced nonadaptive characters that weakened the organisms in the competition with later rivals. Both of these explanations took it for granted that the rise and fall of each group was, in a sense, predetermined. Once on the slippery slope to extinction, nothing could stop the inevitable decline.

The belief that evolution was dominated by parallelism might, in some circumstances, have the opposite implications. Huxley's early suggestion that the various groups of placental mammals might have independently evolved from equivalent marsupial forms would have eliminated the need to postulate a decline of the marsupials in the face of the placentals' advance. But increasingly it was assumed that, for one reason or another, the reptiles, the marsupials, and the archaic placentals had retreated to make room for their successors.

Built-in obsolescence was not, however, the only kind of explanation for the decline of once-dominant groups. Darwin himself accepted that a newly evolved species with some general adaptive improvement would spread out into new territory, diversifying as it did, to produce a new genus and (if really successful) a new order, family, or even a new class. In most cases, however, the group would sooner or later be overtaken by an even more efficient newcomer, which would spread in its turn, gradually wiping out the descendants of the earlier form and driving them toward extinction. There was nothing mysterious or predetermined about the process of rise and fall—it was merely a by-product of the way in which species interact with each other in a competitive world. We shall see in the next chapter how this model was taken up by both biogeographers and paleontologists, especially under the influence of Matthew. The problem was that in a world governed completely by the principle of uniformity, the various cycles, small and large, should overlap to give a relatively continuous process of change. To use the example of the reptiles and mammals again, a steady decline of the great reptiles should have coincided with an equally steady rise of the mammals. This was exactly what the first generation of evolutionists were led to expect when they predicted that the radiation of the mammals would turn out to have taken place in the late Mesozoic. We have already noted that this prediction was essentially falsified by the end of the century. The decline of the reptiles was relatively sudden, if not absolutely catastrophic, and the main episode of mammalian radiation had been held back until the reptiles were out of the way. This encouraged a search for other factors that might have triggered the decline of one class and thus created the conditions favorable to the rise of its successor.

Other suggestions invoked ecological relationships between entirely different types of living organisms. Perhaps the rise of the mammals had had to wait for the appearance of the flowering plants, which provided a new source of food. An even greater sense of the unpredictability of the history of life was implicit in a kind of 'what if' scenario hinted at by paleontologists who wondered if some promising groups had been nipped in the bud by rivals before they got a chance to flourish. Thus Romer envisaged a world that might have been dominated by birds, had not the mammals taken over first. More serious were suggestions that physical changes played a role in extinguishing some groups and stimulating (or creating the opening for) others. Here the renewed interest in the possibility that the earth might have experienced episodes of rapid change brought about by periods of intense geological activity played an important role.

The belief that the history of life on earth was dominated by episodes defined by the rise and fall of particular groups of animals became widespread at the turn of the century. In his survey of ichthylology, Bashford Dean wrote that each geological period had seen the flowering of a particular family, which duplicated the variety of forms displayed by its predecessor.[141] Arthur Smith Woodward carried the metaphor one step further by suggesting that each dominant group had to spend some time in the evolutionary wilderness before it could claim its inheritance:

> All known facts appear to suggest that the processes of evolution have not operated in a gradual and uniform manner, but that there has been a certain amount of rhythm in the course. A dominant old race at the beginning of its greatest vigour seems to give origin to a new type showing some fundamental change; this advanced form then seems to be driven from all the areas where the dominant ancestral race reigns supreme and evolution in this latter becomes comparatively insignificant. Meanwhile the banished type has acquired great developmental energy, and finally spreads over every habitable region, replacing the effete race which originally produced it.[142]

American neo-Lamarckians were enthusiastic about the metaphor of rise and fall because their interpretation of the history of life was strongly influenced by the analogy with the individual life cycle. Thus Alpheus Packard wrote in 1886, "As there is in each individual a youth, manhood

141. Bashford Dean, *Fishes, Living and Fossil*, p. 12.

142. Smith Woodward, *Outlines of Vertebrate Palaeontology*, p. xxi. Woodward claims that the sudden bursts of evolution are what Cope called "expression points," but the latter were, in fact, meant to be jumps between successive species in a single lineage.

and old age, so species and orders rise, culminate and decline, as nations have arisen, reached a maximum of development and decayed."[143] In a study of the archaic mammalian group, the Oreodontidae, Cope expressed his intention of elucidating "the causes of the rise, great development, decadence and extinction of one of the best-marked types of Mammalia the world has seen."[144] In his general survey of evolution, Joseph Le Conte postulated a "law of cyclic movement" which he attributed to the influence of Louis Agassiz.[145] He even gave a diagram illustrating the rise and fall of successive groups as a series of waves pushing the history of life forward.[146]

Similar images continued to be employed by early twentieth-century evolutionists. The image of a rhythm in the development of life was taken up by Richard Swann Lull in his epilogue, "The Pulse of Life": "The great heart of nature beats, its throbbing stimulates the pulse of life, and not until that heart is stilled forever will the rhythmic tide of evolution cease to flow."[147] Lull also drew an explicit comparison with the rise and fall of nations in human history.[148] Osborn wrote of families undergoing a cycle of ascent toward dominance, and identified periods at which the development of each reached its climax.[149] And as a final example, we may quote Arthur Dendy: "Each great group seems to have begun in a small way, then developed rapidly, branching out in many directions, only to dwindle away again and give place to some new and vigorous offshoot.[150]

The popularity of the analogy between ontogeny and phylogeny in the post-Darwinian era predisposed some biologists to explain this phenomenon by assuming that groups of organisms, like individuals, had only a fixed amount of developmental energy. The possibility that species themselves might grow old and die had been considered and rejected by Darwin at an early stage in his career.[151] But others continued to take this idea seriously: it was defended by Charles Naudin in the 1870s, and Naudin's ideas were discussed by Asa Gray.[152] The claim that whole groups might have a life cycle equivalent to that of an individual was developed

143. A. S. Packard, "Geological Extinction and Some of Its Causes," p. 40.

144. Cope, "Synopsis of the Species of Oreodontidae," p. 571.

145. Joseph Le Conte, *Evolution,* p. 16.

146. Ibid., p. 19.

147. Lull, *Organic Evolution,* p. 691.

148. Ibid., pp. 531–532.

149. For example, Osborn, *The Age of Mammals,* pp. 46, 98.

150. A. Dendy, *Outlines of Evolutionary Biology,* p. 304.

151. See, for example, *Charles Darwin's Notebooks,* B notebook, p. 35.

152. Charles Naudin, "Les espèces affines et la théorie de l'évolution"; see Asa Gray, *Darwiniana,* pp. 287–292.

at length in Alpheus Hyatt's orthogenetic theory. Hyatt was particularly associated with the idea of racial senescence or senility. As described above, he believed that once a group's vital energy had been exhausted, it would decline inexorably toward extinction. In his view, the senile phase represented a decline back to the simple form from which the group began. This theme was echoed by H. W. Shimer in 1906.[153] But as we have seen, other invertebrate paleontologists such as C. E. Beecher and W. D. Lang thought the decline to extinction was produced by the development of overelaborate spines or shell structures which eventually choked the organisms within. The overcoiling of the shell of the oyster *Gryphaea* was another popular example.[154]

By the 1890s, paleontologists were divided over the value of the recapitulation theory itself. American neo-Lamarckians such as Beecher and Le Conte retained their commitment to the parallel between the individual and the group's life cycle. In Britain Bather defended it strongly in 1893, but Smith Woodward's textbook of 1898 urged caution.[155] In the early years of the new century the appeal to embryology as a model to help reconstruct the pattern of evolution in a whole group was steadily rejected. Goodrich dismissed the biogenetic law as an incautious generalization that "has perhaps done more to delay the progress of sound views on phylogeny than any other speculation."[156] But rejection of the direct analogy between individual development and evolution did not mean that the whole idea of racial senility had to be abandoned. The 'overdevelopment' theory of extinction postulated by Lang involved no mystical notion of evolutionary energy, only the assumption that harmful trends could eventually become established in successful groups and would then drive them toward extinction.

Vertebrate paleontologists now took up this idea with particular enthusiasm. It was widely believed that the horns and armor of many dinosaurs had developed to the extent that they were a major hindrance. In his article "Momentum in Variation," Loomis gave the example of the plates on the backs of stegosaurs: "With such an excessive load of bony weight entailing a drain on vitality, it is little wonder that the family was short-lived."[157] The same example was used by Ernst Koken and Lull, while Richard Lydekker argued that mammals were also subject to racial old

153. H. W. Shimer, "Old Age in Brachiopoda."

154. See, for example, H. H. Swinnerton, *Outlines of Palaeontology*, p. 222.

155. Bather, "The Recapitulation Theory in Palaeontology," and Smith Woodward, *Outlines of Vertebrate Palaeontology*, p. xxiv. There was also support from James Perrin Smith, "The Biogenetic Law from the Standpoint of Palaeontology."

156. E. S. Goodrich, *Living Organisms*, p. 146.

157. Loomis, "Momentum in Variation," p. 842.

age.[158] Loomis claimed that the horns of mammals such as the Titano-theres and the Irish elk had become so large that they might have contrib-uted to the species' extinction. Depéret also saw the trend toward over-development as characteristic of many vertebrate phyla:

> Each of these branches arrives, with more or less speed, at mutations of great size and of very specialized characteristics, which vanish without leav-ing descendants. When a branch disappears by extinction it is, so to speak, *replaced* by another branch having an evolution until then slower, which in its turn passes through the phases of maturity and old age which conduct it to its end. The species and genera of the present time represent those which have not yet arrived at the senile phase; but it can be foretold that some of them, the Elephants, the Whales, the Ostriches, etc., are approaching this final phase of their existence.[159]

In the postwar years, the claim that the evolution of many vertebrate families was governed by internally programmed trends which drove re-lated species in parallel along the same path was defended by Osborn. His monograph on the Titanotheres also insisted that such trends eventually contributed to the extinction of the group by producing overdeveloped characters.[160] But it was the German paleontologists who remained by far the strongest advocates of orthogenesis. Othenio Abel, professor of paleontology at Vienna and then at Göttingen, retained the analogy be-tween ontogeny and phylogeny, portraying each group as having a life cycle of progress followed by degeneration to extinction.[161] He argued that the postulation of an internally programmed trend would explain both Dollo's law of irreversibility and the law of the progressive reduction of variability within each group postulated by the Italian paleontologist Daniele Rosa.[162]

Similar views were propounded by the invertebrate paleontologist Karl Beurlen in the 1930s, with even greater stress on the explosive radia-tion at the beginning of each group's history being the aftermath of a sud-den macromutation.[163] Beurlen was a strong influence on Otto Schinde-

---

158. Ernst Koken, *Palaeontologie und Descendenzlehre*, pp. 26–27; Lull, "Dinosaurian Cli-matic Response," p. 276; R. Lydekker, *A Geographical History of Mammals*, pp. 18–19. Other examples of racial senility are given in Lull, *Organic Evolution*, pp. 186–191.

159. Depéret, *Transformations of the Animal World*, p. 243.

160. Osborn, *Titanotheres*, II, p. 844.

161. O. Abel, *Paläobiologie und Stammesgeschichte*, pp. 358–375. See Reif, "The Search for a Macroevolutionary Theory in German Paleontology."

162. Abel, ibid., p. 372. Rosa's law was contained in his *La Reduzione Progressiva della Variabilita* of 1899.

163. Karl Beurlen, "Funktion und Form."

wolf, whose "typostrophe" theory based on these principles became infuential in Germany after World War II. Schindewolf stressed that the sudden appearance of vigorous new types, with their subsequent history of rapid expansion followed by gradual orthogenetic decline, imposed a cyclic character on the evolution of life equivalent to that seen in human history.[164] Here the cyclic nature of change is very much a product of internal, biological forces—it is not (in contrast with rival views explored below) imposed on life by cycles of geological change. Both Abel and Beurlen joined the Nazi party, and it has been suggested that their vision of internally programmed trends drew upon a Romantic or vitalistic philosophy that suited the ideology of the 'will to power.'[165] Abel, however, likened his evolutionary trends to the principle of inertia in physics, and thus made at least a pretense of conforming to mechanistic language. Schindewolf's theories were based on the concept of macromutations as defended by geneticists such as Richard Goldschmidt.[166]

The German paleontologists' emphasis on internal forces led them to minimize the role of adaptation in evolution by stressing the effects of nonadaptive orthogenesis. This position was becoming increasingly suspect among English-speaking paleontologists in the interwar years. A favorite example cited by many advocates of orthogenesis was the teeth of the saber-toothed tigers, which were widely believed to have become so big that the animals could hardly have opened their mouths. This interpretation was specifically challenged by Matthew, who argued that the jaws of these felines were designed in such a way that the teeth could have been used very effectively to pierce the thick skins of the many large pachyderms of the period.[167] The saber-tooths would have died out when the prey species declined in numbers, rather than because of any nonadaptive specialization. Matthew noted that there was resistance to his explanation from German paleontolgists, including Abel, who remained an advocate of the claim that characters could develop to harmful extremes.[168] Scott, although a strong advocate of directed evolution, conceded that the saberlike teeth could not have been maladaptive: "No arrangement which was

164. Schindewolf, *Basic Questions in Paleontology*, pp. 193–195, 272–274. Schindewolf explicitly attacked the claim that mass extinctions were due to geological events; see pp. 317–323. This edition is a translation of Schindewolf's *Grundfragen der Paleontologie* of 1950; for background, see the foreword by Stephen Jay Gould and the afterword by Wolf-Ernst Reif; see also Reif's "Evolutionary Theory in German Paleontology."

165. Gould, in Schindewolf, *Basic Questions in Paleontology*, p. xiii.

166. Abel, *Paläobiologie und Stammesgeschichte*, p. 399.

167. Matthew, "Phylogeny of the Felidae," pp. 305–307.

168. Ibid., p. 305. For Abel's later views, see his *Paläobiologie und Stammesgeschichte*, p. 265.

disadvantageous, or even inefficient, could have persisted for such vast periods of time."[169] Romer adopted Matthew's explanation of the saber-tooths' hunting technique.[170]

By the 1920s, then, the excessive claims for the evolution of nonadaptive characters were being greeted with increasing skepticism by English-speaking paleontologists. Even the senile characters postulated by Hyatt for his fossil cephalopods were being interpreted in adaptive terms, although this development seems to have escaped J. B. S. Haldane's attention.[171] This trend clearly predates the emergence of the modern synthetic theory of evolution in the 1940s, showing that paleontologists were playing an independent role in challenging the old non-Darwinian theories. The growing support for an adaptationist, and eventually a selectionist, explanation of the evolutionary mechanism led to a reexamination of the many cases which had been assumed to offer evidence for the development of potentially lethal specializations. In most cases, an adaptive purpose for the allegedly harmful structure could be postulated. Simpson attacked the idea that evolutionary momentum could drive a species to extinction in his *Tempo and Mode in Evolution* of 1944. He argued that even if the senile characters of *Gryphaea* were lethal, this would not matter if the individuals were no longer reproducing.[172] Simpson also devoted two chapters of his popular *The Meaning of Evolution* of 1949 to rebutting the case for both orthogenesis and the theory of racial senility.[173] Evidently Simpson thought that the general public had not yet appreciated the new interpretation.

The collapse of the idea of racial senility did not mean that the whole model based on the cycle of progress and decline had to be adandoned. The claim that a species could develop harmful characters depended on the assumption that there was a fairly loose ecological 'fit' between the organism and its environment: it was only when a character became seriously harmful that it caused extinction. This was directly contrary to the principles of Darwin's theory, which postulated a relentless struggle in which even the slightest disadvantage wiped out the individual. But for all its emphasis on the continuity of transformation, Darwin's theory itself had to accommodate the model of groups undergoing a cycle of expansion and decline in the course of their history. Darwin knew that a successful

169. Scott, *History of the Land Mammals in the Western Hemisphere*, p. 532.

170. Romer, *Vertebrate Paleontology* (1933): p. 294.

171. See, for instance, Edward W. Berry, "Cephalopod Adaptations." J. B. S. Haldane still thought that Hyatt's sequences could not be explained in adaptive terms in 1932; see his *The Causes of Evolution*, pp. 28–29.

172. G. G. Simpson, *Tempo and Mode in Evolution*, pp. 170–179, esp. p. 174.

173. Simpson, *The Meaning of Evolution*, chaps. 11 and 13.

new species, that is, one that had developed a character giving it a significant advantage over its rivals, would expand its territory and diversify to form a genus of closely related species. In most cases, it would eventually be overtaken by a later form with an even greater advantage; it would gradually decline, the individual species in the genus going extinct one by one until eventually it disappeared altogether. "Within the same large group, the later and more highly perfected sub-groups, from branching out and seizing on many new places in the polity of Nature, will constantly tend to supplant and destroy the earlier and less improved sub-groups. Small and broken groups and sub-groups will finally tend to disappear."[174] Even so, Darwin admitted that the disappearance of some groups, such as the ammonites, seemed to have been very abrupt.[175]

The Darwinian model of rise and fall was superior precisely because it could deal with the irregularity and unpredictability of the process observed in the real world. It could explain how a few members of a declining group could survive as 'living fossils' in areas protected from the rise of more highly evolved competitors. Even Smith Woodward, a prominent advocate of orthogenesis, noted this phenomenon. Writing of tapirs, which survive in widely scattered tropical refuges, he noted that "like the surviving Dipnoan fishes . . . they are an illustration of a once-dominant race nearly exterminated but still struggling for existence where competition happens to be least severe in their particular case."[176] Australia was the classic example of a continent that was supposed to have preserved the life typical of an earlier geological period. This model of the history of life was to be developed at length in Matthew's "Climate and Evolution" of 1914, which pictured dominant new groups radiating from a northerly center of progressive evolution, marginalizing or exterminating their predecessors in the process. This model could explain relatively sudden bursts of extinction when a barrier holding back the migration of more highly evolved types was finally removed.

Biogeography forms the theme of the next chapter. In the meantime, we must explore other consequences of the more 'Darwinian' view of the relationship between different branches of the tree of life. It was sometimes argued that certain key steps in evolution had only become possible when suitable conditions were created by the evolution of unrelated types. The land could only be colonized by vertebrates when suitable invertebrate prey had already moved into that environment. In other cases, rival groups evolved at the same time seemed to have struggled to determine

174. Darwin, *Origin of Species*, p. 126. On Darwin's recognition of this point, see Janet Browne, "Darwin's Botanical Arithmetic."

175. Darwin, *Origin of Species*, p. 318.

176. Smith Woodward, *Outlines of Vertebrate Palaeontology*, pp. 321–322.

which should dominate the earth. Sometimes external forces such as geological change played a crucial role in determining the outcome of such contests.

Darwin himself was interested in the possibility that certain key steps in the history of life on earth had been stimulated by developments taking place in other groups. In 1877 he expressed great interest in an idea proposed by the French botanist Gaston de Saporta. Saporta argued that the emergence of the relationship between insects and flowering plants had triggered a rapid expansion of the latter in the Cretaceous period. This in turn had made possible the appearance of the mammalian groups which fed on this type of vegetation.[177] Paleobotanists continued to argue that the appearance of flowering plants had been relatively sudden, although this particular theme was taken up by relatively few paleontologists working on the rise of the mammals. In 1903, however, J. L. Wortman's account of fossils from Marsh's collection noted that the flowering plants may have appeared first in arctic regions and insisted that the evolution of the higher mammals would have been impossible until these plants were widely distributed.[178]

Paleontologists had long been aware of the possibility that trends in the earth's physical development could be correlated with major episodes in the history of life. In the early nineteenth century the pioneering paleobotanist Adolphe Brongniart had suggested that there had been a steady decline in the carbon dioxide level in the atmosphere which might explain the late appearance of the mammals.[179] Brongniart thought the decline was due to the carbon dioxide becoming locked up in coal beds. Some late nineteenth-century evolutionists took up the same theme. In a very conservative survey of the history of life which presented the whole process as a display of divine forethought, the veteran Alfred Russel Wallace explained the origin of the land vertebrates by means of Brongniart's mechanism, while the origin of the mammals was attributed to a further decline due to the carbon dioxide being absorbed to form limestone.[180] It was widely believed that the Mesozoic had been a time of warm, humid conditions ideally suited to the great reptiles. Chamberlin attributed such

177. Saporta to Darwin, 16 December 1877; see Y. Conry, *Correspondence entre Charles Darwin et Gaston de Saporta,* pp. 97–108, esp. pp. 100–101, and Darwin's reply, pp. 109–110. See also *Life and Letters of Charles Darwin,* III, pp. 284–286. Saporta published the idea in his *Le monde des plantes avant l'apparition de l'homme;* see pp. 36–37.

178. J. L. Wortman, "Studies of Eocene Mammalia in the Marsh Collection," p. 419. On the paleobotanists' belief in the sudden appearance of the flowering plants, see, for example, A. C. Seward's 1903 President's address to the Botany Section of the British Association, p. 846.

179. See Bowler, *Fossils and Progress,* pp. 22–25.

180. A. R. Wallace, *The World of Life,* pp. 196, 216–217.

warm periods to an excess of carbon dioxide in the atmosphere and postulated fluctuations in the level of the gas due to climatic changes which affected lime secreting organisms.[181]

The decline of the reptiles was often linked to the emergence of a more varied climate at the end of the Cretaceous. But the climatic transformation was a complex affair, and the subsequent history of the mammals was thought to be linked to a steady decrease in the extent of forested regions. The classic case of the specialization of the modern horse was not a gradual process driven by some inexorable trend adapting the animals to a fixed environment. As Kovalevskii and many others noted, the teeth of *Eohippus* and the early members of the horse family were adapted for browsing on leaves. Evidently, the early horses were adapted to a world that was covered with extensive forests. Only in the Miocene did a new kind of horse emerge, adapted to feeding on the grasses of the open plains—and to the speed necessary to escape predators in this new environment. The expansion of this habitat through into the Pliocene created the world in which the modern horse evolved.[182] The history of the horse family was thus dominated by a change in the environment produced by a gradual dessication in the course of the Cenozoic. W. B. Scott saw the horse family dividing into two groups, the conservative phylum retaining old-fashioned browsing teeth, doomed to eventual extinction, and a more plastic phylum which was able to make use of the new habitat and which flourished through to the present.[183]

Although the language used to describe these events was still redolent of the imagery of progress, the picture of the history of life on earth now emerging was far more 'open-ended' than the more simpleminded progressionism of an earlier age. The fate of many groups could have been different if circumstances had not worked out in exactly the way we observe. The best example of this growing willingness to admit that the progress of life was not predictable on the basis of evolutionary trends was the increasingly common belief that the extinction of the dinosaurs had cleared the way for the expansion of the mammals, discussed below. Most paleontologists dismissed the dinosaurs as sluggish beasts long overdue for replacement, but Lull struck a more positive note by insisting that they should not be regarded as a futile attempt by nature to people the world with insignificant creatures.[184] The general feeling that the mammals got their chance to expand only when some external agency removed the di-

181. T. C. Chamberlin, "The Influence of Great Epochs of Limestone Formation upon the Constitution of the Atmosphere."

182. See, for instance, Osborn, *The Age of Mammals*, p. 44.

183. Scott, *History of the Land Mammals in the Western Hemisphere*, pp. 297–298.

184. Lull, *Organic Evolution*, p. 531.

nosaurs suggests that the image of inevitable progress was now being heavily qualified.

One manifestation of this less deterministic view of evolution was the increasing emphasis on the possibility that the course of life's history was decided by the outcome of a struggle between rival groups. In some cases, the rivalry maintained an equilibrium over a long period of time. In the period just before World War I, Julian Huxley compared the evolution of predators and prey to the arms race between rival naval powers.[185] The same metaphor was used by Huxley and Haldane after the war to account for the improved speed of the horse family:

> In the continual struggle that is going on in mammals and birds between herbivore and carnivore, pursuer and pursued, each new advance in speed and size or strength in one party to the conflict, must call forth a corresponding advance in the other, if it is not to go under in the struggle and become extinct. . . .
>
> A precisely similar state of affairs is often to be seen in the evolution of the tools and weapons and machines of man. For instance, in naval history, the increase throughout the nineteenth century of the range and piercing power of projectiles on the one hand, of the thickness and resistance of armour-plate on the other, provides an exact parallel with the simultaneous increase of speed and strength in both carnivores and their prey.[186]

In such cases the struggle produced progress in both lines—until one could no longer keep up and became extinct.

In other examples the struggle between rivals played a major role in determining which branch of evolution would succeed and which would not. Noting the appearance of giant birds in the early Tertiary, Romer speculated about what would have happened if they had beaten the mammals in the race to dominate the world vacated by the reptiles:

> These giant early birds arouse one to speculation; their presence suggests some interesting possibilities—which never materialized. At the end of the Mesozoic, as we have seen, the great reptiles died off. The surface of the earth was open for conquest. As possible successors there were two groups, the mammals—our own relatives—and the birds. The former group succeeded, but the presence of such forms as *Daitryma* shows that the birds were, at the beginning, their rivals. What would the earth be like today had the birds won and the mammals vanished?[187]

185. Julian Huxley, *The Individual in the Animal Kingdom*, p. 115. I am indebted to Michael Ruse for bringing this analogy to my attention.

186. Haldane and Huxley, *Animal Biology*, p. 237.

187. Romer, *Man and the Vertebrates*, p. 98.

Wells and Huxley wrote of the "Battle of the Giants and Dwarfs," speculating on what might have happened if the arthropods had prevented the vertebrates from dominating the land.[188] They thought that the arthropods' rise had been blocked by the fact that their method of respiration limited their size. But C. Judson Herrick saw this as a battle between instinctive and voluntary behavior: the lower vertebrates had been inferior to the insects, but their form of intelligence had more potential and thus ensured their eventual dominance.[189] Haldane and Huxley shared this view.[190] Alternative histories in which the birds or the insects won the race for dominance are the substance of science fiction, not of science, but the fact that eminent biologists could let their imaginations wander in this way suggests that they were beginning to think of the history of life as a process whose outcome could not easily be predicted. Whatever the pressure for progress, in a slightly different world someone else could have won. Progressive evolution generated a number of rival types and left them to struggle for supremacy, the result determining the future direction of life on earth.

More realistic examples were offered in which the rise of one group had sealed the fate of its rivals, even when the latter were well established. In these cases the emergence of a more powerful rival was the direct cause of a more primitive type's decline. Kovalevskii's inadaptive ungulates had been eclipsed by their more efficiently designed rivals, although the latter were the product of a lucky accident. Many lines of mammalian evolution had been similarly eliminated by the rise of newcomers with bigger brains, most notably the archaic mammals of the Eocene. But in other cases it was difficult to see how the rise of the new group could have been the direct cause of the demise of the previously dominant forms. The decline of the earlier group merely created a vacuum into which a new one could expand. The extinction of the great reptiles was the most obvious instance in which—even if the previously dominant group was weakened by racial senility—the final act of the drama was initiated by some outside agency, not by the rivals who were (in this case, at least) already waiting in the wings. The only suggestion that the mammals may actually have caused the extinction of the dinosaurs was based on the possibility that the primitive mammals may have eaten reptile eggs. This idea was attributed to Cope, but was seldom taken very seriously.[191] Paleontologists were increasingly willing to believe that the reptiles had remained powerful enough to keep the mammals permanently in eclipse. Only when they

188. Wells et al., *The Science of Life,* pp. 465–467.
189. C. Judson Herrick, "The Evolution of Intelligence."
190. Haldane and Huxley, *Animal Biology,* pp. 290–291.
191. See, for example, Lull, *Organic Evolution,* p. 531.

were removed by an external agency did the mammals gain the chance to take over the world. In this case, intergroup struggle could not have been the actual cause of the reptiles' decline.

MASS EXTINCTIONS

Mounting evidence confirmed that the demise of the dinosaurs was too abrupt and the rise of the mammals too late for the latter to be the cause of the former. Late nineteenth-century evolutionists did not concern themselves with the possibility of a mass extinction at the end of the Mesozoic. Haeckel and others seem to have assumed that the reptiles would steadily have given way to the more progressive mammals. We have already noted that it was only with the accumulation of better fossil sequences in the early twentieth century that the abruptness of the transition from the age of reptiles to the age of mammals was confirmed. By the 1920s the combination of the sudden disappearance of the dinosaurs and the rapid expansion of the mammals had come to represent a serious problem. Michael Benton has identified a large number of hypotheses advanced to explain this event, many of which originated in the speculations of the early twentieth century.[192] Some theories attributed the decline of the reptiles to orthogenetic forces, and such ideas were still prevalent in the work of Schindewolf (discussed above). By supposing that the burst of mammalian radiation was a product of the class's youthful vigor following its sudden origin, this approach could account for some aspects of the fossil record. But many paleontologists now believed that the expansion of the mammals was made possible by the elimination of the reptiles—after all, primitive mammals had existed throughout the Mesozoic. The reptiles had not simply declined in numbers over a period of time, even if they had been developing some bizarre specializations. They had disappeared quite suddenly, leaving a world open to mammalian expansion. The abrupt nature of the changeover tempted paleontologists to assume that massive transformations of the environment may have hastened if not actually caused the disappearance of the once dominant group. These early speculations fall into what Benton calls the "dilettante phase" of the debate, because they were seldom subjected to rigorous testing, but they represent a growing awareness that the history of life on earth may have been punctuated by relatively abrupt disturbances.

Although there was a widespread feeling that the dinosaurs had become overspecialized, few were willing to invoke this as the sole cause of

192. Michael Benton, "Scientific Methodologies in Collision: The History of the Study of the Extinction of the Dinosaurs."

their extinction. In his *Age of Mammals* Osborn claimed that the late Cretaceous reptiles were "in the climax of specialization and grandeur" while the mammals were still small and inconspicuous. The sudden extinction of the dinosaurs was "one of the most dramatic moments in the life history of the world," and without it the mammals would not have been able to expand.[193] Osborn admitted that the fossil record for plants showed no sign of any sudden transformation of the climate, but Lull had no doubt that an episode of relatively sudden geological activity had triggered this "dramatic" event, which in turn had made possible the rise of the mammals.[194]

The possibility that major episodes of extinction and evolutionary activity could be initiated by dramatic transformations of the environment brought about by geological forces had begun to be discussed in the late nineteenth century. Even the neo-Lamarckians—who were predisposed to think in terms of a cycle of growth and decline—were keen to exploit the possibility that the rhythms in the history of life were caused by discontinuities in the earth's physical development. Packard thought that mass extinctions were caused when long periods of quiet were succeeded by "more or less sudden crises, or radical changes in the physical structure of continents, resulting in catastrophes, both local and general, to certain faunas or groups of animals as well as individual species."[195] Le Conte was also an early advocate of the view that the sudden development of new forms at certain points in the history of life had been triggered by episodes of rapid geological change.[196]

In the early years of the new century Lull and others developed this as a major theme in their interpretation of the fossil record. The epilogue to Lull's *Organic Evolution,* "The Pulse of Life," was a hymn to the power of the external world as a stimulus to the great surges in the progress of life on earth. Drawing on the geology of Chamberlin, Barrell, and Schuchert, who saw the history of the earth itself as an episodic process, Lull attributed both the creation of new classes and the collapse of old ones to the power of geological events. Nor was this interest in geological discontinuity a purely North American phenomenon. In Britain the physicist John Joly postulated episodes of radioactive heating of the earth's crust, while in Germany a number of geologists led by Hans Stille argued for occasional periods of intense activity. All of these geologists were aware of

193. Osborn, *The Age of Mammals,* pp. 97–98.

194. Lull, *Organic Evolution,* pp. 531, 687–691. But see also the refence to Osborn's more cautious view on the possibility of a climatic change, p. 548.

195. A. S. Packard, "Geological Extinction," p. 39.

196. Le Conte, *Evolution,* pp. 258–258; see also his "On Critical Periods in the History of the Earth."

the potential implications of their work for theories of the development of life on earth.[197] At this point, Schindewolf actively opposed such an interpretation, preferring instead his theory of cyclic change driven by internal biological forces.

The onset of a harsher climate could have many effects. It would stimulate more rapid evolution in those forms that were adaptable enough to cope with the new environment. The early mammals, for instance, had been able to thrive in the harsher conditions that began to emerge at the beginning of the Cenozoic. Lull suggested that the expansion of the modern placental mammals at the expense of their archaic predecessors was a by-product of a later burst of environmental change.[198] Those who responded to the stimulus displaced those who could not. In other cases, however, the capacity of environmental change to cause extinction was equally important. The mammals were only able to flourish when the dinosaurs were swept aside, and it was increasingly being argued that the reptiles' extinction was due to environmental stress. It was not necessary to see the dinosaurs as the victims of racial senility to suppose that they had been too well adapted to the warm conditions of the Mesozoic, and were thus unable to cope with the onset of a harsher environment. Romer noted that an episode of mountain-building at the end of the Cretaceous may have had immense effects on the climate, and commented, "Perhaps the Rocky Mountains killed the dinosaurs!"[199] Charles Knight also saw the dinosaurs—"weird, monstrous and bizarre"—as being wiped out by "a change in climatic conditions, cooling atmosphere, and a gradual drying up of their vast feeding grounds."[200] In the caption to a picture of *Triceratops* and *Tyrannosaurus,* he repeated this conclusion, adding that "Nature itself seemed weary of the great variety of reptiles which had dominated the planet for so long."[201]

Like the notion of rivalry between phyla, the theory of mass extinctions marks the emergence of a new way of thinking about the history of life. The idea of progress had certainly not evaporated, but it was no longer taken for granted that the path of progress was a smooth, unbroken line leading in a predictable direction. The evolution of life was a real historical drama, unpredictable in its outcome, and punctuated by unex-

197. See, for example, Schuchert, *Textbook of Geology,* chap. 34; Barrell, "Rhythms in the Measurement of Geological Time"; John Joly, *The Surface History of the Earth,* pp. 166–168. On the German geologists, see Richard Huggett, *Catastrophism,* pp. 115–117; and on Stille's influence on paleontology, see W.-E. Reif, "The Search for a Macroevolutionary Theory in German Paleontology."

198. Lull, *Organic Evolution,* p. 690.

199. Romer, *Man and the Vertebrates,* p. 89.

200. Knight, *Before the Dawn of History,* p. 8.

201. Ibid., pp. 68–69.

pected events. Other episodes of mass extinction were also coming under scrutiny. The growing interest in ice ages prompted the view that glaciation might have hastened the demise of many large Pleistocene mammals—although the possibility that they might have been hunted to extinction by early humans was also being considered.[202] Indeed, it is probable that the model offered by the ice ages helped to popularize the general belief in episodes of mass extinction brought about by climatic stress.

Wallace gave eloquent support to the theory of glacial extinctions in 1876:

> It is clear, therefore, that we are now in an altogether exceptional period of the earth's history. We live in a zoologically impoverished world, from which all the hugest, and fiercest, and strongest forms have recently disappeared; and it is, no doubt, a much better world for us now that they are gone. Yet it is surely a marvellous fact, and one that has hardly been sufficiently dwelt upon, this sudden dying out of so many large mammalia, not in one place only but over half the land surface of the globe. We cannot but believe that there must have been some physical cause for this great change; and it must have been a cause capable of acting almost simultaneously over large portions of the earth's surface, and which, as far as the Tertiary period at least is concerned, was of an exceptional character. Such a cause exists in the great and recent physical change known as "the Glacial epoch."[203]

Debates over the adequacy of this explanation would rage into the next century. Osborn, for instance, noted that the ice ages seemed the obvious cause of these recent extinctions, but saw problems with the explanation. The American horses were surely mobile enough have escaped south into Mexico, yet they had become extinct along with other species. He suggested other possibile causes of extinction, including sudden epidemics of disease.[204] The causes were uncertain, but the problem posed by mass extinctions seemed inescapable at the time. The recent disappearance of many large mammals was but one example of the comparatively rapid transitions that paleontologists thought they could detect in the history of life.

The issue of mass extinctions was pushed aside during the heyday of modern Darwinism in the 1940s and 1950s. Under the influence of a renewed enthusiasm for the principle of continuity, many paleontologists

202. See Donald Grayson, "Vicissitudes and Overkill: The Development of Explanations of Pleistocene Extinctions."

203. Wallace, *The Geographical Distribution of Animals*, pp. 150–151.

204. Osborn, *The Age of Mammals*, pp. 500–509; see also Osborn's "The Causes of Extinction in the Mammalia."

still believe that there is no need to invoke sudden transitions in the history of life. But in the 1960s Schindewolf fired the first shot in the battle to reawaken interest in mass extinctions by suggesting that they were caused by nearby supernovae. The modern debate over asteroid impacts as the cause of mass extinctions is beyond the scope of our study.[205] The appeal to genuinely catastrophic events of extraterrestrial origin as the cause of mass extinctions is certainly a new feature of the modern debates. But early twentieth-century paleontologists had already confronted the possibility that evolution has been interrupted by sudden transformations of the physical environment, although they tended to prefer slightly less dramatic geological changes as the root cause.

Matthew urged caution in postulating causes for the extinction at the end of the Cretaceous. There was an elevation of the land and a general dessication, but it was difficult to see why the dinosaurs could not have survived in the tropics.[206] He was honest enough to admit that the cause was uncertain, though the effects were obvious enough—a clearing away of one class to leave room for the development of another. In his 1914 study of Tertiary biogeography, "Climate and Evolution," Matthew followed Wallace by invoking the image of successive waves of migration spreading out from a northern center of evolution. In some respects this was a very 'Darwinian' model of the history of life, yet Matthew could not go the whole way with Darwinian gradualism. Like so many North American paleontologists, he saw an episodic character to the process that could not be explained in terms of complete geological uniformitarianism. The trigger for these waves of migration was the relatively sudden appearance of new and more highly evolved groups following episodes of climatic stress induced by geological events. This episodic character of Matthew's theory was directly inspired by Chamberlin's geology.[207] Thus the displacement of the archaic mammals at the end of the Eocene was brought about by the invasion of higher groups evolved in a period of elevation and refrigeration, following which "The northern fauna successively invaded the tropical and southern continents and swept before it nearly all their autochthonic faunae."[208] To explore the implications of this model, we now turn to the whole question of biogeography in the post-Darwinian era.

205. For a survey of the modern debates, see William Glen, ed., *The Mass Extinction Debates: How Science Works in a Crisis.*

206. Matthew, *Outline and General Principles of the History of Life*, pp. 141–142.

207. Matthew's 1914 paper is reprinted in the 1939 edition of *Climate and Evolution*; see p. 3.

208. Ibid., p. 89.

8

THE GEOGRAPHY OF LIFE

As paleontologists extended their knowledge of the history of life on different continents, their work began to interact with another classic line of evidence for evolution: the study of the geographical distribution of species. Darwin had used this extensively in the *Origin of Species* to support his claim that species originated in a single location and then spread out to occupy new territory, often becoming modified in the process. The existing distribution could not be explained in terms of simple adaptation. Only by invoking a historical process was it possible to account for the many apparent anomalies: areas with identical climates but very different floras and faunas, or cases where identical or closely related species lived in widely scattered locations. This style of explanation brought in the geologists and paleontologists, who could chart the migrations of species over periods of time, and thus explain how the various components of each area's flora and fauna had arrived there. The rise and fall of each group of animals and plants was a process that had to be understood in terms of both space and time. The biogeographer worked back from the present distribution, trying to understand how it might have come about. The paleontologist sought evidence of past distributions to throw direct light on where each group had originated and on how it had expanded its range.

Janet Browne has written a valuable survey of biogeography up to the time of Darwin.[1] Martin Fichman has outlined Wallace's biogeography, but little has been written on the explosion of interest in this field which took place in the late nineteenth and early twentieth centuries.[2] The only

1. Janet Browne, *The Secular Ark*.
2. Martin Fichman, "Wallace, Zoogeography, and the Problem of Land Bridges."

exceptions to this are various studies of the origins of the theory of continental drift.[3] These accounts are problematic because they are driven by hindsight: we are expected to know that the various debates over sunken 'land bridges' between continents would eventually be shown to be futile by the belated confirmation of Alfred Wegener's theory of continental drift in the 1950s. The issue is also clouded by the modern debate between biogeographers who emphasize the dispersal of species from localized points of origin and vicariance biogeographers, who argue that many groups have always occupied a wide range of territory. My purpose in addressing the substantial debates of the late nineteenth and early twentieth centuries is not to look forward to the present. I want to show how the first generations of evolutionists attempted to explain the origin and distribution of life, using the ideas and information at their disposal. My account will work forwards from Darwin and Wallace, not backwards from the modern debates.

Darwin's experiences in South America and the Galápagos Islands had taught him the value of biogeography as a testing ground for trying to understand the history of life. This approach had been introduced in the 1840s by Edward Forbes, who had tried to explain the flora of the British Isles in terms of a series of migrations from other parts of Europe.[4] Forbes pioneered an explanation used by Darwin to account for the fact that many mountains support isolated colonies of arctic plants: these populations were relics of the arctic flora that had moved south during the ice ages and then returned north. Forbes also postulated a landmass in what is now the Atlantic ocean, across which plants had migrated from Spain to Ireland. J. D. Hooker built on this model to invoke a great antarctic continent by which plants could have reached scattered locations in the southern hemisphere. These hypothetical ancient landmasses became the stock-in-trade of many later biogeographers. Darwin himself could not accept such massive depressions of the earth's surface in the recent geological past, thus founding a rival school of biogeography.[5] But we should be careful not to dismiss those who invoked sunken land bridges as 'anti-Darwinian.'[6] Darwin only balked at the sinking of continents during the Tertiary, and even he was tempted by the idea of a lost southern continent

---

3. See, for instance, Henry Frankel, "The Paleobiogeographical Debate over the Problem of Disjunctively Distributed Life Forms," and Homer Le Grand, *Drifting Continents and Shifting Theories.*

4. See Browne, *The Secular Ark,* especially chapter 8.

5. Darwin, *Origin of Species,* pp. 357–358.

6. As Adrian Desmond has done in the case of Huxley, for instance; see his *Archetypes and Ancestors,* p. 102.

to explain the apparently sudden appearance of the flowering plants in the fossil record.[7]

Darwin's central purpose in the chapters devoted to biogeography in the *Origin of Species* was to defend the claim that each species appeared in a single locality and then spread outwards.[8] His interest in the alpine plants of mountain ranges was sparked by the fact that here was an anomalous distribution which had to be explained in historical terms. The same was true for his account of oceanic islands: he dismissed the claim that they were the relics of sunken continents by pointing out that most were volcanic, and then offered a complex of processes by which certain types of animals and plants could be accidentally transported from the nearest continent.[9] The chapter on distribution in Haeckel's *History of Creation* is also concerned to defend the single 'creation' of each species, a vital component of his claim that each major group must be seen as monophyletic.[10] Unlike Darwin, Haeckel was prepared to admit recent subsidence of the earth's crust to explain the disappearance of continents.

Darwin was certainly aware that ocean barriers kept the inhabitants of continents isolated from one another, thus accounting for their unique character. He did not, however, become deeply involved in the attempt to divide the world into distinct botanical and zoological provinces. It was Alfred Russel Wallace who took this originally anti-evolutionary approach to biogeography and showed that the composition of each province's flora and fauna had to be explained in terms of how it had received its inhabitants by migration. Naturalists disagreed over the geographical limits of the provinces because different kinds of animals had different means of dispersal, and each expert constructed provinces based on the groups with which he was most familiar. Wallace pioneered the view that the distribution of many groups could be explained in terms of migrations from a northern center of progressive evolution. His *Geographical Distribution of Animals* of 1876 sparked an explosion of interest which continued into the early decades of the new century, although many were critical of Wallace's basic explanatory thesis. Most of the early contributors were field naturalists and museum workers, but paleontolgists became more active as the details of the fossil record for each continent became better known.

Biologists increasingly took it for granted that the population of any

---

7. See, for instance, Darwin to J. D. Hooker, 22 July 1879; *More Letters of Charles Darwin,* II, pp. 20–22.

8. Darwin, *Origin of Species,* p. 353.

9. Ibid., pp. 388–406.

10. Ernst Haeckel, *The History of Creation,* I, chap. 14; see especially pp. 352–353.

species will tend to increase, with the consequence that it will, wherever possible, expand to occupy new territory. For the Darwinians the tendency to expand was a simple consequence of the Malthusian law of population. Yet the specter of population pressure seems increasingly to have haunted the imagination of even those naturalists who rejected the theory of natural selection. In the course of the last decades of the nineteenth century, almost all biogeographers came to assume that species tend to expand their territory. Their work thus began to reflect a Darwinian viewpoint even when they ignored or rejected other aspects of Darwin's thinking. The growing popularity of this assumption marks a distinct phase in the development of biogeography.

The concept of population pressure did not gain immediate popularity following the publication of the *Origin of Species*. One of the first attempts at a global survey, Andrew Murray's *Geographical Distribution of Mammals* of 1866, explicitly denied any automatic tendency to expand. Murray was actually an entomologist, whose interest in biogeography had originally been aroused by problems in the distribution of insects. He now attempted to use the better-known distribution of mammals to solve these problems. Murray thought that geographical provinces were preserved by an *"inertia, or instinctive regard for personal ease, which leads each creature to remain where it is while it is comfortable."* [11] Only when geological forces changed the local conditions did animals move in search of a better environment. Darwin did not like Murray's work, writing to Wallace that he had "a want of all scientific judgement." [12] This was perhaps not surprising, given that Murray was one of the original critics of the *Origin of Species*. The success of at least one aspect of the 'Darwinian' world view can be measured by the almost universal rejection of Murray's concept of inertia in favor of the Malthusian pressure to expand.

Wallace, not surprisingly, took the Darwinian line: "Animals multiply so rapidly, that we may consider them as continually trying to extend their range; and thus any new land raised above the sea by geological causes becomes immediately peopled by a crowd of competing inhabitants, the strongest and best-adapted of which alone succeed in retaining their position." [13] R. F. Scharff, one of the most persistent critics of Wallace's theory of permanent continents, agreed: "It is of course the tendency of every species to spread in all directions from its original home, provided it does not encounter obstacles, such as want of food, unsuitability of cli-

---

11. Andrew Murray, *The Geographical Distribution of Mammals*, p. 10.
12. Darwin to A. R. Wallace, 5 June 1876; *Life and Letters of Charles Darwin*, III, p. 231. On Darwin's response to Murray's criticisms of his theory, see ibid., II, pp. 261, 265.
13. Wallace, *The Geographical Distribution of Animals*, I, p. 7.

mate or soil, or barriers such as mountains, rivers, or the sea."[14] The Royal Geographical Society's *Atlas of Zoogeography* of 1911 concurred: "The natural tendency of any species which is successful in the ever-waging struggle for supremacy is to gradually spread over a wider and wider area."[15] Even the anti-Darwinian J. C. Willis, who used biogeography to support Hugo De Vries's theory of the origin of species by sudden mutation, based his arguments on the claim that species expand at an absolutely regular rate: the older the species, the greater the area it occupies.[16] Small wonder that the literature of biogeography was dominated by metaphors which seem characteristic of an age of imperialism: the rise and fall of species driven by the urge to conquer and dominate new territory.

This emphasis on the power of each species to disperse from a single point of origin makes it impossible to see these early debates as preliminary skirmishes in the modern war between the supporters of dispersal and the vicariance school of biogeography. Dispersalists assume that each group of related species originated in a single home territory. If two related species are found in widely separated locations, they look for ways in which members of the original population could have migrated from the point of origin to new locations. Vicariance biogeographers follow the model proposed by Leon Croizat, which refuses to look for a point of origin and assumes that widely scattered groups have always occupied a large swathe of territory. One supporter of Croizat, Gareth Nelson, has suggested that the whole Darwinian episode in biogeography may turn out to have been a blind alley.[17] The vicariance approach has gained much support from the theory of continental drift, which explains how a population can have been fragmented by the splitting up of its original territory rather than by migration. Vicariance biogeography looks to the same evidence as that used by the early twentieth-century advocates of sunken land bridges. By the standards of the modern debate, however, all of the previous generation of biogeographers were dispersalists—even those who followed the land-bridge school of thought. They all assumed that a species originated in a single location and had a natural tendency to spread. Their debates were merely over the routes available for spreading, not over the

14. R. F. Scharff, *The History of the European Fauna*, p. 12.

15. J. G. Bartholomew, W. Eagle Clarke, and Percy H. Grimshaw, *Atlas of Zoogeography*, p. 3.

16. J. C. Willis, *Age and Area*, for example, pp. 5–6. See also the work of H. B. Guppy, described in Bowler, *Eclipse of Darwinism*, pp. 210–212.

17. See Gareth Nelson, "From Candolle to Croizat." See, for instance, Croizat, *Panbiogeography*, and on the controversy surrounding this area, see Gareth Nelson and Don E. Rosen, *Vicariance Biogeography: A Critique*.

basic question of whether or not one should seek to trace dispersal routes. In this sense, the whole debate was being driven by an evolutionary imperative: the desire to seek the original home where the crucial evolutionary innovation occurred.

Anti-Darwinian evolutionists challenged the assumption that a wide distribution indicated a long-range migration from a single home base. As Darwin and Haeckel both realized, their general thesis that taxonomic relationship is a sign of common descent rested on the belief that each group is monophyletic. If the same or very closely related species could evolve independently in more than one location, the foundations of their whole research program would be threatened. Darwin emphasized the power of dispersal as the basis for his biogeography precisely because he needed to block this challenge. Those evolutionists who postulated massive convergence or parallelism were threatening the foundations of Darwinian biogeography as well as a phylogenetic taxonomy. Tracing migrations would be impossible if the same species could evolve more than once—indeed it would become virtually impossible to distinguish between the effects of parallel evolution and dispersal.

The threat posed by the theory of the independent evolution of identical species was not an empty one, and a number of biogeographers considered the problem, if only to conclude that the possibility was too remote to worry about. Murray raised the problem in the context of a possible multiple origin for the Mammalia, but went on as follows: "We are driven to the conclusion that each class, order, or natural group, started from one parent species, into which had been drawn, as into one focus, all the different rays of previous forms which were afterwards dispersed among the equivalent types which sprang from it."[18] More serious was the possibility that horses had evolved independently on both sides of the Atlantic, making nonsense of any attempt to trace migrations between Eurasia and America. The most positive statement of this idea came from Angelo Heilprin, who also suggested that the same species might evolve independently in different geological epochs.[19] Heilprin's anti-Darwinian views were cited by Frank Beddard in a popular textbook of 1895.[20] The noted mammal expert Richard Lydekker also accepted that true horses had evolved on both sides of the Atlantic.[21] Karl von Zittel argued that migration did not explain the similarities between the Eocene faunas of

18. Murray, *Geographical Distribution of Mammals*, p. 54.

19. Angelo Heilprin, *The Geographical and Geological Distribution of Animals*, pp. 186–187, and on the same species appearing in two geological periods, see pp. 183–184.

20. Frank Beddard, *A Textbook of Zoogeography*, p. 133.

21. Richard Lydekker, *A Geographical History of Mammals*, p. 178. A more restricted problem was the independent evolution of the same blind species in different caves; see ibid., pp. 381–382. See also Lydekker's *Life and Rock*, p. 48.

North America and Europe, postulating instead a parallel evolution of similar forms on both continents.[22] More cautiously, Scharff insisted that nothing ruled out the possibility of dual origin, although he conceded that "almost all species have but one home."[23] It may be significant that Croizat, founder of the modern antidispersal school, was also an advocate of parallel evolution in different regions. Here again, parallelism was used to create an anti-Darwinian model of the history of life which rejected the search for origins.

For the evolutionists of the late nineteenth and early twentieth centuries, this problem could only be solved by better identification of specimens and by the correlation of geological horizons between Europe and America. Tracing past migrations from one continent to the other was clearly impossible if there was no way of being sure which European and North American strata were laid down at the same time. Solving this problem was a major task addressed by a number of paleontologists from both sides of the Atlantic in the last years of the nineteenth century. As the Americans caught up with the Europeans in working out the sequence on their own continent, they began to see the importance of establishing exact correlations between the two. Osborn acted as coordinator for the project, offering trial correlations between the American and European strata which were intended to elicit debate aimed at reaching a consensus.[24] Osborn was led to undertake this task in part because he despaired of working out phylogenies when the possibility of migration from another continent could not be ruled out. Although a leading advocate of parallelism, much of his own work through the early decades of the twentieth century was devoted to the use of paleontology to solve biogeographical problems by tracing migrations through time. Paleontology would answer the questions posed by an earlier generation's study of modern distribution.

Everyone conceded that the hypotheses proposed to explain the facts were somewhat speculative. The Cambridge morphologist Hans Gadow called his little book of 1913, *The Wanderings of Animals*, a "Romance of Land and Water."[25] Gadow saw that "the key to the distribution of any group lies in the geographical configuration of that epoch in which it made its first appearance."[26] When first evolved, a new form was able to supplant any existing rivals and would spread up to the limits imposed by the

22. Karl von Zittel, *Textbook of Palaeontology*, III, p. 295.

23. Scharff, *History of the European Fauna*, p. 38.

24. H. F. Osborn, *Cenozoic Mammal Horizons of Western North America*, and "Correlations between Tertiary Mammal Horizons of Europe and North America." See Ronald Rainger, *An Agenda for Antiquity*, pp. 185–191.

25. Hans Gadow, *The Wanderings of Animals*, p. 3; see also p. vi.

26. Ibid., p. 13.

geographical barriers of the time. Later changes brought about by geological forces would not be so critical because, by then, it would be less able to expand at the expense of others, and might be forced to give up some of its own territory to more highly evolved competitors. The critical question was how to reconstruct the past geography of the earth, and the answers proposed depended on a network of assumptions about the possibilities of both migration and geological change.

One popular assumption was that the large northern continents, Eurasia and North America, constituted a powerhouse of progressive evolution, from which successively higher types migrated out to dominate the rest of the earth. This was Wallace's view, subsequently revived by Matthew. But Hooker's rival idea that there had once existed a great southern continent, now mostly sunk beneath the ocean, served as the basis for a major alternative explanation. This continent not only linked the existing southern landmasses—South America, southern Africa and Australia— but also served as an evolutionary center in its own right, a potential source of newly evolved types that rivaled the northern or holarctic center. Alternatively, the existing southern continents, especially South America and Africa, could have been the original sources for some important groups. The social implications of the claim that the north was the main center for progressive evolution were not lost on some biogeographers, especially as Eurasia was widely seen as the center of human progress. The rival view based on southern continents gained some popularity, however, which must give pause to any historian seeking to argue that the science of the time was completely dominated by an assumption that fitted so well into the white race's ideology of imperialism.

If naturalists disagreed over the relative standing of north and south in the production of higher forms, they also debated the plausibility of the earth movements needed to explain the disappearance of ancient landmasses. Darwin, Wallace, and Matthew accepted that, at least in the Tertiary, the existing outlines of the continents and oceans had remained unchanged. Reductions in the sea level during the glacial epoch could have turned shallow sea into dry land but could not have exposed the deep ocean bed. The rival school of geology led by Eduard Suess held that there has been a steady subsidence in certain areas of the earth's crust throughout its history. Great continents may have existed in the past where now there is deep ocean. Much of the recent literature has assumed that the school of biogeography which appealed to crustal movements to explain past migrations was obsessed with "land bridges." This phrase reflects the disdain of their opponents, however—Matthew called Scharff a "reckless bridge-builder."[27] The term *land bridge* suggests a narrow corridor across

27. W. D. Matthew, *Climate and Evolution*, p. 146.

which migrants pass from one main landmass to another. Such narrow corridors were certainly suggested, but most members of this school of thought held that the sunken lands were themselves very extensive. They were massive continents, not narrow corridors, and far from merely transmitting animals from one still-extant landmass to another, they were the true homes of many types which are now confined to the surviving fragments of their original range.

The biogeographers who insisted on the permanence of continents and oceans explained the existence of widely scattered members of the same group in terms of migration from a center, coupled with subsequent disappearance from the center and all intermediate locations. The only alternative was the kind of accidental transport across oceans by natural rafts, birds, and so on that Darwin used to explain the populations of oceanic islands. The rival school of thought dismissed these dispersal mechanisms as unnecessary: since the islands were themselves the remains of sunken continents, all cases of anomalous distribution could be explained by migration across dry land. Each side accused the other of dreaming up hypotheses that could explain absolutely anything. With sufficient ingenuity, one could postulate a sunken land connection or an 'accidental' migration across the ocean to bridge any two points on the earth's surface. Although geology could be brought in on both sides of the debate on continental permanence, the rival schools of thought disagreed over which kind of evidence was more reliable when assessing the plausibility of a land connection. Darwin and his followers refused to accept the kind of earth movement that would sink land down to deep ocean bed in a geologically short time. For them, accidental dispersal was more plausible than this kind of almost catastrophic geological change. Bridgebuilders such as Scharff tended to ignore geology altogether, except for drawing on the general theory of continental subsidence. They postulated very rapid crustal movements wherever it was necessary to get a species from one point to another and simply assumed that geology would accommodate their ideas. Even Suess and his supporters may not have countenanced some of the more extreme cases of sunken landmasses postulated in their name.

ZOOLOGICAL PROVINCES

In the early nineteenth century it had become accepted that the distribution of species could not be explained by simply assuming that each occupied all the territory to which it was best adapted. Parts of Africa and South America have very similar environments, but the animals and plants inhabiting them are different. Darwin himself had helped to establish the fact that this difference extends back into the fossil record. The South

American fossils he brought back from the voyage of the *Beagle* confirmed the 'law of succession of types' by showing that the unique character of the South American fauna extends back into history. But—as Wallace explained to the lay reader in his *Darwinism* of 1889—the divisions between the regions were by no means self-evident.[28] The animals of western Europe and Japan are very similar despite the enormous distance between them. Yet in other parts of the world a matter of a few miles separates widely different faunas, most obviously in the case of the Asian and Australian faunas, where the junction became known as 'Wallace's line.'

Before the theory of evolution became popular, explanations for these curious facts were sought by postulating 'centers of creation' characteristic of the major continental areas, each somehow generating its own peculiar types of animals and plants. The attempt to divide the earth into self-contained regions became popular in the late nineteenth century, following the publication of Philip Lutley Sclater's scheme in 1857. To some extent this was a very typological view of the situation, an attempt to draw rigid lines across a map when the real situation was much more fluid. It is difficult to escape the feeling that late nineteenth-century Europeans and Americans were fascinated by the attempt to carve the world up into biological provinces which matched their own political and economic zones of influence. The search for rigid boundaries encountered obvious anomalies, however. A popular example exploited by the evolutionists was the distribution of the tapirs, which are found widely separated in South America and Southeast Asia. Furthermore, the maps drawn by specialists in different fields did not always correspond. The evolutionists took over the idea of zoological provinces and used the anomalies as an excuse to modify it in a way that fitted their own requirements. The boundaries were real in the sense that they were defined by major barriers to migration, but they were loosely defined because different types of animals migrated by different means and had originated at different times in the earth's geological history.

Naturalists recognized significant divisions within the great regions defined by climate and environment. Thus Joel Asaph Allen and C. Hart Merriam divided North America into zones corresponding to desert, temperate, and mountainous regions.[29] These zones were important for the emerging science of ecology, but they were of little interest to biogeographers because they merely indicated that the animals characteristic of each province had adapted to the whole range of conditions found there. Of greater concern were anomalies in the flora and fauna of partially isolated

28. Wallace, *Darwinism*, p. 339; see also his *Island Life*, p. 4.

29. J. A. Allen, "The Geographical Distribution of North American Mammals"; C. Hart Merriam, "The Geographical Distribution of Life in North America."

regions, which often lacked some of the species typical for their province. In such cases, as Forbes had demonstrated for the British Isles, it was necessary to treat the problem as a historical one. The inhabitants had crossed the barriers at various times, and the peculiarities of their distribution could be explained by reconstructing the circumstances of their migrations. This type of problem was merely a small-scale version of the larger issues that arose when trying to explain the existence of the major provinces.

Sclater was an ornithologist who became secretary of the Zoological Society of London (he was forced to retire in 1902 following a controversy which also involved his son William, then director of the South African Museum in Cape Town). His scheme was outlined to the Linnean Society of London in 1857. It took the geographical distribution of birds as its primary source of evidence, although Sclater thought this would provide a good indication of the most basic regions. Birds can migrate readily, and if they respect certain boundaries, we can be reasonably certain that most other animals will be similarly confined. Sclater interpreted his regions as a natural consequence of the fact that creation was localized in centers.

> It is a well-known and universally acknowledged fact that we can choose two portions of the globe of which the respective Faunae and Florae shall be so different, that we should not be far wrong in supposing them to have been the result of distinct creations. Assuming then that there are, or may have been, more areas of creation than one, the question naturally arises, how many of them are there, and what are their respective extents and boundaries, or in other words, what are the most natural primary ontological divisions of the earth's surface?[30]

He proposed six regions:

1. Palaearctic (Europe, Asia excluding India, and north Africa).
2. Ethiopian, or Western Palaeotropical (Africa south of the Atlas mountains)
3. Indian, or Central Palaeotropical (India, Southeast Asia, and the islands of what is now Indonesia).
4. Australian, or Eastern Palaeotropical (Australia, New Zealand, Papua, and the Pacific islands).[31]
5. Nearctic or North-American (North America down to Mexico, and Greenland).

30. P. L. Sclater, "On the General Geographical Distribution of the Members of the Class Aves," p. 130. On Sclater's later career, see P. Chalmers Mitchell, *Centenary History of the Zoological Society of London*, pp. 70–75.

31. The original text names both the Ethiopian and Australian regions as "Western Palaeotropical," presumably a misprint.

6. Neotropical, or South American (South America, Southern Mexico, the West Indies).

Writing much later in 1899, Sclater was able to record that his six regions had formed the basis for most of the later debate, especially after they were endorsed by Wallace in 1876.[32] These regions were still employed by the Royal Geographical Society's *Atlas of Zoogeography* in 1911.[33] Various techniques were devised for drawing simplified maps of the regions, which could be modified to show variations in the distibution of particular groups.[34] The simplest was that proposed by Peter Chalmers Mitchell (who succeeded Sclater at the Zoological Society) in 1890:[35]

Nearctic		Palaearctic	
Neotropical	Ethiopian	Oriental	
		Australian	

Mitchell was senior demonstrator in the morphology laboratory at Oxford at the time; evidently he was expected to teach the geographical distribution of the organisms studied.

A series of disagreements emerged around the degree of relationship among the regions, and the status of certain islands such as Madagascar and New Zealand. Some zoologists thought there was so little difference between the Palaearctic and Nearctic regions that they should be united into a single great northern province which was eventually called the holarctic (fig. 8.1). The concept of a unified holarctic region became the basis of the claim that distribution was controlled by the radiation of more highly evolved types from the north. Others noted a surprising degree of similarity between the southern provinces, especially the African and South American, which fueled speculation about a sunken Antarctic continent. Australia, New Zealand, and Madagascar were all seen as isolated regions that had never been stocked by the more highly evolved mammals of the Tertiary.

Sclater himself had admitted that there was less difference between his Palaearctic and Nearctic regions than between any others. American naturalists such as Alpheus Packard and J. A. Allen objected to the term *Nearctic* on the grounds that it gave an unfavorable image of the conti-

32. W. L. and P. L. Sclater, *The Geography of Mammals*, chap. 1.
33. Bartholomew et al., *Atlas of Zoogeography*, pp. 4–12.
34. There is a good survey of the mapping techniques in Beddard, *Textbook of Zoogeography*, pp. 118–123.
35. P. Chalmers Mitchell, "On a Graphic Formula to Express Geographical Distribution."

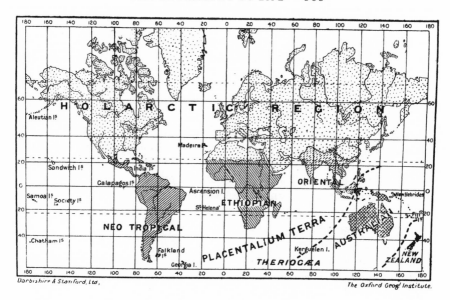

FIGURE 8.1   Zoo-geographical map of the world. From E. Ray Lankester, *Extinct Animals* (London, 1905), p. 63. Like many of his contemporaries, Lankester adopted the convention of linking Eurasia and North America into a single holarctic region.

nent's climate. They wanted to separate temperate North America from the true arctic, which they believed was a uniform circumpolar region covering the northern fringes of both America and Eurasia.[36] A more drastic but ultimately more popular modification was simply to drop the distinction between North America and Eurasia altogether. Writing in 1868, Huxley argued that all of Sclater's provinces except the Neotropical (South America) and the Australasian formed a single great unit, Arctogaea. He also indicated that he would prefer to abandon the distinction between the Palaearctic and Nearctic.[37]

In 1882 Angelo Heilprin formally challenged the status of the Nearctic, suggesting that North America and Eurasia be united as a single province which he at first proposed to call the "Triarctic."[38] This move was disputed by Wallace, who thought that it might establish a trend that would threaten the unity of other regions.[39] In a response to Wallace, Heil-

36. See Allen, "The Geographical Distribution of North American Mammals," pp. 211–212.

37. T. H. Huxley, "On the Classification and Distribution of the Alectoromorphae," pp. 368–370.

38. Heilprin, "On the Value of the 'Nearctic' as One of the Primary Zoological Regions" (1882).

39. Wallace, "On the Value of the 'Nearctic' as One of the Primary Zoological Regions."

prin followed a suggestion made to him by the British ornithologist Alfred Newton, who proposed that the unified region be known as the "Holarctic."[40] Some biogeographers shared Wallace's misgivings and continued to use the old distinction. The tensions were clearly evident in the 1895 textbook written by Frank Beddard, who lectured in biology at Guy's Hospital, London. Beddard noted that the distribution of earthworms supported the idea of a unified holarctic, while that of the Amphibians showed a strong division between Palaearctic and Nearctic.[41] Richard Lydekker supported the concept of Holarctica, and Lankester used it in his popular survey, *Extinct Animals,* of 1905.[42] It became the basis of Matthew's efforts to show that all the southern regions were populated by migrations from a single, unified northern center of evolution.[43] Osborn, who at first opted for more evolutionary centers, still chose to unite the northern continents into a single holarctic region.[44]

Beddard pointed out that those experts who had criticized Sclater's regions were, on the whole, those who dealt with groups that had first appeared far back in geological time.[45] The regions worked best for those groups which had appeared so recently that there had been no subsequent change in the position of land and sea. If a group did conform to Sclater's scheme, then it was not ancient. The wide distribution of the onychophoran *Peripatus* was taken as evidence that this was indeed an ancient form, and hence possibly the ancestor of the later arthropods.[46] Its presence in several different tropical regions suggested land connections in the deep geological past very different from those of today. The most passionate debate in biogeography centered on just how ancient a group had to be for its distribution to be limited by modern barriers.

The claim that modern distribution would have to be explained in historical terms was implicit in Darwin's theory. For Wallace and those who developed an evolutionary biogeography in the late nineteenth century, Sclater's regions were not absolute units but useful heuristic devices for visualizing the major effects of geographical barriers upon the migrations of animals in the course of geological time. Although primarily a morphologist, Huxley made an important step toward a historical ap-

40. Heilprin, "On the Value of the 'Nearctic' as One of the Primary Zoological Regions" (1883); see also his *Geographical and Geological Distribution of Animals,* p. viii.

41. Beddard, *Textbook of Zoogeography,* p. 80.

42. Lydekker, *Geographical Distribution of Mammals,* p. 146; E. Ray Lankester, *Extinct Animals,* pp. 63–65.

43. Matthew, *Climate and Evolution,* for example, p. 11.

44. Osborn, *The Age of Mammals,* for example, pp. 64–65.

45. Beddard, *Textbook of Zoogeography,* p. 73.

46. Ibid., p. 16; see chapter 3 in this text.

proach to biography in his 1868 paper on the classification and distribution of the Alectoromorphae (the game birds, including grouse, etc.). It was in this paper that he suggested that all of Sclater's northern regions should be grouped into a single province, Arctogaea. Only the Neotropical region (South America)—which Huxley preferred to call Austro-Columbia—and Australasia were separated off. The pigeons and parrots were well represented in both Austro-Columbia and Australasia, but how had they become distributed in this way? Huxley argued that the answer to the question lay in whether or not the Miocene rocks of Europe would yield fossils of these birds. If there were no such fossils, then the group must have originated in the southern provinces, and its representatives in the north would be recent immigrants. If the fossils *were* there, then the group had originated in the north, and its currently flourishing state in the southern regions was the result of immigrants there having found the climate highly suitable. He was inclined to suspect that the modern distribution of the parrots favored the latter hypothesis.[47] Huxley was thus prepared to accept that the group had not evolved in the region to which it was best adapted and that its modern distribution was a later phenomenon made possible by its expansion into new territory. Lydekker subsequently noted that Huxley's hypothesis had been confirmed by fossil discoveries.[48]

Something of the same concerns enter into the last chapter of Huxley's classic introductory biology text, *The Crayfish*. In his chapter on comparative morphology, he showed that the modern crayfish must be divided into two groups, a northern hemisphere family, the Potamobiidae, and a southern hemisphere group, the Parastacidae. Embryological evidence suggested that both groups had evolved from a common parent form, the Protostacidae, now extinct. When he turned to geographical distribution, Huxley noted that the pattern for the crayfish is quite different from that observed in birds and mammals, since the regions are determined by different factors. The European crayfish can be interpreted as the products of a series of westward migrations of an Aralo-Caspian stock following the ice age.[49] The distinction between the northern and southern types can be explained on the assumption that the crayfish are the descendants of marine ancestors which migrated into rivers and lakes and adapted to fresh water. The situation is similar to that of the freshwater prawns, but the marine ancestor of the crayfish has vanished, leaving two groups of its

47. Huxley, "On the Classification and Distribution of the Alectoromorphae," pp. 372–373.

48. Lydekker, *Geographical History of Mammals*, p. 130.

49. Huxley, *The Crayfish*, pp. 321–322.

descendants isolated at opposite ends of the globe. Huxley noted that the absence of crayfish from certain apparently suitable regions can be accounted for when we note that fluvatile crabs would be powerful rivals for this group.[50] Wherever the crabs had flourished, the crayfish could not gain a foothold.

Here we see Huxley being prepared to admit that the present distribution must be seen as the product of a historical process governed by an interaction between the possibilities of evolution and the limitations of the environment. Although primarily a morphologist, and strongly attracted to the idea of parallelism, he was nevertheless prepared to follow Darwin in seeing the history of life as a process that could not altogether be predicted on the basis of purely biological trends. Hans Gadow is another example of a morphologist who developed an interest in geographical distribution.[51] As Wallace noted, accurate taxonomy was vital to the assessment of biogeographical problems. The Mexican *Bassaris astula* had been classified as a civet, a group which was normally confined to Africa and Asia—but W. H. Flower had shown that it was actually a form of raccoon, thus eliminating what had otherwise been a serious anomaly in distribution.[52] Morphologists such as Huxley, Gadow, and Mitchell were concerned about taxonomic relationships, and they could also appreciate the problems created by the geographical locations from which their specimens came.

Not surprisingly, though, field naturalists were most active in taking up the study of biogeography. It was Wallace who did most to convince biologists that Sclater's abstract division of the globe into regions must be seen as a problem that had to be solved in historical terms. Wallace had spent considerable amounts of time collecting in both South America and the Malay archipelago (modern Indonesia). He was responsible for establishing the exact position of the line dividing the Asian from the Australian faunas, which in the Malay archipelago runs between the two adjacent islands of Bali and Lombock.[53] In 1870 Darwin and Newton encouraged

50. Ibid., p. 337.

51. Gadow was lecturer in morphology at Cambridge when he published his *Wanderings of Animals*.

52. Wallace, *Geographical Distribution of Animals*, I, chap. 6; see pp. 83–84.

53. Wallace, *The Malay Archipelago*, chap. 10. On Wallace's line, and more generally on the importance of biogeographical maps, see Jane Camerini, "Evolution, Biogeography, and Maps: An Early History of Wallace's Line." On later developments, see Ernst Mayr, "Wallace's Line in the Light of Recent Zoogeographic Studies." On Wallace's work, see his autobiography, *My Life*, and Alfred Marchant, *Alfred Russel Wallace*. Modern studies include Barbara G. Beddall, *Wallace and Bates in the Tropics;* Martin Fichman, *Alfred Russel Wallace;* Wilma George, *Biologist-Philosopher;* H. Lewis McKinney, *Wallace and Natural Selection;* and Amabel Williams-Ellis, *Darwin's Moon.*

him to undertake a more general study of geographical problems in the light of evolution theory. Although put off at first by the lack of accurate information, Wallace was at last able to put together tolerable lists for the distribution of many genera and was able to publish his monumental *Geographical Distribution of Animals* in 1876. A host of authorities including Newton, Flower, Gunther, and Mivart helped him to process the information—an interesting list since it includes one specialist who was indifferent to the whole idea of evolution (Gunther) and the archetypical anti-Darwinian, Mivart. Wallace expressed the hope that his work would encourage others to join in and help create a field of research equivalent to "the lofty heights of transcendental anatomy, or the bewildering mazes of modern classification."[54] There can be little doubt that his book triggered an explosion of interest in biogeography over the next several decades. By the 1890s Newton could write of morphology, paleontology, and biogeography as the "triple alliance" that guided attempts to reconstruct phylogeny.[55]

Wallace argued that the distribution of mammals offered the best guide to dividing the earth up into regions, since they were least likely to be transported accidentally across the oceans and thus blur the lines defined by major geographical barriers.[56] Birds differed only marginally in this respect, so it was not surprising that Sclater's regions worked for mammals too. Wallace noted Huxley's argument that there was an overarching similarity between all the northern regions but held that this was far too great an area to make a useful region.[57] He would not go along with Murray, who had grouped North America and Eurasia into a single unit, and Africa and India into another. Huxley's claim that New Zealand should be treated as a region on its own (discussed below) was dismissed on the ground that some of its characteristic species might yet be discovered in Australia.[58] In the end, Wallace came firmly down on the side of Sclater's regions as the most useful and natural divisions of the earth's surface. His subsequent opposition to a unified holarctic region has already been noted.

Yet this analysis of the biogeographic regions came in the fourth chapter of Wallace's book, following an extensive discussion of the factors which might have created the present distribution. He was interested in *why* the regions existed as they did, and was convinced that this question

54. Wallace, *The Geographical Distribution of Animals*, II, p. 553. The circumstances leading to Wallace's undertaking the project are described in the book's preface.

55. Newton, *A Dictionary of Birds*, p. 120.

56. Wallace, *Geographical Distribution of Animals*, I, p. 57.

57. Ibid., I, p. 59.

58. Ibid., I, p. 62.

could only be answered by considering the history of the earth and its inhabitants. Wallace took it for granted that animals have a natural tendency to spread out from their home territory, "to roam in every direction in search of fresh pastures and new hunting grounds."[59] Regions would thus be defined by geographical barriers, and by the changes in such barriers brought about by geological forces. Wallace himself had once been tempted by the idea of land bridges, and he still believed that geographical distribution would reveal the previous existence of islands and continents now sunk.[60] He now insisted that, in the Tertiary period at least, it was unrealistic to postulate the existence of landmasses occupying the positions of the modern oceans.

The great ocean basins were the most fundamental barriers to dispersal. Insects and plants might be occasionally transported across the oceans, accounting for some anomalies, but the higher animals were confined to those areas they could reach by land. Shallow seas could be exposed as dry land by a relatively slight depression of the sea level such as might have taken place in the ice age, thus opening temporary migration routes.[61] Wallace was prepared to accept that extensive tracts of land had once existed in what are now the Indian and western Pacific oceans.[62] But he insisted that deep oceans were permanent barriers, at least on the timescale of the Tertiary period. The crucial barrier marking 'Wallace's line' was a strait of exceptionally deep water between Bali and Lombock that could not have been exposed as dry land by any imaginable reduction in the sea level. For Wallace, as for Darwin, those who conjured up land bridges to solve every anomaly of distribution were simply letting their imaginations run riot:

> If we once admit that continents and oceans have changed places over and over again (as many writers maintain), we lose all power of reasoning on the migrations of ancestral forms of life, and are at the mercy of every wild theorist who chooses to imagine the former existence of a now submerged continent to explain the existing distribution of a group of frogs or a genus of beetles.[63]

Curiously, this was exactly the argument used by the supporters of land bridges against the mechanisms of accidental transportation postulated by

59. Ibid., I, p. 10

60. For example, ibid., I, pp. v–vi, 8. See Fichman, "Wallace, Zoogeography, and the Problem of Land Bridges."

61. Wallace, *Geographical Distribution of Animals*, I, pp. 36–37.

62. Ibid., I, pp. 358–360.

63. Wallace, *Island Life*, p. 10.

Darwin and Wallace to explain anomalies in the distribution of lower animals.

Barriers could also exist even where there was dry land: deer and bear could not enter Africa from Eurasia because they were woodland creatures, and the deserts of North Africa are impassable to them. The absence of elephants and other typically 'African' creatures from Eurasia was a puzzle, given that the fossil record confirmed their presence in the recent past. Wallace here appealed to the ice age as the agent that had both driven these creatures south into Africa and wiped them out in their original home.[64] Some of the existing regions were thus geologically very recent. As we shall see below, Wallace extended his use of the fossil evidence to create a whole theory of the origin and dispersal of species based on migration from a northern center of evolution.

LOST WORLDS

Previous chapters have noted the evolutionists' continued fascination with the concept of the 'living fossil'—a modern species that has somehow preserved the character of a long-vanished ancestral stage in the development of life. The same motivation seems to have supported the popular assumption that certain regions of the world have preserved populations characteristic of an earlier phase in the history of life. Some regions were, in effect, 'older' than others; they had been protected from the invasion of more highly evolved forms and retained a fauna typical of what had once been worldwide in an earlier geological epoch. Australia was the classic example, and many scientists seem to have welcomed the idea that this continent's animals—like its human inhabitants—were 'primitive' and hence unable to compete with European invaders once they had gained access to the continent. No one seriously expected to find dinosaurs still alive in South America, but Sir Arthur Conan Doyle's story *The Lost World* of 1912 built upon a popular fascination with the idea that the past might still linger on in some remote corner of the globe.

Another example of such thinking is the expectation that the deep sea might yield evidence of animals dating back to the Mesozoic or even the Paleozoic era. Louis Agassiz had given voice to this possibility in a letter of 1871 to Benjamin Peirce, superintendent of the U.S. Coast Survey.[65] Charles Wyville Thomson shared the same hopes, and there is no doubt that this prediction helped to generate enthusiasm for the *Challenger* ex-

---

64. Wallace, *Geographical Distribution of Animals,* I, pp. 149–150.

65. Louis Agassiz, "A Letter Concerning Deep-Sea Dredgings."

pedition of 1872–76.[66] Moseley, the expedition's naturalist, recalls how at first each cuttlefish was squeezed to see if it contained a belemnite bone, and trilobites were eagerly looked for in the dredgings.[67] The hopes were soon dashed, of course, although the same idea resurfaced in another form in the 'Bipolar hypothesis,' much debated in the 1890s. This theory assumed that the Arctic and Antarctic oceans shared many identical species, presumably the relics of a once world-wide marine fauna.[68] Not even Agassiz expected the seas to yield ichthyosaurs or plesiosaurs, but the efforts of Richard Owen and others to combat popular belief in sea serpents suggests that the public would take some convincing that ancient monsters were not still alive.[69] H. G. Wells and Julian Huxley still thought it necessary to include a short chapter rejecting such beliefs in their popular survey of 1931.[70]

Adrian Desmond has commented on the emergence in the early nineteenth century of the view that the Australian marsupials were relics of the Mesozoic mammalian fauna being found in the European fossil record.[71] Many geologists assumed that the earth had enjoyed a warm and essentially uniform climate throughout the Mesozoic. Significant cooling and regionalization of the climate was a product of the Tertiary. Coupled with the widespread belief that there had been a much more extensive land surface at the time, especially in the southern hemisphere, it was easy to imagine that a universal fauna had penetrated into Australia and then been cut off by a later geographical barrier. The early population would have been protected from the impact of later mammalian evolution in the north. As early as 1862 Frederick M'Coy, professor of natural science at the University of Melbourne, was led to protest against this almost universal assumption of his adopted continent's primitive character. The geological survey of Victoria revealed that, far from having been in existence since the Mesozoic, Australia had been under the sea for much of the Tertiary. Any previous inhabitants had been destroyed, and the continent had been repeopled in relatively modern times.[72]

66. C. Wyville Thomson, "On Deep Sea Climates."

67. H. N. Moseley, *Notes by a Naturalist*, p. 509.

68. See, for example, John Murray, "On the Deep and Shallow-Water Marine Faunas of the Kerguelen Region." For opposition to the theory, see D'Arcy Wentworth Thompson, "On a Supposed Resemblance between the Marine Faunas of the Arctic and Antarctic Regions"; and Louis Dollo, *Resultats du Voyage du S.Y. Belgica: Zoologie: Poissons*, pp. 191–194. See also the anonymous article "The Bipolar Theory."

69. On Owen and the sea-serpent sightings, see Nicolaas Rupke, *Richard Owen: Victorian Naturalist*, pp. 324–332.

70. H. G. Wells et al., *The Science of Life*, pp. 187–188.

71. Desmond, *Archetypes and Ancestors*, pp. 101–107.

72. Frederick M'Coy, "Note on the Ancient and Recent Natural History of Victoria."

M'Coy's views were challenged in Murray's pioneering survey of bio-geography of 1866, which reinforced most of the traditional prejudices.[73] Murray believed in the gradual cooling of the earth since the Mesozoic and in the existence of great southern landmasses at that time.[74] He argued that Australia has "preserved the general type of the eocene life down to the present day."[75] Murray drew upon a new line of evidence suggested by Franz Unger, professor of botany at Vienna, whose lecture "New Holland in Europe" had drawn parallels between the fossil plants of the Eocene in Europe and the gum-trees of modern Australia.[76] Unger thought that Europe had received some of its plants by migration from Australia, but his theory could be taken as evidence for the land link that had allowed Australia to host the worldwide Mesozoic fauna before being cut off.

Unlike Darwin, T. H. Huxley was sympathetic to Murray's "thoughtful and ingenious work."[77] In his 1870 address to the Geological Society on paleontology and evolution, Huxley addressed the problem posed by the unusual character of the inhabitants of southern regions, especially Australia and New Zealand. In a paper of the previous year on the fossil reptile *Hyperodapedon*, he had already noted the close resemblance between this and the living New Zealand reptile *Sphenodon* (the tuatara). Coining the term *Poikilitic* to denote the Permian and Trias from which *Hyperodapedon* dated, he asked, "What if this present New Zealand fauna, so remarkable and so isolated from all other faunae, should be a remnant, as it were, of the life of the Poikilitic period which has lingered on isolated, and therefore undisturbed, down to the present day?"[78] In his 1868 paper on geographical distribution, he was even inclined to take seriously the suggestion that New Zealand might count as a full province in its own right.[79]

The 1870 address extends the same line of argument to the inhabitants of Africa and India, as well as those of Australia. Huxley believed that Africa and India had once been united by dry land. Their present population of ungulates was derived from the European Miocene fauna, which had migrated south after a sea barrier had been removed and had

73. Murray, *Geographical Distribution of Mammals*, p. 284.

74. Ibid., pp. 1–12.

75. Ibid., p. 23.

76. Unger's lecture "Neu Holland in Europa" was translated as "New Holland in Europe" for the *Journal of Botany*.

77. Huxley, "Palaeontology and the Doctrine of Evolution," p. 367.

78. Huxley, "On Hyperodapedon," p. 388. On the tuatara as a relic of the Secondary era, see also Beddard, *Textbook of Zoogeography*, p. 14, and Lydekker, *Geographical History of Mammals*, p. 8.

79. Huxley, "On the Classification and Distribution of the Alectoromorphae," p. 369.

flourished in their new home.[80] Subsequently, these "lost tribes of the Miocene Fauna were hemmed by the Himalayas, the Sahara, the Red Sea and the Arabian deserts, within their present boundaries," while their northern relatives were exterminated by the ice ages.

The same applied to the Australian marsupials, which Huxley assumed to be relics of a once widely distributed population of Mesozoic mammals. He took it for granted that there was a relationship between the faunas of the European Eocene and modern Australia.[81] This similarity implied that the Australasian province was once much larger, including even Europe. He postulated an ancient Mesozoic continent, now sunk beneath the waters of the North Pacific and Indian oceans.[82] This continent had once been linked to Australia, and its inhabitants lingered on there still, protected from the more highly developed placental mammals which had expanded their territory after the land bridge had sunk. Following his general tendency to assume that most types had evolved long before their first appearance in the fossil record, Huxley thought that the higher mammals must already have been in existence in the Mesozoic, but confined to a much smaller Arctogaeal province whose remains are yet unexplored. For Huxley, the abrupt appearance of new types in the fossil record was the result of geological changes which permitted the invasion of new territory by forms already evolved in unknown areas.

Huxley's views on the antiquity of the vertebrate classes were steadily abandoned in the later nineteenth century. His appeals to Tertiary land connections were repudiated by Wallace, although many field naturalists preferred Huxley's approach. Even Wallace admitted wider land connections in the Mesozoic, and was quite happy with the idea that the marsupials had migrated to Australia from the north at that time, where they had subsequently been cut off.[83] In his *Island Life,* Wallace also singled out Madagascar as a classic example of an island cut off from the later stages of evolution: "Such islands preserve to us the record of a by-gone world,—of a period when many of the higher types had not yet come into existence and when the distribution of others was very different from what prevails at the present day."[84] The possibility that remote refuges might protect the products of earlier evolutionary phases from the advance of their successors was also an integral part of the theory advanced in Matthew's "Climate and Evolution" of 1915.

80. Huxley, "Palaeontology and the Doctrine of Evolution," p. 374.

81. Ibid., pp. 378–379.

82. Ibid., pp. 380, 387.

83. Wallace, *Geographical Distribution of Animals,* I, p. 399 and p. 465.

84. Wallace, *Island Life,* p. 411.

The image of Australia and Madagascar as refuges for ancient types of life remained popular through the period of most intense debate of biogeography at the turn of the century. Richard Lydekker intepreted them in this way in his *Geographical Distribution of Mammals* of 1896.[85] The Australian fauna in particular was seen as a relic of the Mesozoic. Flower and Lydekker's *Introduction to the Study of Mammals* pointed to the isolation of Australia and argued, "Consequently Australia has never been able to receive an influx of Eutherian [placental] orders, which have probably swept away all the Marsupials except the small American Opossums from the rest of the globe."[86] Beddard was puzzled by the inability of the placentals to penetrate Australia despite their superior powers.[87]

The possibility of a connection between the marsupials of Australia and South America was problematic for Wallace's thesis of a northerly origin for higher types. It was precisely this kind of evidence that led Gadow and others to argue that the southern continents had once been united in a great Antarctic landmass. Gadow accepted that the marsupials had had a worldwide distribution until wiped out from most areas by the placentals.[88] But he did not agree that they had spread down to Australia from the north. The Australian marsupials had evolved from opossumlike ancestors migrating from South America via Antarctica, although the opossums themselves had disappeared from Australia after having served this pioneering function.[89]

Arthur Bensley accepted the same explanation for the origin of the Australian marsupials. He also noted an important link between this view of their geographical origin and the ongoing reassessment of the marsupials' taxonomic status. The great radiation that had produced the kangaroos and others must have been a comparatively recent event, since the opossums themselves had not appeared until the Oligocene.[90] The image of a continent teeming with Mesozoic life became somewhat tarnished, a change of emphasis consistent with the growing belief that the marsupials were a specialized and fairly late offshoot of mammalian evolution, not a stage in the development toward the placentals. By the 1920s it appeared increasingly unlikely that any of the Mesozoic mammals were marsupials (see chapter 6). Australia remained unique, a center of evolutionary radia-

85. Lydekker, *Geographical History of Mammals*, pp. 6, 9.

86. Flower and Lydekker, *Introduction to the Study of Mammals*, p. 95.

87. Beddard, *Textbook of Zoogeography*, p. 223.

88. Gadow, *The Wanderings of Animals*, pp. 14–16.

89. Ibid., pp. 119–120.

90. Arthur Bensley, "On the Evolution of the Australian Marsupalia," pp. 206–207, and his "A Theory of the Origin and Evolution of the Australian Marsupials," p. 260.

tion unlike that of any other region, but its range of marsupial types did not offer a clue as to the state of the rest of the world in Mesozoic times. The image of a lost world had to be abandoned.

NORTHERN ORIGINS

The theory that distribution could be explained in terms of more advanced types migrating from a wide northern center to the isolated southern continents was a powerful but controversial explanatory tool. Its resonance with the assumption that Europeans were destined to rule the world is obvious enough. But its origins as a biogeographical theory interact with other problems debated in paleontology and geology. An important source of inspiration was the belief that the earth has been cooling down in the course of the Tertiary era. This theory was at first linked to the more general claim that the earth has been steadily cooling from an originally molten state. But by the end of the nineteenth century this association was being challenged, and there was evidence for cooler climates much earlier, in the Paleozoic. We have already seen how it became popular to associate major progressive steps in evolution with the periodic onset of harsher climates. The theory of a uniformly warm Mesozoic remained popular, but it was now assumed that the cooling that began with the Tertiary was part of an endless cycle of alternating climates. The theory of northern origins can be seen as a by-product of the growing willingness to assume that critical episodes in the evolution of life on earth have been triggered by climatic stress.

The theory of a cooling earth dominated nineteenth-century geology, despite the efforts of Charles Lyell to promote the steady-state alternative implied by his uniformitarian principles. Historians of evolutionism are well aware of the efforts made by the physicist William Thomson, Lord Kelvin, to undermine Darwin's theory by claiming that the cooling did not allow enough time for natural selection to bring about the evolution of higher forms of life.[91] But this attack did not necessarily imply a cooling of the climate, because Kelvin admitted from the beginning of the debate that once a solid crust was formed, the amount of heat reaching the earth's surface from the interior was insignificant compared to that from the sun.[92] There were efforts to revive the claim that the cooling affected the surface later in the century. Joseph Le Conte rebutted one effort in this

---

91. See Joe D. Burchfield, "Darwin and the Dilemma of Geological Time."
92. William Thomson, "On the Secular Cooling of the Earth," p. 8. This point was also made in William Hopkins's Presidential Address to the Geological Society of London in 1858; see p. lviii.

direction by Samuel Haughton, professor of geology at Dublin, in 1878.[93] Le Conte explained the cooling of the climate by invoking what we now call the greenhouse effect, coupled with a decline in the carbon dioxode level of the atmosphere. Efforts to explain the ice ages centered on astronomical causes, especially the theory of James Croll.[94] From the middle of the century, A. C. Ramsay of the Scottish Geological Survey had begun to argue that there was evidence of glaciation as far back as the Permian.[95] Although this was disputed at first, by the end of the century it was widely accepted. The cyclic theories of geological change developed by Chamberlin and others took it for granted that there was no overall cooling, only a fluctuation between mild and harsh climates. We have already seen how these theories had a major effect on thinking about earlier phases in the evolution of life, although the basic idea that there had been a cooling since the end of the Mesozoic remained intact.

Geologists naturally assumed that when the uniformly warm climate of the Mesozoic began to break down, the polar regions would be the first to experience colder conditions. These regions would inevitably form the center for a new burst of evolution in the Cenozoic. The logic of this argument could be developed in two ways, however. In its most basic form, the cooling-earth theory would imply a 'bipolar' origin for new forms of life, since both north and south polar regions would experience the harsher climate. The bipolar theory would only make sense in practical terms, however, if there was a large southern landmass which could serve as the locus for evolution. The enlarged Antarctica would also have to be linked by land connections to the existing continents of the equatorial region. Those who followed the geological theory of Suess, in which there had been depression of the land surface throughout the earth's history, could argue that at least some of these hypothetical southern lands had lasted into relatively recent times. A whole network of theories based on such southern continents and land connections was developed to exploit the evidence which Alfred Wegener was to explain in terms of the breakup of a great southern landmass, Pangaea, by the process of continental drift. One ornithologist, Charles Dixon, tried to challenge the fundamental basis of the argument by suggesting that it was the warm, stable climate of the equatorial region that had been most productive of evolutionary

93. Joseph Le Conte, "Geological Climates and Geological Time"; see also Samuel Haughton, "Notes on Physical Geology."

94. James Croll, *Climate and Time*. See Christopher Hamlin, "James Croll, James Geikie, and the Eventful Ice Age."

95. Ramsay began to make this point as early as 1854; see A. Geikie, *Memoir of Sir Andrew Crombie Ramsay*, pp. 228–229. Geikie himself, however, remained skeptical; see pp. 362–363.

change, and hence the center of dispersal.[96] This suggestion was seldom taken seriously, although there were efforts to show that equatorial continents could have acted as minor centers of evolution when isolated from the rest of the world.

To those who preferred to believe that the continents and oceans were essentially stable, the bipolar hypothesis was redundant. There was simply not enough land in the southern hemisphere for the overall cooling to have the same evolutionary effects as in the north. It made much more sense to visualize the great northern landmasses of Eurasia and North America as a partially unified center of evolution from which more advanced types made their way by existing land connections to southern Asia, Africa, and South America. Complications might arise if the land routes were temporarily blocked. Thus South America seemed to have acquired some primitive mammals before being isolated by a break in the isthmus of Panama. It had then functioned as a subsidiary center of adaptive radiation, protected from further invasion from the north. When the isthmus was reconnected, animals flooded north and south across it, with the northerners being more successful in the long run.

The debate over the relative evolutionary significance of north and south was a complex one, because similar arguments were used to postulate extended landmasses in both the northern and the southern hemispheres. The first evidence for temperate and even tropical climates in the arctic during the early Tertiary came from paleobotanists who wanted to unify the northern region even more by postulating a great continent across what is now the Atlantic ocean. Not surprisingly, the name 'Atlantis' was revived, and a great deal of work was done trying to explain the flora and fauna of isolated regions such as the British Isles in terms of transatlantic land connections. But the speculations about a Miocene Atlantis did not necessarily lend support to the theory of northern origins. The idea of an extended northern landmass inevitably fostered expectations that the same argument could be applied further south. Most supporters of Atlantis also favored the existence of Gondwana, Lemuria, and other hypothetical southern continents, and were thus deflected toward a bipolar, or a more complex, multicentered theory of progressive evolution. In contrast, the proponents of fixed continents argued that the Bering Strait was shallow enough to explain the interconnection between Eurasia and North America in most geological periods. They argued that there was no more need to invoke an Atlantic bridge than to postulate any other sunken landmass. In rejecting Atlantis, they also rejected the far

96. Charles Dixon, "A New Law of Geographical Dispersal Based on Ideas Developed in the Study of the Migration of Birds."

greater southern landmasses that would be needed to make a bipolar theory plausible.

Paleobotany provided the best evidence for climatic cooling during the Tertiary, but also some of the most active support for the Atlantic land bridge. In 1865, Oswald Heer, professor of botany at Zurich, published his *Die Urwelt der Schweiz,* a book notable for its plates depicting scenes from earlier geological periods. Heer noted that the plants of Miocene Europe were equivalent to those of the modern tropics.[97] A few years later the first volume of his *Flora Fossilis Arctica,* based especially on material from Spitzbergen, argued that the arctic itself had enjoyed a tropical climate in the Miocene.[98] Asa Gray noted that the plants of both America and Eurasia had moved south from a once-tropical arctic during the ice ages.[99] Joseph Dalton Hooker supported Heer's views in his presidential address to the Royal Society in 1878, and followed Le Conte in postulating a decline in the carbon dioxide level of the atmosphere to account for the cooling climate.[100] The French paleobotanist Gaston de Saporta intepreted the whole history of vegetable life in terms of a gradual cooling from the poles.[101]

The botanists' ideas encouraged Wallace and others who wished to explain animal distribution in terms of migrations from the north. The situation was complicated, however, by the fact that some of the botanists were leading supporters of the theory that the north itself had once been a more continuous geographical area. They interpreted the similarities between the fossil floras of North America and Europe as evidence that a continuous landmass had linked the continents across what is now the north Atlantic during the Miocene. Franz Unger, professor of botany at Vienna, gave a lecture on "The Sunken Island of Atlantis" in 1860.[102] Heer supported the idea, and Murray devoted a chapter of his *Geographical Distribution of Mammals* to the botanists' theory of a Miocene Atlantis.[103] The American botanist Asa Gray, a strong supporter of Darwin, and Daniel Oliver of University College, London, were critical. They argued

97. Oswald Heer, *Die Urwelt der Schweiz,* pp. 588–589.

98. Heer, *Flora Fossilis Arctica,* I, pp. 53–77, esp. p. 72.

99. See Gray's article "Sequoia and Its History," reprinted in his *Darwiniana,* esp. pp. 224–225.

100. Hooker, "Royal Society, President's Anniversary Address," p. 112.

101. Gaston de Saporta and A.-F. Marion, *L'évolution du règne végétal,* II, chap. 2.

102. Unger's lecture "Die versunken Insel Atlantis" was translated as "The Sunken Island of Atlantis" in 1865.

103. See, for example, Heer, *Die Urwelt der Schweiz,* pp. 588–589; Murray, *Geographical Distribution of Mammals,* chapter 4.

that all the evidence could be explained by transmission across the Bering Strait.[104] Others, however, saw that the botanical evidence was consistent with the reconstruction of the earth's past geography endorsed by Edouard Suess, the influential professor of geology at Vienna. Suess was convinced that both a northern continent of Atlantis and a southern one he called "Gondwana land" had disappeared in the course of the Tertiary.[105]

Belief in the existence of a North Atlantic continent lasting until relatively recent times (by geological standards) remained popular well into the twentieth century. One of its chief supporters was Scharff (Matthew's "reckless bridge-builder"), who used it as part of a general campaign against the Wallace-Matthew theory of dispersal from a northern center. Scharff (who was born in Britain of German parents) was keeper of the natural history collections in the Science and Art Museum in Dublin. He rejected the idea of dispersal by 'accidental' means such as ocean currents as both implausible and open to abuse by overenthusiastic supporters.[106] He agreed that the idea of a land connection across the main body of the Atlantic could not be sustained, but followed the geologist and paleoanthropologist William Boyd Dawkins in arguing that a bridge via Greenland had existed until recently. He even suggested that the bridge may have caused the more severe climate of the so-called ice ages by deflecting the ocean currents.[107] Similar views were articulated by Theodor Arldt in his *Entwicklung der Kontinente* of 1907 and a later handbook of paleogeography. Arldt thought a North Atlantic bridge via Greenland had existed up until the Miocene, the Greenland-North American link being broken in the Pliocene.[108]

Scharff was also a member of a small but vocal group convinced that the idea of an ice age in which large portions of North America and Eurasia had been covered with an ice sheet was a gross exaggeration.[109] He accepted a less severe cooling and postulated rapid elevations and depressions of the earth's surface to account for the evidence usually attributed to glaciation. This approach allowed him to develop the position first suggested by Forbes in the 1840s. The flora and fauna of the British Isles were

104. Gray, *Darwiniana*, p. 225; Daniel Oliver, "The Atlantis Hypothesis in Its Botanical Aspect."

105. Edouard Suess, *The Face of the Earth*, II, for example, p. 254.

106. Scharff, *History of the Europeam Fauna*, chap. 1.

107. Ibid., pp. 168–171. See also Scharff's "Some Remarks on the Atlantis Problem." See William Boyd Dawkins, *Early Man in Britain*, pp. 22, 44. Scharff disagreed with Dawkins on the question of whether Iceland was once part of the land bridge.

108. Theodor Arldt, *Handbuch der Palaeogeographie*, pp. 82–98 and maps, pp. 413–422.

109. Scharff, *History of the European Fauna*, pp. 64, 71; see also his *Distribution and Origin of Life in America*, pp. 41–47. Scharff supported the extreme opposition to the ice age concept articulated in Sir Henry Howorth's *The Glacial Nightmare and the Flood*.

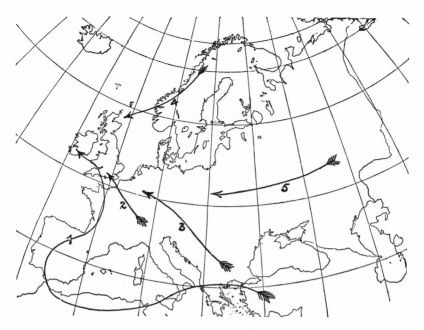

FIGURE 8.2   Migration routes to the British Isles. From R. F. Scharff, *The History of the European Fauna* (London, 1899), p. 117. This map uses existing coastlines and does not show the additional land areas postulated by Scharff both to the north and south of Britain.

composed of several different populations brought in by migration from different sources at different times (fig. 8.2). "Just as we distinguish in the British Isles the parts inhabited by Englishmen, Scotchmen, and Irishmen, so we can recognize these divisions in the animal world, and these roughly correspond to the boundaries of England, Scotland, and Ireland." [110] Ireland contained a predominance of species derived from the south, the Lusitanian fauna, and this must have come not via England (since the species were not found there) but via a land bridge to Spain. Some mammals had come down from the north (which was linked to America via the Atlantean continent), but others had come from the east, ultimately from Siberia. Scharff opposed the theory of the ice age because the massive glaciations envisaged by some geologists would have wiped out the earlier populations which he claimed to distinguish within the fauna of the British Isles. His approach to biogeography did not envisage species being displaced from any part of the territory they had once occupied—except by submergence of the land. Thus the center of dispersal for any group was always the center of its existing range.

110. Scharff, *History of the European Fauna*, p. 99.

Scharff's views were extremely controversial. One anonymous comment on his theory of the Irish fauna worried that his speculation would bring the whole of biogeography into disrepute, although Scharff was immediately defended by his assistant, George H. Carpenter.[111] Matthew's opinion has already been noted, but Osborn adopted a far more positive attitude toward Scharff's land bridges. A northern Atlantic continent was endorsed by the geologist J. W. Gregory in 1929 and by the Welsh naturalist H. Edward Forrest in 1933, the latter using it to explain the origin of the ice sheets that Scharff rejected.[112] Most supporters of Atlantis, however, including Arldt and Gregory, were equally interested in a south Atlantic connection. The theories of southern continents and centers of dispersal (discussed below) all reflect the kind of thinking that Scharff employed to deal with distributional problems. The assumption that there had once been a more continuous area of land in the northern hemisphere was by no means part of the general theory of northern origins. Instead, it reflected a preference for the creation of hypothetical ancient land masses that was to have its main impact on those who denied the claim that the main flow of migration was always from north to south.

The botanists' evidence for a cooling of the earth during the Tertiary predicted a general movement of life from poles to equator. To those who accepted the permanence of continents, geography itself dictated a theory of dispersal from the almost continuous holarctic region southwards to the isolated peninsulas of South America, Africa, and Southeast Asia. But to anyone who accepted great southern continents in the past, the general cooling would imply a bipolar theory. Ludwig Rütimeyer adopted this view in his *Ueber die Herkunft unserer Thierwelt* of 1867. Rütimeyer was professor of Anatomy at Basel in Switzerland, and had gained his reputation by proposing some of the earliest phylogenies for mammalian groups. Although convinced of the ability of northern types to succeed in the struggle for existence, he also postulated an antarctic center of which Australia was a remnant.[113] Huxley envisaged a southward migration of Miocene mammals in his 1870 address, "Palaeontology and the Doctrine of Evolution," but he too favored sunken continents to the south. The alternative theory based on fixed continents was articulated most effectively in Wallace's *Geographical Distribution of Animals* of 1876. Wallace sought to explain the biogeographical regions as the consequence of a historical process, and his major explanatory tool was southward migration cou-

111. Anon., "The Origin of the Irish Fauna"; George H. Carpenter, "The Problem of the British Fauna."

112. J. W. Gregory, Anniversary Address of the President: "The Geographical History of the Atlantic Ocean." H. E. Forrest, *The Atlantean Continent* (first edition published 1933).

113. Ludwig Rütimeyer, *Ueber die Herkunft unserer Thierwelt*, pp. 15, 46.

pled with the subsequent extinction of many northern types in their original homeland.

Wallace was certainly prepared to accept that a continent like South America could act as a minor evolutionary center during those periods when it was isolated from the northern landmass. He insisted, though, that the general pressure of life was always from the northern center southwards. Darwin himself supported this idea, writing of "the more efficient workshops of the north." He attributed the superior powers of the northern types to the sheer extent of the landmass on which they evolved, which encouraged more active competition.[114] Wallace believed that evolution acted more effectively in a harsh and constantly changing environment; thus it was North America and more especially Eurasia that tended to witness the appearance of higher types as the climate cooled: "All the chief types of animal life appear to have originated in the great north temperate or northern continents." He went on as follows:

> In the northern, more extensive, and probably more ancient land, the process of development has been more rapid, and has resulted in more varied and higher types; while the southern lands, for the most part, seem to have produced numerous diverging modifications of the lower grades of organization, the original types of which they derived either from the north, or from some of the ancient continents in Mesozoic or Palaeozoic times.[115]

Cases of anomalous distribution could be explained either by accidental transportation across oceans, or more often by noting the originally much wider distribution of genera revealed by the fossil record. The tapirs of Southeast Asia and South America could be seen as the relics of a fauna once continuous across Eurasia and later North America.

> During Miocene and Pliocene times tapirs abounded over the whole of Europe and Asia, their remains having been found in the tertiary deposits of France, India, Burmah, and China. In both North and South America, fossil remains of tapirs only occur in caves and deposits of post-Pliocene age, showing that they are comparatively recent immigrants to that continent. . . . In Asia, they were driven southwards by the competition of numerous higher and more powerful forms, but have found a last resting-place in the swampy forests of the Malay region.[116]

Wallace was unwilling to accept the idea of a unified holarctic region. For him, Eurasia was the real center of progressive evolution, with occasional

114. Darwin, *Origin of Species,* pp. 379–380.
115. Wallace, *Geographical Distribution of Animals,* I, pp. 173–174.
116. Wallace, *Darwinism,* pp. 352–353.

contacts allowing the types developed there to penetrate across to North America, and hence eventually to South America.

Wallace's views were applied in the field of botany by W. D. Thiselton-Dyer in 1878. Thiselton-Dyer was Hooker's assistant and later his successor at the Royal Botanical Gardens at Kew (he was also his son-in-law). Rejoicing that Britain was the leading country in the study of geographical distribution, he argued that the evidence drove him to the opinion "that the northern hemisphere has always played the most important part in the evolution and distribution of new vegetable types."[117] His words were quoted with approval by Hooker, who (despite his early support for antarctic land bridges) thought it was significant that paleontology led to the same conclusions as biogeography on the predominance of the north.[118]

Wallace's position gained some support from zoologists in the 1880s and 1890s. In 1886 Wilhelm Haacke proposed an extreme version of the theory in which the actual north polar region had been the center for the evolution of higher animals when it had enjoyed a more temperate climate.[119] In the following year the ornithologist H. B. Tristram argued that the theory of northern dispersal explained patterns of bird migration.[120] He noted that on the theory of general cooling, the south ought to have served as a center too, but pointed out that any higher types evolved on the small antarctic continent would have been trapped there and perished as the climate became intolerable to life. Another expert on bird distribution, Alfred Newton, argued that the holarctic region (whose name he had coined) "seems to have the most highly developed fauna, in that it is one from which the weakest types have generally been eliminated, though that result is seen chiefly in its Palaearctic area, and perhaps especially in the western part of this, shewing the truth of the poet's line happily applied by Mr. Sclater in his classical essay—'Better fifty years of Europe than a cycle of Cathay.' "[121]

A popular metaphor invoked by the supporters of this theory was that of successive waves of migration spreading out from the north, displacing and marginalizing their predecessors in the southern regions. Wallace

---

117. W. D. Thiselton-Dyer, "Lecture on Plant Distribution as a Field of Geographical Research," p. 441. On Britain's leading role in biogeography, see p. 412.

118. Hooker, "President's Anniversary Address," p. 113.

119. Wilhelm Haacke, "Der Nordpol als Schopfungscentrum der Landfauna."

120. H. B. Tristram, "The Polar Origin of Life." Tristram thus took up exactly the opposite position on bird migration to that adopted by Dixon; see note 96 above.

121. Alfred Newton, *A Dictionary of Birds*, p. 328. For the Sclater reference, see the latter's "On the General Geographical Distribution of the Members of the Class Aves," p. 134—although Sclater actually uses the quotation to highlight biogeography's departure from the idea that Europe constitutes a distinct region.

identified Australia, southern Africa, and South America as the three ancient southern continents, and wrote, "Into these flowed successive waves
of life, as they each in turn became temporarily united with some part of
the northern land." [122] For Beddard, the Australian marsupials were the
most ancient relic of this process: their presence "is due to the fresh development of types of life in the polar regions which have forced the older
and less vigorous forms of life to emigrate; a continuous efflux of waves of
life spreading out from the place of origin push further away the races
which have the start." [123] The following year Lydekker wrote:

> There is a considerable probability that at least a very large proportion of
> the animals that have populated the globe in the later geological epochs
> originated high up in the northern hemisphere, if not, indeed, in the neigh-
> bourhood of the pole itself (which is known to have enjoyed a genial climate
> during the Tertiary period), and that they gradually migrated southwards in
> a series of waves, probably under pressure of the development of new and
> higher types in high latitudes; and it is to such southerly migrations that the
> present marked differentiation of the fauna of different parts of the earth's
> surface is chiefly due. [124]

In 1903 J. L. Wortman of Yale interpreted the succession of North American faunas as "new-comers representing so many waves or impulses of
migration" driven south by the gradual cooling. [125]
    This image of successive waves of migration sweeping their predeces-
sors southwards to the margins of the earth was built into the theory proposed in Matthew's "Climate and Evolution" of 1915 (fig.8.3). [126] In 1902
Matthew still accepted an antarctic link between South America and Australia at the beginning of the Tertiary. He also postulated a narrow sea
dividing Europe from Asia, suggesting that the disappearance of this barrier would account for the sudden appearance of the Eocene placentals in
Europe. [127] The antarctic links were soon abandoned, however, and the
1915 paper offered a strong defense of the permanence of continents based
on the different character of continental and ocean-bed rocks. Chamber-

122. Wallace, *Geographical Distribution of Animals*, II, p. 155.

123. Beddard, *Textbook of Zoogeography*, p. 227.

124. Lydekker, *Geographical History of Mammals*, pp. 7–8.

125. J. L. Wortman, "Studies of Eocene Mammalia in the Marsh Collection," p. 432.

126. The original 1915 text of "Climate and Evolution" was published in the *Annals of the New York Academy of Sciences*, vol. 24, pp. 171–318, and is reprinted in the 1939 edition, *Climate and Evolution*, cited here. For details of Matthew's work see Rainger, *An Agenda for Antiquity*, chap. 8.

127. Matthew, "Hypothetical Outlines of the Continents in Tertiary Times," p. 359. See also *Climate and Evolution*, p. 139, quoting H. G. Stehlin's similar opinion on the source of the Eocene placentals.

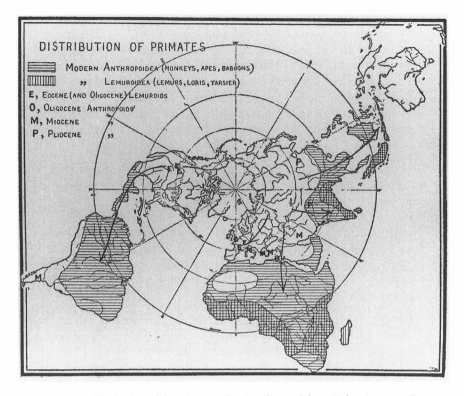

DISTRIBUTION OF PRIMATES

■■■■ Modern Anthropoidea (monkeys, apes, baboons)

▥▥▥▥ „ Lemuroidea (lemurs, loris, tarsier)

E, Eocene (and Oligocene) Lemuroids
O, Oligocene Anthropoids
M, Miocene
P, Pliocene   „

FIGURE 8.3   Distribution of the primates, showing dispersal from Holarctic center. From W. D. Matthew, *Climate and Evolution* (New York, 1939), p. 46.

lin's geological theories were used to explain the cycle of change between warm, uniform conditions and harsh, zonal ones.[128] The distinct episodes of harsher conditions stimulated bursts of progressive evolution, each of which generated new types spreading out in a wave of migration. Like Wallace, Matthew emphasized the possibility of accidental transportation across oceans to explain occasional anomalies. His main explanatory tool was the spreading out of higher types from a northern center by overland migration. He argued that those who appealed to land bridges often worked from incomplete 'book knowledge' of the fossil record.[129] The more the record was explored, the more it confirmed that most groups originally had a distribution across the north at some early point in geological time.

128. Matthew, *Climate and Evolution*, pp. 4–8.
129. Ibid., p. 142.

The increasing harshness of the Tertiary climate spread outwards from the poles, ensuring that higher types would evolve first in the north and then spread southwards. The fauna of modern Africa, for instance, is "the late Tertiary fauna of the northern world, driven southward by climatic change and the competition of higher types."[130] The Australian and South American marsupials were relics of an even more ancient holarctic fauna. For Matthew, the common assumption the the center of a group's modern range was once its ancestral home is the very reverse of the truth. For most older types, the present distribution represents merely the refuges to which it has been driven by the outward pressure from the center. Matthew applied the same arguments to the primates and to the various branches of the human species. Like many of his contemporaries, he thought that central Asia was the evolutionary home of humankind, and that the more primitive races had always been displaced to the southern regions.[131] His biogeography endorsed the arguments of Osborn and many others who were convinced that the white races were the highest products of evolution. The visual impact of the maps that Matthew provided to illustrate his thesis rests on his readers' assumption that the landmasses of the north define the center of power and inflence in a world dominated by America and Europe.

Matthew's views were a deliberate assault on a rival tradition which had grown up in opposition to Wallace's work and which depended on the postulation of hypothetical land connections in the past. This rival school reacted strongly to "Climate and Evolution," although modern analysis of the resulting debate has tended to focus on the reasons why both sides denied the third option of continental drift, which was first put forward by Alfred Wegener in 1912. Matthew's reputation has also been caught up in the debate between modern supporters of dispersal and the rival school of vicariance biogeography founded by Croizat. The arguments pioneered in "Climate and Evolution" were certainly taken up by George Gaylord Simpson, Ernst Mayr, and other founders of the modern synthetic theory of evolution. Then and now, Darwinism is taken to imply that species originate at a single locality and then spread outward. Accidental transportation is used to explain many cases of disjointed distribution (closely related species in widely separated localities). The vicariance biogeographers distrust the unpredictability of accidental migration and prefer to assume that species evolve *in situ*. They assume that a disjointed distribution is produced by a once widespread population being broken up by the interspersal of geographical barriers. The theory of continental drift has made it possible to see how a continuous population

130. Ibid., p. 11.

131. Ibid., pp. 40–45. See Bowler, *Theories of Human Evolution*, pp. 174–176.

spread over the old supercontinent of Pangaea has been fragmented as the continuous landmass separated into the southern continents of today.

One modern opponent of dispersal biogeography has suggested that Matthew enjoyed 'cult status' among biologists, which allowed his views to serve as an irrational barrier to serious consideration of the drift hypothesis.[132] This claim reflects the sense that dispersal became the orthodoxy of post-1940s Darwinism, but ignores the strength of the opposition to Matthew in the early decades of the century. Matthew's emphasis on migration is seen as a target by those who accuse modern Darwinian biogeographers of being similarly infatuated by the image of territorial expansion. This use of a historical figure as a symbol in the modern debate can impede any attempt to understand the theoretical issues as they were perceived in the early twentieth century. Many appeals to hypothetical land briges used the evidence now explained in terms of drift. But drift was bound to be greeted with suspicion at a time when the prevailing geological theories allowed only vertical, not horizontal movements of the earth's crust.

The modern debate does not map neatly back onto that of the early twentieth century. The union between vicariance biogeography and continental drift does not provide an exact equivalent of the old land-bridge school of thought, although both share a suspicion of accidental dispersal. The advocates of sunken land bridges still supported the 'Darwinian' idea of migration from a home territory—they merely had more dispersal centers and routes available than Wallace and Matthew. Neither side in the earlier debate anticipated the modern vicariance biogeographers' reluctance to concede any dispersal, however far back one traces the origins of a group. Even Scharff wanted to identify the centers from which groups radiated out, and would have regarded vicariance biogeography as a deliberate evasion of the question of origins. Croizat's approach has at least some resemblance to that of the exponents of parallelism such as Heilprin, who thought that the same forms could evolve independently in separate locations from very distantly related ancestors. Vicariance theory thus revives a mode of explanation that is even more deliberately anti-Darwinian than that of the bridge-builders. We should also remember that Matthew's chief purpose was to explain Tertiary biogeography; he deliberately avoided detailed discussion of the evidence for Mesozoic and Paleozoic links which are now explained by continental separation. To understand the source of the opposition to Matthew at the time, we need to trace the

132. Robin Craw, "'Conservative Prejudice' in the Debate over Disjunctively Distributed Life Forms." This was written to oppose Frankel's "The Paleobiogeographical Debate over Disjunctively Distributed Life Forms"; see also Frankel's reply: "Biogeography before and after the Rise of Sea-Floor Spreading." For a more balanced account of Matthew's impact, see Rainger, *An Agenda for Antiquity*, chap. 8.

origins of the school of biogeography which took the geologically recent sinking of continents and land bridges for granted in its search for the ancestral home of each species.

SOUTHERN CONTINENTS

The fact that so influential a geologist as Suess had given his support to the former existence of great continents where now there is deep ocean was taken by some field naturalists as a license to postulate a land connection wherever a disjointed distribution seemed to require it. In some cases these were indeed merely 'bridges'—narrow corridors of land linking, for instance, the tips of South America, Australia, and perhaps even South Africa to Antarctica. Even such limited connections allowed Antarctica to function as a major center of evolution and dispersal when the climate was warmer, thus balancing the northern center of creation. Other theories postulated major landmasses in what are now the Indian and Pacific oceans, including the hypothetical 'Lemuria.' Again, these were supposed to function as centers of evolution, perhaps explaining the origin of groups whose ancestry could not be found in the known fossil record. Oceanic islands might be the relics of these ancient continents, undermining Darwin's theory that they had to receive their inhabitants by accidental transportation. As a variant on Suess's theory of constant depression of the crust, some efforts were made to suggest that modern continents had once been broken up by seas. Even if extensive land movements were not thought plausible, it was possible to challenge the theory of northern origins by arguing that South America and Africa had served as major centers of evolution and were thus the original homes of some important groups.

The evidence for bridges that was based on groups evolved before the Tertiary can still be taken seriously today, thanks to the revolution in the earth sciences that made continental drift plausible. Wegener certainly appealed to this evidence, arguing that drift was a better explanation than bridges, but the geological thinking of the time was against him. Some advocates of the permanence of continents, including Wallace, did not extend this principle back into the Mesozoic and Paleozoic, and were by no means obliged to reject the evidence for connections so far back in time. The real debate, as in the case of Atlantis, was between the Wallace-Matthew school and those who extended the hypothetical southern landmasses into the early or mid-Tertiary, thereby providing an alternative explanation for some episodes of mammalian distribution. Scharff, for instance, explained the westward dispersal of tapirs from a European center via a mid-Atlantic link to South America in the early Tertiary, not via North America.[133]

133. Scharff, *Distribution and Origin of Life in America*, p. 353.

The simplest alternative to the theory of northern origins was based on the possibility that the existing southern continents had served as important evolutionary centers. In human terms, Africa had traditionally been dismissed as an unprogressive continent, and the geology of its central regions was unexplored in the late nineteenth century. The mammal-like reptiles discovered by Broom and others suggested that important evolutionary events had taken place in southern Africa in the Paleozoic, but this was too long ago to be relevant for biogeography. Huxley and Wallace's assumption that the Miocene mammals of Europe had migrated to Africa to escape the growing cold meant that the continent was seen as essentially a refuge for outdated forms. This view was challenged in 1876 by W. T. Blanford, who argued that the present fauna of India is mostly African rather than Asian in origin.[134] The same theme was taken up at the turn of the century by Tycho Tullberg in his work on the origin of rodents, by H. G. Stehlin in his work on pigs, and by H. F. Osborn. Tullberg linked his hypothesis to a full-blown bipolar theory in which Africa had once been linked to a great antarctic continent, but Stehlin and Osborn adopted a more limited view in which Africa itself was "a great center of independent evolution" from which the antelopes, giraffes, elephants, and hippos were derived.[135] Osborn later noted that expeditions to the fossil beds of Fayum in Egypt had confirmed that Africa was "a great breeding place not only of animals which subsequently wandered into Europe, but of animals belonging to types hitherto unknown and undreamed of."[136] The unique fauna of Lake Tanganyika was thought to be a relic of marine life which was trapped and forced to adapt to fresh water as central Africa was elevated. Lankester organized a series of expeditions led by J. E. S. Moore to test this theory in the 1890s.[137]

South America was always seen as something of an exception. Like Australia, it harbored marsupials, especially the opossums, but it was also the home of a unique fauna of edentates, or toothless mammals: the anteaters, armadillos, and sloths. On the voyage of the *Beagle,* Darwin himself had discovered remains of the many giant forms that had once existed

---

134. W. T. Blanford, "The African Element in the Fauna of India." Blanford accepted a direct link via the Indian ocean continent of Lemuria, but only up to the end of the Cretaceous—so this could not have been a route for more recent migrations; see his "Anniversary Address of the President" to the Geological Society of London, pp. 88–89.

135. Osborn, "The Geological and Faunal Relations of Europe and America," pp. 568–569. Extracts from Tullberg's *Über das System der Nagethiere* of 1899 and Stehlin's "Über die Geschichte des Suiden-Gebisses" of 1899–1900 are translated in Osborn, *The Age of Mammals,* pp. 70–71.

136. Osborn, *Age of Mammals,* pp. 72–73.

137. See James R. Troyer, "On the Name and Work of J. E. S. Moore."

alongside the surviving members of this group. It was almost universally accepted that the wide range of edentates had evolved in South America from primitive mammals that had reached the continent very early in the Tertiary. Protected by the subsequent breaking of the isthmus of Panama, they had undergone a considerable radiation. When the isthmus was reconnected in recent times, probably at the time of the ice ages, northern types had flooded into South America—but, as Wallace noted, some of the southern types had also moved north, despite the great carnivores which ought to have found them an easy prey.[138] The South American radiation was thus by no means as unsuccessful as was once assumed. Lankester suggested that some of the bizarre giant edentates might have survived almost to the present, since there was evidence that the Indians had kept ground sloths captive in caves.[139] The scenario he depicts offers some parallels to that presented in Conan Doyle's *The Lost World,* although the latter has even more ancient animals isolated on a plateau. The fictitious Professor Challenger of the story is said to be a friend of Lankester, suggesting that Conan Doyle's inspiration may well have come from this source.

Whatever their ultimate fate, the edentates must have evolved from primitive mammals that had migrated into South America during the early Tertiary. Unless one accepted a southern origin for the mammals themselves, these early representatives of the class must have entered from the north before South America became an island. In 1896 Jacob L. Wortman described an early North American mammal, *Psittacotherium,* which both he and Osborn thought might be the ancestor from which the edentates had sprung.[140] By the 1920s several fossils had been found in the North American Eocene which seemed more likely candidates for primitive edentates.[141] Although they were soon swept away in the north, it was assumed that their relatives in the south had prospered with the severing of the land connection.

A much-discussed alternative to this conventional interpretation was proposed by the Argentinian paleontologists Carlos and Florentino Ameghino. Carlos collected, and Florentino described, a great range of South

---

138. Wallace, *Geographical Distribution of Animals,* I, pp. 131–132. See also Osborn, *Age of Mammals,* pp. 78–79; A. Smith Woodward, "The Evolution of Mammals in South America"; K. von Zittel, "The Geological Development, Descent, and Distribution of the Mammalia," p. 457.

139. Lankester, *Extinct Animals,* pp. 174–179.

140. J. L. Wortman, "Psittacotherium, A Member of a New and Primitive Suborder of the Edentata." See Rainger, *An Agenda for Antiquity,* p. 93 and p. 200, note 88.

141. See, for instance, A. S. Romer, *Vertebrate Paleontology* (1933), pp. 382–383.

American fossil mammals.[142] They also tried to work out the stratigraphy of the continent, but although they got the sequence of beds right, they systematically overestimated their age with respect to the rest of the world. As a result, they were able to claim that the South American mammals predated, and hence were ancestral to all others: "At the end of the Secondary period there lived in Argentine Territory not only the ancestors of the mammals which inhabit it now, but also of those which live in all parts and all climates of the world."[143] Many paleontologists, including Gaudry and Lydekker, went to South America to see for themselves. By the 1890s a major debate was raging, with most authorities attacking the Ameghinos' dates. In order to settle the matter, John Bell Hatcher led three expeditions organized by Princeton University to Patagonia. Many fossil mammals were unearthed, but studies of the invertebrate fossils by A. E. Ortmann were conclusive in showing that the Ameghinos' correlations based solely on the mammals were in error. Louis Dollo also refuted the Ameghinos' chronology as developed by Santiago Roth.[144]

A more serious challenge to Wallace's interpretation came from those naturalists who thought that ancient landmasses once linked the surviving southern continents. The most widely discussed version of this approach centered on the attempt to reconstruct a giant antarctic continent which had thrust up spurs to link with the tips of South America, Africa, and Australia. Links were also proposed across the oceans closer to the equator: a direct south Atlantic bridge between Africa and South America, a continent (Lemuria) in what is now the Indian ocean, and another continent in the Pacific. Coupled with such theories was the claim that oceanic islands such as the West Indies and the Galápagos were not recently upraised from the seabed (and hence colonized accidentally as Darwin supposed), but were the mountain peaks of once-continuous landmasses that had gradually sunk. Scharff supported the idea that the Carribean was originally dry land, and Thomas Barbour of Harvard's Museum of Comparative Zoology conducted a campaign against Matthew on this issue.[145]

---

142. On the Ameghinos and other fossil hunters in South America, see George Gaylord Simpson, *Discoverers of the Lost World*; also Rainger, *An Agenda for Antiquity*, pp. 189–190.

143. F. Ameghino, "South America as the Source of the Tertiary Mammalia," p. 260.

144. W. B. Scott, ed., *Reports of the Princeton University Expeditions to Patagonia, 1896–99*, see, for example, I, p. vii, and Ortmann's work comprising volume 4 of the series. Matthew translated Dollo's "Patagonia and the Pampas Cenozoic: A Critical Review of the Correlations of Santiago Roth" for the New York Academy of Sciences.

145. Scharff, *Distribution and Origin of Life in America*, chap. 2. On Barbour, see Mary P. Winsor, *Reading the Shape of Nature*, pp. 256–261. Even Wallace did not deny that the West Indies were once connected by dry land; see his *Geographical Distribution of Animals*, II, p. 78.

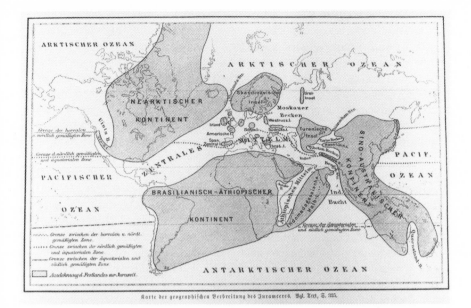

FIGURE 8.4   Map of continents in the Jurassic, showing a Brazilian-Ethiopian and a Sino-Australasian continent in the southern hemisphere. From M. Neumayr, *Erdgeschichte* (Vienna, 1890), I, p. 336.

Scharff was also impressed with Georg Baur's attack on the Darwinian view of the Galápagos islands and his claim that they too were the relics of an ancient continent in the Pacific.[146]

The possibility of a direct link between Africa and South America had been raised as early as 1862 by Murray, who drew attention to the existence of the same genera of beetles on the two continents.[147] Gunther made the same point in 1880 on the basis of similarities in the freshwater fish.[148] But the idea really took shape when it was endorsed in Melchior Neumayr's authoritative *Erdgeschichte* of 1887 (fig. 8.4). Neumayr had studied under Wilhelm Waagen and became professor of paleontology at Vienna. He had already drawn attention to the existence of climatic provinces in the marine fossils of the Jurassic, but now he postulated a Brazilian-Ethiopian continent lying right across what has become the

146. Scharff, *Distribution and Origin of Life in America,* chap. 12; see G. Baur, "On the Origin of the Galápagos Islands" and "New Observations of the Origin of the Galápagos Islands."

147. Murray, "On the Geographical Relations of the Coleoptera of Old Calabar"; see also Murray's *Geographical Distribution of Mammals,* p. 29.

148. A. Gunther, *An Introduction to the Study of Fishes,* p. 233.

south Atlantic ocean.[149] This was taken seriously by both Scharff and Arldt, and was also incorporated into Hermann von Ihering's interpretation of South American biogeography.[150] Von Ihering had originally made important morphological studies of molluscs, but later moved to South America and became interested in the distribution of the invertebrates there. Wallace had noted the possibility that the modern continent might be a composite of several smaller landmasses welded together by the formation of the Amazon basin.[151] Von Ihering exploited the same idea, creating two South American provinces that were once supposed to have been separated by water: Archiplata and Archamazonia (later renamed Archhelenis). In the early Tertiary, the former was linked to a great antarctic continent, the latter directly to Africa across the south Atlantic.[152] The English translation of von Ihering's views was greeted favorably by Ortmann, who also endorsed them in the report of the Princeton expeditions to Patagonia.[153] Gadow thought that most experts accepted a south Atlantic link at least up to the Mesozoic; the real debate was over the possibility that it might have survived long enough to affect the distribution of mammals in the early Tertiary.[154]

The belief that the Polynesian islands were relics of a great Pacific continent was implied in Heer's linking of the floras of Australia and Miocene Europe.[155] Murray supported this view, noting that Darwin's own theory of coral reefs implied that the floor of the Pacific was still sinking.[156] Haeckel and Huxley both argued for a great Mesozoic landmass in the north Pacific and Indian oceans, of which Australia was a remnant.[157] The hypothetical continent of the Indian ocean was often called Lemuria on the grounds that Madagascar, with its unique fauna that included the lemurs, was a relic of it. Some regarded Lemuria as equivalent to the Indo-Madagascan peninsula shown to the east of Neumayr's Brazilian-

149. M. Neumayr, *Erdgeschichte*, II, pp. 333–335, and the map on p. 306. On climatic zones, see his "Über klimatische Zonen wahrend der Jura und Kreidzeit."

150. Scharff, *Distribution and Origin of Life in America*, chap. 14. Arldt, *Handbuch der Palaeogeographie*, pp. 195–224.

151. Wallace, *Geographical Distribution of Animals*, II, p. 27.

152. H. von Ihering, "The History of the Neotropical Region." For his later views, see "Land Bridges across the Atlantic and Pacific Oceans during the Kainozoic Era."

153. A. E. Ortmann, "Von Ihering's Archiplata and Archhelenis Theory"; see also Ortmann's work in volume 4 of Scott, ed., *Reports of the Princeton University Expeditions to Patagonia*, pp. 313–324.

154. Gadow, *The Wanderings of Animals*, pp. 79–80.

155. Heer, "New Holland in Europe," pp. 47–48.

156. Murray, *Geographical Distribution of Mammals*, pp. 25–27.

157. Haeckel, *History of Creation*, I, p. 361; Huxley, "Palaeontology and the Doctrine of Evolution," pp. 380, 387.

Ethiopian continent.[158] Haeckel thought that the human race itself might have evolved in Lemuria, a possibility that might explain the absence of human fossils in other parts of the world. The existence of Lemuria was endorsed in a study of the birds of Madagascar by G. Hartlaub and later by Beddard.[159] Blanford thought there might once have been a girdle of land three-quarters of the way around the southern hemisphere, from Peru to New Zealand.[160]

The possibility of a greatly extended Antarctica attracted most attention, however, because such a continent would provide a near equivalent of the northern landmass as a center for evolution. Hooker had argued for such a continent to explain the similarities in the floras of the southern regions, although by 1881 he was having second thoughts.[161] By the early twentieth century, paleobotanists had stronger evidence for a continuous southern-hemisphere flora in the late Paleozoic in the form of the fossil seed fern *Glossopteris,* found in South America, Africa, Australia, and even Antarctica itself.[162] Most zoologists and paleontologists accepted that these links had survived into the Mesozoic, but again the crucial question was, Had they also survived into the early Tertiary? If so, the marsupials and early mammals of the southern hemisphere might not be relics of northern types, but the survivors of an evolutionary radiation stimulated by the gradual cooling of the great southern landmass.

This bipolar interpretation was suggested in Rütimeyer's sketch of 1867, in which Australia was seen as the relic of a lost world of antarctic life.[163] In 1874 the New Zealand naturalist F. W. Hutton explained the distribution of *Peripatus,* the frogs, and the struthious birds by postulating an antarctic continent lasting until the lower Cretaceous, and a lesser landmass reemerging in the Eocene.[164] In a popular article of 1894 the ornithologist Henry O. Forbes attacked Wallace's theory and invoked the lost continent of Antarctica to explain the southern distribution of birds. He argued that the southern hemisphere must also have experienced the refrigeration leading to the ice ages, and saw the inhabitants of Antarctica

158. See, for instance, C. Schuchert, "Gondwana Land Bridges."

159. G. Hartlaub, *Die Vögel Madagascars,* pp. x–xi; see also Beddard, *Textbook of Zoogeography,* pp. 176–182.

160. Blanford, "Anniversary Address of the President," p. 106.

161. For Hooker's retraction of his earlier ideas, see his "Opening Address, Geographical Section" [of the British Association for the Advancement of Science], p. 448.

162. See E. W. Berry, "Scientific Results of the Terra Nova Expedition"; also T. C. Chamberlin and A. Salisbury, *Geology: Earth History,* II, p. 645.

163. Rütimeyer, *Ueber die Herkunft unserer Thierwelt,* p. 15.

164. F. W. Hutton, "The Geographical Relations of the New Zealand Fauna." But note Hutton's subsequent withdrawal of this interpretation in favor of a south Pacific continent in his 1884 paper "On the Origin of the Fauna and Flora of New Zealand."

being "expelled from their southern paradise" by the onset of the cold.[165] Beddard's textbook added the evidence of the distribution of earthworms to support the antarctic continent, while von Ihering and Ortmann used the distribution of land molluscs for the same purpose.[166] Arldt's textbook of paleogeography also contained a substantial discussion of the antarctic links.[167]

The evidence from the distribution of invertebrates was widely taken to confirm the existence of an antarctic continent at least up to the early part of the Tertiary. This in turn made it possible to suppose that the continent played a role in the transmission of the early mammals into South America and Australia, if it was not actually a center for mammalian evolution. Von Zittel and Smith Woodward favored this view, as did Gadow, although the latter made it clear that the ultimate source of the marsupials was from the north via South America.[168] A 1905 study of the fossil marsupials of the Santa Cruz beds in Patagonia by William J. Sinclair noted their relationship to the Australian forms and suggested that South America was the actual center from which the ancestors of the Australian marsupials had migrated via Antarctica.[169] Even Matthew accepted this link in his 1906 reconstructions of the early Tertiary continents, although he soon repudiated it.[170] Osborn followed the same line, reconstructing the antarctic continent on the basis of the 3040-metre soundings of the ocean (fig. 8.5), an approach favored also by Dollo.[171] In his *Age of Mammals* Osborn at first appeared to support the bipolar theory, writing that "one of the greatest triumphs of recent biological investigation is the reconstruction of a great southern continent." A few pages later, however, he reversed this opinion, stating that he now accepted that the north was

165. H. O. Forbes, "Antarctica: A Vanished Austral Land," p. 211.

166. Beddard, *Textbook of Zoogeography,* pp. 60, 164–176; also the same author's "The Former Northward Extension of the Antarctic Continent." See von Ihering and Ortmann's work cited above, notes 152 and 153; also Ortmann's "The Geographical Distribution of the Freshwater Decapods" and "The Theories of the Origin of the Antarctic Faunas and Floras." Ortmann was a strong opponent of Weismann and a supporter of neo-Lamarckism; see his "Facts and Theories in Evolution" and "Weismannism: A Critique of Die Selektionstherie."

167. Arldt, *Handbuch der Palaeogeographie,* pp. 255–277. See also E. L. Trouessart, *Le géographie zoologique,* pp. 49–50.

168. Von Zittel, "The Geological Development, Descent, and Distribution of the Mammalia," p. 457; Smith Woodward, "The Evolution of Mammals in South America"; Gadow, *The Wanderings of Animals,* pp. 119–120.

169. W. J. Sinclair, "The Marsupial Fauna of the Santa Cruz Beds," pp. 80–81.

170. Matthew, "Hypothetical Outlines of the Continents in Tertiary Times," for example, see the map, p. 356.

171. Osborn, "The Geological and Faunal Relations of Europe and North America," pp. 565–566; Dollo, *Voyage du S. Y. Belgica . . . Poissons,* p. 223.

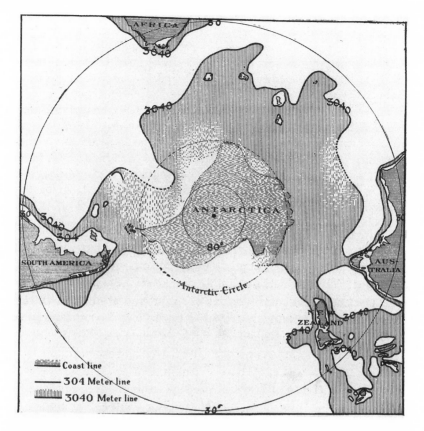

FIGURE 8.5   Map showing the extent of Antarctica if the sea level were to drop by 3040 meters. Note the links to South America, Australia, New Zealand, and almost to Africa. From H. F. Osborn, *The Age of Mammals* (New York, 1910), p. 77.

the center of all mammalian dispersal.[172] As Ronald Rainger observes, it appears that Matthew's new geography of dispersal was already exerting an influence on his superior at the American Museum.[173]

Nevertheless, the land-bridge school of thought defended itself strongly against Matthew's attack, and also formed a main line of resistance to Wegener's theory of continental drift. Wegener reinterpreted the evidence for connections between the southern continents up to the Mesozoic in terms of his own theory. As a geophysicist, he shared Matthew's

172. Osborn, *Age of Mammals*, pp. 74, 80.
173. Rainger, *An Agenda for Antiquity*, p. 192.

distrust of those who imagined that land and deep ocean bed were interchangable. The floor of the south Atlantic simply could not have been dry land at an earlier period because it was formed of a different, denser type of rock.[174] The biogeographers who made these claims were simply ignorant of the basic explanatory tools of geology—yet the evidence for intercontinental links was incontrovertible. For Wegener the conclusion was obvious: the continents had been linked, but by actual contact rather than by bridges between their modern positions. Thus he attacked the work of Arldt and von Ihering, arguing that all the evidence was better explained by continental drift.[175] The *Glossopteris* flora and the discovery of freshwater Carboniferous reptiles, the Mesosauridae, in South America and Africa, were evidence that the two continents had once fitted together like the pieces of a jigsaw puzzle.

The bridge-builders fought back. Von Ihering dismissed Wegener's claimed 'fit' between the coastlines of South America and Africa as naive and reiterated his arguments for bridges across the Atlantic.[176] In America, one of Wegener's leading opponents was Charles Schuchert, who also insisted that land bridges were the most plausible explanation.[177] Osborn depicted a Gondwana land bridge across the south Atlantic up to the Cretaceous in his *Origin and Evolution of Life* of 1917.[178] But Schuchert continued the hypothetical bridges into the Cenozoic, thus coming more directly into conflict with Matthew. In a note published posthumously in the 1939 edition of *Climate and Evolution,* Matthew himself admitted that if a geophysically plausible mechanism for Wegener's hypothesis could be found, it would resolve the problems created by the distribution of species in the Paleozoic and Mesozoic.[179] Other vertebrate paleontologists saw no need to follow Wegener on this score, however. Romer noted the occurrence of the freshwater Mesosauridae in South America and Africa, but argued that they may have been distributed worldwide, and had simply not been found in other areas because the fossil deposits there were of a different kind.[180] This would have resolved the problem in very much the same way that Matthew used for the mammals. As far as the Cenozoic

174. Alfred Wegener, *The Origin of Continents and Oceans,* p. 96.

175. Ibid., pp. 98–101.

176. Von Ihering, "Land Bridges across the Atlantic and Pacific Oceans during the Kainozoic Era"; see pp. 366–367.

177. Schuchert, "The Hypothesis of Continental Displacement" and "Gondwana Land Bridges."

178. Osborn, *The Origin and Evolution of Life,* pp. 217.

179. Matthew, *Climate and Evolution,* pp. 164–165.

180. Romer, *Vertebrate Paleontology* (1933), pp. 141–142.

was concerned, Matthew's argument for the dispersal of mammals from a northern center remained plausible.

The consolidation of modern neo-Darwinism in the 1940s prompted a reaction against the bridge-builders in favor of Matthew's theory. Simpson and Mayr intepreted the new theory in a way that reemphasized the origin of each species in a single, often very restricted, locality. Their emphasis on geographically localized speciation made dispersal a major factor in the history of life on earth and revived interest in Matthew's approach to biogeography. The revolution in geophysics during the 1950s which made continental drift plausible had an impact on this field because it gave new weight to the evidence that widely scattered groups might once have had a continuous range. Croizat's assault on the new dispersalist orthodoxy was incorporated into the alternative school of vicariance biogeography. The debate between those who favor dispersal and those who see all widely scattered groups as having an extensive range as far back as we can trace them continues today.

The land-bridge school has now been partially vindicated, since the evidence it employed is taken seriously, even if it has had to be reinterpreted in terms of the breakup of a great southern continent. But this leaves much of Matthew's argument for the dispersal of mammals in the later part of the history of life intact. Suggestions that his approach might be extended further back in geological time are no longer plausible, but they were always the more speculative part of the thesis in any case. The theory of northern origins made good sense, given the evidence for the gradual cooling of the earth in Tertiary times. But any extension of this into a more general theory of polar origins based on a steady cooling of the earth throughout its history was already problematic even in Matthew's time. Paleontologists such as Lull were stressing the periodic occurrence of episodes of harsher climates, in line with the geological theories which predicted distinct phases of continental elevation. As we have seen in earlier chapters, the belief that the onset of a cooler climate might trigger a burst of progressive evolution had become widespread in the early twentieth century. Matthew's theory can best be understood as an application of this wider theory to the field of biogeography. Those who opposed him had no intention of denying the dispersal of species from restricted origins, nor were they necessarily challenging the view that a more stimulating climate, perhaps on an antarctic continent, would trigger progressive evolution.

The complex debates over biogeography that took place in the decades around 1900 were thus no mere prelude to later developments, although they allowed the exploration of some lines of evidence that we can now reinterpret profitably in the light of modern geophysics. The almost universal tendency to search for a home territory within which the first members of a group had evolved, and the automatic assumption that any successful species would expand its territory as far as possible, show that these debates were stimulated by the more general interest in evolutionary origins. The disagreements over the location of the most important phylogenetic innovations were all played out within an emerging consensus that areas which experience harsher climates have offered the most stimulating locations for bursts of progressive evolution. And finally, the widespread assumption that successive episodes of climatic stress had triggered waves of invasion spreading out from the most active areas allowed biogeography to be incorporated into the attempt to synthesize the theories of geology and evolution into a comprehensive history of the earth and its inhabitants.

9

THE METAPHORS OF EVOLUTION

Surveys of the history of life on earth traditionally end with the emergence of the human species, in effect presenting us as the final act in the splendid drama. This is as true for Stephen Gould's recently published *Book of Life* as it is for Matthew's *Outline and General Principles of the History of Life* or even Haeckel's *History of Creation*. Modern accounts may stress that this is a sign of popular interest in our own origins, not an attempt to portray the history of life as the ascent of a ladder toward the perfection of the human species. The previous generation of surveys, however, followed Haeckel in conveying the image of a main line of progress. Few early twentieth-century evolutionists doubted that—however diverse the branches of the evolutionary tree—one branch had advanced much further than any other. The very structure of such accounts, along with the phylogenetic trees used to illustrate them, thus constitute a model which invites us to read a certain significance into the events described.

So far, this account of the history of the genre has followed the orthodox pattern. One reader of the original manuscript suggested moving the chapter on arthropod origins to a later position in the book precisely to show the extent to which the linear model is imposed by hindsight. Such a structure would be anachronistic in any account which intends to convey an impression of how the topic was (and often still is) approached. The linear sequence still has a role to play—especially if readers are warned of the hidden assumptions underlying the model. The one difference from the orthodox pattern is that this book does not end with a chapter on human origins. This strategy is possible only because the author has already devoted a whole book to that topic; in effect, my *Theories of Human Evolution* should be read as the final chapter of a truly comprehensive account of the way life's splendid drama has been portrayed.[1] One purpose of that book was to demonstrate how accounts of human origins in the late nineteenth and early twentieth centuries followed the models used to describe earlier phases of the history of life on earth. Depicting the emergence of the human species as a continuation of the whole process of evolution, encouraged readers to see human progress as the inevitable consequence of processes built into the very nature of life on earth.

In this final chapter, I want to look more directly at the styles of explanation adopted by these scientists in order to illuminate the preconceptions that they built into their understanding of how evolution must work. I hope that we shall learn something of value about the culture within which the scientists tried to articulate their theories, both from the preconceptions that seem characteristic of a bygone age, and from those areas where their language seems to reflect a harsher, less optimistic viewpoint more in line with our current way of thinking. The metaphors and visual images they used to describe the history of life may reflect their feelings about human history and hence about human nature and society. Since the survey covers a period of eighty years, it is hardly surprising that the metaphors should change to reflect changing values. In conclusion, I want to examine the overall effect of these changes in preparing the way for the advent of modern Darwinism in the mid-twentieth century.

Thanks to the writings of Stephen Jay Gould and others, we are now alert to the hidden messages which can be conveyed by evolutionary trees. The structure of the tree is intended to make us think that the relationships between species must be understood in a certain way. Phylogenetic trees are not the only source of visual influences: think for instance of the way in which a map such as Matthew's projection of the world viewed as from the North pole (fig 8.3) seeks to persuade us that the northern regions really are the center of evolutionary activity. During the decades around 1900, many artists developed the ability to depict the past inhabitants of the earth in apparently realistic settings, thus helping to create imaginary worlds in the past. Martin Rudwick has surveyed the origins of this tradition, but its true flowering began in the period we have surveyed.[2] The paintings of Charles Knight, but poorly reproduced in his *Before the Dawn of History,* are classic examples of this genre (e.g., fig. 6.7). By constructing a sequence of such scenes, an artist such as Knight could draw the viewer along a scale of increasing organic complexity. The transition from the Paleozoic invertebrates to the lumbering dinosaurs and on to the brutish Neanderthals encouraged a sense of steady progress in the his-

1. Osborn's *Origin and Evolution of Life* does not include human origins, but he wrote two books on that topic: *Men of the Old Stone Age* and *Man Rises to Parnassus.*
2. M. J. S. Rudwick, *Scenes from Deep Time.*

tory of life. Visual images thus joined with literary metaphors to create the cultural framework within which evolution was understood and popularized.

By incorporating human origins, and often comments on at least the early phases of human history, into the account of life's development on earth, biologists promoted the belief that humans were products of the natural laws of evolution. They also legitimized the use of models and metaphors derived from human history in the attempt to understand the earlier phases of evolution. This survey has shown how concepts defined originally in terms of human affairs were employed to illustrate what writers thought was going on at various stages in the ascent of life. Many technical discussions of particular episodes did not employ such socially loaded metaphors, of course. One could write about the morphological changes that created the mammalian ear bones, for instance, without invoking images of global progress. Other topics lent themselves much more readily to the use of value-laden terminology. Even detailed analyses of problems in biogeography used language that reflected the territorial concerns of the age of European colonization and conquest. Whenever the scientists wrote on a larger scale, or for a more popular audience, the use of socially derived metaphors became all too apparent.

Exploration of this topic is potentially controversial. Scientists are often unwilling to admit that their thinking is influenced by factors derived from other areas of human culture. This issue has been debated at great length, for instance in the case of Darwin's possible use of the metaphors of free-enterprise liberalism in the construction of his theory of natural selection.[3] I have no wish to reopen this question at the theoretical level; it seems to me self-evident that many of the biologists I have been discussing did use metaphors that are derived from human affairs, and I do not think that we should worry too much about whether or not this disqualifies them from being counted as true scientists. If we accept Ernst Mayr's point that evolutionary biology has to employ different modes of argument than those used in the physical sciences, we can concede that the discussion of historical events in the development of life on earth almost inevitably requires the use of language that has already acquired meaning from its application to human history.[4] More generally, writers such as George Lakoff and Mark Johnson have stressed the extent to which all human understanding is based on the use of metaphors.[5]

When we find that the language used in the late nineteenth and early

---

3. See, for instance, R. M. Young, *Darwin's Metaphor,* and for a survey of the debate, Bowler, *Evolution: The History of an Idea,* chap. 6.

4. Mayr, *Toward a New Philosophy of Biology.*

5. George Lakoff and Mark Johnson, *Metaphors We Live By.*

twentieth centuries is quite self-consciously redolent of the ideology of, say, Western capitalism or imperialism, we may suspect that the postulation of hypotheses in this area was to some extent shaped by those ideologies. But we might also ask how on earth the first generation of evolutionists could have tackled the problem without borrowing metaphors from the existing areas of historical discourse. The hypotheses counted as scientific if they were capable of being tested. Some critics argue that this requirement is precisely where the whole enterprise fails as science—but they would apply this argument equally to modern explanations which are far more carefully constructed in an attempt to circumvent the problem. My purpose here is not to debate the extent to which the reconstruction of phylogenies was or was not scientific. It was counted as scientific at the time—although everyone accepted that it was at the most speculative end of the spectrum of scientific activities. If the biologists had not tackled it, they would have been accused of leaving the territory to the theologians. Their willingness to try betokens a level of self-confidence built on the successes of the original Darwinian debates. If they consciously or unconsciously used ideologically loaded language, we should not automatically reject their work as unscientific.

Scientists creating hypotheses in any new area of study may have to seek inspiration in models and metaphors borrowed from other fields. If these models yield useful results, then the borrowing has paid off. Problems arise only if bad models are retained for ideological reasons—but who decides which are the bad models? In some areas we are still debating which are the good and bad theories today, as in the clash between vicariance and dispersalist biogeographers. Where one side sees a bad theory being retained for nonscientific reasons, the other sees a model that is doing good work whatever its origins. In such areas, historians cannot evaluate the scientific credentials of their subjects without taking one side or the other in an ongoing debate. This chapter explores how the science of the period around 1900 reflected the wider culture within which it was articulated, but without passing such judgments. That the science was influenced by external factors seems evident, but I certainly do not wish to imply that it was therefore bad science. Anyone who is tempted to argue that the whole enterprise was nothing more than the imposition of social metaphors onto nature should note that one of the most crucial issues debated by evolutionists had no social analog. The problem of identifying similarities produced by convergence, rather than by common descent, does not map onto any popular social debate (although the possibility of the 'independent invention' of similar artifacts and cultural traits did worry contemporary anthropologists).

If the scientists themselves used language which reflects assumptions about human affairs, it is hardly surprising that writers on social and political questions used the science in an effort to show that their policies were merely reflections of what was natural—and hence inevitable. These applications raise the thorny question of 'social Darwinism,' the alleged use of scientific theories to bolster political systems in which struggle or other biological necessities are given free reign. Many books have been written on this topic, and I do not intend to explore the wide range of nonscientific writing which may have used the theories of biological evolutionism to make its point.[6] I am more concerned with cases where the scientists themselves not only use ideologically loaded metaphors, but also extend these explicitly to human affairs by treating human nature as a product of the evolutionary process they have described. At this point the incorporation of paleoanthropology into the story of life on earth becomes crucial. Scientists such as Lankester, Matthew, and Osborn took an active interest in hominid fossils and the story of human origins. Where we can see parallels between their accounts of evolution and their views on human nature, and hence on social issues, we must be willing to admit that science was being exploited in the service of ideology. We are also left with a strong impression that the science was being constructed on foundations which incorporated elements of contemporary cultural values.

The one point that emerges very clearly from the exploration of these metaphors of evolution is that the phrase *social Darwinism* is singularly misleading. I do not mean to imply that Darwinism was not employed to promote the ideologies of individual, national, or racial struggle. Darwin's theory certainly helped to incorporate the phrases *struggle for existence* and *survival of the fittest* into everyday language. But I do want to urge that the use of metaphors which convey the impression that progress is achieved only through struggle was far more widespread than we might expect if we took the phrase *social Darwinism* to mean that Darwin's theory of natural selection was the sole or even the primary agent on the scientific side. I have argued elsewhere that Herbert Spencer's philosophy of progress through individual struggle was based as much on Lamarckian as Darwinian principles.[7] Struggle was important because it encouraged individual self-development, not just because it eliminated the congenitally unfit. Lamarckians were as active as Darwinians in promoting the view that a harsh environment was the greatest stimulus to evolutionary progress. Here we see clear evidence that even in discussions of the evolutionary mechanism, natural selection was not the only hypothesis which exploited the metaphor of struggle.

6. The classic account is Richard Hofstadter, *Social Darwinism in American Thought.* For a critique of this, see Robert Bannister, *Social Darwinism: Science and Myth in Anglo-American Social Thought.* See also Greta Jones, *Social Darwinism and English Thought.*

7. Bowler, *Non-Darwinian Revolution,* pp. 156–158.

Many of the debates over the exact nature of steps in the history of life were conducted by biologists who did not need to commit themselves regarding the evolutionary mechanism. Yet their language invoked images of progress through struggle, of the conquest of lower by higher types, and of the invasion of territory by newcomers, which could be used to promote the ideologies of national or racial rivalry. In some cases that language was explicitly applied to human affairs, as in the increasingly popular view that the Neanderthal race (or species) was wiped out by our own ancestors—and the claim that this event served as a warning for modern nations engaged in the worldwide struggle for dominance.[8] Some of the scientists who employed this language of struggle, invasion, and conquest were Darwinians only by the loosest definition of the term: they did not believe that natural selection acting within a population was the mechanism of evolutionary change. Some were active opponents of selectionism. In these circumstances, we should be very careful not to use the term *social Darwinism* in a way that would encourage the belief that Darwin's selection theory was the only source of the struggle metaphor in biology. The reconstruction of life's splendid drama by a generation of scientists who were not primarily concerned with the nature of the evolutionary mechanism created an arena in which ideologically loaded discourses could be developed without reference to the debates over selectionism which have been the principal source of information used by historians of social Darwinism. Some of the metaphors used, especially those centered on images of conquest and invasion, may reflect a more broadly defined Darwinian viewpoint that derives only indirectly from the selection theory. The purpose of this chapter is to provide an overview of the metaphors encountered in the course of this survey, with some hints at how those metaphors could have been used by writers seeking a scientific justification for social policies.

TREES AND LADDERS

The fact that most surveys of the history of life—even today—start with the invertebrates, run through the vertebrate classes to the mammals, and then end with the emergence of humankind suggests the strength of the image in which evolution is reduced to a ladder, a sequence of stages leading up to ourselves as the pinnacle of creation. The image gives purpose to human existence by implying that we are in some sense the goal of the evolutionary process, an inevitable outcome of the most fundamental laws of nature. The ladder model predates Darwin himself, and can be seen in

8. See Bowler, *Theories of Human Evolution,* chap. 9.

the temporalized version of the old chain of being invoked by Lamarck and Robert Chambers.[9] In theory, at least, Darwin exploited a very different image, that of the tree whose branches diverge in many different directions at once (see fig. 2.2). As Stephen Gould has argued, this tree (or more properly, bush) of evolution prevents us from seeing the human species as the goal of creation.[10] All branches are developing in their own direction, and it is illegitimate to treat any other form of life as merely a step on the way toward becoming human. As noted in chapter 2, however, the post-Darwinian evolutionists had difficulty in coming to terms with the tree metaphor. Very few nineteenth- or early twentieth-century evolutionists could tolerate the idea that the human species was not, in some important respects, raised above all the other animals. The most sophisticated way of coming to terms with this assumed superiority was to suppose that many of the main branches in the tree of life had progressed toward higher levels of complexity, each in its own way, but ours had advanced further than any other. All too often a simpler approach was adopted, however: the tree of life was given a main stem or trunk leading toward humans, while all other forms of life were depicted as the products of side branches that did not progress once they had separated from the main stem.

The tree of evolution was thus redrawn to hide the ladder within, providing an axis for the progress that was widely hailed as a necessary feature of the history of life. This can be seen in the tree with a central trunk used to illustrate Haeckel's *History of Creation* (see fig. 2.5) In his 1880 account of vertebrate evolution, Huxley excused the lack of any discussion of the Teleosts, or bony fish, by noting that they lay "off the main line of evolution."[11] Dohrn visualized a single, progressive animal phylum with degenerate side branches, while Lankester defined the main grades of animal organization in terms of the stages at which they branched off the main line of progress.[12] Gregory's *Our Face from Fish to Man* shows that even a sophisticated evolutionist of the 1920s could find it opportune to treat many forms of vertebrate life as merely stages on the way toward the superior human form.

Gregory's book illustrates the tendency to describe certain modern forms of life as "living fossils," so that they can be seen a part of a chain of being leading toward ourselves. In theory, of course, a living fossil could

9. On pre-Darwinian ideas of progress, see Bowler, *Fossils and Progress*.

10. See Gould, *Wonderful Life*, and *Eight Little Piggies*, chap. 30.

11. Huxley, "On the Application of the Laws of Evolution to the Arrangement of the Vertebrata," p. 471.

12. Dohrn, *Ursprung der Wirbelthiere*, pp. xi, 74; Lankester, "Notes on the Embryology and Classification of the Animal Kingdom," p. 440. See Bowler, "Development and Adaptation."

be an unchanged relic of the early stage of any branch in the history of life, as in the case of *Peripatus*. Darwinians were supposed to be aware of the dangers of this approach, since they knew that each phylum continues to change even after a 'higher' form has branched off. We have seen that Lankester warned against the tendency to treat the gorilla as a human ancestor on these grounds.[13] George Gaylord Simpson accepted the possibility of "bradytely," or very slow rates of evolution, where species were adapted to a very stable environment, but modern evolutionists reject the whole idea of branches that have changed at so slow a rate that they can be used to illustrate earlier stages of evolution.[14] From Haeckel into the early twentieth century, however, the temptation to use living forms as models for earlier stages in the history of life proved too tempting, especially when trying to illustrate the process for a nonspecialist audience. And never was this temptation greater than when trying to provide an overview of the sequence leading to the human form. Gregory's use of living marsupials as illustrations of the early stage of mammalian evolution has been noted—despite his awareness of growing evidence that in fact they were a degenerate side branch. Cambridge University Press published a popular book, *Primitive Animals,* in 1911 that included *Peripatus, Amphioxus,* the lungfish, and other living fossils.[15] The image of the 'lost world,' where relics of the past have been protected by isolation, was an important by-product of the hope that nature will help us to reconstruct the past by preserving ancestral forms through to the present. The publication in 1912 of Conan Doyle's story *The Lost World,* in which the irascible Professor Challenger leads an expedition to discover dinosaurs and ape-men trapped on a South American plateau, suggests that popular interest in these topics was still strong.

The problem with Haeckel's tree was that it tried to represent two different variables along its vertical axis: geological time and level of organic development. It made sense only if one assumed that any side branch from the main stem continued to the present retaining the level of development characteristic of the period in which it split off. This point is made even clearer in another tree used by Haeckel (fig. 9.1) This type of tree at least has the merit of allowing one to see that some side branches, such as the dinosaurs, did not survive through to the present. The Darwinian kind of tree (fig. 2.2) does not allow the terminal points of any branch to serve as the node from which others diverge: no living species can be the ancestor of any other. Some early twentieth-century evolutionary classifications

13. Lankester, *Zoological Articles,* p. 175; see chap. 2 in this text.

14. Simpson, *Tempo and Mode in Evolution,* chap. 4; see Niles Eldredge and Stephen M. Stanley, eds., *Living Fossils.*

15. Geoffrey Smith, *Primitive Animals.*

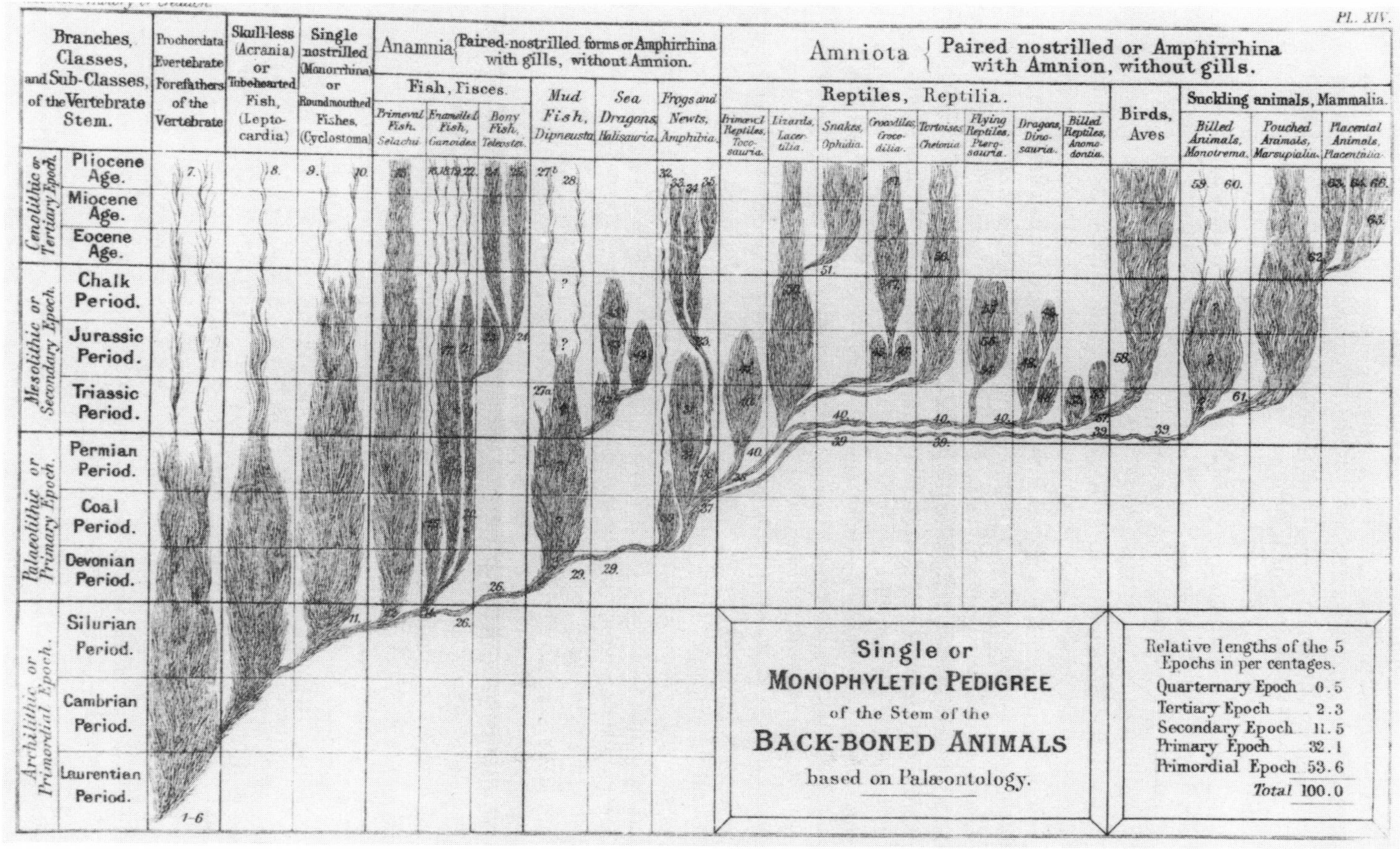

FIGURE 9.1   Monophyletic pedigree of the back-boned animals. From Ernst Haeckel, *History of Creation* (New York, 1883), II, plate XIV.

sented at the same level; by refusing to depict some as 'lower' than others, the temptation to say that primitive forms are ancestral to those which are more highly developed is blocked. This mode of representation discourages its users from thinking in terms of 'missing links' or 'living fossils'— although it is still possible to identify primitive characters and thus build up an image of what the common ancestor of a group might have looked like. The 'living fossil' simply has more of these characters than most other members of the group. In the most extreme manifestation of this technique, fossil species are represented at the same level as living ones, thus preventing even an extinct form from being be treated as the 'missing link' in the group's ancestry. No early twentieth-century paleontologist would have used such a technique, although all were aware of the extreme unlikelihood of a fossil representing the direct common ancestor of a group. The best one could hope for was a close relative of the common ancestor, which would at least share most of the ancestral features. Deciding which were the primitive characters and which were secondary specializations was a problem, even with fossils. Huxley nevertheless used the term *missing link* when describing *Compsognathus* as the kind of reptile from which the birds had sprung,[16] and we have seen over and over again how fossils were used to fill in the important steps in the ascent of life.

Haeckel's popular descriptions of the history of life tend to mention only those forms which can be assigned to a place on the ascending scale toward humankind. In more careful discussions it was accepted, even by Haeckel, that most of the evolutionary trends within major groups were not concerned with the ascent toward the next highest stage. The wide acceptance of Cope's "law of the unspecialized" suggests that everyone recognized this point. Most branches of the tree led toward specialization, not progress in any absolute sense, and specialized forms were not flexible enough to be reorganized as the basis for a major step forward. The increasing emphasis on paleontology as a guide to the history of life meant that most popular accounts devoted a great deal of space to exploring what were, in effect, the side branches of the tree. The small number of events in which new classes emerged were thus increasingly seen as unique: each had to be described in its own terms, as the product of a particular set of circumstances. If there were laws of evolution, as many biologists assumed, they applied to the general trends within classes, not to these unique events. The strong emphasis given to the divergent specializations within each class by early twentieth-century paleontologists such as Matthew and Osborn represents a significant break with the tradition established in Haeckel's time.[17] Instead of treating everything as a

16. Huxley, "On the Animals Which Are Most Nearly Intermediate between Birds and Reptiles," p. 311.

step in the main ascent, biologists deliberately emphasized the diversity of life in each geological period. Progress was no longer the inevitable consequence of a steadily operating trend: it was the sum total of a series of individual events by which unspecialized animals had been restructured into something quite new. The story becomes more difficult to tell coherently, since the narrative has to switch back and forth between two modes of evolution, progressive and specialized. Evolutionary trees for particular groups tend to stress the divergent nature of most evolution, as Osborn's do (fig. 7.2). Various techniques were developed for illustrating how successful some of the side branches had become in terms of the sheer number of species they had generated (fig. 9.2). Osborn's emphasis on the multiple branches that evolved side by side even within a single family provided a foundation for his views on the human races, which he insisted were distinct species that had evolved separately over a vast period of time.[18] Osborn had no interest in the image of a single ladder of progress precisely because it would have undermined his sense of the white race's unique character and destiny.

One consequence of this willingness to emphasize the diversity within each class was the admission that the history of life on earth could not be depicted as the product of a continuous evolutionary trend. Most authorities now believed that the ascent of life occurred in a sequence of distinct episodes, each initiated by the emergence of the primitive members of a class exhibiting a new grade of organization. The subsequent evolution of the class constituted an episode resembling the rise and fall of one of the great ancient civilizations, a topic we shall return to in the next section. It seemed unlikely that such a process could be the product of a "continuous perfecting principle" such as that advocated by Carl von Nägeli, or of the *élan vital* postulated by Henri Bergson.[19] The almost teleological progressionism of William Patten's *The Evolution of the Vertebrates and Their Kin* of 1912 was already out of date.[20] Osborn explicitly rejected such ideas when he proposed his theory in which evolution was the product of an interaction between the germ plasm, the organism, and the environment.[21]

The role of the environment could be seen as passive or active. Most evolutionists saw the ever-changing nature of most physical environments

17. See Bowler, "Darwinism and Modernism."

18. Osborn, *Man Rises to Parnassus;* see Rainger, *An Agenda for Antiquity,* pp. 145–151, and Bowler, *Theories of Human Evolution,* pp. 125–128, 176–179.

19. On Nägeli and Bergson, see Bowler, *Eclipse of Darwinism,* pp. 56–57, 149–150.

20. Patten, *The Evolution of the Vertebrates and Their Kin,* for example, p. 472; see above, chap. 4.

21. Osborn, *Origin and Evolution of Life,* p. 278.

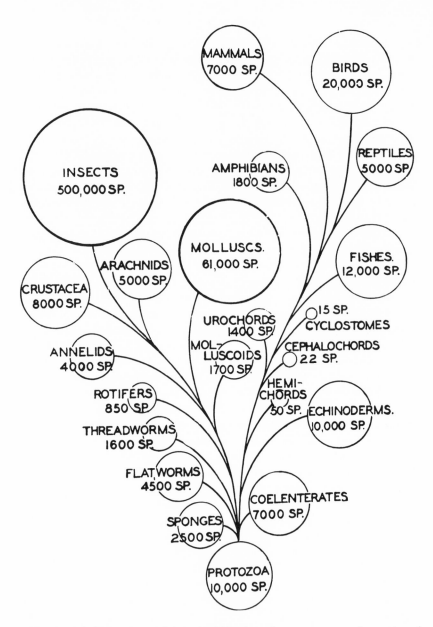

FIGURE 9.2 Phylogenetic tree of the animal kingdom, drawn to compare the number of living species in each branch (although the circle for the insects is still far too small). From H. V. Neal and H. W. Rand, *Comparative Anatomy* (London, 1936), p. 665. Another popular technique was to vary the thickness of the branch to indicate the number of species, which could be used to indicate the rise and fall of groups in geological time (see figs. 4.6 and 9.1).

as a challenge to their inhabitants. The struggle for existence was not between the individuals making up each population, but between the population and the limitations imposed by an unstable environment. O. C. Marsh insisted that natural selection was defined by most American evolutionists to include "the equally important contest with the elements, and all surrounding nature. By changes in the environment, migrations are enforced, slowly in some cases, rapidly in others, and with change of locality must come adaptation to new conditions, or extinction."[22] Progress would always take place more rapidly in a harsh environment, especially in one subject to rapid fluctuations. The greatest challenge, presumably, would be posed in an entirely new environment, as in the transition from water to land. Many late nineteenth-century accounts seem to take it for granted that new environments simply waited for life to invade them as soon as it could. The seabed, the shoreline, the dry land, the air, each would be exploited as soon as animals were able to acquire the characters needed to exploit them. Naturally, those which arose to the challenge first were those which had already advanced further than their rivals. According to such a model, progressive evolution was still an active force even though it depended upon an external stimulus: animals voluntarily invaded new environments to claim their resources. A parallel in theories of human origins was Grafton Elliot Smith's claim that our ancestors moved out of the trees as soon as they had enough intelligence to exploit the environment of the open plains.[23]

Many evolutionists accepted that certain environments were more stimulating than others, so that those groups which made the transition into them were more likeley to progress. Other environments were less stimulating and might even lead to degeneration. The emphasis on degeneration by Dohrn and Lankester illustrates a growing realization that the appeal to an environmental stimulus was double-edged. Parasitism was an obvious case where the move to a less stimulating environment led to a degeneration that might produce superficially 'primitive' organisms. Dohrn's ideas on progress and degeneration were inspired by the model of human mental development. Lankester was firm in his warning that human civilizations had degenerated, when—like Rome—they had left themselves with no further challenges.[24] There was much concern in the late nineteenth century over the possibility that civilization was leading to degeneration. The picture of a degenerate humanity in the distant future invoked in H. G. Wells's popular story "The Time Machine" is almost certainly derived from Lankester.[25] E. W. MacBride's theory that inverte-

22. Marsh, "Origin and Succession of Vertebrate Life in North America," p. 376.
23. See Bowler, *Theories of Human Evolution*, pp. 161–173.
24. Lankester, *Degeneration*, pp. 33, 59–60.

brates were degenerate vertebrates was founded on the idea that crawling on the seabed was inherently less stimulating than swimming in the open waters.[26] MacBride was also a strong exponent of eugenics: he wanted to restrict the breeding of the Irish and other races which he assumed were inferior to the Anglo-Saxons as a result of their evolution in a less stimuating environment.[27]

Early twentieth-century paleontologists placed increasing emphasis on the active power of a changing environment to trigger changes in animal life. The dry land did not simply await invasion, at least as far as vertebrates were concerned. Only when populated by plants and invertebrates did it offer an environment fit to be colonized, although once that had been achieved, the air-breathing fish had not been slow to exploit it. In some cases, changes in the physical environment became the imperative that forced animals to adapt. Romer's claim that the amphibians had been forced to walk on land by the drying up of their ancestral pools is a classic example of this new style of explanation.[28] Here living things are no longer seen as active agents, triumphantly invading a new territory in order to exploit its resources. They have become the hapless victims of climatic changes that might easily have led to extinction, were it not for their ability to innovate in the face of adversity. The human parallel is the theory that our ancestors were driven out of the trees by a climatic change which forced them to adapt to the open plains by standing upright and inventing tools.[29] This style of explanation emphasizes the unpredictable nature of each major step in the ascent of life.

The unique character of the events which introduced the pioneers of each class made it possible to imagine that they were stimulated by some external force which acted upon life sporadically in the course of the earth's history. The possibility that new classes could have been founded by large-scale saltations ("hopeful monsters," to use the term later introduced by Richard Goldschmidt) was increasingly discounted. External triggers thus seemed all the more likely, especially when geologists such as Chamberlin began to see the history of the earth itself as an episodic process capable of generating occasional periods of climatic stress. The in-

25. Wells's "The Time Machine" is reprinted in *The Short Stories of H. G. Wells;* see esp. pp. 36–37, 52–53. See also his article "Zoological Retrogression," reprinted in *H. G. Wells: Early Writings in Science and Science Fiction*, pp. 158–168. See Bowler, "Holding Your Head Up High," and on the subsequent friendship of Lankester and Wells, see Lester, *E. Ray Lankester*, chap. 17.

26. MacBride, *Textbook of Embryology*, p. 472.

27. For details, see Bowler, "E. W. MacBride's Lamarckian Eugenics."

28. Romer, *Vertebrate Paleontology* (1933), p. 105.

29. Bowler, *Theories of Human Evolution*, pp. 173–185; see also Misia Landau, *Narratives of Human Evolution*.

ability of the mammals to exploit their superiority until the extinction of the dinosaurs reinforced the feeling that the history of life could not be understood in terms of progressive forces arising from within the organism. We have seen how the possibility of an environmental trigger was called upon to explain the origin of the vertebrates and of each vertebrate class. The image of the "pulse of life" invoked by Lull provides the best expression of the growing conviction that the episodic character of evolution followed the episodic character of the earth's physical history.[30]

The role of the environment in stimulating progressive evolution had been seen as important even before the reemergence of discontinuous theories of geological change. Darwin had, of course, rejected any notion of a continuous internal trend toward perfection; for him, evolution was essentially a response to the environment (although the environment included other species which were themselves evolving). Lamarckians too saw the stimulus of the environment as the key to progress. Whether in the form of Spencer's almost mechanistic version of self-improvement, or Samuel Butler's quasi-mystical vision of life responding creatively when challenged, the role of the environment as stimulus was crucial.[31] Neo-Lamarckians such as Cope could stress the specialization of each group for its chosen way of dealing with the environment. It was also possible to imagine how, in rare combinations of circumstances, a population of organisms might 'invent' a new structure that was a significant improvement on anything that had gone before, thus pioneering a new wave of evolutionary development. All evolutionists, Darwinian and non-Darwinian, came to accept that major innovations were followed by a process of perfection and refinement which generated the diverse specializations within each class.

An important product of the belief that the course of life's evolution was conditioned by a haphazardly changing environment was a growing willingness to admit that the actual course of evolution was unpredictable. Most modern Darwinians would see this as an integral part of their world view, but the idea met a good deal of resistance at first. Patten's insistence that the human form was the goal of evolution was perhaps the last expression of a once-common belief that we are the inevitable outcome of a purposeful trend built into nature. Some biologists had already begun to doubt this in the late nineteenth century. In the conclusion to his account of the *Challenger* voyage, Mosely stressed that the dependence of life on its environment meant that evolution was essentially unpredictable.[32] Few would admit that such a position undermined the idea of progress itself: it

30. Lull, *Organic Evolution*, epilogue.

31. On Lamarckism, see Bowler, *Eclipse of Darwinism*, chap. 4.

32. Mosely, *Notes by a Naturalist*, p. 518.

was still possible to imagine many lines of evolution progressing in their own ways. The 'main line' leading to humankind was identifiable only by hindsight, now that the product of the most advanced line had gained self-awareness. We saw in chapter 7 how even working biologists were led to speculate on the possibility that, under slightly different circumstances, progress would have taken a different course. Romer's vision of a world dominated by giant birds is a classic example of this.[33] Matthew accepted that if the human race were wiped out, progress would be set back millions of years—yet eventually the world would be dominated by "some super-intelligent dog or bear or glorified weasel," or even by intelligent reptiles.[34] The image invoked here, and also in the theories of a multiple-branched tree of hominid evolution, was of nature constantly experimenting to see which of her products would do best in the long run.[35]

In 1923 Julian Huxley complained that professional biologists no longer believed in evolutionary progress.[36] He must have been thinking of the new breed of experimentalists, because we have seen little evidence that those actually involved in the reconstruction of life's ancestry had abandoned the idea—although they certainly saw it as a complex and unpredictable process. Huxley himself claimed that natural selection would produce an overall, if irregular, increase in intelligence and the ability of organisms to cope with their environment.[37] Matthew drew inspiration from the fact that progress was always due to "little steps forward achieved by individuals in each generation."[38] This almost Lamarckian emphasis on self-improvement is all the more remarkable coming from a paleontologist who did much to promote the resurgence of Darwinism in the 1920s and 1930s. It suggests that for some, the Spencerian philosophy of progress was still alive. But for many others, including Osborn (Matthew's superior at the American Museum of Natural History), the message of the new progressionism was harsher, less accommodating to the individual's aspirations. For Osborn, people were locked into a way of life defined by their racial ancestry: some branches of the tree of life had always developed further than others, and nature's guarantee of progress was built on the extinction of the less effective products of the

33. Romer, *Man and the Vertebrates*, p. 98.

34. Matthew, *Outline and General Principles of the History of Life*, p. 234.

35. Bowler, *Theories of Human Evolution*, pp. 87–104.

36. Julian Huxley, *Essays of a Biologist*, p. 10. This passage was brought to my attention by Mark Swetlitz, who also wondered whom Huxley might be referring to.

37. Wells and Huxley, *The Science of Life*, pp. 477–480. On Huxley's views, see C. Kenneth Waters and Albert Van Helden, eds., *Julian Huxley: Biologist and Statesman of Science*.

38. Matthew, *Outline and General Principles*, p. 235.

evolutionary experiment. The fact that Matthew's own biogeography was an integral part of this new philosophy suggests that the ideological factors at work in early twentieth-century science were many and various.

THE BIOLOGY OF IMPERIALISM?

A significant proportion of the images and metaphors used to explain the history of life seem to resonate with the cultural values of an age in which Western societies self-consciously set out to dominate the rest of the world. The assumption that a more stimulating environment led to faster progress seemed to vindicate the claim that the white race had a natural superiority by virtue of its ancestry in harsh northern regions. Most paleo-anthropologists believed that the human race had originated in the plains of Eurasia, so that the races which had migrated to Africa and other tropical locations would have stagnated. The model of evolution in which each family split into many parallel lines of development fostered the belief that the human races were distinct species, perhaps even distinct genera. Beyond these obvious appeals to the white race's sense of superiority, we can observe other ways in which the description of life's ancestry mirrored accounts of human history in which the dominant feature was the rise and fall of great empires. The claim that history had a cyclic rhythm marked by the succession of dominant races and civilizations was a central feature of conservative historiography in the nineteenth century.[39] This cyclic model of progress certainly had analogs in those anti-evolutionary models of the fossil record which stressed the distinct character of the successive episodes in the ascent of life. Ostensibly, at least, the Darwinian revolution represented the triumph of the rival, liberal model of history in which progress was the sum total of many individual acts of self-improvement. The renewed emphasis on the claim that the history of life does exhibit distinct phases marked a reemergence of the cyclic model of progress. Some versions of this model may have reflected an idealist view of the world similar to that which had inspired the conservative models of human history. But others seem to represent an adaptation of the cyclic model to the demands of a new, more materialistic imperialism which still needed to stress the divisions and rivalries between groups of animals and humans.

Many scientists were themselves aware of the fact that their work was part of the imperial enterprise. From Richard Owen onward, British naturalists saw themselves as playing a vital role in the process by which the untamed parts of the world were brought under European control.[40]

39. See Bowler, *The Invention of Progress.*

The great museums of natural history which adorned the principal cities of the Western world symbolized the scientists' participation in the drive to understand and control the world of nature.[41] The universities, botanical and zoological gardens, and geological surveys of Europe and America—along with those established in the colonies—also created a framework within which the scientists could self-consciously participate in the work of empire. In some respects, the additional evidence brought to light by these institutions played a vital role in extending the explanatory power of science. It became possible to describe the history of life in terms analogous to the succession of human civilizations precisely because domination of the world allowed a new precision in the study of biogeography and the geological structure of remote areas. The metaphors of imperialism were products not only of cultural influence, but of the kind of evidence that an imperial world power alone could command.

Throughout our survey we have seen examples of the history of life being described in terms of the rise and fall of particular groups of living things. Each family, each class has its origin, its rise to dominance, its decline and, often, its extinction. Continuity is maintained only by the fact that a small offshoot of one class occasionally acquires a significant new character that will eventually enable it to supplant its parents. The rise and fall of groups such as the Ammonites was widely seen as a parallel to the life cycle of the individual: birth, youth, maturity, senility, and death. In Matthew's words, life's splendid drama was composed of many acts: "Race after race of animals and plants arises, flourishes, and spreads widely over the earth, and then disappears, or gives rise to new races which in turn play their part and vanish from the scene."[42] Osborn wrote of various groups rising to a "climax" and then declining, while Othenio Abel referred to the *Blütezeit* (blossom or bloom-time) of each group preceding the time of decline.[43] The assumption that the main eras of the geological record were marked by the dominance of certain groups (the "Age of Reptiles," etc.) was widespread. The dinosaurs had been the "lords of creation" before an impatient nature had swept them away.[44] The growing concentration of effort on the origin of major groups and the growing debate over the causes of their decline and extinction both mark the re-

40. This is a major theme in Nicolaas Rupke's *Richard Owen: Victorian Naturalist.* On the parallel influence of Osborn at the American Museum of Natural History, see Ronald Rainger, *An Agenda for Antiquity.*

41. Susan Sheets-Pyenson, *Cathedrals of Science.*

42. Matthew, *Outline and General Principles,* p. 6.

43. Osborn, *The Age of Mammals,* pp. 46, 198; Abel, *Paläobiologie und Stammesgeschichte,* p. 358. The term *climax* is also used in the context of the rise and fall of the trilobites by Chamberlin and Salisbury, *Geology: Earth History,* II, pp. 347–348.

44. For example, Knight, *Before the Dawn of History,* p. 7.

vival of interest in a model of history that has strong parallels with the rise and fall of the great empires of human civilization.

One way of explaining this phenomenon was to invoke a direct parallel with the life cycle of the individual and assume that each group was originally endowed with a certain amount of evolutionary energy. At first the group would have enough energy to respond to a wide range of environmental challenges, but when the supply of energy eventually ran out, the group would no longer be able to adapt and would decline to extinction. Such an explanation was preferred by Hyatt and other neo-Lamarckians, and seems to reflect a direct continuation of the idealist world view of Louis Agassiz.[45] German paleontologists continued this way of thinking into the Nazi era. Abel thought that the period of decline was marked by a growing number of degenerate individuals who could not respond to environmental challenges.[46] By this time, English-language paleontologists were less likely to invoke actual degeneration as a cause of extinction, although the theory of overdevelopment as the result of orthogenetic trends retained some popularity. Here too there was a sense of restricted adaptability becoming eventually so rigid that the results were positively harmful. In all of these explanations there was an internal cause of decline built into each group from its inception. Evolutionary trends or laws were so rigorously enforced by nature that no group of animals could escape the consequences in the long run. Not surprisingly, there were few attempts to apply so pessimistic a philosophy to the human species, although Hyatt did try to use it as an argument against female emancipation by claiming that a reduction in sexual dimorphism was a sign of racial senility.[47]

In the English-speaking world especially, evolutionists turned away from this interpretation in the early twentieth century. Increasingly, they adopted a more Darwinian emphasis on success or failure in the struggle between different kinds of animals. Newly evolved forms with some general advantage would spread out to occupy more territory, displacing and eventually exterminating members of previously successful groups. The element of 'rise and fall' in the cyclic model of development was provided by the inexorable competition between the new groups constantly generated by evolution. Biogeography was linked to paleontology because the history of life could only be explained in terms of an "ever-waging struggle for supremacy" in which successful species had a natural tendency "to spread over a wider and wider area."[48] Virtually all biogeographers took

---

45. See Bowler, *Eclipse of Darwinism*, pp. 121–133.

46. Abel, *Paläobiologie und Stammesgeschichte*, pp. 365–367.

47. Hyatt, "The Influence of Woman on the Evolution of the Human Race." See Bowler, "Holding Your Head Up High."

it for granted that a species would spread as widely as possible, and that the consequences were often disastrous for the previous inhabitants of the areas they invaded. Initial success inevitably led to overspecialization for the environments conquered, and thus to stagnation and an inability to face the challenges posed by geographical changes or the influx of newcomers. As Matthew wrote, the eventual fall of each group was a consequence of "the inexorable law of the survival of the fittest, of adaptive radiation leading into a specialization of habits and characters that results in prosperity and progress so long as the environment remains favorable, but leads in the end to extinction when the conditions of life are changed."[49] The overspecialized dinosaurs were wiped out by a change in the environment, thus leaving room for the archaic mammals to expand—but the latter were wiped out directly by the modern mammals who became their successors. In either case, the decline of one group was intimately bound up with the rise of another. In other areas rivalry had more positive effects, as in Julian Huxley's metaphor of the "arms race" spurring the development of weapons and defense mechanisms in predators and prey.[50]

The effects of the natural tendency to expand were particularly obvious when a geographical barrier was eventually broken down. A 'lost world' of primitive types could then be swept away by more highly evolved newcomers. The influx of new inhabitants did not have to be described in imperialist terms; at one point Wallace wrote of the "extensive immigration" of northern forms into South America, a form of words which avoids the imagery of invasion and conquest.[51] Elsewhere he employed more violent imagery: there was an "incursion" of modern mammals into Africa, where they "overran the whole continent."[52] The metaphor of an invasion became commonplace in paleontological and biogeographical descriptions around 1900. In his discussion of South America, Matthew wrote, "The great invasion of northern animals swept away all the native groups of hoofed animals or ungulates and all of the marsupial carnivors."[53] Elsewhere he emphasized that new types spread out from the limited center where they first appeared, "displacing the old

48. Bartholomew et al., *Atlas of Zoogeography,* p. 3; see also other examples of this language given in chapter 8 of this text.

49. Matthew, *Outline and General Principles,* pp. 230–231.

50. Huxley, *The Individual in the Animal Kingdom,* p. 115; Haldane and Huxley, *Animal Biology,* p. 237.

51. Wallace, *The Geographical Distribution of Animals,* I, pp. 131–132.

52. Ibid., I, p. 288.

53. Matthew, *Outline and General Principles,* p. 92. The same metaphor is used in Lydekker's *Geographical History of Mammals,* p. 119.

and driving them before them into more distant regions."[54] Obviously these were not planned invasions of the kind organized by William the Conqueror: they resembled those unconscious expansions of human populations that had occurred whenever a strong emerging race had pushed into its neighbors' territory. The expansion of the white settlers in America and Australia would have seemed familiar examples. The extermination of the Neanderthals by early modern humans was seen as a classic parallel in prehistory.[55]

For those who believed that the harsh climate of the northern regions made them the powerhouse of progressive evolution, the model of invasion could be generalized. Most invasions were of northern types displacing or exterminating the relics of earlier phases which lingered in the south. Gadow quoted Darwin on "the efficient workshops of the North" which generated most of the forms that have dominated the rest of the earth.[56] In the *Origin of Species* Darwin had indeed stressed the ability of northern plants and animals, introduced by white colonists in Australia and elsewhere, to displace their southern counterparts.[57] The dominance, he claimed, arose because the greater extent of land area in the north fostered a more intense struggle for existence, which advanced the native forms to a "higher stage of perfection or dominating power." Matthew argued that the northern animals "have invaded other regions and, competing with the animals of those regions, they have almost always gained the upper hand, and tend to displace them."[58] We have also seen how this model was often extended to include the image of "successive waves of life" flowing south.[59]

The human races too were seen as waves of invaders spreading out from central Asia, and then the imagery could become quite biblical, as when Boyd Dawkins wrote of "the mysterious birthplace of successive races, the Eden of mankind, Central Asia."[60] More realistically, W. J. Sollas warned of the need for Europeans to be on their guard: progress was always accompanied by the sweeping away of the old by the new, and any nation that did not keep itself up to the mark would soon suffer the penalty imposed by natural selection, "the stern but beneficient tyrant of

54. Matthew, *Climate and Evolution,* p. 32.

55. See Bowler, *Theories of Human Evolution,* chap. 4, and *The Invention of Progress,* chaps. 2 and 4.

56. Gadow, *The Wanderings of Animals,* pp. 141–142.

57. Darwin, *Origin of Species,* pp. 379–380.

58. Matthew, *Outline and General Principles,* p. 85.

59. Wallace, *Geographical Distribution of Animals,* II, p. 155, and other examples given in chapter 8 of this text.

60. Boyd Dawkins, *Early Man in Britain,* p. 306.

the organic world."[61] The Darwinian source of this imagery is obvious enough: it was now taken for granted that any successful new type would expand its territory and displace its now-stagnant predecessors. Sollas was no Darwinian, however, and elswhere in his book he ridiculed the idea that natural selection was the agent which actually produced new forms of life.[62] The belief that selection acted as a negative force to weed out the unsuccessful was widely accepted even by those who dismissed it as a positive evolutionary force. The role of Darwin's theory in promoting this imperialist form of 'social Darwinism' must thus be evaluated with care. Everyone now accepted the Malthusian pressure to expand. But supporters of Lamarckism and other non-Darwinian mechanisms could exploit the assumption that a stimulating environment was the spur to progress, appealing to natural selection only as a means of explaining the eventual elimination of those who fell behind in nature's race.

The assumption that there were distinct waves of invaders following each other in succession across the globe dovetailed neatly with the growing belief that periodic episodes of climatic stress triggered bursts of progressive evolution. According to this model, one would not expect to see a constant rate of replacement of old types by new. The apparent fact that the history of life on earth could be divided into a series of discrete epochs became part of a more general philosophy of evolution which stressed its discontinuous or cyclic nature. The dinosaurs may have been the lords of creation, but nature had eventually swept them away through an environmental revolution. The rise and fall of the dinosaurs and other once-dominant groups could be seen as elements in a process by which life periodically surged forward in response to a climatic stimulus. Lull's purple prose thus makes a fitting conclusion to this survey of the metaphors used to depict the cyclic nature of evolutionary progress:

> Thus time has wrought great changes in earth and sea, and these changes, acting directly or through climate, have always found somewhere in the unending chain of living beings certain groups whose plasticity permitted their adaptation to the newly arising conditions. The great heart of nature beats, its throbbing stimulates the pulse of life, and not until that heart is stilled for ever will the rhythmic tide of evolution cease to flow.[63]

61. Sollas, *Ancient Hunters*, p. 385. See Bowler, *Theories of Human Evolution,* pp. 223–230, and *The Invention of Progress,* chap. 4.

62. Sollas, ibid., p. 405.

63. Lull, *Organic Evolution,* p. 691.

PHYLOGENY AND MODERN DARWINISM

By exposing the metaphors used to describe the history of life on earth, we uncover the extent to which science reflects the values of the culture in which it is conducted. We also expose a systematic pattern in the changing use of particular metaphors over the period under study. The new metaphors also tell us something about the changing resources and priorities of the scientists themselves. The relative decline of morphology led to a diminished use of models that related solely to the taxonomic relationships between different animal groups (trees and ladders). At the same time, developments in paleontology and the rise of historical biogeography provided new sources of evidence that allowed the articulation of new images of how life might evolve. The metaphors of imperialism may represent a form of social Darwinism, but they also illustrate the evolutionists' need to address questions posed by the flood of information brought in from conquered territories. Whatever the source of their new insights, the evolutionists of the 1920s and 1930s perceived the history of life on earth in a significantly different way than their predecessors who worked in the immediate aftermath of the *Origin of Species*.

Throughout this study I have argued that the debates of the early twentieth century reflected an increasingly Darwinian way of thinking about the history of life on earth. In conclusion, I want to summarize what I mean by by this claim. In what ways did the new phylogenetic studies reflect what we now recognize as key aspects of modern Darwinism—and to what extent did these studies actually help to shape the synthetic Darwinism that emerged in the mid-twentieth century?

At first sight, my argument seems to violate traditional ideas about the origins of the modern synthesis. Many interpretations have focused on population genetics and the emergence of a new generation of biologists committed to the theory of natural selection. But Ernst Mayr has challenged this orthodox historiography, arguing that field naturalists had independently begun to develop an approach that reflected Darwinian ideas about the role of adaptation and geographical isolation in the formation of new species.[64] It might seem perverse to extend this kind of argument to include paleontology and even morphology. After all, morphologists had no direct interest in the geographical dimension of evolution, and paleontology remained a hotbed of support for nonselectionist theories through to the 1940s. In the conventional view of the origins of modern Darwinism, Simpson's *Tempo and Mode in Evolution* of 1944 used the

64. This point was first developed in Mayr's "Where Are We?" and is continued in his contributions to Mayr and Provine, *The Evolutionary Synthesis*.

new broom of the genetic theory of natural selection to sweep away all this dead wood. My contention is that such an interpretation ignores a number of developments in phylogenetic research that paved the way for a more Darwinian way of thinking, here as much as in the field studies emphasized by Mayr.

The argument falls into two stages, one passive and one active. The passive, and hence perhaps less interesting, phase merely notes the existence of a number of phylogenetic researchers in the early twentieth century who participated in the growing awareness that natural selection might, after all, be the most effective explanation of how evolution works. These biologists may not have been without influence, even if they did not actively engage in the synthesis of genetics and Darwinism. But the more active phase of my argument depends upon the claim that modern Darwinism reflects something more than a mere loyalty to the selection theory. After all, some pioneers of the genetical theory of natural selection—R. A. Fisher is a good example—paid little attention to the element of geographical isolation that Mayr sees as central to the new Darwinism. Modern Darwinism extends certain key ideas that were developed, either explicitly or implicitly, by Darwin himself, and which were ignored by many biologists of the immediately post-Darwinian era. Changing styles of phylogenetic research helped to articulate these more generally Darwinian insights quite independently of the rise of the new selection theory. Even those evolutionists who still accepted a role for nonselectionist mechanisms could thus participate in the formulation of a Darwinian world view. The new paleontology of the early twentieth century shaped the climate of opinion within which modern Darwinism was born, even before Simpson began his assault on the surviving aspects of the old way of thinking.

This reassessment of the history of evolutionism draws upon the claim that much late nineteenth-century evolutionism was developmental rather than Darwinian in character.[65] The early evolutionists presented the development of life on earth as a largely predetermined process, unfolding almost inevitably toward obvious goals such as the emergence of humankind. Such a historicist perspective missed many of the crucial insights proclaimed by Darwin—insights based on the recognition that evolution was driven by the organisms' response to an ever-changing environment and was thus both divergent and largely unpredictable or open-ended. If we accept that the phylogenetic research of the early twentieth century was increasingly able to incorporate these aspects of the Darwinian program, then we must recognize that the biologists involved may have participated in the articulation of the modern Darwinian world view

65. See Bowler, *The Non-Darwinian Revolution.*

even when they did not see natural selection as the sole mechanism of evolution.

The background to this process is provided by the decline of evolutionary morphology and its replacement by paleontology as the primary focus of phylogenetic research. The first post-Darwinian generation— Haeckel, Dohrn, Lankester, and others who became active in the 1860s— drew upon the experience of predecessors such as Gegenbaur and Huxley, who had already achieved fame by pre-evolutionary efforts to represent the natural arrangement of species. They used anatomical and embryological evidence to reconstruct the major points of divergence in the tree of life. By the last decade of the nineteenth century, their approach was already seen as a wasted effort, since the evidence stubbornly refused to admit clear evaluation of the various schemes which had been proposed. A new generation of evolutionists drew upon the increasingly rich fossil record in an effort to reconstruct the later phases in the history of life— although their efforts too were plagued by the constant realization that some of the most crucial breakthroughs seem to have left no direct evidence in the rocks. Paleontologists drew upon some aspects of the old morphology, especially the functional analysis of how organic structures actually work, in an effort to understand transitions such as the origin of the tetrapod limb. They exploited not only the fossils, but also other evidence from the geological record to throw light on the changing conditions which might have stimulated important transformations. The attempt to understand the origin, rise and fall of groups such as the dinosaurs began to take on the form that we still see today. The paleontologists also linked up with field naturalists seeking to explain geographical distribution as the outcome of a historical process of evolution and migration.

Many early twentieth-century paleontologists retained an interest in mechanisms such as Lamarckism and orthogenesis. Osborn in America and D. M. S. Watson in Britain are good examples of this conservatism. In Germany, the rising influence of Schindewolf meant that developmental ideas successfully resisted the rising Darwinism of the English-speaking world. But in America especially, a new generation of paleontologists was throwing off the shackles of the old non-Darwinian approach. Matthew and Gregory both turned against the anti-Darwinian elements in Osborn's thought even though they worked in the institution he controlled. Romer also adopted an explicitly pro-selectionist stance in the 1930s. These paleontologists may not have anticipated Simpson's imaginative challenge to the old theories of parallel evolution, but they had already given up much of the old anti-selectionist way of thinking.

Nor can we dismiss the influence of surviving evolutionary morphologists. Lankester retired from his last official position in 1907, but he continued to write about science and to participate in debates about evolution. He had no patience with the new genetics, but was an active supporter of the selection theory. Lankester advised H. G. Wells when the latter was writing the introductory sections of his *Outline of History,* first published in 1920, where a brief outline of the history of life was used to expound the claim that natural selection could explain the whole process.[66] Wells's account evoked outrage among literary figures such as Hilaire Belloc, who were still under the impression that Darwinism had been dealt its death-blow by an earlier generation of biologists.[67] A recent study suggests that J. B. S. Haldane, one of the founders of the genetic theory of natural selection, was encouraged to work on the theory by the strength of Belloc's response.[68] Thus it was possible for Lankester to play an indirect role in the revival of the selection theory, even though he distrusted the new genetics. His one-time student E. S. Goodrich continued in the morphological tradition, yet stayed abreast of the new evolutionism and directly influenced the work of Julian Huxley and Gavin De Beer, both architects of the new Darwinian synthesis.[69] We have made frequent references to the popular *Science of Life* published by Wells and Huxley in 1931, which also linked phylogenetic studies to the new developments in thinking about the evolutionary mechanism.

Some morphologists and paleontologists were thus able to play a role in stimulating the revival of the selection theory. But it would be wrong to base an argument for the significance of these fields purely on this evidence. For every forward-looking student of phylogenetic reseach, there were many others who retained the old faith in nonselectionist mechanisms. A more substantive argument can, however, be made by stretching our definition of Darwinism to include factors that go beyond mere faith in the selection theory. The Darwinian world view represents a far more basic challenge to the old world of developmental thinking. It rests on acceptance of the possibility that there is no predetermined pattern of evolution, no goal toward which everything must strive, and no internal force pushing life through a developmental sequence. These insights are a consequence of Darwin's recognition (1) that species evolve in response to challenges from their environments, both inorganic and organic, and (2) that the environment is constantly changing, both as a result of geological forces and because animals and plants can move or be moved to

66. Wells, *The Outline of History,* book 1. On Lankester's involvement, see Lester, *E. Ray Lankester,* p. 178.

67. See Hilaire Belloc, *Mr. Belloc Objects to "The Outline of History."*

68. Gordon McQuat and Mary P. Winsor, "J. B. S. Haldane's Darwinism in Its Religious Context."

69. Stephen Waisbren, "The Importance of Morphology in the Evolutionary Synthesis."

new parts of the globe. As a result, evolution is branching, divergent, open-ended, and unpredictable.

My argument for a positive contribution from phylogenetic research rests on the demonstration that in the first half of the twentieth century, paleontogists and biogeographers moved a significant way along the path from a developmental toward a Darwinian way of representing the history of life on earth. Developmental ways of thought did not disappear: think of Patten's progressionism, Broom's theistic evolutionism, and Osborn's continued support for orthogenesis. In some areas, biologists retaining these outdated ways of thought made important contributions, as with Broom's work on the mammal-like reptiles. But even where older ways of thinking were retained, they were increasingly overlaid by concerns about new questions that would demand new kinds of answers. The answers offered may not always seem plausible by modern standards; indeed they may seem downright speculative. But scientists cannot develop the techniques needed to solve problems until they have begun to ask the right questions. By the 1920s and 1930s paleontologists were beginning to wonder about the forces that caused the rise and fall of different groups of animals in the course of the earth's history. In so doing, I maintain, they were taking the first, and perhaps crucial, steps on the way toward a more open-ended view of how life has developed.

Examples of this new way of thinking can be seen in the range of new metaphors outlined above, those invoking images of geographical conquests leading to the steady elimination of inferior types. Equally striking is the growing interest in the possibility of environmental triggers for key developments in the history of life. Other factors are more subtle. The willingness of most twentieth-century evolutionists to accept that the mammals could not dominate the earth until the reptiles had been removed marks a major qualification of the old, simpleminded progressionism. The pioneering efforts of Romer and others to explain the invasion of the land as an unanticipated by-product of the early amphibians' desperate attempts to reach more water reflects a way of thought that could not have been entertained by an evolutionist of the previous generation, yet which fits naturally into the viewpoint of modern Darwinism. Romer's willingness to imagine a world in which the mammals never got their chance to expand is another example of this more modern way of thinking about evolution. Faith in the inevitability of progress remained, but no one saw the exact course of evolution as predetermined. Even conservatives such as Osborn began to ask new kinds of questions about the origin, spread, and decline of classes in the fossil record.

I do not want to claim that, by themselves, these initiatives could have triggered Simpson and others to develop the very different style of paleontological research that has become common today. The chief input into the modern synthesis clearly came from outside the realm of phylogenetic research, as recognition of the range of variation within living species destroyed the old argument for parallelism. But I do want to argue that the path toward a wider acceptance of these initiatives was paved by developments taking place in paleontology and biogeography during the early decades of the twentieth century. Biologists thought about the history of life on earth in a significantly different way than their forerunners of the 1870s and 1880s, and the new questions they asked were important for creating a climate of opinion in which the last vestiges of the old system—the theories of Lamarckism and orthogenesis—could be dismantled under pressure from the new generation of selectionists. In effect, the areas of biology devoted to phylogenetic research had become preadapted to the worldview of the new Darwinism. Naturalists such as Huxley and paleontologists such as Simpson could appreciate the initiatives coming from the geneticists because they were already used to thinking about evolution in terms that made it easy to apply the new Darwinian theory to their problems.

Outside the realms of professional biology, the paleontologists had already helped to popularize the new kinds of questions that were being asked about the history of life. Popular modern accounts such as Gould's *Book of Life* are recognizably part of the same genre as that established by the earlier surveys of Osborn and Matthew. The techniques used by the specialists are far more sophisticated, but the questions that the readers expect to see addressed are those formulated in the early twentieth century—including the much-debated topic of the dinosaurs' extinction. Given that, for most ordinary people, the story of life's splendid drama represents the key focus of evolutionary science, the steady developments taking place in phylogenetic research from 1860 to 1940 represent a significant factor in the consolidation of the Darwinian revolution.

The following biographical summaries cover most of the figures who appear more than once in the text, and those minor contributors who played a prominent role in other fields. Some less well-known figures are not included, partly to keep the list to manageable proportions, but also because it has sometimes proved impossible to track down information beyond what is mentioned where they appear in the text.

ABEL, OTHENIO (1875–1946). Studied under E. Suess and was strongly influenced by the work of L. Dollo. Professor of paleontology and paleobiology at Vienna, then at Göttingen. Published extensive surveys of the fossil record supporting non-Darwinian ideas of evolution.

AGASSIZ, ALEXANDER (1835–1910). Son of Louis Agassiz; studied under his father at Harvard; curator at Museum of Comparative Zoology from 1874 and director from 1902. Worked on deep-sea animals, espcially echinoids.

AGASSIZ, JEAN LOUIS RODOLPHE (1807–1873). Trained in France and Germany; produced first comprehensive study of fossil fish; professor of natural history at Harvard University from 1847; established Museum of Comparative Zoology there, 1859. Prominent opponent of evolutionism, and pioneer advocate of the ice-age theory.

AMEGHINO, FLORENTINO (1853–1911). Born in Italy and taken to Argentina as child; after little formal education, he began collecting fossils in Argentina with help of his brother Carlos. Held several minor posts and also worked in business; secretary to La Plata Museum from 1886 to 1890. Published theory deriving all mammals from South American ancestors.

BAER, KARL ERNST VON (1792–1876). Of Baltic German origin; became librarian of the Academy of Natural Sciences, St. Petersburg. Discovered the mammalian ovum and published the classic embryological text, *Über Entwickelungsgeschichte der Thiere* (1828–1873).

BALFOUR, FRANCIS MAITLAND (1851–1882). Studied under Michael Foster at Cambridge, then at Naples; gained a fellowship in natural sciences at Trinity College, Cambridge, in 1874, and subsequently became university lecturer, then reader, in animal morphology. Studied the embryology of the elasmobranch fish and produced textbook of vertebrate embryology. Appointed to a personal chair in 1882, the year of his death in a climbing accident in the Alps.

BARRELL, JOSEPH (1869–1919). Worked with the U.S. Geological Survey in Montana from 1901, where he developed the theory that surface features have been shaped by the intrusion of magma. Appointed professor of geology at Yale in 1908. Pioneered the view that a number of sedimentary rocks have been laid down by rivers, winds, and ice.

BATESON, WILLIAM (1861–1926). Worked under Sedgwick at St. John's College, Cambridge, and under Brooks at Johns Hopkins (1883–1884). Held the Balfour studentship at Cambridge, 1887–1890, and became professor of biology there in 1908. In 1910 he became director of the John Innes Horticultural Institution. Worked first on the origin of the chordates, but turned from Darwinism to saltative evolution and then to genetics; published the first English translation of Mendel's papers and coined the term *genetics*.

BATHER, FRANCIS (1863–1934). Trained at Oxford, where he was awarded a D.Sc. in 1900. In 1887 became assistant in Department of Geology, British Museum (Natural History); keeper of geology there from 1924. Worked on echinoderm morphology and paleontology.

BAUR, GEORG (1859–1898). Studied anatomy and paleontology at Munich and Leipzig, and became assistant to K. Kuppfer at Leipzig. In 1882 moved to Yale University to work with O. C. Marsh; became assistant professor of osteology and paleontology at Chicago in 1892. Worked on fossil amphibians and on mammal-like reptiles; also organized expedition to Galápagos islands.

BEEBE, CHARLES WILLIAM (1877–1962). Curator of ornithology (1899) and director of tropical research (1919) at the New York Zoological Society. Well known as explorer and author of books on birds. Made record-breaking descent in bathysphere in 1934.

BEECHER, CHARLES EMERSON (1856–1904). Yale Ph.D. in paleontology; worked mainly on brachiopods and trilobites. Professor of paleontology and curator of the geological collections at Yale.

BLANFORD, WILLIAM THOMAS (1832–1905). Studied at the Royal School of Mines, London (1852–54) and also at Frieberg in Saxony. On staff of Geological Survey of India from 1854, then geologist on expedition to Abyssinia. President of Geological Society of London, 1888–1890.

BROOKS, WILLIAM KEITH (1848–1908). Wrote thesis under Alexander Agassiz at Harvard; with H. Newell Martin founded graduate program in biology at Johns Hopkins University and became professor of biology there. Worked on the morphology of the ascidian *Salpa* and on crustaceans.

BROOM, ROBERT (1866–1951). Trained in medicine at Glasgow and practiced in Australia and South Africa. Professor of zoology at Victoria College, Stellenbosch, 1903–1910; appointed curator of Transvaal Museum, Pretoria, 1934. Studied morphology of marsupials, and the fossils of mammal-like reptiles and Australopithecines.

CHAMBERLIN, THOMAS CHROWDER (1843–1928). University of Michigan Ph.D. in geology; worked with Wisconsin State Geological Survey and U.S. Geological Survey (chief geologist, 1871–1882). President of University of Wisconsin, 1887–1892, then professor of geology at University of Chicago. Developed theories of the origin of the earth and of episodic geological change.

CLAUS, CARL (1835–1899). Professor of zoology and director of the Institute of Comparative Physiology and Anatomy, Vienna. Published extensive work on invertebrate morphology, especially crustaceans, and an important textbook of zoology (1866). Criticized by Haeckel for limited support of evolutionism. Taught zoology to Sigmund Freud.

COPE, EDWARD DRINKER (1840–1899). Studied paleontology under Joseph Leidy; worked with U.S. Geological Survey from 1872 until Marsh took over. Linked to the Academy of Natural Sciences, Philadelphia; founder of the *American Naturalist*. Great rival of Marsh in collection of vertebrate fossils from the American west. Prominent neo-Lamarckian.

DARWIN, CHARLES ROBERT (1809–1882). Naturalist on voyage of H.M.S. *Beagle*, 1831–36; discovered principle of natural selection 1837; published *Origin of Species* (1859) and *Descent of Man* (1871).

DEAN, BASHFORD (1867–1928). Ph.D. from Columbia University, 1889; appointed lecturer (1886), then professor of vertebrate zoology (1904) at College of City of New York. Prominent ichthyologist; served on U.S. government fish commissions. Also expert on arms and armor.

DEPÉRET, CHARLES (1854–1929). Wrote thesis on geology of Perpignan region; became professor (1889) and subsequently Dean in Faculty of Science at Lyon. Influential expert on Tertiary geology and paleontology.

DOHRN, FELIX ANTON (1840–1909). Studied under Haeckel and Gegenbaur at Jena, but later was bitter opponent of both. Left academic life and founded Zoological Station at Naples, 1874, funded at first by his wealthy father, then by Prussian government. Worked on arthropod and chordate morphology, and proposed annelid theory of vertebrate origins.

DOLLO, LOUIS ANTOINE MARIE JOSEPH (1857–1931). Studied at Lille under Giard; moved to Brussels and became engineer, but took up study of paleontology after reading V. Kovalevskii. Appointed to staff of National Museum, Brussels, in 1882 and became Curator of the Vertebrate Section there in 1891. Also interested in linguistics. Restored fossil *Iguanodon* specimens from Bernissart, and worked on evolution of many groups in fossil record, including lungfish, turtles, dinosaurs, and marsupials.

FLOWER, WILLIAM HENRY (1831–1899). Studied zoology at University College, London; conservator of the museum of the Royal College of Surgeons, 1861–1884, and Hunterian professor of comparative anatomy, 1870–1884. Director of the British Museum (Natural History), 1884–1898. Expert on mammals, especially whales, and on museum techniques.

FÜRBRINGER, MAX (1846–1920). Studied under Gegenbaur at Jena; appointed professor at Amsterdam and then at Jena before succeeding Gegenbaur to the chair at Heidelberg. Published influential study of anatomy and systematics of birds.

GADOW, HANS FRIEDRICH (1855–1928). Studied zoology at Berlin, Jena, and Heidelberg; worked at British Museum (Natural History), 1880–1882; reader in vertebrate morphology at Cambridge from 1884.

GASKELL, WALTER HOLBROOK (1847–1914). Originally trained in mathemetics at Cambridge, then studied physiology at Leipzig, 1874–75. Obtained Cambridge M.D., 1878, and appointed university lecturer in physiology there in 1883. Revolutionized ideas on the action of the heart and on cardiac disease.

GAUDRY, ALBERT (1827–1908). Appointed assistant to A. D'Orbigny and then professor of paleontology at the Muséum d'histoire naturelle, Paris. Published studies of Tertiary mammals discovered in Greece supporting the idea of evolution.

GAUPP, ERNST (1865–1916). Studied at Breslau and became assistant in the Breslau Anatomical Insitute. In 1895 became prosector, then in 1897 professor of anatomy at Freiburg-im-Breslau. Worked on homologies of bones in reptilian jaw and mammalian ear.

GEGENBAUR, CARL (1825–1903). Studied under R. A. Kölliker at Würzburg; became professor of zoology at Jena in 1855 and professor of anatomy at Heidelberg in 1872. Worked especially on the structure of fish, and published several influential textbooks of vertebrate anatomy.

GIARD, ALFRED (1846–1908). Studied at the École normale supérieure; produced doctoral thesis on ascidians in 1872. Became profesor of science at Lille, then lecturer at the École normal supérieure, and finally professor in the Faculty of Science at Paris. Noted link between articulates and molluscs later promulgated by Hatschek; strongly supported the Lamarckian mechanism of evolution.

GILL, THEORORE NICHOLAS (1837–1914). Trained in law, but turned to study of ichthyology with scholarship from Wagner Free Institute of Science, Philadelphia. Librarian at Smithsonian Institution and later at Library of Congress; subsequently appointed professor of zoology at Columbia College.

GOODRICH, EDWIN STEPHEN (1868–1946). Originally studied at Slade Art School, but persuaded by Lankester to take up zoology at University College, London. Assistant to Lankester at Oxford, 1892, and himself appointed to the Linacre chair of comparative anatomy there in 1921. Made extensive studies of vertebrate anatomy.

GREGORY, WILLIAM KING (1876–1970). Assistant to Osborn at American Museum of Natural History; appointed assistant curator in Department of Vertebrate Paleontology there (1911) and curator of the Department of Comparative Anatomy (1921); also professor of vertebrate paleontology at Columbia University. Studied functional morphology of vertebrates and vertebrate fossils.

HAECKEL, ERNST (1834–1919). Professor of zoology at Jena from 1862. Important work on calcareous sponges led to proposal of gastrea theory and the "biogenetic law" (recapitulation theory). Campaigned vigorously in support of evolutionism; published *Generelle Morphologie* (1866) and many popular works on evolutionism.

HATSCHEK, BERTHOLD (1854–1941). Studied under Claus at Vienna, R. Leuckhart at Leipzig, and Haeckel at Jena; became professor of zoology at the German University of Prague, then in 1896 professor in the Zoological Insitute, Vienna. Published important textbook of zoology. Later studies interrupted by nervous depression. Worked on invertebrate morphology, and proposed trochophore theory uniting articulates and molluscs.

HEER, OSWALD (1809–1883). Professor of botany at Zurich. Published extensively on fossil plants of Switzerland and the arctic regions.

HEILPRIN, ANGELO (1853–1907). Born in Hungary and brought to the United States as a child. Studied under T. H. Huxley at the Royal School of Mines, London, 1876, and also at Paris and Geneva. Appointed professor of vertebrate paleontology at the Academy of Natural Sciences, Philadelphia, 1879, and professor of geology at Wagner Free Institute of Science, 1885. Became well known as explorer of Mexico, South America, and the Arctic.

HERTWIG, WILHELM AUGUST OSCAR (1849–1922). Studied under Haeckel and Gegenbaur at Jena, and went on to become assistant (1878) and then professor (1881) of anatomy there. In 1888 appointed director of the Anatomical and Biological Institute, Berlin. Worked on the origin of the body cavities, and proposed the coelom theory.

HILGENDORF, FRANZ (1839–1904). Curator of the Zoological Museum, Berlin. Published studies of the Tertiary snails in deposits at Steinheim showing gradual evolution.

HUBRECHT, AMBROSIUS ARNOLD WILHELM (1853–1915). Studied at Delft, Utrecht, Leiden and Erlangen; obtained D. Phil. from Utrecht in 1874, then became professor of zoology and entomology there. Worked on nemertine worms and their relationship to chordates; also worked on primates. Visited Dutch East Indies in 1890 to study tree shrews.

HUXLEY, JULIAN SORRELL (1877–1975). Grandson of T.H. Huxley and brother of Aldous. Studied at Naples and Oxford; taught biology at Rice Institution, Huston (1912–1916), Oxford (1919–1925), and King's College, London (1926–1929). Published important work on comparative growth rates and on Darwinism, but later accepted many nonscientific appointments, including director of UNESCO.

HUXLEY, THOMAS HENRY (1825–1895). Naturalist on H.M.S. *Rattlesnake* (1846–1850) studying marine invertebrates; appointed lecturer in paleontology at Royal College of Mines, London, 1854. Founded important course in biology that included laboratory work at Royal College of Science, 1872, with Lankester as assistant. Best known for defense of Darwinism, but also campaigned for reform of education, and served on many government commissions. President of the Royal Society, 1883–1885.

HYATT, ALPHEUS (1838–1902). Studied under Louis Agassiz; worked at the Peabody Academy, Salem, Mass., and later as custodian of the Boston Society of Natural History. Professor of zoology and paleontology at the Massachusetts Institute of Technology, 1870–1888. Worked on fossil cephalopods; advocated theories of Lamarckian and orthogenetic evolution.

KINGSLEY, JOHN STIRLING (1853–1929). Princeton D.Sc., 1885; studied at Freiburg, 1891–92. Taught at Indiana University (1887–1889), University of Nebraska (1889–1891), Tufts College (1892–1913), and University of Illinois (1913–1921). Worked on both vertebrate and invertebrate morphology.

KOKEN, ERNST (1860–1912). Professor of mineralogy, geology, and paleontology at Tübingen; specialist in invertebrate paleontology.

KOVALEVSKII, ALEXANDR ONUFRIEVICH (1840–1901). Studied biology at St. Petersburg, then at Heidelberg under Ludwig Carus and at Tübingen under F. Leidig. Worked on *Amphioxus* and tunicates at Naples in 1864; subsequently developed theory of vertebrate origins. Taught at universities in Kazan, Kiev, Odessa, and finally St. Petersburg.

KOVALEVSKII, VLADIMIR (1842–1883). Brother of Alexandr; studied law in Russia before working with Haeckel. Published studies of evolution in fossil ungulates in 1870s which gained him wide recognition except in Russia, where he found it impossible to obtain an academic post. After several abortive efforts to start another career he committed suicide.

LANG, ARNOLD (1855–1914). Studied zoology at Geneva and under Haeckel at Jena. Worked briefly at Naples before becoming professor of phylogenetic zoology at Jena in 1886, then professor of zoology at Zurich. Worked on annelid phylogeny and origin of body cavities, and published textbook of invertebrate anatomy. Translated Lamarck into German.

LANKESTER, EDWIN RAY (1847–1929). Studied zoology at Cambridge, Leipzig, Jena, and Naples. Appointed fellow of Exeter College, Oxford (1872); professor of zoology, University College, London (1874); Linacre professor of comparative anatomy, Oxford (1891); director of the British Museum (Natural History) (1898). Worked widely on morphology and paleontology, both vertebrate and invertebrate. Campaigned for improvements in British science education, and wrote many popular articles on science. He was knighted in 1907.

LULL, RICHARD SWAN (1867–1957). Assistant professor of zoology, Massachusetts State College, 1894–1906; assistant professor, then professor (1911) of vertebrate paleontology at Yale University, and curator of vertebrate paleontology at the Peabody Museum, Yale.

LYDEKKER, RICHARD (1849–1915). Studied at Trinity College, Cambridge; paleontologist on staff of Geological Survey of India, 1874–1882; produced ten volumes of catalogs of the reptiles, birds, and mammals in the collection of the British Museum (Natural History) and many popular books on mammals.

MACBRIDE, ERNEST WILLIAM (1866–1940). Studied under Sedgwick at Cambridge, spent a year a Naples, then became demonstrator in animal morphology at Cambridge and a fellow of St. John's College (1897). Subsequently appointed first professor of zoology at McGill University, Montreal, then assistant to Sedgwick at Imperial College, London; finally succeeded to the chair at Imperial College in 1914. Published textbook on invertebrate embryology; prominent defender of Lamarckism and supporter of eugenic policies.

MARSH, OTHNIEL CHARLES (1831–1899). Graduate of Yale University; became professor of paleontology there in 1886. Appointed vertebrate paleontologist of U.S. Geological Survey in 1892. Organized fossil-hunting expeditions to the west, and described many new species of extinct reptiles and mammals. Became bitter rival of E. D. Cope.

MATTHEW, WILLIAM DILLER (1871–1930). Born in Canada; obtained Ph.D. in geology from Columbia University, 1895. Worked under Osborn in Department of Vertebrate Paleontology at American Museum of Natural History; in 1927 appointed professor in the Department of Paleontology, University of California, Berkeley.

MINOT, CHARLES SEDGWICK (1852–1914). Studied at Massachusetts Institute of Technology and Harvard Graduate School, and then at Leipzig under K. Ludwig and R. Leuckart. Taught at Harvard Dental School and then in the university's Department of Histology and Embryology. Built up an outstanding collection of vertebrate embryos, and invented the automatic rocking microtome for producing sections. Published a textbook, *Human Embryology,* in 1892.

MITCHELL, PETER CHALMERS (1864–1945). Studied science at Oxford; appointed university demonstrator in animal morphology, 1888. Lecturer in biology at Charing Cross Hospital, London, from 1894. Secretary of the Zoological Society of London, 1903–1935; responsible for establishment of Whipsnade zoological park.

MIVART, ST. GEORGE JACKSON (1827–1900). Worked under Huxley in 1860s on primates and later on many other kinds of vertebrates. Lectured at St. Mary's Hospital Medical School, London, 1862–1884. Proposed theories on origin of fish fins and origin of mammals, but increasingly based his work on an anti-Darwinian view of evolutionary relationships, leading to a feud with Darwin's supporters.

MORGAN, THOMAS HUNT (1866–1945). Did graduate work at Johns Hopkins University on morphology of sea spiders, but turned to experimental biology under influence of Jacques Loeb and Hans Driesch. Appointed professor of experimental zoology at Columbia University, 1904. In 1928 organized the division of biology of the California Insitute of Technology. Founder of the modern theory of the gene.

MOSELEY, HENRY NOTTIDGE (1844–1891). Studied at Oxford and in Germany; naturalist on the voyage of H.M.S. *Challenger,* 1872–1876. Fellow of Exeter College, Oxford; appointed to Linacre chair of comparative anatomy, Oxford, 1881.

MÜLLER, FRITZ (1822–1897). Left Germany in 1852 for Brazil, where he supported himself by various jobs, including teaching school. Inspired by reading

Darwin to undertake study of crustacean ontogeny in the hope of reconstructing the group's phylogeny.

MURRAY, ANDREW (1812–1878). Abandoned a career in the law for botany and entomology; president of Edinburgh Botanical Society, 1848, and of Royal Horticultural Society, 1860. Scientific director of the Linnean Society of London, 1877.

NAEF, ADOLPH (1883–1949). Born in Switzerland; studied in Germany and at Naples Zoological Station. Published on idealistic morphology, but was unable to obtain permanent position in Europe; became professor of zoology at Cairo. Became known for work on ontogeny and phylogeny of cephalopods.

NEUMAYR, MELCHIOR (1845–1890). Studied paleontology under W. Waagen at Munich and became professor of paleontology at Vienna. Studied Miocene Gastropods and noted geographical variations in fossil populations; became strong advocate of Darwinian selectionism.

NEWTON, ALFRED (1829–1907). Ornithologist with interest in migration and distribution of birds; despite lack of academic training in biology appointed to chair of zoology and comparative anatomy at Cambridge, 1866.

OSBORN, HENRY FAIRFIELD (1857–1935). Studied morphology with Huxley and Balfour but turned to vertebrate paleontology. Appointed assistant professor of natural science at Princeton, 1881; in 1890 moved to New York to take up position at Columbia College and the Department of Vertebrate Paleontology at the American Museum of Natural History.

OWEN, RICHARD (1804–1892). Curator of Hunterian Museum, Royal College of Surgeons, London, from 1827; appointed superintendant of natural history collections at the British Museum, 1856, and created the present Natural History Museum at South Kensington. Worked extensively on both vertebrate and invertebrate morphology and paleontology; accepted evolution, but was critical of Darwin, and fell out with Huxley over the relationship of humans to apes.

PACKARD, ALPHEUS SPRING (1839–1905). Student of Louis Agassiz; professor of zoology and paleontology at Brown University from 1878. Entomologist; also studied blind cave-animals; prominent neo-Lamarckian.

PATTEN, WILLIAM (1861–1932). Studied at Harvard and at Leipzig; appointed professor of zoology at Dartmouth College, New Hampshire, 1893. Made extensive studies of the horseshoe crab, *Limulus*, and proposed theory of origin of vertebrates from arthropods.

RABL, CARL (1853–1917). Originally influencd by Haeckel, but then studied medicine at Vienna and Leipzig. Taught at Vienna and Prague; obtained a chair at Leipzig in 1904. Studied the form of the germ layers, and later established the continuity of the chromosomes through cellular division.

ROLLESTON, GEORGE (1829–1881). Trained as physician; appointed Regius professor of physic, Oxford, 1857, and to Linacre chair of anatomy and physiology, 1860. Published textbook of vertebrate anatomy.

ROMER, ALFRED SHERWOOD (1874–1973). Ph.D. from Columbia University, 1921; appointed associate professor (1923), then professor (1931) of vertebrate paleontology, University of Chicago. Professor of zoology and curator of vertebrate paleontology at Harvard, 1934–1965.

RÜTIMEYER, KARL LUDWIG (1825–1895). Professor of anatomy at Basel; published early studies of mammalian paleontology supporting evolution.

SCHARFF, ROBERT FRANCIS (1858–1934). Born in England of German parents; studied at University College, London, Heidelberg, and Naples. Keeper of the natural history collections in the National Museum of Science, Dublin. Expert on geographical distribution of mammals, especially in western Europe.

SCHINDEWOLF, OTTO H. (1896–1971). Associate professor at Marburg, then director of the Geological Survey in Berlin; from 1948, professor of geology and paleontology at Tübingen. Studied stratigraphy and evolution of ammonoids and corals, and developed "typostrophe theory."

SCHUCHERT, CHARLES (1858–1942). Curator at U.S. National Museum, 1894–1904, then appointed professor of vertebrate paleontology at Yale University. Regarded as a founder of paleobiogeography.

SCLATER, PHILLIP LUTLEY (1829–1913). Studied mathematics at Oxford; qualified as lawyer, 1855. Secretary of the Zoological Society of London, 1859–1902. Ornithologist and founder of *Ibis*; best known for work on geographical distribution.

SCOTT, WILLIAM BERRYMAN (1858–1947). Studied under Huxley and Balfour; obtained Heidelberg Ph.D. under Gegenbaur in 1880. Made extensive studies of American fossil mammals. Appointed assistant in geology at Princeton, 1880, then professor of geology and paleontology, 1884.

SEDGWICK, ADAM (1854–1913). Studied under Balfour and succeeded him as Cambridge reader in animal morphology, 1882; appointed to chair of zoology at Imperial College, London, in 1909. Studied embryology, and proposed theories of early differentiation of animal kingdom, but subsequently turned away from phylogenetic speculation. Not to be confused with the Adam Sedgwick (1785–1873) who was professor of geology at Cambridge, a prominent catastrophist, and later opponent of Darwinism.

SEELEY, HARRY GOVIER (1839–1909). Studied at Woodwardian Museum of Geology, Cambridge, under Adam Sedgwick (1785–1873); appointed professor of geology, King's College, London. Worked on pterosaurs; visited South Africa to study fossils of mammal-like reptiles.

SEMON, RICHARD (1859–1918). Studied under Haeckel at Jena and under Gegenbaur at Heidelberg. Became assistant to Oscar Hertwig at the Jena Anatomical Institute and subsequently became professor of normal and comparative anatomy there. Visited Australia to study lungfish and monotremes; also published theory of heredity in *Die Mneme* (1904).

SIMPSON, GEORGE GAYLORD (1902–1984). Obtained Yale Ph.D. under R. S. Lull; worked on Mesozoic mammals at British Museum (Natural History), 1926–

1927. Appointed assistant curator of fossil vertebrates, American Museum of Natural History, 1927. Made expeditions to South America in search of fossil mammals. Published *Tempo and Mode in Evolution,* 1944, and many later books on Darwinism. Appointed Alexander Agassiz Professor at the Museum of Comparative Zoology, Harvard, 1958.

SUESS, EDUARD (1831–1914). Appointed assistant at the Hofmuseum, Vienna, 1852, then professor of geology at University of Vienna, 1857. Wrote on origin of Alps (1875) and published major survey of geology, *Der Antlitz der Erde* (1883–1888). Liberal deputy in lower house of Austrian parliament.

WAAGEN, WILHELM (1841–1900). Worked on staff of Geological Survey of India studying fossil cephalopods, then became professor of paleontology successively at Munich, Prague, and Vienna. Proposed theory of evolution by "mutations," or directed variations.

WALCOTT, CHARLES DOOLITTLE (1850–1923). Assistant to James Hall of New York Geological Survey, then moved to U.S. Geological Survey (1879) and became director (1894). Secretary of the Smithsonian Insitution, 1907–1927. Worked on Cambrian fossils; first to describe Burgess Shale fossils.

WALLACE, ALFRED RUSSEL (1823–1913). Originally worked as surveyor in England; organized zoological collecting expedition to South America with H. W. Bates, then to Southeast Asia. Worked on biogeography and relationship between species and varieties; discovered principle of natural selection in 1858. Supported himself by writing on zoology, evolution theory, and religious issues, and was eventually awarded a small government pension.

WATSON, DAVID MEREDITH SEARS (1886–1973). Studied at the University of Manchester and was appointed lecturer in vertebrate paleontology at University College, London; subsequently became Jodrell Professor of Zoology and Comparative Anatomy there. Worked on fossil amphibians; supported theory of orthogenetic evolution.

WEGENER, ALFRED LOTHAR (1880–1930). Geophysicist and meteorologist; appointed professor of geophysics at Graz, 1924. Worked on thermodynamics of atmosphere, but best known for theory of continental drift advanced in *Die Entstehung der Kontinente und Ozeane* (1915). Died while on expedition to Greenland.

WIEDERSHEIM, ROBERT ERNST (1848–1923). Studied at Lausanne, Tübingen, Würzburg, and Freiburg-im-Breslau; became assistant to R. A. Kölliker at the Anatomical Insitute, Wurzburg, and later professor of comparative anatomy at Freiburg. Worked on many problems in vertebrate anatomy, and published important textbooks of comparative anatomy.

WILLEY, ARTHUR (1867–1942). Studied at University College, London, and was then appointed to Balfour Studentship at Cambridge. Director of the Colombo Museum in Ceylon (Sri Lanka), 1902–1910, then appointed professor of zoology at McGill University, Montreal.

WOODWARD, ARTHUR SMITH (1864–1944). Studied at Owen's College, Manchester; appointed to Department of Geology, British Museum (Natural History) in 1882 and became assistant Keeper (1892) and finally Keeper (1901). Worked on fossil fish, and produced catalog of Museum's collection; also the discoverer of the notorious Piltdown hominid remains (1912), subsequently shown to be fraudulent.

ZITTEL, KARL ALFRED VON (1839–1904). Professor of geology and paleontology at Munich. Worked on vertebrate paleontology and produced massive textbook in that field.

TIMELINES

The timelines given below are meant to help the reader visualize how the more significant personalities might have interacted with one another. Dates of birth and death are indicated to the nearest five years. The personalities are divided into groups based on location:

America (including South America and those who moved to America for most of their career)

Britain (including British Empire)

Continental Europe (excluding those who moved to America or Britain for most of their career).

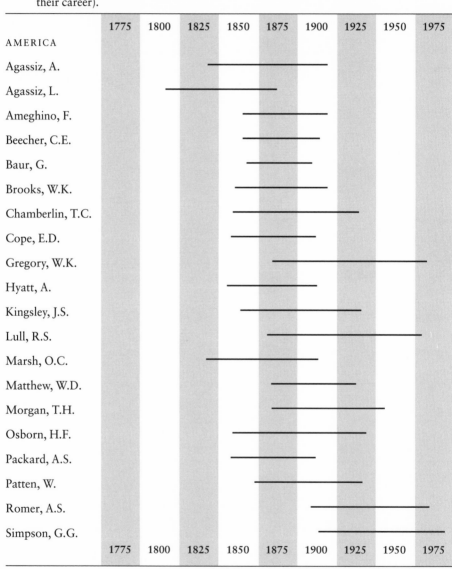

	1775	1800	1825	1850	1875	1900	1925	1950	1975
AMERICA									
Agassiz, A.				————————————					
Agassiz, L.			————————————						
Ameghino, F.					—————				
Beecher, C.E.					————————				
Baur, G.					—————				
Brooks, W.K.				————————					
Chamberlin, T.C.				—————————————					
Cope, E.D.				—————————					
Gregory, W.K.					————————————————————				
Hyatt, A.				————————					
Kingsley, J.S.				—————————					
Lull, R.S.					————————————————				
Marsh, O.C.			————————————						
Matthew, W.D.				—————————————					
Morgan, T.H.					————————————				
Osborn, H.F.				—————————————					
Packard, A.S.				—————————					
Patten, W.					———————————				
Romer, A.S.						————————————————			
Simpson, G.G.						————————————————			
	1775	1800	1825	1850	1875	1900	1925	1950	1975

	1775	1800	1825	1850	1875	1900	1925	1950	1975
BRITAIN									
Balfour, F.M.					——				
Bateson, W.					————————				
Broom, R.					—————————————			
Darwin, C.			————————						
Gadow, H.					———————————			
Goodrich, E.S.					—————————————			
Huxley, J.S.					———————————————————		
Huxley, T.H.				—————————					
Lankester, E.R.				——————————————				
MacBride, E.W.					————————————			
Mivart, St.G.J.				————————					
Newton, A.				—————————					
Owen, R.			———————————						
Scharff. R.F.					—————————			
Sclater, P.L.				———————————				
Sedgwick, A.					————				
Seeley, H.G.				——————————————				
Wallace, A.R.				———————————————				
Watson, D.M.S.					———————————————		
Willey, A.					—————————			
Woodward, A.S.					———————————			

	1775	1800	1825	1850	1875	1900	1925	1950	1975

	1775	1800	1825	1850	1875	1900	1925	1950	1975

EUROPE

Abel, O.

Baer, K.E. von

Claus, C.

Depéret, C.

Dohrn, A.

Dollo, L.

Fürbringer, M.

Gegenbaur, C.

Gaupp, E.

Gaudry, A.

Giard, A.

Haeckel, E.

Hatschek, B.

Hertwig, O.

Hubrecht, A.A.W.

Kovalevskii, A.

Kovalevskii, V.

Lang, A.

Müller, F.

Neumayr, M.

Rabl, C.

Rütimeyer, K.

Schindewolf, O.

Semon, R.

Wiedersheim, R.

Zittel, K.A. von

	1775	1800	1825	1850	1875	1900	1925	1950	1975

BIBLIOGRAPHY

Abel, Othenio. *Paläobiologie und Stammesgeschichte.* Jena: Gustav Fischer, 1929.

Adams, Mark B. "Svertsov and Schmalhausen: Russian Morphology and the Evolutionary Synthesis." In Ernst Mayr and William Provine, eds., *The Evolutionary Synthesis.* Cambridge, Mass.: Harvard University Press, 1980. pp. 193–225.

Agassiz, Alexander. "Paleontological and Embryological Development." *Proceedings of the American Association for the Advancement of Science,* 29 (1880): 389–414; also *American Journal of Science,* 20 (1880): 294–302, 325–389.

Agassiz, Louis. "A Letter Concerning Deep-Sea Dredgings, Addressed to Prof. Benjamin Peirce, Superintendent, United States Coast Survey." *Annals and Magazine of Natural History,* ser. 4, 9 (1872): 169–173.

Allen, Garland E. *Life Science in the Twentieth Century.* New York: Wiley, 1975.

———. *Thomas Hunt Morgan: The Man and His Science.* Princeton: Princeton University Press, 1978.

Allen, Joel Asaph. "The Geographical Distribution of North American Mammals." *Bulletin of the American Museum of Natural History,* 4 (1892): 199–243.

Allen, K. C., and D. E. G. Briggs. *Evolution and the Fossil Record.* London: Balhaven Press, 1989.

Allis, Edward Phelps. "The Lips and the Nasal Apertures in the Gnathostome Fishes." *Journal of Morphology,* 32 (1919): 145–197.

———. "Concerning the Nasal Apertures, the Lachrymal Canal and the Bucco-Pharyngeal Upper Lip." *Journal of Anatomy,* 66 (1932): 650–658.

Ameghino, Florentino. "South America as the Source of the Tertiary Mammalia." *Natural Science,* 11 (1897): 256–264.

Anderson, D. T. *Embryology and Phylogeny in Annelids and Arthropods.* Oxford: Pergamon Press, 1973.

Andrews, Ethan Allen. "Affinities of Annelids to Vertebrates." *American Naturalist*, 19 (1885): 767–774.

Anon., "The Kinship of Ascidians and Vertebrates." *Quarterly Journal of Microscopical Science*, 10 (1870): 59–69.

———. "The Genetic Relationship between Ascidians and Vertebrates." *Quarterly Journal of Microscopical Science*, 10 (1870): 299.

———. "The History of the Founding of the Marine Biological Association." *Journal of the Marine Biological Association*, 1 (1887): 17–44.

———. "The Evolution of the Mammalia." *Natural Science*, 1 (1892): 489.

———. "The Evolution of Life." *Natural Science*, 3 (1893): 222–225.

———. Obituary of Henry Bargman Pollard. *Nature*, 54 (1896): 183.

———. "The Origin of the Irish Fauna." *Natural Science*, 11 (1897): 223–224.

———. Obituary of Georg Baur. *Nature*, 58 (1898): 350.

———. "The Bipolar Hypothesis." *Natural Science*, 14 (1899): 348–349.

———. Review of W. Patten, *The Evolution of Vertebrates and their Kin. Nature*, 91 (1913): 79–80.

Appel, Tobey. *The Cuvier-Geoffroy Debate: French Biology in the Decades before Darwin*. New York: Oxford University Press, 1987.

Argyll, Dowager Duchess of, ed. *George Douglas Campbell, Eighth Duke of Argyll . . . Autobiography and Memoirs*. London: John Murray, 1906. 2 vols.

Argyll, George Douglas Campbell, Duke of. *The Reign of Law*. 5th ed. London: Alexander Strahan, 1868.

Arkell, W. J., and J. A. Moy-Thomas. "Palaeontology and the Taxonomic Problem." In J. Huxley, ed., *The New Systematics*, pp. 395–410. Oxford: Oxford University Press, 1940.

Arldt, Theodor. *Handbuch der Palaeogeographie*. Leipzig: Gebrüder Borntraeger, 1919.

Ayers, Howard. "The Unity of the Gnathostome Type." *American Naturalist*, 40 (1906): 75–94.

Baer, Karl Ernst von. "Entwickelt sich die Larve der einfachen Ascidien in der ersten Zeit nach dem Typus der Wirbelthiere?" *Mémoires de l'Académie des Sciences de St.-Pétersbourg*, 7th ser., 19, no. 8 (1873).

Baitsell, G. A., ed. *The Evolution of the Earth and Man*. New Haven: Yale University Press, 1929.

Balfour, Francis Maitland. "A Comparison of the Early Stages in the Development of Vertebrates." *Quarterly Journal of Microscopical Science*, 15 (1875): 207–226.

———. "A Monograph on the Development of the Elasmobranch Fishes." Reprinted from *Quarterly Journal of Microscopical Science* (1878) in Balfour, *Works*, I, pp. 203–520.

————. "On Certain Points in the Anatomy of Peripatus capensis." Reprinted from *Proceedings of the Cambridge Philosophical Society* (1879) in Balfour, *Works*, I, pp. 657–660.

————. "On the Development of the Skeleton of the Paired Fins of Elasmobranchii, Considered in Relation to Its Bearing on the Nature of the Limbs of the Vertebrata." Reprinted from *Proceedings of the Zoological Society of London* (1881) in Balfour, *Works*, I, pp. 714–737.

————. *A Treatise on Comparative Embryology* (London, 1880–81). Reprinted in Balfour, *Works*, vols. II and III.

————. *The Works of Francis Maitland Balfour.* Ed. M. Foster and A. Sedgwick. London: Macmillan, 1885. 4 vols.

Balfour, F. M., and W. N. Parker. "On the Structure and Development of Lepidosteus." *Philosophical Transactions of the Royal Society,* 173 (1883): 359–442.

Bannister, Robert C. *Social Darwinism: Science and Myth in Anglo-American Social Thought.* Philadelphia: Temple University Press, 1979.

Barrell, Joseph. "Influences of Silurian-Devonian Climates on the Rise of Air-Breathing Vertebrates." *Bulletin of the Geological Society of America,* 27 (1916): 387–436.

————. "Rhythms and the Measurement of Geological Time." *Bulletin of the Geological Society of America,* 28 (1917): 745–904.

Barrell, Joseph, et al. *The Evolution of the Earth and Its Inhabitants: A Series of Lectures Delivered before the Yale Chapter of the Sigma Xi during the Academic Year 1916–17.* New Haven: Yale University Press, 1918.

Bartholomew, J. G., W. Eagle Clarke, and Percy H. Grimshaw. *Atlas of Zoogeography.* London: John Bartholomew for the Royal Geographical Society, 1911.

Bates, Henry Walter. "Contributions to an Insect Fauna of the Amazon Valley: Lepidoptera: Heliconidae." *Transactions of the Linnean Society of London,* 23 (1862): 495–566.

————. *The Naturalist on the River Amazons.* London: John Murray, 1863. 2 vols.

Bateson, Beatrice. *William Bateson: Naturalist.* Cambridge: Cambridge University Press, 1928

Bateson, William. "The Later Stages in the Development of Balanoglossus Kowalevskii, with a Suggestion as to the Affinities of the Enteropneusta." *Quarterly Journal of Microscopical Science,* 25 (1885) supplement: 81–122.

————. "Continued Account of the Later Stages in the Development of Balanoglossus Kowalevskii, and of the Morphology of the Enteropneusta." *Quarterly Journal of Microscopical Science,* 26 (1886): 511–533.

————. "The Ancestry of the Chordata." Reprinted from *Quarterly Journal of Microscopical Science,* (1886), in Bateson, *Scientific Papers,* I, 1–31.

―――. *Materials for the Study of Variation; Treated with Especial Regard to Discontinuity in the Origin of Species.* London: Macmillan, 1894. Reprinted with an introduction by Peter J. Bowler and Gerry Webster. Baltimore: Johns Hopkins University Press, 1992.

―――. *The Scientific Papers of William Bateson.* Ed. R. C. Punnett. Cambridge: Cambridge University Press, 1928. 2 vols.

Bather, Francis. Review of "The Perisomic Plates of the Crinoids" by C. Wachsmuth and F. Springer. *Geological Magazine,* 4 (1891): 219–224.

―――. "The Recapitulation Theory in Palaeontology." *Natural Science,* 2 (1893): 275–281.

―――. "Fossils and Life." *Report of the British Association for the Advancement of Science,* 1910 meeting: 61–86.

―――. Anniversary Address of the President: "Biological Classification: Past and Present." *Quarterly Journal of the Geological Society of London,* 83 (1927): lxii–civ.

Baur, Georg. "Note on the Pelvis in Birds and Dinosaurs." *American Naturalist,* 18 (1884): 1273–1275.

―――. "On the Phylogenetic Arrangement of the Sauropsida." *Journal of Morphology,* 1 (1887): 93–104.

―――. "Über die Kanäle im Humerus der Amnioten." *Morphologisches Jahrbuch,* 12 (1887): 299–305.

―――. "Ueber die Abstammung der amnioten Wirbelthiere." *Sitzungsberichte der Gessell für Morphologie und Physiologie im München,* 3 (1888): 46–61.

―――. "On the Origin of the Galápagos Islands." *American Naturalist,* 25 (1891): 217–229, 307–326.

―――. "The Stegocephali: A Phylogenetic Study." *Anatomischer Anzeiger,* 11 (1896): 657–673.

―――. "New Observations on the Origin of the Galápagos Islands, with Remarks on the Geological Age of the Pacific Ocean." *American Naturalist,* 31 (1897): 661–680, 864–890.

Baur, G., and E. C. Case. "On the Morphology of the Skull of the Pelycosauria and the Origin of the Mammals." *Science,* n.s., 5 (1897): 592–594.

Beard, John. "The Old Mouth and the New." *Nature,* 37 (1888): 224–227.

―――. "Some Annelidan Affinities in the Ontogeny of the Vertebrate Nervous System." *Nature,* 39 (1889): 259–261.

Beddall, Barbara G. *Wallace and Bates in the Tropics: An Introduction to the Theory of Natural Selection.* London: Macmillan, 1969.

Beddard, Frank E. *A Textbook of Zoogeography.* Cambridge: Cambridge University Press, 1895.

―――. "The Former Northward Extension of the Antarctic Continent." *Nature,* 53 (1895–96): 129.

Beebe, C. William. "A Tetrapteryx Stage in the Ancestry of Birds." *Zoologica: Scientific Contributions of the New York Zoological Society*, 2 (1915): 39–52.

Beecher, C. E. "Larval Forms of Trilobites from the Lower Heidelberg Group." *American Journal of Science*, ser. 3, 46 (1893): 142–147.

———. "The Morphology of Triarthus." *American Journal of Science*, ser. 4, 1 (1896): 251–256.

———. "Outline of a Natural Classification of Trilobites." *American Journal of Science*, ser. 4, 3 (1897): 89–106.

———. *Studies in Evolution: Mainly Reprints of Occasional Papers Selected from Publications of the Laboratory of Invertebrate Paleontology, Peabody Museum, Yale University*. New York: Scribners/London: Edward Arnold, 1901.

Beeson, Roberta Jane. *Bridging the Gap: The Problem of Vertebrate Ancestry, 1859–1875*. Oregon State University Ph.D. thesis, 1978.

Belloc, Hilaire. *Mr. Belloc Objects to "The Outline of History."* London: Watts, 1926.

Bendall, D. S., ed. *Evolution from Molecules to Men*. Cambridge: Cambridge University Press, 1983.

Bensley, B. Arthur. "On the Question of the Arboreal Ancestry of the Mammalia, and the Interrelationships of the Mammalian Subclasses." *American Naturalist*, 35 (1901): 117–138.

———. "A Theory of the Origin and Evolution of the Australian Marsupalia." *American Naturalist*, 35 (1901): 245–269.

———. "On the Evolution of the Australian Marsupalia: with Remarks on the Relationships of the Marsupials in General." *Transactions of the Linnean Society*, 9 (1903–1907): 83–217.

Benson, Keith R. "From Museum Research to Laboratory Research: The Transformation of Natural History into Academic Biology." In R. Rainger, K. R. Benson, and J. Maienschein, eds., *The American Development of Biology*. Philadelphia: University of Pennsylvania Press, 1988, pp. 49–83.

Benton, Michael J. "Scientific Methodologies in Collision: The History of the Study of the Extinction of the Dinosaurs." *Evolutionary Biology*, 24 (1990): 371–400.

Bernard, Henry Meyners. *The Apodidae: A Morphological Study*. London: Macmillan, 1892.

———. "The Comparative Morphology of the Galeodidae." *Transactions of the Linnean Society of London (Zoology)*. ser. 2, 6 (1894–97): 305–417.

———. "The Endosternite of *Scorpio* Compared with the Homologous Structures in Other Arachnida." *Annals and Magazine of Natural History*, ser. 6, 6 (1894): 18–26.

———. "The Systematic Position of the Trilobites." *Quarterly Journal of the Geological Society of London*, 50 (1894): 411–434.

――――. "Supplementary Notes on the Systematic Position of the Trilobites." *Quarterly Journal of the Geological Society of London*, 51 (1895): 352–360.

Bernard, H. M., et al. "Are the Arthropoda a Natural Group?" *Natural Science*, 10 (1897): 97–117, 264–268.

Berrill, N. J. *The Origin of Vertebrates*. Oxford: Clarendon Press, 1955.

Berry, Edward W. "Scientific Results of the Terra Nova Expedition." *Science*, n.s., 41 (1915): 830–831.

――――. "Cephalopod Adaptations—The Record and Its Interpretation." *Quarterly Review of Biology*, 3 (1928): 92–108.

Beurlen, Karl. "Funktion und Form in der organischen Entwicklung." *Naturwissenschaft*, 20 (1932): 73–80.

Blanford, W. T. "The African Element in the Fauna of India: A Criticism of Mr. Wallace's Views as Expressed in the 'Geographical Distribution of Animals.'" *Annals and Magazine of Natural History*, ser. 4, 18 (1876): 277–294.

――――. "Anniversary Address of the President." *Quarterly Journal of the Geological Society of London*, 46 (1890): 43–110.

Boore, Jeffrey L., Timothy M. Collins, David Stanton, L. Lynne Daehler, and Wesley Brown. "Deducing the Pattern of Arthropod Phylogeny from Mitochondrial DNA Rearrangements." *Nature* 376 (1995): 163–165.

Boulenger, G. A. "On Reptilian Remains from the Trias of Elgin." *Philosophical Transactions of the Royal Society*, 196 B (1903): 175–189.

Bowler, Peter J. *Fossils and Progress: Paleontology and the Idea of Progressive Evolution in the Nineteenth Century*. New York: Science History Publications, 1976.

――――. *The Eclipse of Darwinism: Anti-Darwinian Evolution Theories in the Decades around 1900*. Baltimore: Johns Hopkins University Press, 1983.

――――. "E. W. MacBride's Lamarckian Eugenics and Its Implications for the Social Construction of Scientific Knowledge." *Annals of Science*, 41 (1984): 245–260.

――――. *Theories of Human Evolution: A Century of Debate, 1844–1944*. Baltimore: Johns Hopkins University Press/Oxford: Basil Blackwell, 1986.

――――. *The Non-Darwinian Revolution: Reinterpreting a Historical Myth*. Baltimore: Johns Hopkins University Press, 1988.

――――. *Evolution: The History of an Idea*. 2d ed. Berkeley: University of California Press, 1989.

――――. "Holding Your Head Up High: Degeneration and Orthogenesis in Theories of Human Evolution." In James R. Moore, ed., *History, Humanity, and Evolution: Essays for John C. Greene*, pp. 329–353. Cambridge: Cambridge University Press, 1989.

――――. *The Mendelian Revolution: The Emergence of Hereditarian Concepts in Modern Science and Society*. London: Athlone/Baltimore: Johns Hopkins University Press, 1989.

————. "Development and Adaptation: Evolutionary Concepts in British Morphology, 1870–1914." *British Journal for the History of Science,* 22 (1989): 283–297.

————. *The Invention of Progress: The Victorians and the Past.* Oxford: Basil Blackwell, 1990.

————. "Darwinism and Modernism: Genetics, Paleontology, and the Challenge to Progressionism, 1880–1930." In Dorothy Ross, ed., *Modernist Impulses in the Human Sciences, 1870–1930,* pp. 236–254. Baltimore: Johns Hopkins University Press, 1994.

Brooks, William Keith. "Salpa in Its Relation to the Evolution of Life." *Studies from the Biological Laboratory, Johns Hopkins University,* 5 (1893), No. 3: 129–211.

————. "The Origin of the Oldest Fossils and the Discovery of the Bottom of the Ocean." *Journal of Geology,* 2 (1894): 455–479.

————. *The Foundations of Zoology.* 2d ed. New York: Columbia University Press, 1915.

Broom, Robert. "On the Development and Morphology of the Marsupial Shoulder Girdle." *Transactions of the Royal Society of Edinburgh,* 39 (1899): 749–770.

————. "On the Early Condition of the Shoulder Girdle in the Polyprotodont Marsupials *Dasyurus* and *Perameles.*" *Journal of the Linnean Society (Zoology),* 28 (1902): 449–454.

————. "A Comparison of the Permian Reptiles of North America with Those of South Africa." *Bulletin of the American Museum of Natural History,* 28 (1910): 197–234.

————. "On the Structure of the Internal Ear and the Relations of the Basicranial Nerves in *Dicynodon,* and on the Homology of the Mammalian Auditory Ossicles." *Proceedings of the Zoological Society of London,* 1912: 419–425.

————. "On the Origin of the Cheiropterygium." *Bulletin of the American Museum of Natural History,* 32 (1913): 265–272.

————. "On the South-African Pseudosuchian *Euparkeria* and Allied Genera." *Proceedings of the Zoological Society of London,* 1913: 619–633.

————. "On the Structure and Affinities of the Multituberculata." *Bulletin of the American Museum of Natural History,* 33 (1914): 115–134.

————. Croonian Lecture: "On the Origin of Mammals." *Philosophical Transactions of the Royal Society,* 206 B (1915): 1–48.

————. "On the Classification of the Reptiles." *Bulletin of the American Museum of Natural History,* 51 (1924–25): 39–65.

————. *The Mammal-Like Reptiles of South Africa and the Origin of Mammals.* London: H. F. & G. Witherby, 1932.

————. *The Coming of Man: Was It Accident or Design?* London: H. F. & G. Witherby, 1933.

————. *Finding the Missing Link*. London: Watts, 1950.

Browne, Janet. "Darwin's Botanical Arithmetic and the 'Principle of Divergence,' 1854–1858." *Journal of the History of Biology*, 13 (1980): 53–89.

————. *The Secular Ark: Studies in the History of Biogeography*. New Haven: Yale University Press, 1983.

Buffetaut, Eric. *A Short History of Vertebrate Paleontology*. Beckenham: Croom Helm, 1986.

Burchfield, Joe D. "Darwin and the Dilemma of Geological Time." *Isis*, 65 (1974): 301–321.

————. *Lord Kelvin and the Age of the Earth*, New York: Science History Publications, 1975.

Burian, R. M., J. Gayon, and D. Zallen. "The Singular Fate of Genetics in the History of French Biology." *Journal of the History of Biology*, 21 (1988): 357–402.

Cain, A. J. "On Homology and Convergence." In K. A. Joysey and A. E. Friday, eds., *Problems of Phylogenetic Reconstruction*. Systematics Association Special Volume 21. New York: Academic Press, 1982, pp. 1–19.

Caldwell, W. H. "The Embryology of the Monotremata." *Philosophical Transactions of the Royal Society*, 178 B (1887): 463–486.

Calman, W. T. *Crustacea*. In E. Ray Lankester, ed., *A Treatise on Zoology*, vol. 7, 3d fascicule. London: A. & C. Black, 1909.

————. "Ernest William MacBride." *Obituary Notices of the Fellows of the Royal Society of London*, 3 (1940): 747–759.

Camerini, Jane R. "Evolution, Biogeography, and Maps: An Early History of Wallace's Line." *Isis*, 84 (1993): 700–727.

Cannon, H. G. "On the Feeding Mechanism of the Branchiopoda." *Philosophical Transactions of the Royal Society*, 222 B (1933): 267–352.

————. *Lamarck and Modern Genetics*. Manchester: Manchester University Press, 1959.

Cannon, H. G., and S. M. Manton. "On the Feeding Mechanism of a Mysid Crustacean, Hemimysis lamornae." *Transactions of the Royal Society of Edinburgh*, 55 (1927): 219–253.

Cardwell, D. S. L. *The Organization of Science in England*. Rev. ed. London: Heinemann, 1972.

Caron, Joseph A. "'Biology' in the Life Sciences: A Historiographical Reappraisal." *History of Science*, 26 (1988): 223–268.

Carpenter, George H. "The Problem of the British Fauna." *Natural Science*, 11 (1897): 375–386.

————. "On the Relations of the Classes of Arthropoda." *Proceedings of the Royal Irish Academy*, 24 (1902–1904), section B, pp. 320–360.

Case, E. C. "The Development and Geological Relations of the Vertebrates." *Journal of Geology,* 6 (1898): 393–416, 622–646, 816–839.

Chamberlin, Thomas C. "A Systematic Source of the Evolution of Provincial Faunas." *Journal of Geology,* 6 (1898): 597–608.

———. "The Influence of Great Epochs of Limestone Formation upon the Constitution of the Atmosphere." *Journal of Geology,* 6 (1898): 609–621.

———. "On the Habitat of the Early Vertebrates." *Journal of Geology,* 8 (1900): 400–412.

Chamberlin, Thomas C., and Rollin D. Salisbury. *Geology: Earth History.* London: John Murray, 1908. 3 vols.

Chambers, Robert. *Vestiges of the Natural History of Creation and Other Evolutionary Writings.* Introduction by James Secord. Chicago: University of Chicago Press, 1994

Claus, Carl. *Untersuchungen zur Erforschung der genealogischen Grundlage des Crustaceen-Systems.* Vienna: Carl Gerolds Sohn, 1876.

———. *Elementary Textbook of Zoology.* Adam Sedgwick, trans. London: W. Swan Sonnenschein, 1884. 2 vols.

———. "Neue Beiträge zur Morphologie der Crustaceen." *Arbeiten der Zoologischen Institut, Wein,* 6 (1886): 1–108.

———. "On the Heart of *Ganasidae* and Its Significance in the Phylogenetic Consideration of the Acaridae and Arachnoidea, and the Classification of the Arthropoda." *Annals and Magazine of Natural History,* ser. 5, 17 (1886): 168–170.

———. "Professor E. Ray Lankester's Memoir 'Limulus an Arachnid' and the Pretensions and Charges Founded upon It." *Annals and Magazine of Natural History,* ser. 5, 18 (1886): 55–65.

———. "Reply to Professor E. Ray Lankester's 'Rejoinder.' " *Annals and Magazine of Natural History,* ser. 5, 18 (1886): 467–470.

Cohen, I. Bernard. "Three Notes on the Reception of Darwin's Ideas on Natural Selection (Henry Baker Tristram, Alfred Newton, Samuel Wilberforce)." In D. Kohn, ed., *The Darwinian Heritage,* pp. 589–607. Princeton: Princeton University Press, 1985.

Colbert, Edwin H. *Men and Dinosaurs: The Search in Field and Laboratory.* Harmondsworth: Penguin Books, 1971.

———. *Evolution of the Vertebrates.* 3d ed. New York: Wiley, 1980.

———. *William Diller Matthew, Paleontologist: The Splendid Drama Observed.* New York: Columbia University Press, 1992.

Coleman, William, *Georges Cuvier: Zoologist.* Cambridge, Mass.: Harvard University Press, 1964.

———. *Biology in the Nineteenth Century: Problems of Form, Function, and Transformation.* New York: Wiley, 1971.

————. "Morphology between Type Concept and Descent Theory." *Journal of the History of Medicine,* 31 (1976): 149–175.

————, ed. *The Interpretation of Animal Form.* New York: Johnson Reprint Corporation, 1967.

Conry, Yvette. *Correspondence entre Charles Darwin et Gaston de Saporta.* Paris: Presses Universitaires de France, 1972.

Cope, Edward Drinker. "An Account of the Extinct Reptiles which Approached the Birds." *Proceedings of the Academy of Natural Sciences, Philadelphia,* 1867: 234–235.

————. "Second Contribution to the History of the Vertebrates of the Permian Formation in Texas." *Proceedings of the American Philosophical Society,* 19 (1882): 38–58.

————. "The Relations between the Theromorphous Reptiles and the Monotreme Mammalia." *Proceedings of the American Association for the Advancement of Science,* 33 (1884): 471–482.

————. "On the Trituberculate Type of Molar Tooth in the Mammalia." *Proceedings of the American Philosophical Society,* 21 (1884): 324–326.

————. "Synopsis of the Species of Oreodontidae." *Proceedings of the American Philosophical Society,* 21 (1884): 503–572.

————. "Fifth Contribution to the Knowledge of the Fauna of the Permian Formation of Texas and the Indian Territory." *Proceedings of the American Philosophical Society,* 22 (1884): 28–47.

————. "The Batrachia of the Permian Period of North America." *American Naturalist,* 18 (1884): 26–38.

———— "Position of Pterichthys in the System." *American Naturalist,* 19 (1885): 289–291.

————. *The Origin of the Fittest: Essays in Evolution.* New York: Macmillan, 1887.

————. "The Mechanical Causes of the Development of the Hard Parts of the Mammalia." *Journal of Morphology,* 3 (1889): 137–290.

————. "On the Phylogeny of the Vertebrata." *Proceedings of the American Philosophical Society,* 30 (1892): 278–281.

————. "On the Homologies of the Posterior Cranial Arches in the Reptilia." *Transactions of the American Philosophical Society,* 17 (1893): 11–26.

————. *The Primary Factors of Organic Evolution.* Chicago: Open Court, 1896.

Craw, Robin. "'Conservative Prejudice' in the Debate over Disjunctively Distributed Life Forms." *Studies in the History and Philosophy of Science,* 15 (1984): 131–140.

————. "Margins of Cladistics: Identity, Difference, and Place in the Emergence of Phylogenetic Systematics, 1864–1975." In P. Griffiths, ed., *Trees of Life: Essays on the Philosophy of Biology,* pp. 65–107. Dordrecht: Kluwer Academic Publishers, 1992.

Croizat, Leon. *Panbiogeography*. Caracas: The author, 1958. 3 vols.

————. *Space, Time, Form: The Biological Synthesis*. Caracas: The author, 1964.

Croll, James. *Climate and Time in Their Geological Relations*. London: Daldy, Ibister, 1875.

Dames, Wilhelm. "Ueber Archaeopteryx." In W. Dames and E. Kayser, eds., *Palaeontologische Abhandlungen*. Vol. 2, part 3. Berlin: G. Reimer, 1884, pp. 119–196 (also separate pagination as 1–79).

Darwin, Charles Robert. *On the Origin of Species by Means of Natural Selection; or the Preservation of Favoured Races in the Struggle for Life*. London: John Murray, 1859. Reprint, ed. Ernst Mayr. Cambridge, Mass.: Harvard University Press, 1964.

————. *The Descent of Man and Selection in Relation to Sex*. 2d ed. London: John Murray, 1874.

————. *The Life and Letters of Charles Darwin*. Ed. Francis Darwin. London: John Murray, 1887. 3 vols.

————. *More Letters of Charles Darwin*. Ed. Francis Darwin. New York: Appleton, 1903. 2 vols.

————. *Charles Darwin's Notebooks, 1836–1844*. Ed P. H. Barrett et al. London: British Museum (Natural History)/Cambridge: Cambridge University Press, 1987.

Davidoff, M. von. "Beiträge zur vergleichenden Anatomie der hinteren Gliedmassen der Fische." *Morphologisches Jahrbuch*, 5 (1879): 450–520.

Dawkins, W. Boyd. *Early Man in Britain and His Place in the Tertiary Period*. London: Macmillan, 1880.

Dean, Bashford. *Fishes, Living and Fossil: An Outline of their Forms and Probable Relationships*. New York: Macmillan, 1895.

————. "Sharks as Ancestral Fishes." *Natural Science*, 8 (1896): 245–253.

De Beer, Gavin. *Embryology and Evolution*. Oxford: Clarendon Press, 1930.

Delsman, Hendricus C. *The Ancestry of Vertebrates*. Weltervreden, Java: Visser & Co., and Amersfoort, Holland: Valkhoff, 1922.

Dendy, Arthur, ed. *Animal Life and Human Progress*. London: Constable, 1919.

————. *Outlines of Evolutionary Biology*. London: Constable, 1919.

Depéret, Charles. *The Transformations of the Animal World*. London: Kegan Paul, Trench, Trubner, 1909.

Desmond, Adrian. *The Hot-Blooded Dinosaurs: A Revolution in Palaeontology*. London, 1975, reprinted London: Futura, 1977.

————. *Archetypes and Ancestors: Palaeontology in Victorian London, 1850–1875*. London: Blond and Briggs, 1982.

————. *The Politics of Evolution: Morphology, Medicine, and Reform in Radical London*. Chicago: University of Chicago Press, 1982.

Di Gregorio, Mario. *T. H. Huxley's Place in Natural Science*. New Haven: Yale University Press, 1984.

———. "A Wolf in Sheep's Clothing: Carl Gegenbaur, Ernst Haeckel, the Vertebrate Theory of the Skull, and the Survival of Richard Owen." *Journal of the History of Biology* 28 (1955): 247–280.

Dixon, George. "A New Law of Geographical Dispersal Based on Ideas Developed in the Study of the Migration of Birds." *Fortnightly Review*, n.s., 57 (1895): 640–657.

Dohrn, Anton. "On Eugereon Boeckingi and the Genealogy of Insects." *Annals and Magazine of Natural History*, ser. 4, 1 (1868): 448–455.

———. "Die Ueberreste des Zoëa-Stadiums in der ontogenetischen Entwickelung der verschiedenen Crustaceen-Familien." *Jenaische Zeitschrift für Medicin und Naturwissenschaft*, 5 (1870): 54–81.

———. *Untersuchungen über Bau und Entwickelung der Arthropoden*. Leipzig: Wilhelm Engelmann, 1870.

———. "Geschichte des Krebsstammes, nach embryologischen, anatomischen, und palaeontologischen Quellen. Ein Versuch" *Jenaische Zeitschrift für Medicin und Naturwissenschaft*, 6 (1871): 96–156.

———. *Der Ursprung der Wirbelthiere und das Princip des Functionswechsels: Genealogische Skizzen*. Leipzig: Wilhelm Englemann, 1875.

———. *Die Pantopoden des Golfes von Neapel und der angrezenden Meeres-Abschnitte. Flora und Fauna des Golfes von Neapel*, vol. 3. Leipzig: Wilhelm Engelmann, 1881.

———. "Studien zur Urgeschichte der Wirbelthierkörpers." I. "Der Mund der Knockenfische." *Mitheilungen aus der Zoologische Stazion zu Neapel*, 3 (1881–82): 252–279.

———. "Studien . . ." VI. "Die paarigen und unpaaren Flossen der Selachier." *Mitheilungen aus der Zoologische Stazion zu Neapel*, 5 (1884): 161–195.

———. "Studien . . ." VIII. "Die Thyreoidea bei *Petromyzon, Amphioxus* und der Tunicaten." *Mitheilungen aus der Zoologische Stazion zu Neapel*, 6 (1886–86): 49–92.

———. "Studien . . ." XII. "Thyroidea und Hypobranchialrinne, Spritzlochsack und Pseudo-branchialrinne bei Fischen, *Ammocoetes* und Tunicata." *Mitheilungen aus der Zoologische Stazion zu Neapel*, 7 (1887): 301–337.

———. "Studien . . ." XXI. "Theoretisches über Occipitalsomite und Vagus: Competenzkonflikt zwischen Ontogenie und vergleichender Anatomie." *Mitheilungen aus der Zoologische Stazion zu Neapel*, 15 (1901–2): 186–279.

———. "The Origin of Vertebrates and the Principle of Succession of Functions: Genealogical Sketches." Translated and introduced by Michael T. Ghiselin. *History and Philosophy of the Life Sciences*, 16 (1993): 1–98.

Dollo, Louis. "Première note sur les dinosauriens de Bernissart." *Bulletin du Musée Royal de l'Histoire Naturelle de Belgique*, 1 (1882): 161–178.

————. "Les Lois d'évolution." *Bulletin de la Société Belgique de Géologie, de Paléontologie et d'Hydrogéologie*, 7 (1893): 164–166.

————. "Sur la phylogenie des Dipneustes." *Bulletin de la Société Belgique de Géologie, de Paléontologie et d'Hydrogéologie*, 9 (1895): 79–128.

————. "Les ancêstres des Marsupiaux étaient-ils arboricoles?" *Traveaux de L'Institut Zoologique de Lille*, 7 (1899): 199–203.

————. *Resultats du Voyage du S. Y. Belgica en 1897–1898–1899. Rapports Scientifiques. Zoologie: Poissons.* Anvers: J. E. Buschmann, 1904.

————. "Patagonia and the Pampas Cenozoic of South America: A Critical Review of the Correlations of Santiago Roth." *Annals of the New York Academy of Sciences*, 19 (1910): 149–160.

Doyle, Arthur Conan. *The Lost World.* Reprinted, London: John Murray, 1960.

Dunbar, Carl Owen. "Phases of Cephalopod Adaptation." In M. R. Thorpe, ed., *Organic Adaptation to the Environment*, pp. 188–223. New Haven: Yale University Press, 1924.

Duncan, P. Martin. Anniversary Address of the President. *Quarterly Journal of the Geological Society of London*, 33 (1877): 41–88.

Egerton, Frank N. "Refutation and Conjectures: Darwin's Response to Sedgwick's Attack on Chambers." *Studies in History and Philosophy of Science*, 1 (1970): 176–183.

Eigenmann, Carl. "The American Characidae." *Memoirs of the Museum of Comparative Zoology*, 43 (1917–27).

Eiseley, Loren. *Darwin's Century: Evolution and the Men who Discovered it.* New York: Doubleday, 1958.

Eisig, Hugo. *Monographie der Capitelliden der Golfes von Neapel. Flora und Fauna des Golfes von Neapel*, vol. 16. Berlin: R. Friedlander, 1887.

Eldredge, Niles, and Steven M. Stanley, eds. *Living Fossils.* New York and Berlin: Springer-Verlag, 1984.

Ellegard, Alvar. *Darwin and the General Reader: The Reception of Darwin's Theory of Evolution in the British Periodical Press, 1859–1872.* Göteburg: Acta Universitatis Gothenburgenis, 1958.

Falconer, Hugh. *Palaeontological Memoirs and Notes of the Late Hugh Falconer.* Ed. Charles Murchison. London: Robert E. Hardwicke, 1868. 2 vols.

Farley, John. *The Spontaneous Generation Controversy: From Descartes to Oparin.* Baltimore: Johns Hopkins University Press, 1974.

Fernald, H. T. "The Relationship of Arthropods." *Studies from the Biological Laboratory of Johns Hopkins University*, 4 (1887–90): 431–513.

Fichman, Martin. "Wallace, Zoogeography, and the Problem of Land Bridges." *Journal of the History of Biology*, 10 (1977): 45–63.

————. *Alfred Russel Wallace.* New York: Twayne, 1981.

Findlay, G. H. *Dr. Robert Broom, F. R. S.: Palaeontologist and Physician, 1866–1951*. Cape Town: A. A. Balkema, 1972.

Flower, William Henry. *Essays on Museums and Other Subjects Connected with Natural History*. London: Macmillan, 1898.

Flower, W. H., and Richard Lydekker. *An Introduction to the Study of Mammals, Living and Extinct*. London: A. & C. Black, 1891.

Flynn, T. Thomson. "The Phylogenetic Significance of the Marsupial Allanto-placenta." *Proceedings of the Linnean Society of New South Wales*, 47 (1922): 541–544.

Forbes, Henry O. "Antarctica: A Vanished Austral Land." *Fortnightly Review*, n.s., 55 (1894): 194–214.

Forrest, H. Edward. *The Atlantean Continent: Its Bearing upon the Great Ice Age and the Distribution of Species*. 2d ed. London: Witherby, 1935.

Frankel, Henry. "The Paleobiogeographical Debate over the Problem of Disjunctively Distributed Life Forms." *Studies in the History and Philosophy of Science*, 12 (1981): 211–259.

———. "Biogeography before and after the Rise of Sea-Floor Spreading." *Studies in the History and Philosophy of Science*, 15 (1984): 141–168.

Friedmann, Hermann. *Die Konvergenz der Organismen: Eine empirische begründete Theorie als Ersatz für die Abstammungslehre*. Berlin: Gebrüder Patel, 1904.

Fürbringer, Max. *Untersuchungen zur Morphologie und Systematik der Vögel*. Amsterdam: T. van Holkema, 1888.

———. "Ueber die spino-occipitalen Nerven der Selachier und Holocephalen und ihre vergleichenden Morphologie." In *Festchrift zur siebenzigsten Geburtstag von Carl Gegenbaur*. Leipzig: Wilhelm Engelmann, 1896–97, vol. III, pp. 351–788. 3 vols.

———. "Morphologisches Streitfragen." *Morphologisches Jarhbuch*, 30 (1902): 85–274.

Gadow, Hans. "The Morphology of Birds." *Nature*, 39 (1888): 150–152, 172–181.

———. "The Evolution of the Auditory Ossicles." *Anatomischer Anzeiger*, 19 (1901): 396–411.

———. *The Wanderings of Animals*. Cambridge: Cambridge University Press, 1913.

Gardiner, Brian G. "Tetrapod Ancestry: A Reappraisal." In A. L. Panchen, ed. *The Terrestrial Environment and the Origin of Land Vertebrates*, pp. 177–185. London: Academic Press, 1980.

Garstang, Walter. "Preliminary Note on a New Theory of the Phylogeny of the Chordata." *Zoologisches Anzeiger*, 17 (1894): 122–125.

———. "The Theory of Recapitulation: A Critical Restatement of the Biogenetic Law." *Journal of the Linnean Society (Zoology)*, 35 (1922): 81–101.

————. "The Morphology of the Tunicata, and Its Bearings on the Phylogeny of the Chordata." *Quarterly Journal of Microscopical Science,* 72 (1928): 51–187.

————. *Larval Forms and Other Zoological Verses.* Oxford: Basil Blackwell, 1951.

Gaskell, Walter Holbrook. "On the Relation between the Structure, Function, Distribution, and Origin of the Cranial Nerves, Together with a Theory of the Origin of the Nervous System of the Vertebrates." *Journal of Physiology,* 10 (1889): 153–211.

————. "On the Origin of Vertebrates from a Crustacean-like Ancestor." *Quarterly Journal of Microscopical Science,* n.s., 31 (1890): 379–444.

————. "President's Address, Physiology Section." *Report of the British Association for the Advancement of Science,* 1896 meeting: 942–972.

————. "On the Origin of Vertebrates, Deduced from the Study of Ammocoetes." *Journal of Anatomy and Physiology,* 32 (1898): 513–581.

————. *The Origin of Vertebrates.* London: Longmans, Green, 1908.

————, et al. "Discussion 'Origin of the Vertebrates.' " *Proceedings of the Linnean Society of London,* 122nd session (1909–10): 9–50.

Gaudry, Albert. *Animaux fossiles et géologie de l'Attique d'après les recherches faites en 1855–56 et 1860 sous les auspices de l'Académie des Sciences.* Paris, 1862–67.

————. *Essai de paléontologie philosophique.* Paris: Masson, 1896.

Gaupp, Ernst. "Das Lakrimale des Menschen und der Säuger und seine morphologische Bedeutung. *Anatomischer Anzeiger,* 36 (1910): 529–555.

————. "Die Verwandtschaftsbeziehungen der Säuger, vom Standpunkte der Schädelmorpholgie aus erörtert." *Verhandlung der VIII International Zoolgie Kongress zu Graz, 1910,* pp. 215–240.

————. "Beiträge zur Kenntnis des Unterkiefers der Wirbeltiere. I. Der Processus anterior (Folii) des Hammers der Säuger und das Goniale der Nichtsäuger." *Anatomischer Anzeiger,* 39 (1911): 97–135.

————. "Beiträge . . . II. Die Zusammensetzung des Unterkiefers der Quadrupeden." *Anatomischer Anzeiger,* 39 (1911): 433–473.

————. "Beiträge . . . III. Das Probleme der Entstehung eines 'sekundären' Kiefergelenkes bei den Säugern." *Anatomischer Anzeiger,* 39 (1911): 609–666.

Gegenbaur, Carl. *Untersuchungen zur vergleichenden Anatomie der Wirbelthiere.* Leipzig: Wilhelm Engelmann, 1865–72. 3 vols.

————. *Grundzüge der vergleichenden Anatomie.* 2d ed. Leipzig: Wilhelm Engelmann, 1870.

————. "Ueber das Skelet der Gliedmassen der Wirbelthiere im Allgemeinen und der Hintergliedmassen der Selachier inbesondere." *Jenaische Zeitschrift für Medicin und Naturwissenschaft,* 5 (1871): 397–447.

———. "Ueber das Archipterygium." *Jenaische Zeitschrift für Medicin und Naturwissenschaft*, 7 (1872): 131–134.

———. *Grundriss der vergleichenden Anatomie*. Leipzig: Wilhelm Englemann, 1874; 2d ed., 1878.

———. "Zur Morphologie der Gliedmaassen der Wirbelthiere." *Morphologisches Jahrbuch*, 2 (1876): 396–420.

———. *Elements of Comparative Anatomy*. Trans. F. Jeffrey Bell. London: Macmillan, 1878.

———. "Die Stellung und Bedeutung der Morphologie." *Morphologisches Jahrbuch*, 1 (1879): 1–19.

———. "Zur Gliedmassenfrage: An die Untersuchungen von Davidoff's angeknüpfte Bemerkungen." *Morphologisches Jahrbuch*, 5 (1879): 521–525.

———. "Das Flossenskelet der Crossopterygier und das Archipterygium." *Morphologisches Jahrbuch*, 22 (1895): 119–160.

———. *Vergleichenden Anatomie der Wirbelthiere*. Leipzig: Wilhelm Engelmann, 1898. 2 vols.

Geikie, Archibald. *Memoir of Sir Andrew Crombie Ramsay*. London: Macmillan, 1895.

Geison, Gerald. *Michael Foster and the Cambridge School of Physiology*. Princeton: Princeton University Press, 1978.

George, Wilma. *Biologist-Philosopher: A Study of the Life and Writings of Alfred Russel Wallace*. New York: Abelard-Schumann, 1964.

Gemmill, James F. "The Development of Certain Points in the Adult Structure of the Starfish *Asterias rubens, L.*" *Philosophical Transactions of the Royal Society*, 205 B (1914): 213–294.

Ghiselin, Michael T. *The Triumph of the Darwinian Method*. Berkeley: University of California Press, 1969.

———. "The Origin of Molluscs in the Light of Molecular Evidence." In P. H. Harvey and L. Partridge, eds., *Oxford Surveys in Evolutionary Biology*, 5 (1988): 66–95.

———. "Classical and Molecular Phylogenetics." *Bolletino Zoologica*, 58 (1991): 289–294.

Giard, Alfred, "Étude critique des traveaux d'embryogénie relatifs a la parenté des vertébrés et des tuniciers." *Archives des zoologie experimentale et générale*, 1 (1872): 233–288.

———. "Deuxième étude critique des traveaux d'embryogénie relatifs a la parenté des vertébrés et des tuniciers." *Archives des zoologie experimentale et générale*, 1 (1872): 397–428.

Gill, Theodore. "Arrangement of the Families of Fishes, or Classes Pisces, Marsipobranchii, and Leptocardii." *Smithsonian Miscellaneous Collections*, no. 247, 5 (1872): xlvi + 45 pp.

————. "On the Geographical Distribution of Fishes." *Annals and Magazine of Natural History,* ser. 4, 15 (1875): 251–255.

————. "Fishes, Living and Fossil." *Science,* n.s., 3 (1896): 909–917.

Gilmour, J. S. L. "Taxonomy and Philosophy." In Julian Huxley, ed., *The New Systematics,* pp. 461–470. Oxford: Oxford University Press, 1940.

Gislen, Torsten. "Affinities between the Echinodermata, Enteropneusta, and Chordonia." *Zoologiska Bidrag, Uppsala,* 12 (1930): 199–304.

Glen, William, ed. *The Mass Extinction Debates: How Science Works in a Crisis.* Stanford: Stanford University Press/Cambridge: Cambridge University Press, 1994.

Goode, George Brown, "The Principles of Museum Administration." *Annual Report of the Board of Regents of the Smithsonian Insitution, Report of the U.S. National Museum.* Part 2. Washington, D.C.: U.S. Government Printing Office, 1895–1901, pp. 193–240.

Goodrich, Edwin S. "On the Fossil Mammalia from the Stonesfield Slate." *Quarterly Journal of Microscopical Science,* 35 (1894): 407–432.

————. "On the Coelom, Genital Ducts, and Nephridia." *Quarterly Journal of Microscopical Science,* 37 (1895): 477–510.

————. "On the Pelvic Girdle and Fin of Eusthenopteron." *Quarterly Journal of Microscopical Science,* 45 (1901): 311–324.

————. "Notes on the Development, Structure, and Origin of the Median and Paired Fins of Fish." *Quarterly Journal of Microscopical Science,* 50 (1906): 333–376.

————. "On the Systematic Position of Polypterus." *Report of the British Association for the Advancement of Science,* 1907 meeting: 545–546.

————. *Vertebrata Craniata (Cyclostomes and Fishes).* Part 9 of E. Ray Lankester, ed., *A Treatise on Zoology.* London: A. & C. Black, 1909.

————. "Metameric Segmentation and Homology." *Quarterly Journal of Microscopical Science,* 59 (1913): 227–248.

————. "On the Classification of the Reptilia." *Proceedings of the Royal Society,* 89 B (1917): 261–276.

————. "On the Development of the Segments of the Head in *Scyllium.*" *Quarterly Journal of Microscopical Science,* 63 (1918): 1–30.

————. *Living Organisms: An Account of their Origin and Evolution.* Oxford: Clarendon Press, 1924.

————. "The Origin of Land Vertebrates." *Nature,* 114 (1924): 935–936.

————. "Polypterus a Palaeoniscid?" *Palaeobiologica,* 1 (1927): 87–92.

————. *Studies on the Structure and Development of Vertebrates.* London: Macmillan, 1930. Reprint introd. Keith S. Thomson. Chicago: University of Chicago Press, 1986.

———. "Syndactyly in Marsupials." *Proceedings of the Zoological Society of London*, 1935: 175–178.

Gould, Stephen Jay. "Dollo on Dollo's Law: Irreversibility and the Status of Evolutionary Laws." *Journal of the History of Biology*, 1 (1970): 189–212.

———. *Ontogeny and Phylogeny*. Cambridge, Mass.: Harvard University Press, 1977.

———. *The Panda's Thumb: More Reflections on Natural History*. New York: Norton, 1980.

———. "Irrelevance, Submission, and Partnership: The Changing Role of Paleontology in Darwin's Three Centennials, and a Modest Proposal for Macroevolution." In D. S. Bendall, ed., *Evolution from Molecules to Men*, pp. 347–366. Cambridge: Cambridge University Press, 1983.

———. "The Fossil Fraud That Never Was." *New Scientist*, 12 March 1987: 32–36.

———. *Wonderful Life: The Burgess Shale and the Nature of History*. London: Hutchinson Radius, 1989.

———. *Eight Little Piggies*. London: Jonathan Cape, 1993.

———, ed. *The Book of Life*. London: Ebury-Hutchinson, 1993.

Gray, Asa. *Darwiniana: Essays and Reviews Pertaining to Darwinism*. New York: Appleton, 1876.

Grayson, Donald K. "Vicissitudes and Overkill: The Development of Explanations of Pleistocene Extinctions." *Advances in Archaeological Method and Theory*, 3 (1980): 357–403.

Greene, John C. *The Death of Adam: Evolution and Its Impact on Western Thought*. Ames: Iowa State University Press, 1959.

Gregory, John Walter. "Contributions to the Palaeontology and Physical Geology of the West Indies." *Quarterly Journal of the Geological Society of London*, 51 (1895): 255–312.

———. Anniversary Address of the President: "The Geographical History of the Atlantic Ocean." *Quarterly Journal of the Geological Society of London*, 85 (1929): lvi–cxxii.

Gregory, William King. "Application of the Quadrate-Incus Theory to the Conditions in Theriodont Reptiles and the General Relations of the Latter to the Mammalia." *Science*, n.s., 31 (1910): 600.

———. "The Orders of Mammals." *Bulletin of the American Museum of Natural History*, 27 (1910): 1–524.

———. "Critique of Recent Work on the Morphology of the Vertebrate Skull, Especially in Relation to the Origin of Mammals." *Journal of Morphology*, 24 (1913): 1–42.

———. "Present Status of the Problem of the Origin of the Tetrapoda, with Special Reference to the Skull and Paired Limbs." *Annals of the New York Academy of Sciences*, 26 (1915): 317–383.

————. "Theories of the Origin of Birds." *Annals of the New York Academy of Sciences*, 27 (1916): 31–38.

————. *Our Face from Fish to Man: A Portrait Gallery of our Ancient Ancestors and Kinsfolk, Together with a Concise History of Our Best Features.* New York: G. P. Putnam's Sons, 1929.

————. "Further Observations on the Pectoral Girdle and Fin of Sauripterus Taylori, Hall, A Crossopterygian Fish from the Upper Devonian of Pennsylvania, with Special Reference to the Origin of the Pentadactylate Extremities of Tetrapods." *Proceedings of the American Philosophical Society*, 75 (1935): 673–690.

————. "Biographical Memoir of Henry Fairfield Osborn." *Biographical Memoirs of the National Academy of Sciences*, 19 (1938): 53–119.

Gregory, W. K., and L. A. Adams. "The Temporal Fossae of Vertebrates in Relation to the Jaw Muscles." *Science*, n.s., 41 (1915): 763–765.

Gregory, W. K., and C. L. Camp. "Studies in Comparative Myology and Osteology, No. 3." *Bulletin of the American Museum of Natural History*, 38 (1918): 447–563.

Gregory, W. K., and Henry C. Raven. "Studies on the Origin and Early Evolution of Paired Fins and Limbs." *Annals of the New York Academy of Sciences*, 42 (1941): 273–360.

Griesemer, James R., and Elihu M. Gerson. "Collaboration at the Museum of Vertebrate Zoology." *Journal of the History of Biology*, 26 (1993): 185–203.

Griffiths, P., ed. *Trees of Life: Essays on the Philosophy of Biology.* Dordrecht: Kluwer Academic Publishers, 1992.

Grinnell, Joseph. "The Uses and Methods of a Research Museum." *Popular Science Monthly*, 77 (1910): 163–169.

Grobben, Karl. "Die Systematische Einteilung des Thierreiches." *Verhandlung der Kaiserlich-Königlichen Zoologischen-Botanischen Gesselschaft in Wein*, 58 (1908): 491–511.

Groeben, Christiane, ed. *Charles Darwin, 1809–1882, Anton Dohrn, 1840–1909. Correspondence.* Naples: Macchiaroli, 1982.

Groeben, Christiane, et al. "The Naples Zoological Station and the Marine Biological Laboratory: One Hundred Years of Biology." *Biological Bulletin*, 168 (1985): supplementary issue.

Gruber, Jacob W. "From Myth to Reality: The Case of the Moa." *Archives of Natural History*, 14 (1987): 339–352.

————. "Does the Platypus Lay Eggs? The History of an Event in Science." *Archives of Natural History*, 18 (1991): 51–123.

Gunther, Albert. "Description of *Ceratodus*, a Genus of Ganoid Fishes, Recently Discovered in Rivers of Queensland, Australia." *Philosophical Transactions of the Royal Society*, 161 (1871): 511–571.

———. *An Introduction to the Study of Fishes.* Edinburgh: A. & C. Black, 1880.

Gunther, Albert E. *A Century of Zoology at the British Museum through the Lives of Two Keepers, 1815–1914.* London: Dawson, 1975.

Guppy, H. B. "Plant Distribution from the Standpoint of an Idealist." *Journal of the Linnean Society of London (Botany),* 44 (1917–20): 439–472.

Gupta, A. P., ed. *Arthropod Phylogeny.* New York: Van Nostrand Reinhold, 1979.

Haacke, Wilhelm. "Der Nordpol als Schöpfungscentrum der Landfauna." *Biologisches Centralblatt,* 6 (1886–87): 363–370.

Haeckel, Ernst. *Generelle Morphologie der Organismen.* 2 vols. Berlin: Georg Reimer, 1866. Reprinted, Berlin: Walter de Gruyter, 1988.

———. *Natürliche Schöpfungsgeschichte.* Berlin: Georg Reimer, 1868.

———. "The Calcispongiae: Their Position in the Animal Kingdom, and Their Relation to the Theory of Descendence." *Annals and Magazine of Natural History,* ser. 4, 11 (1873): 241–262, 421–430.

———. *Anthropogenie oder Entwickelungsgeschichte des Menschen.* Leipzig: Wilhelm Englemann, 1874.

———. *The History of Creation: Or the Development of the Earth and Its Inhabitants by the Action of Natural Causes.* New York: Appleton, 1876. 2 vols.

———. *The Evolution of Man: A Popular Exposition of the Principal Points of Human Ontogeny and Phylogeny.* New York: Appleton, 1879. 2 vols.

Haldane, J. B. S. *The Causes of Evolution.* London: Longmans, 1932.

Haldane, J. B. S., and Julian Huxley. *Animal Biology.* Oxford: Clarendon Press, 1927.

Hamlin, Christopher. "James Geikie, James Croll, and the Eventful Ice Age." *Annals of Science,* 39 (1982): 565–583.

Hansen, H. C. "A Contribution to the Morphology of the Limbs and Mouth Parts of Crustaceans and Insects." *Annals and Magazine of Natural History,* ser. 6, 12 (1893): 417–434.

Hardy, A. C. "Edwin Stephen Goodrich, 1868–1946." *Quarterly Journal of Microscopical Science,* 87 (1946): 317–355.

Harmer, S. F., and A. E. Shipley, eds., *The Cambridge Natural History.* London: Macmillan, 1906–30. 10 vols.

Hartlaub, G. *Die Vögel Madagascars und der benachbarten Inselgruppen. Ein Beitrag zur Zoologie der äthiopischen Region.* Halle: H. W. Schmidt, 1877.

Harwood, Jonathan. *Styles of Scientific Thought: The German Genetics Community, 1900–1933.* Chicago: University of Chicago Press, 1993.

Haseman, John D. "Some Factors of Geographical Distribution in South America." *Annals of the New York Academy of Sciences,* 22 (1912): 9–112.

Hatschek, Berthold. "Beiträge zur Entwicklungsgeschichte der Lepidopteren." *Jenaische Zeitschrift für Naturgeschichte,* 11 (1877): 115–148.

———. "Beiträge zur Entwicklungsgeschichte und Morphologie der Anneliden." *Sitzungsberichte der Mathematische-Naturwissenschaftlichen Classe der kaiserlichen Akademie der Wissenschaften, Wein,* 74 (1877): 443–463.

———. "Studien über Entwicklungsgeschichte der Anneliden: Ein Beitrag zur Morphologie der Bilateren." *Arbeiten der Zoologischen Institut, Wein,* 1 (1878): 1–128.

———. *The Amphioxus and Its Development.* Trans. James Tuckey. London: Swan Sonnenschein, 1893.

Haughton, Samuel. "Notes on Physical Geology." *Nature,* 18 (1878): 266–268.

Heer, Oswald. *Die Urwelt der Schweiz.* Zurich: Friedrich Schultheis, 1865.

———. *Flora Fossilis Arctica: Die fossile Flora der Polarlander.* Zurich: Friedrich Schultheis/J. Wurster, 1868–75. 4 vols.

Heider, Karl. "Phylogenie der Wirbellosen." In R. Hertwig and R. von Wettstein, eds., *Abstammunslehre, Systematik, Palaeontologie, Biogeographie.* Vol. 4, part 3 of P. Hinneberg, ed., *Die Kultur der Gegenwart.* Leipzig and Berlin: B. G. Teubner, 1914, pp. 453–529.

Heilmann, Gerhard. *The Origin of Birds.* London: H. F. & G. Witherby, 1926.

Heilprin, Angelo. "On the Value of the Nearctic as One of the Primary Zoological Regions." *Proceedings of the Academy of Natural Sciences, Philadelphia,* 1882: 316–334, and 1883: 266–275.

———. "On the Value of the 'Nearctic' as One of the Primary Zoological Regions." *Nature,* 27 (1883): 606.

———. *The Geographical and Geological Distribution of Animals.* London: Kegan Paul, Trench & Co., 1887.

Henson, Pamela M. "The Comstock Research School in Evolutionary Entomology." *Osiris,* 8 (1993): 1–19.

Herrick, C. Judson. "The Evolution of Intelligence and Its Origins." *Science,* n.s., 31 (1910): 7–18.

Hertwig, Oscar, and Richard Hertwig. "Die Actinien anatomisch und histologisch mit besonderer Berücksichtigung der Nervenmuskelsystems untersucht." *Jenaische Zeitschrift für Naturwissenschaft,* 13 (1879): 457–640.

———. "Die Coelomtheorie. Versuch einer Erklärung des mittleren Keimblattes. *Jenaische Zeitschrift für Naturwissenschaft,* 15 (1882): 1–150.

———. *The Chaetognatha: A Monograph.* Trans. Susan M. Egan and Paul N. Sund. La Jolla: Scripps Institution of Oceanography, 1959.

Heuss, Theodor. *Anton Dohrn: A Life for Science.* Trans. Liselotte Dieckmann. Berlin: Springer-Verlag, 1991.

Hill, J. P. "The Placentation of Perameles." *Quarterly Journal of Microscopical Science,* 40 (1897): 385–442.

Himmelfarb, Gertrude. *Darwin and the Darwinian Revolution.* New York: Norton, 1959.

Hodge, M. J. S. "The Universal Gestation of Nature: Chambers' *Vestiges* and *Explanations*." *Journal of the History of Biology*, 5 (1972): 127–152.

Hoenigswald, Henry M., and Linda F. Wiener. *Biological Metaphor and Cladistic Classification: An Interdisciplinary Perspective.* Philadelphia: University of Pennsylvania Press, 1987.

Hofstadter, Richard. *Social Darwinism in American Thought.* Rev. ed. New York: George Braziller, 1959.

Holmgren, N. "On the Origin of the Tetrapod Limb." *Acta Zoologica, Stockholm,* 14 (1933): 185–295.

Hooker, Joseph Dalton. "Royal Society. President's Anniversary Address." *Nature,* 19 (1878): 109–113, 132–135. Also *Proceedings of the Royal Society,* 28 (1878–79): 43–68.

———. Opening Address [Geography Section, British Association for the Advancement of Science], *Nature,* 24 (1881): 443–448.

Hopkins, William. Anniversary Address of the President. *Quarterly Journal of the Geological Society of London,* 8 (1852): xxiv–lxxx.

Howarth, Janet. "Scientific Education in Late Victorian Oxford: A Curious Case of Failure?" *English Historical Review,* 102 (1987): 334–371.

Howes, G. B. "On a Hitherto Unrecognized Feature in the Larynx of the Anurous Amphibia." *Proceedings of the Zoological Society of London,* 1887: 491–510.

Howorth, Henry H. *The Glacial Nightmare and the Flood.* London: Sampson Low, Marston & Co., 1893. 2 vols.

Hoyle, Fred. *Archaeopteryx: The Primordial Bird.* Swansea: Christopher Davies, 1986.

Hubrecht, A. A. W. "On the Ancestral Form of the Chordata." *Quarterly Journal of Microscopical Science,* 23 (1883): 349–368.

———. "The Relation of the Nemertea to the Vertebrata." *Quarterly Journal of Microscopical Science,* 27 (1887): 605–644.

———. "Die Phylogenie des Amnions und die Bedeutung des Trophoblastes." *Verhandelingen der Koninklijke Akademie van Wetenschappen, Amsterdam,* 4, no. 5 (1896): 66 pp.

———. *The Descent of the Primates. Lectures Delivered on the Occasion of the Sesquicentennial Celebration of Princeton University.* New York: Charles Scribner's Sons, 1897.

———. "The Gastrulation of the Vertebrates." *Quarterly Journal of Microscopical Science,* 49 (1906): 403–419.

Huggett, Richard. *Catastrophism: Systems of Earth History.* London: Edward Arnold, 1990.

Hull, David L. *Darwin and His Critics: The Reception of Darwin's Theory of Evolution by the Scientific Community.* Cambridge, Mass.: Harvard University Press, 1973.

———. "Darwinism as a Historical Entity: A Historiographical Proposal." In

David Kohn, ed., *The Darwinian Heritage*, pp. 773–812. Princeton: Princeton University Press, 1985.

——. *Science as Process: An Evolutionary Account of the Growth and Conceptual Development of Science*. Chicago: University of Chicago Press, 1988.

Hutchinson, G. Evelyn. "Restudy of Some Burgess Shale Fossils." *Proceedings of the U. S. National Museum*, 78 (1931): 1–24.

Hutchinson, H. N. *Extinct Monsters and Creatures of Other Days: A Popular Account of Some of the Larger Forms of Ancient Animal Life*. New ed. London: Chapman and Hall, 1910.

Hutton, F. W. "The Geographical Relations of the New Zealand Fauna." *Annals and Magazine of Natural History*, ser. 4, 13 (1874): 25–39, 85–102.

——. "On the Origin of the Fauna and Flora of New Zealand." *Annals and Magazine of Natural History*, ser. 5, 13 (1884): 425–448.

Huxley, Julian. *The Individual in the Animal Kingdom*. Cambridge: Cambridge University Press, 1912.

——. *Essays of a Biologist*. London: Chatto and Windus, 1923.

——. *Evolution: The Modern Synthesis*. London: Allen and Unwin, 1942.

——, ed. *The New Systematics*. Oxford: Oxford University Press, 1940.

Huxley, Leonard, ed. *The Life and Letters of Thomas Henry Huxley*. 2d ed. London: Macmillan 1908. 3 vols.

——. *The Life and Letters of Joseph Dalton Hooker*. London: John Murray, 1918. 2 vols.

Huxley, Thomas Henry. "Preliminary Essay upon the Systematic Arrangement of the Fishes of the Devonian Epoch." Reprinted from *Memoirs of the Geological Survey of the United Kingdom* (1861) in Huxley, *Scientific Memoirs*, II, pp. 417–460.

——. "Geological Contemporaneity and Persistent Forms of Life." Reprinted from *Quarterly Journal of the Geological Society of London* (1862) in Huxley, *Collected Essays*, VIII, pp. 272–304.

——. *Lectures on the Elements of Comparative Anatomy*. London: Churchill, 1864.

——. "Remarks upon Archaeopteryx lithographica." Reprinted from *Proceedings of the Royal Society* (1868) in Huxley, *Scientific Memoirs*, III, pp. 340–345.

——. "On the Animals Which are Most Nearly Intermediate between Birds and Reptiles." Reprinted from *Annals and Magazine of Natural History* (1868) in Huxley, *Scientific Memoirs*, III, 303–313.

——. "On the Classification and Distribution of the Alectoromorphae and Heteromorphae." Reprinted from *Proceedings of the Zoological Society of London* (1868) in Huxley, *Scientific Memoirs*, III, pp. 346–373.

——. *An Introduction to the Classification of Animals*. London: Churchill, 1869.

———. "On Hyperodapedon." Reprinted from *Quarterly Journal of the Geological Society of London* (1869) in Huxley, *Scientific Memoirs*, III, pp. 374–390.

———. "Further Evidence of the Affinities between the Dinosaurian Reptiles and Birds." Reprinted from *Quarterly Journal of the Geological Society of London* (1870) in Huxley, *Scientific Memoirs*, III, pp. 465–486.

———. "On the Classification of the Dinosauria, with Observations on the Dinosaurs of the Trias." Reprinted from *Quarterly Journal of the Geological Society of London*, (1870) in Huxley, *Scientific Memoirs*, III, pp. 487–509.

———. "Palaeontology and the Doctrine of Evolution." Reprinted from *Quarterly Journal of the Geological Society of London* (1870) in Huxley, *Collected Essays*, VIII, pp. 340–388.

———. "On Stagonolepis Robertsoni and the Evolution of the Crocodilia." Reprinted from *Quarterly Journal of the Geological Society of London* (1875) in Huxley, *Scientific Memoirs*, IV, pp. 66–83.

———. "On the Classification of the Animal Kingdom." Reprinted from *Journal of the Linnean Society (Zoology)* (1876) in Huxley *Scientific Memoirs*, IV, pp. 35–60.

———. Contributions to Morphology. Ichthyopsida: On Ceratodus forsteri, with Observations on the Classification of Fishes." Reprinted from *Proceedings of the Zoological Society of London* (1876) in Huxley, *Scientific Memoirs*, IV, pp. 84–124.

———. "Lectures on Evolution." Reprinted from Huxley, *American Addresses* (1877) in Huxley, *Collected Essays*, IV, pp. 46–138.

———. "Evolution in Biology." Reprinted from *Encyclopaedia Britannica* (1878) in Huxley, *Collected Essays*, II, pp. 187–226.

———. "On the Character of the Pelvis in the Mammalia, and the Conclusions Respecting the Origin of Mammals Which May be Based on Them." Reprinted from *Proceedings of the Royal Society* (1879) in Huxley, *Scientific Memoirs*, IV, pp. 345–356.

———. *The Crayfish: An Introduction to the Study of Zoology*. London: Kegan Paul, 1880.

———. "On the Application of the Laws of Evolution to the Arrangement of the Vertebrata, and Particularly of the Mammalia." Reprinted from *Proceedings of the Zoological Society of London* (1880) in Huxley, *Scientific Memoirs*, IV, pp. 457–472.

———. "On the Respiratory Organs of Apteryx." Reprinted from *Proceedings of the Zoological Society of London* (1882) in Huxley, *Scientific Memoirs*, IV, pp. 529–539.

———. *Collected Essays*. London: Macmillan, 1893–94. 9 vols.

———. *The Scientific Memoirs of Thomas Henry Huxley*. Ed. M. Foster and E. Ray Lankester. London: Macmillan, 1898–1902. 5 vols.

Hyatt, Alpheus. "On the Parallelism between the Different Stages of Life in the Individual and Those in the Entire Group of the Molluscous Order Tetrabranchiata." *Memoirs of the Boston Society of Natural History,* 1 (1866): 193–209.

———. "The Genesis of the Tertiary Species of Planorbis at Steinheim." *Anniversary Memoir of the Boston Society of Natural History, 1830–1880.* Boston, 1880: 1–114. Summarized in *American Naturalist,* 16 (1882): 441–455.

———. "Evolution of the Cephalopoda." *Science,* 3 (1884): 122–127.

———. *Genesis of the Arietidae.* Washington: Smithsonian Contributions to Knowledge, no. 673, 1889.

———. "The Influence of Woman in the Evolution of the Human Race." *Natural Science,* 11 (1897): 89–93.

Ihering, Hermann von. *Vergleichenden Anatomie des Nervensystems und Phylogenie der Mollusken.* Leipzig: Engelmann, 1877.

———. "The History of the Neotropical Region." *Science,* n.s., 12 (1900): 857–864.

———. "Land Bridges across the Atlantic and Pacific Oceans during the Kainozoic Era." *Quarterly Journal of the Geological Society of London,* 87 (1931): 376–391.

Jarvik, Erik. "On the Structure of the Snout in Crossopterygians and in Lower Gnathostomes in General." *Zoologica Bidrag, Uppsala,* 21 (1942): 235–675.

Jefferies, R. P. S. *The Ancestry of the Vertebrates.* London: British Museum (Natural History), 1986.

Joly, John. *The Surface History of the Earth.* Oxford: Oxford University Press, 1925.

Jones, Frederic Wood. "The Origin of Man." In Arthur Dendy, ed., *Animal Life and Human Progress,* pp. 99–131. London: Constable, 1919.

———. *Man's Place among the Mammals.* London: Edward Arnold, 1929.

———. *Design and Purpose.* London: Kegan Paul, 1942.

———. *Habit and Heritage.* London: Kegan Paul, 1943.

Jones, Greta. *Social Darwinism and English Thought.* London: Harvester, 1980.

Joysey, K. A., and A. E. Friday, eds. *Problems of Phylogenetic Reconstruction.* Systematics Association Special Volume 21. New York: Academic Press, 1982.

Kerr, J. Graham. "The Origin of the Paired Limbs of Vertebrates " *Report of the British Association for the Advancement of Science,* 1901 meeting: 693–695.

———. *Textbook of Embryology.* Vol. II. *Vertebrata, with the Exception of Mammalia.* London: Macmillan, 1919.

———. "Archaic Fishes—Lepidosiren, Protopterus, Polypterus—and their Bearing upon the Problems of Vertebrate Morphology." *Jenaisch Zeitschrift für Naturwissenschaft,* 67 (1932): 419–433.

Kesteven, H. L. "Contributions to the Cranial Osteology of the Fishes: The Skull of Neoceratodus fosteri: A Study in Phylogeny." *Records of the Australian Museum*, 18 (1931): 236–265.

Kingsley, J. S. "Is the Group Arthropoda a Valid One?" *American Naturalist*, 17 (1883): 1034–1037.

———. "The Head of an Embryo Amphiuma." *American Naturalist*, 26 (1892): 671–680.

———. "The Embryology of Limulus." *Journal of Morphology*, 7 (1892): 35–68, and 8 (1893): 195–268.

———. "The Classification of the Arthropoda." *American Naturalist*, 28 (1894): 118–135, 220–235.

———. "The Origin of the Mammals." *Science*, n.s., 14 (1901): 193–205.

———. "The Bones of the Reptilian Lower Jaw." *American Naturalist*, 39 (1905): 59–64.

Kingsley, J. S. and W. H. Ruddick. "The Ossicula Auditus and Mammalian Ancestry." *American Naturalist*, 33 (1899): 219–230.

Klaatsch, Hermann. "Zur Morphologie der Mesenterialbildungen am Darmkanal der Wirbelthiere." *Morphologisches Jarhbuch*, 18 (1892): 386–450.

———. "Die Brutflosse der Crossopterygier: Ein Beitrag zur Anwendung der Archipterygium-Theorie auf die Gliedmassen der Landwirbelthiere." in *Festschrift zur siebenzigsten Geburstag von Carl Gegenbaur*. Leipzig: Wilhelm Engelmann, 1896–97, III, pp. 259–391. 3 vols.

———. *The Evolution and Progress of Mankind*. Ed. Adolph Heilborn; trans. Joseph McCabe. London: T. Fisher Unwin, 1923.

Knight, Charles R. *Before the Dawn of History*. New York: Whittlesey House/ McGraw-Hill, 1935.

Kobelt, W. *Studien zur Zoogeographie: Die Mollusken der Palaearktischen Region*. Weisbaden: C. W. Kreidel, 1897.

Koestler, Arthur. *The Case of the Midwife Toad*. London: Hutchinson, 1971.

Kohn, David, ed. *The Darwinian Heritage*. Princeton: Princeton University Press, 1985.

Koken, Ernst. *Palaeontologie und Descendenzlehre*. Jena: G. Fischer, 1902.

Korschelt, E., and K. Heider. *Textbook of the Embryology of Invertebrates*. Trans. Matilda Bernard. London: Swan Sonnenschein, 1895–1900. 4 vols.

Kowalevsky, Alexander [Kovalevskii, Alexandr], "Entwickelungsgeschichte der einfachen Ascidien." *Mémoires de l'Académie des Sciences de St.-Pétersbourg*, 7th ser, 10, no. 4 (1866).

———. "Entwickelungsgeschichte des *Amphioxus lanceolatus*." *Mémoires de l'Académie des Sciences de St.-Pétersbourg*, 7th ser, 11, no. 4 (1867).

———. "Weiterer Studien über die Entwickelung der einfachen Ascidien." *Archiv für mikroskopische Anatomie*, 6 (1871): 101–130.

———. "Weiterer Studien über die Entwickelungsgeschichte des *Amphioxus lan-*

ceolatus, nebst einem Beiträge zur Homologie des Nervensystems der Würmer und Wirbelthiere." *Archiv für mikroskopische Anatomie,* 13 (1877): 181–204.

Kowalevsky, W. [Kovalevskii, Vladimir]. "On the Osteology of the Hyopotamidae" (abstract). *Proceedings of the Royal Society of London,* 21 (1873): 147–165.

———. "Sur l'*Anchitherium aurelianse* Cuv. et sur l'histoire paléontologique des chevaux." *Mémoires de l'Académie des Sciences de St. Pétersbourg,* ser. 7, 20, no. 5 (1873).

———. "On the Osteology of the Hyopotamidae." *Philosophical Transactions of the Royal Society,* 163 (1874): 19–94.

———. *The Complete Works of Vladimir Kovalevsky.* Ed. Stephen Jay Gould. New York: Arno Press, 1980.

Kühn, Alfred. *Anton Dohrn und die Zoologie seiner Zeit.* Naples: Publicazioni della Stazione Zoologica, Supplemento 1950.

Kuhn, Thomas S. *The Structure of Scientific Revolutions.* Chicago: University of Chicago Press, 1962.

Kükenthal, Willy. "Ueber die Enstehung und Entwicklung des Säugetierstammes." *Biologisches Centralblatt,* 12 (1892): 400–413.

Kuppfer, Carl Ritter von. "Die Stammverwandschaft zwischen Ascidien und Wirbelthieren." *Archiv für mikroskopische Anatomie,* 5 ((1869): 459–463.

———. "Die Stammverwandschaft zwischen Ascidien und Wirbelthieren." *Archiv für mikroskopische Anatomie,* 6 (1870): 115–172.

———. "Zur Entwicklung der einfachen Ascidien." *Archiv für mikroskopische Anatomie,* 8 (1872): 358–396.

Lakoff, George, and Mark Johnson. *Metaphors We Live by.* Chicago: University of Chicago Press, 1980.

Landau, Misia. *Narratives of Human Evolution.* New Haven: Yale University Press, 1991.

Lang, Arnold. *Textbook of Comparative Anatomy.* Trans. H. M. Bernard and Matilda Bernard. London: Macmillan, 1891. 2 vols.

Lang, W. D. *Catalogue of the Fossil Bryozoa (Polyzoa) in the Department of Geology, British Museum (Natural History). The Cretaceous Bryozoa.* Vol. 3. London, 1921.

Lanham, Url. *The Bone Hunters.* New York: Columbia University Press, 1973.

Lankester, E. Ray. *A Monograph: The Fishes of the Old Red Sandstone of Britain. Pt. 1. The Cephalaspidae.* London: Palaeontographical Society, 1868.

———. "On the Use of the Term Homology in Modern Zoology and the Distinction between Homogenetic and Homoplastic Agreements." *Annals and Magazine of Natural History,* 4th ser., 6 (1870): 34–43.

———. "On the Primitive Cell-Layers of the Embryo as the Basis of Genealogical Classification of Animals." *Annals and Magazine of Natural History,* 4th ser., 11 (1873): 321–338.

————. "Observations on the Development of the Pond Snail (*Lymneus stagnalis*) and on the Early Stages of Other Mollusca." *Quarterly Journal of Microscopical Science*, 14 (1874): 365–391.

————. "Contribution to the Developmental History of the Mollusca." *Philosophical Transactions of the Royal Society*, 165 (1875): 1–48.

————. "Dohrn on the Origin of the Vertebrata and the Principle of Succession of Functions." *Nature*, 12 (1875): 479–481.

————. Preface. In Carl Gegenbaur, *Elements of Comparative Anatomy*. London: Macmillan, 1878.

————. "Notes on the Embryology and Classification of the Animal Kingdom: Comprising a Revision of Speculations Relative to the Origin and Significance of the Germ Layers." *Quarterly Journal of Microscopical Science*, 17 (1877): 399–454.

————. *Degeneration: A Chapter in Darwinism*, London: Macmillan, 1880.

————. "Limulus an Arachnid." *Quarterly Journal of Microscopical Science*, new ser., 21 (1881): 504–649.

————. "New Hypothesis as to the Relation of the Lung-Book of Scorpio to the Gill-Book of Limulus." *Quarterly Journal of Microscopical Science*, new ser., 25 (1885): 339–342.

————. "Professor Claus and the Classification of the Arthropoda." *Annals and Magazine of Natural History*, ser. 5, 17 (1886): 364–372.

————. "Professor Claus: A Rejoinder." *Annals and Magazine of Natural History*, ser. 5, 18 (1886): 179–182.

————. "Zoology." *Encyclopaedia Britannica*. 9th ed. Vol. 24 (1888): 799–820.

————. "The Coelom and the Vascular System of Mollusca and Arthropoda." *Nature*, 37 (1888): 498.

————. "The Taxonomic Position of the Pteraspidae, Cephalaspidae, and Asterolepidae." *Natural Science*, 11 (1897): 45–47.

————. *Extinct Animals*. London: Constable, 1905.

————. "Arthropoda." *Encyclopaedia Britannica*, 11th edn, Cambridge: Cambridge University Press, 1910–11, 29 vols. Vol. 2, pp. 673–681.

————, ed. *Zoological Articles Contributed to the Encyclopaedia Britannica*. London: A. & C. Black, 1891.

————, ed. *A Treatise on Zoology*. London: A. & C. Black, 1900–1909 (9 vols.).

Laporte, Leo F. "Simpson on Species." *Journal of the History of Biology*, 27 (1994): 141–160.

Laudan, Larry. *Progress and Its Problems: A Theory of Scientific Growth*. Berkeley: University of California Press, 1977.

Laurie, Malcolm, "The Anatomy and Relations of the Eurypteridae." *Transactions of the Royal Society of Edinburgh*, 37 (1892–93): 509–528.

Le Conte, Joseph. "On Critical Periods in the History of the Earth and Their Relation to Evolution, and on the Quaternary as Such a Period." *American Journal of Science,* ser. 3, 14 (1877): 99–114.

———. "Geological Climates and Geological Time." *Nature,* 18 (1875): 668.

———. "Critical Periods in the History of the Earth." *Journal of Geology,* 3 (1895): 869–870.

———. *Evolution: Its Nature, Its Evidences, and Its Relation to Religious Thought.* 2d ed. New York: Appleton, 1899.

Le Grand, Homer. *Drifting Continents and Shifting Theories* Cambridge: Cambridge University Press, 1989.

Lester, Joseph. *E. Ray Lankester and the Making of Modern British Biology.* Ed. Peter J. Bowler. British Society for the History of Science Monographs, 1995.

Linklater, Eric. *The Voyage of the Challenger.* London: Murray, 1972.

Loomis, F. B. "Momentum in Variation." *American Naturalist,* 39 (1905): 839–843.

Lowe, Percy Roycroft. "Studies and Observations on the Phylogeny of the Ostrich and Its Allies." *Proceedings of the Zoological Society of London,* (1928): 185–247.

———. "On the Relationship of the Struthiones to the Dinosaurs and the Rest of the Avian Class, with Special Reference to the Position of Archaeopteryx." *Ibis,* 5 (1935): 398–432.

Lubbock, John. *On the Origin and Metamorphoses of Insects.* London: Macmillan, 1895.

Lull, Richard Swan. "S. W. Williston, *Water Reptiles of Past and Present.*" *Science,* n.s., 61 (1915): 391–392.

———. *Organic Evolution.* New York: Macmillan, 1917.

———. "The Pulse of Life." In Joseph Barrell et al., *The Evolution of the Earth and Its Inhabitants: A Series of Lectures Delivered Before the Yale Chapter of the Sigman Xi During the Academic Year 1916–1917.* New Haven: Yale University Press, 1918, pp. 109–146. Reprinted in G. A. Baitsell, ed., *The Evolution of the Earth and Man,* pp. 110–147. New Haven: Yale University Press, 1929.

———. "Dinosaurian Climatic Response." In M. R. Thorpe, ed., *Organic Adaptation to the Environment,* pp. 225–279. New Haven: Yale University Press, 1924.

Lydekker, Richard. *Phases of Animal Life: Past and Present.* London: Longman, Green, 1892.

———. *Life and Rock: A Collection of Zoological and Geological Essays.* London: Universal Press, 1894.

———. *A Geographical History of Mammals.* Cambridge: Cambridge University Press, 1896.

MacBride, Ernest William. "Sedgwick's Theory of the Embryonic Phase of Ontogeny as a Guide to Phylogenetic Speculation." *Quarterly Journal of Microscopical Science,* 37 (1895): 325–342.

———. "The Development of Asterina gibbosa." *Quarterly Journal of Microscopical Science,* 38 (1896): 339–411.

———. "The Early Development of Amphioxus." *Quarterly Journal of Microscopical Science,* 40 (1898): 589–612.

———. "On the Origin of Echinoderms." *Proceedings of the 4th International Congress of Zoology, Cambridge, 1898.* London, 1899: 142–148.

———. *Textbook of Embryology.* Vol. I. *Invertebrata.* London: Macmillan, 1914.

Macfarlane, John Muirhead. *The Causes and Course of Organic Evolution: A Study in Bioenergetics.* New York: Macmillan, 1918.

McKinney, H. Lewis. *Wallace and Natural Selection.* New Haven: Yale University Press, 1972.

MacLeod, J. "Recherches sur la structure et la signification de l'appareil respiratoire des Arachnides." *Archives de biologie,* 5 (1884): 1–34.

MacLeod, Roy. "Embryology and Empire: The Balfour Students and the Quest for Intermediate Forms in the Laboratory of the Pacific." In Roy MacLeod and P. F. Rehbock, eds., *Darwin's Laboratory: Evolutionary and Natural History in the Pacific,* pp. 140–165. Honolulu: University of Hawaii Press, 1994.

McQuat, Gordon, and Mary P. Winsor. "J. B. S. Haldane's Darwinism in Its Religious Context." *British Journal for the History of Science,* in press.

Maienschein, Jane. *Transforming Traditions in American Biology, 1880–1915.* Baltimore: Johns Hopkins University Press, 1991.

———. "Cutting Edges Cut Both Ways." *Biology and Philosophy,* 9 (1994): 1–24.

Manton, S. M. *The Arthropoda: Habits, Functional Morphology, and Evolution.* New York: Van Nostrand Reinhold, 1979.

Manton, S. M., and D. T. Anderson. "Polyphyly and the Evolution of the Arthropods." In M. R. House, ed., *The Origin of Major Invertebrate Groups,* pp. 269–322. London: Academic Press for the Systematics Association, 1979.

Marchant, James. *Alfred Russel Wallace: Letters and Reminiscences.* New York: Harper, 1916.

Marsh, Othniel Charles. "Introduction and Succession of Vertebrate Life in America." *American Journal of Science,* 3rd ser., 9 (1877): 337–378.

———. "Fossil Mammalia from the Jurassic of the Rocky Mountains." *American Journal of Science,* ser. 3, 15 (1878): 459.

———. *Odontornithes: A Monograph on the Extinct Toothed Birds of North America.* Report of the Geological Exploration of the Fortieth Parallel. Vol. VII. (Washington: Government Printing Office, 1880).

———. "Notice of Jurassic Mammals Representing Two New Orders." *American Journal of Science,* ser. 3, 20 (1880): 235–239.

———. "American Jurassic Mammals." *American Journal of Science,* ser. 3, 33 (1887): 327–348.

———. *Dinocerata: A Monograph of an Extinct Order of Gigantic Mammals.* Memoirs of the U.S. Geological Survey. Vol. X. (Washington: Government Printing Office, 1886).

———. "Discussion on the Origin of Mammals." *Proceedings of the Fourth International Congress of Zoology, Cambridge, 1898.* London: 1899, pp. 71–74.

Matthew, William Diller. "On the Antennae and Other Appendages of Trilobites." *American Journal of Science,* ser. 3, 46 (1893): 121–125.

———. "The Arboreal Ancestry of the Mammalia." *American Naturalist,* 38 (1904): 811–818.

———. "Hypothetical Outlines of the Continents in Tertiary Times." *Bulletin of the American Museum of Natural History,* 22 (1906): 353–383.

———. "The Phylogeny of the Felidae." *Bulletin of the American Museum of Natural History,* 28 (1910): 289–316.

———. "The Evolution of the Mammals in the Eocene." *Proceedings of the Zoological Society of London,* 1927: 947–985.

———. *Outline and General Principles of the History of Life (Synopsis of Lectures in Paleontology).* Vol. I. Berkeley: University of California Press, 1928. Reprinted, New York: Arno Press, 1980.

———. *Climate and Evolution.* 2d ed. New York: New York Academy of Sciences, 1939.

Maurer, Friedrich. "Zur Phylogenie der Säugethierhaare." *Morphologisches Jahrbuch,* 20 (1893): 260–275.

Mayr, Ernst. "The Role of Systematics in Biology." Reprinted in Mayr, *Evolution and the Diversity of Life,* pp. 416–424.

———. "Wallace's Line in the Light of Recent Zoogeographical Studies." Reprinted in Mayr, *Evolution and the Diversity of Life,* pp. 626–645.

———. "Where Are We?" Reprinted in Mayr, *Evolution and the Diversity of Life,* pp. 307–328.

———. *Evolution and the Diversity of Life.* Cambridge, Mass.: Harvard University Press, 1976.

———. *The Growth of Biological Thought: Diversity, Evolution, and Inheritance.* Cambridge, Mass.: Harvard University Press, 1982.

———. *Toward a New Philosophy of Biology.* Cambridge, Mass.: Harvard University Press, 1988.

———. *One Long Argument: Charles Darwin and the Genesis of Modern Evolutionary Thought.* Cambridge, Mass.: Harvard University Press, 1991.

Mayr, Ernst, and William B. Provine. *The Evolutionary Synthesis: Perspectives on the Unification of Biology.* Cambridge, Mass.: Harvard University Press, 1980.

M'Coy, Frederick. "Note on the Ancient and Recent Natural History of Victoria." *Annals and Magazine of Natural History,* ser. 3, 9 (1862): 137–150.

Mehnert, Ernst. "Untersuchungen über die Entwicklung des Os Pelvis der Vögel." *Morphologisches Jahrbuch,* 13 (1888): 259–295.

Merriam, C. Hart. "The Geographical Distribution of Life in North America, with Special Reference to the Mammalia." *Proceedings of the Biological Society of Washington,* 7 (1892): 1–64.

Merriam, John Campbell. *Published Papers and Addresses.* Washington: Carnegie Institution, 1938. 4 vols.

Meyerhof, Otto. *Die chemischen Vorgänge im Muskel und ihr Zusammenhang mit Arbeitsleitung und Warmbildung.* Berlin: J. Springer, 1930.

Meyrick, Edward. *A Revised Handbook of British Lepidoptera* (1927) Reprinted, Hampton, Middlesex: E. W. Classey, 1968.

Miall, L. C. "The Transformations of Insects." *Nature,* 53 (1895): 152–158.

Millhauser, Milton. *Just before Darwin: Robert Chambers and Vestiges.* Middletown, Conn.: Wesleyan University Press, 1959.

Mills, Eric L. *Biological Oceanography: An Early History, 1870–1960.* Ithaca, N.Y.: Cornell University Press, 1989.

Miner, Roy Waldo. "The Pectoral Limb of *Eryops* and Other Primitive Tetrapods." *Bulletin of the American Museum of Natural History,* 51 (1924–25): 145–312.

Minot, Charles Sedgwick. "Cephalic Homologies: A Contribution to the Determination of the Ancestry of Vertebrates." *American Naturalist,* 31 (1897): 927–943.

Mitchell, Peter Chalmers. "On a Graphic Formula to Express Geographical Distribution." *Proceedings of the Zoological Society of London,* 1890: 607–609.

———. "On the Intestinal Tract of Birds; with Remarks on the Valuation and Nomenclature of Zoological Characters." *Transactions of the Linnean Society of London (Zoology),* 2d ser., 8 (1901): 173–275.

———. "On the Intestinal Tract of Mammals." *Transactions of the Zoological Society of London (Zoology),* 17 (1905): 437–536.

———. *Centenary History of the Zoological Society of London.* London: Zoological Society, 1929.

Mitman, Gregg. "Evolution as Gospel: William Patten, the Language of Democracy, and the Great War." *Isis,* 81 (1990): 446–463.

Mivart, St. George Jackson. *On the Genesis of Species.* London: Macmillan, 1871.

———. "Notes on the Fins of Elasmobranches, with Considerations on the Nature and Homologues of Vertebrate Limbs." *Transactions of the Zoological Society of London,* 10 (1879): 439–484.

————. "A Popular Account of Chamaeleons." *Nature,* 24 (1881): 309–312 and 335–338.

————. "On the Possible Dual Origin of the Mammalia." *Proceedings of the Royal Society,* 43 (1888): 372–379.

Moore, James R., ed. *History, Humanity and Evolution: Essays for John C. Greene.* Cambridge: Cambridge University Press, 1989.

Morgan, Thomas Hunt. "The Growth and Metamorphosis of Tornaria." *Journal of Morphology,* 5 (1891): 407–447.

————. "A Contribution to the Embryology and Phylogeny of the Pycnogonids." *Studies from the Biological Laboratory of the Johns Hopkins University,* 5 (1893): 1–76.

————. "The Development of Balanoglossus." *Journal of Morphology,* 9 (1894): 1–86.

Morris, Charles. "The Origin of Lungs." *American Naturalist,* 26 (1892): 975–986.

Moseley, Henry N. "On the Structure and Development of *Peripatus capensis.*" *Philosophical Transactions of the Royal Society,* 164 (1874): 757–782.

————. *Notes by a Naturalist: An Account of Observations Made During the Voyage of H.M.S. 'Challenger' Round the World in the Years 1872–1876.* New ed. London: John Murray, 1892.

Moy-Thomas, J. A. "The Early Evolution and Relationships of the Elasmo-branches." *Biological Reviews of the Cambridge Philosophical Society,* 14 (1939): 1–26.

————. "The Devonian Fish *Palaeospondylus gunni,* Traquair." *Philosophical Transactions of the Royal Society,* 230 B (1940): 391–413.

Müller, Fritz. *Facts and Arguments for Darwin: With Additions by the Author.* Trans. W. S. Dallas. London: John Murray, 1869.

Murray, Andrew. "On the Geographical Relations of the Coleoptera of Old Cala-bar." *Transactions of the Linnean Society of London,* 23 (1862): 449–455.

————. *The Geographical Distribution of Mammals.* London: Day & Son, 1866. Facsimile reprint, New York: Arno Press, 1978.

Murray, John. "On the Deep and Shallow-Water Marine Fauna of the Kerguelen Region of the Great Southern Ocean." *Transactions of the Royal Society of Edinburgh,* 38 (1894–95): 343–500.

————. *Report of the Scientific Results of the Voyage of H.M.S. Challenger. . . . A Summary of the Scientific Reports.* London: HMSO, 1895. Reprinted New York: Arno, 1977.

Naef, Adolph. *Idealistische Morphologie und Systematik: Zur Methodik der sys-tematischen Morphologie.* Jena: Gustav Fischer, 1919.

————. "Notizen zur Morphologie und Stammesgeschichte der Wirbeltiere. 6: Zur Stammesgeschichte der Säuger-Molaren." *Biologisches Zentralblatt,* 45 (1925): 668–676.

————. "Notizen zur Morphologie und Stammesgeschichte der Wirbeltiere. 7: Das Verhältnis der Chordaten zu niederen Tierformen und der typische Verlauf ihre frühen Entwicklung." *Biologisches Zentralblatt,* 46 (1926): 39–50.

————. "Notizen zur Morphologie und Stammesgeschichte der Wirbeltiere." *Pubblicazioni della Stazione Zoologica di Napoli,* 7 (1926): 299–333.

————. "Allgemeine Morphologie. I. Die Gestalt als Begriff und Idee." In L. Bolk, E. Goppert, E. Kallius and W. Lubosch, eds., *Handbuch der vergeleichenden Anatomie der Wirbeltiere.* Berlin: Urban and Schwarzenberg, 1931, pp. 77–118.

Naudin, Charles. "Les espèces affines et la théories de l'évolution." *Bulletin de la Société Botanique de France,* 21 (1874): 240–272.

Neal, Herbert V. and Herbert W. Rand. *Comparative Anatomy.* London: H. K. Lewis, 1936.

Needham, Dorothy, Josheph Needham, Ernest Baldwin, and John Yudkin. "A Comparative Study of the Phosphagens, with Some Remarks on the Origin of the Vertebrates." *Proceedings of the Royal Society,* 110 B (1932): 260–294.

Nelson, Gareth. "From Candolle to Croizat: Comments on the History of Biogeography." *Journal of the History of Biology,* 11 (1978): 269–305.

Nelson, Gareth, and Don E. Rosen, eds. *Vicariance Biogeography: A Critique.* New York: Columbia University Press, 1981.

Neumayr, Melchior. "Über klimatische Zonen wahrend der Jura- und Kreidzeit." *Denkschriften der Akademie der Wissenschaft, Wein, Mathematisch-naturwissenschaftliche Klasse,* 47 (1883): 277–310.

————. *Erdgeschichte.* New ed. Leipzig and Vienna: Bibliographischen Institut, 1890. 2 vols.

Newton, Alfred. *A Dictionary of Birds.* London: A. & C. Black, 1893–96.

Newton. E. T. "Reptiles from the Elgin Sandstone: Descriptions of Two New Genera." *Philosophical Transactions of the Royal Society,* 185 B (1894): 573–607.

Nicholson, Alleyne. *The Ancient Life History of the Earth.* Edinburgh: William Blackwood, 1877.

Nopsca, Francis. "Ideas on the Origin of Flight." *Proceedings of the Zoological Society of London,* 1907: 223–236.

Norman, David B. "On the History of the Discovery of Fossils at Bernissart in Belgium." *Archives of Natural History,* 14 (1987): 59–75.

Norton, Bernard. "The Biometric Defence of Darwinism." *Journal of the History of Biology,* 6 (1973): 283–316.

Nübler-Jung, K., and D. Arendt. "Is Ventral in Insects Dorsal in Vertebrates? A History of Embryological Arguments Favouring Axis Inversion in Chordate Ancestors." *Roux's Archives in Developmental Biology,* 203 (1994): 357–366.

Nyhart, Lynn K. *Morphology and the German University, 1860–1900.* University of Pennsylvania Ph.D. thesis, 1986.

———. "The Disciplinary Breakdown of German Morphology, 1870–1900." *Isis*, 78 (1987): 365–389.

———. *Biology Takes Form: Animal Morphology and the German Universities, 1800–1900*. Chicago: University of Chicago Press, 1995.

Nuttall, George H. F. *Blood Immunity and Blood Relationships: A Demonstration of Certain Blood Relationships Amongst Animals by Means of the Precipitin Test for Blood*. Cambridge: Cambridge University Press, 1904.

O'Brien, Charles F. "*Eozoön canadense:* The Dawn Animal of Canada." *Isis*, 61 (1970): 200–223.

O'Hara, Robert J. "Diagrammatic Classification of Birds, 1819–1901: Views of the Natural System in 19th-century British Ornithology." In H. Ouellet, ed. *Acta XIX Congressus Ornithologici*, pp. 2746–2759. Ottawa: National Museum of Natural Science, 1988.

———. "Representations of the Natural System in the Nineteenth Century." *Biology and Philosophy*, 6 (1991): 255–274.

———. "Telling the Tree: Narrative Representation and the Study of Evolutionary History." *Biology and Philosophy*, 7 (1992): 135–160.

Olby, Robert C. *Origins of Mendelism*. Rev. ed. Chicago: University of Chicago Press, 1985.

———. "Scientists and Bureaucrats in the Establishment of the John Innes Horticultural Institution under William Bateson." *Annals of Science*, 46 (1989): 497–510.

Oliver, Daniel. "The Atlantis Hypothesis in Its Botanical Aspect." *Natural History Review*, 1862: 149–170.

Oppenheimer, Jane. *Essays in the History of Embryology*. Cambridge, Mass.: MIT Press, 1967.

Ortmann, Arnold E. "Von Ihering's Archiplata and Archhelennis Theory." *Science*, n.s., 12 (1900): 929–930.

———. "The Theories of the Origin of the Antarctic Faunas and Floras." *American Naturalist*, 35 (1901): 139–142.

———. "The Geographical Distribution of Freshwater Decapods and Its Bearing on Ancient Geography." *Proceedings of the American Philosophical Society*, 41 (1902): 267–400.

———. "Facts and Theories in Evolution." *Science*, n.s., 23 (1906): 947–952.

———. "Weismannism: A Critique of Die Selektionstheorie." *Science*, n.s., 31 (1910): 815–819.

Orvig, Tor, ed. *Current Problems of Lower Vertebrate Phylogeny*. Stockholm: Almqvist and Wiksell/New York: Interscience Publishers, 1968.

Osborn, Henry Fairfield. "The Rise of the Mammalia in North America." *American Journal of Science*, ser. 3, 46 (1893): 373–392, 448–466.

———. "The Origin of the Mammalia." *American Naturalist*, 32 (1898): 309–334.

———. "Discussion on the Origin of Mammals." *Proceedings of the Fourth International Congress of Zoology, Cambridge, 1898.* London: 1899.

———. "The Origin of Mammals." *Proceedings of the Fourth International Congress of Zoology, Cambridge, 1898.* London: 1899. Appendix C, pp. 415–419.

———. "The Origin of Mammals." *American Journal of Science,* ser. 4, 7 (1899): 92–96.

———. "Reconsideration of the Evidence for a Common Dinosaur-Avian Stem in the Permian." *American Naturalist,* 34 (1900): 777–799.

———. "Origin of the Mammalia. III. Occipital Condyles of Reptilian Tripartite Type." *American Naturalist,* 34 (1900): 943–947.

———. "Correlations between Tertiary Mammal Horizons of Europe and America." *Annals of the New York Academy of Sciences,* 13 (1900): 1–64.

———. "Phylogeny of the Rhinoceroses of Europe." *Bulletin of the American Museum of Natural History,* 13 (1900): 229–267.

———. "The Geological and Faunal Relations of Europe and America during the Tertiary Period and the Theory of the Successive Invasion of an African Fauna." *Science,* n.s., 11 (1900): 561–574.

———. "Homoplasy as a Law of Latent or Potential Homology." *American Naturalist,* 36 (1902): 259–271.

———. "The Causes of Extinction of Mammalia." *American Naturalist,* 40 (1906): 829–859.

———. *The Evolution of Mammalian Molar Teeth to and from the Triangular Type.* New York: Macmillan, 1907.

———. "The Four Inseperable Factors of Evolution." *Science,* n.s., 27 (1908): 148–150.

———. *Cenozoic Mammal Horizons of Western North America.* Washington: U.S. Geological Survey Bulletin 361, 1909.

———. *The Age of Mammals in Europe, Asia, and North America.* New York: Macmillan 1910; reissue 1921.

———. "The Continuous Origin of Certain Unit Characters as Observed by a Paleontologist." *American Naturalist,* 46 (1912): 185–206, 249–278.

———. *Men of the Old Stone Age.* London: George Bell, 1916.

———. *The Origin and Evolution of Life: On the Theory of the Action, Reaction, and Interaction of Energy.* New York: Charles Scribner's Sons, 1917.

———. "Recent Discoveries Relating to the Origin and Antiquity of Man." *Proceedings of the American Philosophical Society,* 66 (1927): 373–389.

———. *Man Rises to Parnassus.* 2d ed. Princeton: Princeton University Press, 1928.

———. *The Titanotheres of Ancient Wyoming, Dakota, and Nebraska.* Washington: U.S. Geological Survey Monograph 55, 1929. 2 vols.

———. *Cope: Master Naturalist. The Life and Writings of Edward Drinker Cope.* Princeton: Princeton University Press, 1931.

———. *Proboscidea: A Monograph of the Discovery, Evolution, Migration, and Extinction of the Mastodonts and Elephants of the World.* New York: American Museum Press for the Trustees of the American Museum of Natural History, 1936. 2 vols.

Osborn, H. F., and J. L. Wortman. "Fossil Mammals of the Wasatch and Wind River Formations." *Bulletin of the American Museum of Natural History,* 4 (1892): 81–148.

Ospovat, Dov. "The Influence of Karl Ernst von Baer's Embryology, 1828–1859: A Reappraisal in the Light of Richard Owen's and William B. Carpenter's Paleontological Application of von Baer's Law." *Journal of the History of Biology,* 9 (1976): 1–28.

Owen, Richard. *On the Archetype and Homologies of the Vertebrate Skeleton.* London: Van Voorst, 1848.

———. *On the Nature of Limbs.* London: Van Voorst, 1849.

———. *On the Classification and Geographical Distribution of the Mammalia.* London: John Parker, 1859.

———. "On the Archaeopteryx of von Meyer." *Philosophical Transactions of the Royal Society,* 153 (1864): 33–47.

———. *A Monograph of the Fossil Reptilia of the Liassic Formations.* London: Palaeontographical Society, 1865–81.

———. *On the Anatomy of Vertebrates.* London: Longmans, Green, 1866–68. 3 vols.

———. *A Monograph on the Fossil Reptilia of the Mesozoic Formations.* London: Palaeontographical Society, 1874–89.

———. *Descriptive and Illustrated Catalogue of the Fossil Reptilia of South Africa in the Collection of the British Museum.* London: Trustees of the British Museum, 1876.

———. "Evidence of a Carnivorous Reptile (Cynodraco major, Ow.) about the Size of a Lion, with Remarks Thereon." *Quarterly Journal of the Geological Society of London,* 32 (1876): 95–101.

———. "Evidence of Theriodonts in Permian Deposits Elsewhere Than in South Africa." *Quarterly Journal of the Geological Society of London,* 32 (1876): 352–368.

———. "On the Influence of the Advent of a Higher Form of Life in Modifying the Structure of an Older and Lower Form." *Quarterly Journal of the Geological Society of London,* 34 (1878): 421–430.

———. "Description of Parts of the Skeleton of an Anomodont Reptile (Platypodosaurus robustus, Ow.) from the Trias of Graaf Reinet, S. Africa." *Quarterly Journal of the Geological Society of London,* 36 (1880): 414–425.

———. "On the Order Theriodontia, with a Description of a New Genus and

Species (Aeluosaurus felinus, Ow.). *Quarterly Journal of the Geological Society of London*, 37 (1881): 261–265.

———. "On the Homology of the Conario-hypophysial Tract, or the So-called Pineal and Pituitary Glands." *Journal of the Linnean Society (Zoology)*, 16 (1882): 131–149.

———. "On Cerebral Homologies in Vertebrates and Invertebrates." *Journal of the Linnean Society (Zoology)*, 17 (1883): 1–13.

———. "On the Answerable Divisions of the Brain in Vertebrates and Invertebrates." *Annals and Magazine of Natural History*, ser. 5, 12 (1883): 303–307.

Packard, Alpheus. "On the Embryology of Limulus polyphemus." *Proceedings of the American Association for the Advancement of Science*, 21 (1870): 247–255.

———. "On the Development of Limulus polyphemus." *Memoirs of the Boston Society of Natural History*, 2, part. 2, no. 1, 1872.

———. "Geological Extinction and Some of Its Causes." *American Naturalist*, 20 (1886): 29–40.

———. "Hints on the Classification of the Arthropoda: The Group a Polyphyletic One." *Proceedings of the American Philosophical Society*, 42 (1903): 142–161.

Panchen, A. L., ed. *The Terrestrial Environment and the Origin of Land Vertebrates*. London: Academic Press, 1980.

Parker, W. N. "On the Anatomy and Physiology of Protopterus annectans." *Transactions of the Royal Irish Academy*, 30 (1892): 109–226.

———. "On the Anatomy and Physiology of Protopterus annectans." *Proceedings of the Royal Society*, 49 (1893): 549–554.

Patten, William. "Studies on the Eyes of Arthropods. I. The Development of the Eyes of Vespa." *Journal of Morphology*, 1 (1887): 193–226.

———. "Studies on the Eyes of Arthropods. II. Eyes of Acilius." *Journal of Morphology*, 2 (1888): 97–190.

———. "On the Origin of Vertebrates from Arachnids." *Quarterly Journal of Microscopical Science*, n.s., 31 (1890): 317–378.

———. "Gaskell's Theory of the Origin of Vertebrates from Crustacean Ancestors." *American Naturalist*, 33 (1899): 360–369.

———. *The Evolution of the Vertebrates and their Kin*. Philadelphia: P. Blakiston's Sons, 1912.

Patterson, Colin. "The Origin of the Tetrapods: Historical Introduction to the Problem." In A. L. Panchen, ed., *The Terrestrial Environment and the Origin of Land Vertebrates*, pp. 159–175. London: Academic Press, 1980.

Pauly, Philip J. "Summer Resort and Scientific Discipline: Woods Hole and the Structure of American Biology, 1882–1925." In R. Rainger, K. R. Benson and J. Maienschein, eds., *The American Development of Biology*. Philadelphia: University of Pennsylvania Press, 1988, pp. 121–150.

Pelseneer, Paul. *Mollusca.* Vol. 5 of E. Ray Lankester, ed., *Treatise of Zoology.* London: A. & C. Black, 1906.

Plate, Robert. *The Dinosaur Hunters: Othniel C. Marsh and Edward D. Cope.* New York: D. McKay, 1964.

Pocock, R. I. "On Some Points in the Morphology of the Arachnida (s.s.), with Notes on the Classification of the Group." *Annals and Magazine of Natural History,* ser. 6, 11, (1893): 1–19.

———. "The Scottish Silurian Scorpion." *Quarterly Journal of Microscopical Science,* ser. 3, 22 (1900–1901): 291–311.

———. "On the Geographical Distribution of Spiders of the Order Mygalomorphae." *Proceedings of the Zoological Society of London,* 1903 (1): 340–368.

Pollard, H. B. "On the Anatomy and Phylogenetic Position of Polypterus." *Anatomische Anzeiger,* 6 (1891): 338–344.

Provine, William B. *The Origins of Theoretical Population Genetics.* Chicago: University of Chicago Press, 1971.

Pycraft, W. P. "The Wing of Archaeopteryx." *Natural Science,* 5 (1894): 350–360, 437–448.

———. *A History of Birds.* London: Methuen, 1910.

Rabl, Carl. "Ueber die Bildung des Herzens der Amphibien." *Morphologisches Jahrbuch,* 12 (1887): 252–274.

———. "Theorie des Mesoderms." *Morphologisches Jahrbuch,* 15 (1889): 113–252; 19 (1893): 65–144.

———. "Ueber den Bau und die Entwickelung der Linse. 3 Teil." *Morphologisches Jarhbuch,* 67 (1898): 1–138.

———. "Gedanken und Studien über den Ursprung der Extremitäten." *Zeitschrift für wissenschaftliche Zoologie,* 70 (1901): 476–557.

Rackoff, Jerome S. "The Origin of the Tetrapod Limb and the Ancestry of Tetrapods." In A. L. Panchen, ed., *The Terrestrial Environment and the Origin of Land Vertebrates,* pp. 255–292. London: Academic Press, 1980.

Raff, R., and T. C. Kaufman. *Embryos, Genes, and Evolution: The Developmental Genetic Basis of Evolutionary Change.* London: Macmillan, 1983.

Rainger, Ronald. "The Continuation of the Morphological Tradition in American Paleontology." *Journal of the History of Biology,* 14 (1981): 129–158.

———. *An Agenda for Antiquity: Henry Fairfield Osborn and Vertebrate Paleontology at the American Museum of Natural History, 1890–1935.* Tuscaloosa: University of Alabama Press, 1991.

———. "Biology, Geology, or Neither, or Both: Vertebrate Paleontology at the University of Chicago, 1892–1950." *Perspectives on Science,* 1 (1993): 478–519.

Rainger, Ronald, Keith R. Benson, and Jane Maienschein, eds., *The American Development of Biology.* Philadelphia: University of Pennsylvania press, 1988.

Rasmussen, Nicolas. "The Decline of Recapitulationism in Early Twentieth-Century Biology: Disciplinary Conflict and Consensus on the Battleground of Theory." *Journal of the History of Biology,* 24 (1991): 51–89.

Raymond, Percy E. *The Appendages, Anatomy, and Relationships of Trilobites.* Memoirs of the Connecticut Academy of Arts and Sciences, 7 (1920).

Rehbock, Philip F. "Huxley, Haeckel, and the Oceanographers: The Case of *Bathybius haeckelii.*" *Isis,* 66 (1975): 504–533.

Reif, Wolf-Ernst. "Evolutionary Theory in German Paleontology." In M. Grene, ed., *Dimensions of Darwinism,* pp. 173–204. Cambridge: Cambridge University Press, 1983.

———. "The Search for a Macroevolutionary Theory in German Paleontology." *Journal of the History of Biology,* 19 (1986): 79–130.

Richards, Robert J. *The Meaning of Evolution: The Morphological Construction and Ideological Reconstruction of Darwin's Theory.* Chicago: University of Chicago Press, 1992.

Ridley, Mark. "Embryology and Classical Zoology in Britain." In T. J. Horder et al., *A History of Embryology.* Cambridge: Cambridge University Press, 1986, pp. 35–67.

Ritvo, Lucille B. *Darwin's Influence on Freud.* New Haven: Yale University Press, 1990.

Rolleston, George. *Scientific Papers and Addresses.* Oxford: Clarendon Press, 1884. 2 vols.

———. *Forms of Animal Life.* 2d ed. Oxford: Clarendon Press, 1888.

Romer, Alfred S. *Vertebrate Paleontology.* Chicago: University of Chicago Press, 1933. 2d ed., 1945.

———. *Man and the Vertebrates.* Chicago: University of Chicago Press, 1933.

———. *Notes and Comments on Vertebrate Paleontology.* Chicago: University of Chicago Press, 1968.

Romer, Alfred S., and Frank Byrne. "The Pes of Diadectes: Notes on the Primitive Tetrapod Limb." *Palaeobiologica,* 4 (1931): 24–48.

Rosa, Daniele. *La Reduzione progressiva della Variabilita a i suoi rapporti coll'Eztinzione e coll'Origine della Specie.* Turin: Carlo Clausen, 1899.

Rudwick, Martin J. S. *The Meaning of Fossils: Episodes in the History of Paleontology.* 2d ed. New York: Science History Publications, 1976.

———. *Scenes from Deep Time: Early Pictorial Representations of the Prehistoric World.* Chicago: University of Chicago Press, 1992.

Rupke, Nicolaas A. "The Road to Albertopolis: Richard Owen (1804–1892) and the Founding of the British Museum of Natural History." In Rupke, ed., *Science, Politics and the Public Good,* pp. 63–89. London: Macmillan, 1988.

———. "Richard Owen's Vertebrate Archetype." *Isis,* 84 (1993): 231–251.

──────. *Richard Owen: Victorian Naturalist.* New Haven: Yale University Press, 1994.

Ruse, Michael. *The Darwinian Revolution: Science Red in Tooth and Claw.* Chicago: University of Chicago Press, 1979.

──────. "Molecules to Men: The Concept of Progress in Evolutionary Biology." Cambridge, Mass.: Harvard University Press, in preparation.

Russell, E. S. *Form and Function: A Contribution to the History of Animal Morphology.* London: John Murray, 1916.

Russell, F. S. "The Plymouth Laboratory of the Marine Biological Association." *Proceedings of the Royal Society of London,* 135 B (1947–48): 12–25.

Rütimeyer, Ludwig. *Ueber die Herkunft unserer Thierwelt: Eine Zoogeographische Skizze.* Basel: H. Georg, 1867.

Saporta, Gaston de. *Le Monde des plantes avant l'apparition de l'homme.* Paris: Masson, 1879.

Saporta, Gaston de, and A.-F. Marion. *L'évolution du règne vegetal: Les Phanerogames.* Paris: Felix Alcan, 1885. 2 vols.

Säve-Soderburgh, G. "Preliminary Note on Devonian Stegocephalians from East Greenland." *Meddelelser om Gronland,* 94 (1932): 1–107.

──────. "Some Points of View Concerning the Evolution of the Vertebrates and the Classification of the Group." *Arkiv for Zoologi,* 26 (1934): 1–20.

Schaeffer, B. "The Evolution of Concepts Related to the Origin of the Amphibia." *Systematic Zoology,* 14 (1965): 115–118.

Scharff, Robert Francis. *The History of the European Fauna.* London: Walter Scott, 1899.

──────. "Some Remarks on the Atlantis Problem." *Proceedings of the Royal Irish Academy,* 24 B (1902–4): 268–302.

──────. *European Animals: Their Geological History and Geographical Distribution.* London: Constable, 1907.

──────. "On an Early Tertiary Land Connection between North and South America." *American Naturalist,* 42 (1909): 513–531.

──────. *Distribution and Origin of Life in America,* London: Constable, 1911.

Schindewolf, Otto H. *Basic Questions in Paleontology: Geologic Time, Organic Evolution, and Biological Systematics.* Trans. Judith Schaefer. Foreword by Stephen Jay Gould. Afterword by W.-E. Reif. Chicago: University of Chicago Press, 1993.

Schmalhausen, I. I. *The Origin of Terrestrial Vertebrates.* Trans. Leo Kelso. New York: Academic Press, 1968.

Schuchert, Charles. "Paleogeography of North America." *Bulletin of the Geological Society of America,* 20 (1910): 427–606.

──────. "Climates of Geological Time." In E. Huntingdon, ed., *The Climatic Factor,* pp. 265–298. Washington: Carnegie Insitute Publication No. 192, 1914.

———. Review of W. D. Matthew, "Climate and Evolution." *American Journal of Science,* ser. 4, 40 (1915): 83–85.

———. *A Textbook of Geology.* Part II, *Historical Geology.* 2d ed. rev., New York: John Wiley, 1924.

———. "The Hypothesis of Continental Displacement." In W. A. J. M. van Waterschoot van der Gracht et al., *Theory of Continental Drift: A Symposium on the Origin and Movement of Land Masses both Inter-Continental and Intra-Continental, as Proposed by Alfred Wegener.* Tulsa, Oklahoma: American Association of Petroleum Gelogists, 1928, pp. 104–144.

———. "Gondwana Land Bridges." *Bulletin of the Geological Society of America,* 43 (1932): 875–916.

Schuchert, Charles, and Clara Mae Levene. *O. C. Marsh: Pioneer in Paleontology.* New Haven: Yale University Press, 1940.

Sclater, Philip Lutley. "On the General Geographical Distribution of the Members of the Class Aves." *Journal of the Linnean Society of London, Zoology,* 2 (1857): 130–145.

Sclater, William Lutley and Philip Lutley Sclater. *The Geography of Mammals.* London: Kegan Paul, Trench, Trubner, 1899.

Scott, William Berryman. "On the Osteology of *Poebrotherium:* A Contribution to the Phylogeny of the Tylopoda." *Journal of Morphology,* 5 (1891): 1–78.

———. "On the Osteology of *Mesohippus* and *Leptomeryx,* with Observations on the Modes and Factors of Evolution in the Mammalia." *Journal of Morphology,* 5 (1891): 310–406.

———. "On Variations and Mutations." *American Journal of Science,* ser. 3, 48 (1894): 355–374.

———. "The Palaeontological Record. I. Animals." In A. C. Seward, ed., *Darwin and Modern Science,* pp. 200–222. Cambridge: Cambridge University Press, 1909.

———. *The Theory of Evolution, with Special Reference to the Evidence upon Which It Is Founded.* New York: Macmillan, 1917.

———. *A History of Land Mammals in the Western Hemisphere.* New York: Macmillan, 1929.

———. *Memoirs of a Paleontologist.* Princeton: Princeton University Press, 1939.

———, ed. *Reports of the Princeton University Expeditions to Patagonia, 1896–1899.* Princeton: Princeton University Press, 1903–6. 8 vols.

Secord, James. "Behind the Veil: Robert Chambers and *Vestiges.*" In J. R. Moore, ed., *History, Humanity and Evolution,* pp. 165–194. Cambridge: Cambridge University Press, 1989.

Sedgwick, Adam. "The Origin of Metameric Segmentation." *Quarterly Journal of Microscopical Science,* 24 (1884): 43–82.

———. "The Development of Peripatus Capensis." *Quarterly Journal of Micro-*

scopical Science, 25 (1885): 449–468; 26 (1886): 175–212; 27 (1887): 467–550; and 28 (1888): 373–396.

———. "On the Law of Development Commonly Known as Von Baer's Law; and on the Significance of Ancestral Rudiments in Embryonic Development." *Quarterly Journal of Microscopical Science*, new ser., 36 (1894): 35–52.

———. *Student's Textbook of Zoology*. London: George Allen, 1898–1909. 4 vols.

———. "The Influence of Darwin on the Study of Embryology." In A. C. Seward, ed., *Darwin and Modern Science*, pp. 171–184. Cambridge: Cambridge University Press, 1909.

Seeley, Harry Govier. "Remarks on Prof. Owen's Monograph on Dimorphodon." *Annals and Magazine of Natural History*, ser. 4, 6 (1870): 129–152.

———. *Ornithosauria*. Cambridge: Deighton Bell, 1870.

———. "Note on a Femur and a Humerus of a Small Mammal from the Stonesfield Slate." *Quarterly Journal of the Geological Society of London*, 35 (1879): 456–463.

———. "Professor Carl Vogt on the Archaeopteryx." *Geological Magazine*, decade 2, 8 (1881): 300–309.

———. "On the Classification of the Fossil Animals Commonly Named Dinosauria." *Proceedings of the Royal Society*, 42 (1887–88): 165–171.

———. Croonian Lecture: "Researches on the Structure, Organization, and Classification of the Fossil Reptilia. II. On Pareiasaurus bombidens (Owen), and the Significance of Its Affinities to Amphibians, Reptiles, and Mammals." *Philosophical Transactions of the Royal Society*, 179 B (1888): 59–109.

———. "Researches . . . III. On Parts of the Skeleton of a Mammal from the Triassic Rocks of Klipfontein, Fraserburg, South Africa (Theriodesmus phylarchus), Illustrating the Reptilian Inheritance of the Mammalian Hand." *Philosophical Transactions of the Royal Society*, 179 B (1888): 141–155.

———. "Researches . . . IV. On the Anomodont Reptilia and Their Allies." *Philosophical Transactions of the Royal Society*, 180 B (1889): 215–296.

———. "The Ornithosaurian Pelvis." *Annals and Magazine of Natural History*, ser. 6, 7 (1891): 237–255.

———. "Researches . . . VI. Further Observations on Pareiasaurus." *Philosophical Transactions of the Royal Society*, 183 B (1892): 311–370.

———. "Researches . . . IX. On the Therosuchia." *Philosophical Transactions of the Royal Society*, 185 B (1894): 987–1018.

———. "Researches . . . IX, pt.2. The Reputed Mammals from the Karoo Formation of the Cape Colony." *Philosophical Transactions of the Royal Society*, 185 B (1894): 1019–1028.

———. "Researches . . . IX (abstract). On the Complete Skeleton of an Anomodont Reptile (Aristodesmus Rutimeyeri, Wiedersheim) from the Bunter Sandstone of Reichen, near Basel, Giving Evidence of the Relation of the Anomodontia to the Monotremata." *Proceedings of the Royal Society,* 59 (1896): 167–169.

———. "Discussion on the Origin of Mammals." *Proceedings of the Fourth International Congress of Zoology, Cambridge, 1898.* London, 1899, pp. 68–70.

———. *Dragons of the Air: An Account of Extinct Flying Reptiles.* London: Methuen, 1901.

Semon, Richard, ed. *Zoologische Forschungsreisen in Australien und dem Malayischen Archipel.* Jena: G. Fischer, 1893–1915. 5 vols. (I. Ceratodus; II and III, Monotremen und Marsupalier.)

———. *In australischen Busch und an den Küsten des Korallenmeeres: Reiseerlebnisse und Beobachtungen einer Naturforschers in Australien, Neu-Guinea und den Molukken.* Leipzig: W. Engelmann, 1896.

———. *In the Australian Bush and on the Coast of the Coral Sea: being the Experiences and Observations of a Naturalist in Australia, New Guinea, and the Molucas.* London: Macmillan, 1899.

———. "Über das Verwandschaftverhältnis der Dipnoer und Amphibien." *Zoologische Anzeiger,* 24 (1901): 180–188.

———. *Die Mneme als erhaltendis Prinzip im Wechsel des organischen Geschehens.* 3d ed. Leipzig: W. Engelmann, 1911.

———. *The Mneme.* London: Allen & Unwin/New York: Macmillan, 1921.

Semper, Carl. "Die Stammesverwandtschaft der Wirbelthiere und Wirbellosen," *Arbeiten aus dem Zoologisch-zootomischen Institut in Würzburg,* 2 (1875): 25–76.

———. "Die Verwandtschaftsbeziehungen der gegliederten Thiere," *Arbeiten aus dem Zoologisch-zootomischen Institut in Würzburg,* 3 (1876): 9–404.

Seward, A. C. President's Address, Botany Section. *Report of the British Association for the Advancement of Science,* 1903 meeting: 824–849.

———, ed. *Darwin and Modern Science.* Cambridge: Cambridge University Press, 1909.

Sharov, A. G. *Basic Arthropodan Stock.* Oxford: Pergamon Press, 1966.

Sharpe, R. Bowdler. *A Review of Recent Attempts to Classify Birds.* Budapest: Second International Orthithological Congress, 1891.

———. "On the Zoo-Geographical Areas of the World Illustrating the Distribution of Birds." *Natural Science,* 3 (1893): 100–108.

Sheets-Pyenson, Susan. *Cathedrals of Science: The Development of Colonial Natural History Museums in the Late Nineteenth Century.* Montreal: McGill-Queens University Press, 1989.

Shimer, H. W. "Old Age in Brachiopoda—A Preliminary Study." *American Naturalist,* 40 (1906): 95–121.

Shor, Elizabeth. *The Fossil Feud between E. D. Cope and O. C. Marsh.* Hicksville, N.Y.: Exposition Press, 1974.

Simpson, George Gaylord. *A Catalogue of the Mesozoic Mammalia in the Geological Department of the British Museum*. London: British Museum, 1928.

——. *Tempo and Mode in Evolution*. New York: Columbia University Press, 1944.

——. "Fossil Penguins." *Bulletin of the American Museum of Natural History*, 87 (1946): 1–99.

——. *The Meaning of Evolution*. New Haven: Yale University Press, 1949.

——. *Concession to the Improbable*. New Haven: Yale University Press, 1978.

——. *Discoverers of the Lost World: An Account of Some of Those who Brought Back to life South American Mammals Long Buried in the Abyss of Time*. New Haven: Yale University Press, 1984.

——. *Simple Curiosity: Letters from George Gaylord Simpson to His Family, 1921–1970*. Ed. Leo F. Laporte. Berkeley: University of California Press, 1987.

Simroth, Heinrich. *Die Entstehung der Landthiere: ein biologische Versuch*. Leipzig: Wilhelm Engelmann, 1891.

Sinclair, William J. "The Marsupial Fauna of the Santa Cruz Beds." *Proceedings of the American Philosophical Society*, 44 (1905): 73–81.

Smith, Andrew B. *Systematics and the Fossil Record: Documenting Evolutionary Patterns*. Oxford: Blackwell Scientific, 1994.

Smith, Geoffrey. *Primitive Animals*. Cambridge: Cambridge University Press, 1911.

Smith, James Perrin. "The Biogenetic Law from the Standpoint of Paleontology." *Journal of Geology*, 8 (1900): 413–425.

Snodgrass, R. E. "Evolution of the Annelida, Onychophora, and Arthropoda." *Smithsonian Miscellaneous Collections*, 97, no. 6 (1938).

Sober, Elliot. *Reconstructing the Past: Parsimony, Evolution, and Inference*. Cambridge, Mass.: MIT Press, 1989.

Sollas, W. J. "On the Origin of Freshwater Faunas: A Study in Evolution." *Scientific Transactions of the Royal Dublin Society*, ser. 2, 3, (1883–87): 87–118.

——. *Ancient Hunters and Their Modern Representatives*. London: Macmillan, 1911.

Spengel, J. W. *Die Enteropneusten des Golfes von Neapel und der angrezenden Meeres-Abschnitte. (Flora und Fauna des Golfes von Neapel*, vol. 18) Berlin: R. Friedlander, 1893.

Stearn, William T. *The Natural History Museum at South Kensington*. London: Heinemann, 1981.

Steinmann, Gustav. *Die geologischen Grundlagen der Abstammungslehre*. Leipzig: Wilhelm Engelmann, 1908.

Stenhouse, John. "Darwin's Captain: F. W. Hutton and the Nineteenth-Century Darwinian Debates." *Journal of the History of Biology*, 23 (1990): 411–442.

Stensiö, Erik Andersson. *The Downtonian and Devonian Vertebrates of Spitsbergen. Pt. 1. Family Cephalaspidae.* Oslo: Jacob Dybwad, 1927. 2 vols.

Suess, Eduard. *The Face of the Earth.* Trans. Hertha B. C. Sollas. Oxford: Clarendon Press, 1904–1924. 5 vols.

Swetlitz, Marc. "Julian Huxley and the End of Evolution." *Journal of the History of Biology,* 28 (1995): 181–217.

Swinnerton, Henry Hurd. *Outlines of Palaeontology.* London: Edward Arnold, 1923.

Swinton, W. E. "Harry Govier Seeley and the Karoo Reptiles." *Bulletin of the British Museum (Natural History). Historical Series,* 3, no. 1 (1962): 1–39.

Tauber, Alfred I., and Leon Chernyak. *Metchnikoff and the Origins of Immunology: From Metaphor to Theory.* New York: Oxford University Press, 1991.

Thatcher, James K. "Median and Paired Fins: A Contribution to the History of Vertebrate Limbs." *Transactions of the Connecticut Academy of Arts and Sciences,* 3 (1874–78): 281–308.

Theobald, F. V. *A Monograph on the Culcidae, or Mosquitoes.* London: British Museum (Natural History), 1901.

Thiele, Johannes. "Die Stammesverwandtschaft der Mollusken." *Jenaische Zeitschrift für Naturwissenschaft,* 25 (1891): 480–543.

Thiselton-Dyer, W. T. "Lecture on Plant Distribution as a Field for Geographical Research." *Proceedings of the Royal Geographical Society,* 22 (1878): 412–445.

Thompson, D'Arcy Wentworth. "On a Supposed Resemblance between the Marine Faunas of the Arctic and Antarctic Regions." *Proceedings of the Royal Society of Edinburgh,* 22 (1897–99): 311–349.

Thompson, Keith Stewart. "A Critical Review of the Diphyletic Theory of Rhipidistrian-Amphibian Relationships." In T. Orvig, ed., *Current Problems in Lower Vertebrate Phylogeny,* pp. 285–305. Stockholm: Almqvist and Wiksell/New York: Interscience Publishers, 1968.

————. *Living Fossil: The Story of the Coelacanth.* New York: Norton, 1991.

Thomson, Charles Wyville. "On Deep Sea Climates." *Nature,* 2 (1870); 257–261.

Thomson, J. Arthur. *Outline of Zoology.* 7th ed. Edinburgh: Frowd, Hodden, and Stoughton, 1921.

Thomson, William. "On the Secular Cooling of the Earth." *Philosophical Magazine,* ser. 4, 25 (1863): 1–14.

Thorpe, M. R., ed. *Organic Adaptation to the Environment.* New Haven: Yale University Press, 1924.

Todes, Daniel P. "V. O. Kovalevskii: The Genesis, Content, and Reception of His Paleontological Work." *Studies in History of Biology,* 2 (1978): 99–165.

Tothill, John T. "The Ancestry of Insects with Particular Reference to Chilopods and Trilobites." *American Journal of Science,* ser. 4, 42 (1916): 373–383.

Traquair, Ramsay H. "A Further Description of *Palaeospondylus gunni*, Traquair." *Proceedings of the Royal Physical Society of Edinburgh,* 12 (1894): 87–94, 312–321.

————. "Report on Fossil Fishes Collected by the Geological Survey of Scotland in the Silurian Rocks of the South of Scotland." *Transactions of the Royal Society of Edinburgh,* 39 (1899): 827–864.

————. "The Bearing of Fossil Ichthyology on the Problem of Evolution." *Geological Magazine,* ser. 4, 7 (1900): 463–470, 516–524.

Tristram, H. B. "The Polar Origin of Life considered in Its Bearing on the Distribution and Migration of Birds." *Ibis,* ser. 5, 5 (1887): 236–242.

Trouessart, E.-L. *La géographie zoologique.* Paris: J.-B. Ballière, 1890.

Troyer, James R. "On the Name and Work of J. E. S. Moore (1870–1947)." *Archives of Natural History,* 18 (1991): 31–50.

Unger, Franz. "The Sunken Island of Atlantis." *Journal of Botany,* 3 (1865): 12–26.

————. "New Holland in Europe." *Journal of Botany,* 3 (1865): 39–70.

Uschmann, Georg. *Geschichte der Zoologie und der zoologischen Anstalten in Jena, 1779–1919.* Jena: Gustav Fischer, 1959.

Vogt, Carl. "L'Archaeopteryx macroura—un intermédiare entre les oiseaux et les reptiles." *Révue Scientifique,* 17 (1879): 241–248.

Waagen, Wilhelm. "Die Formenreihe des *Ammonites subradius*." *E. W. Beneke's Geognostische-Paleontologische Beiträge,* 2 (1869): 179–256.

————. *Jurassic Fauna of Kutch. I. Cephalopoda. Palaeontologica Indica (Memoirs of the Geological Survey of India).* Calcutta. 1875.

Wachsmuth, F., and C. Springer. "The North American Crinoidea Camerata." *Memoirs of the Museum of Comparative Zoology,* 20 and 21 (1897).

Waisbren, Stephen James. "The Importance of Morphology in the Evolutionary Synthesis as Demonstrated by the Contributions of the Oxford Group: Goodrich, Huxley, and de Beer." *Journal of the History of Biology,* 21 (1988): 291–330.

Walcott, Charles Doolittle. "The Trilobite: New and Old Evidence Relating to Its Organization." *Bulletin of the Museum of Comparative Zoology at Harvard College,* 8 (1880–81): 190–214.

————. "Cambrian Geology and Paleontology." I, no. 1. "Abrupt Appearance of the Cambrian Fauna on the North American Continent." *Smithsonian Miscellaneous Collections,* 53 (1910): 1–14.

————. "Cambrian Geology and Paleontology." I, no. 3. "Holothurians and Medusae." *Smithsonian Miscellaneous Collections,* 53 (1910): 41–68.

————. "Cambrian Geology and Paleontology." I, no. 4. "Middle Cambrian Annelids." *Smithsonian Miscellaneous Collections,* 53 (1910): 109–144.

————. "Cambrian Geology and Paleontology." I, no. 6. "Middle Cambrian

Branchiopoda, Malacostraca, Trilobita, and Merostoma." *Smithsonian Miscellaneous Collections*, 53 (1910): 145–228.

———. "Cambrian Geology and Paleontology." IV, no. 4. "Appendages of Trilobites." *Smithsonian Miscellaneous Collections*, 67 (1924): 115–216.

Wallace, Alfred Russel. *The Malay Archipelago: The Land of the Orang-Utan and the Bird of Paradise.* London, 1869. Reprinted, Singapore: Oxford University Press, 1989.

———. *Contributions to the Theory of Natural Selection.* London: Macmillan, 1870.

———. *The Geographical Distribution of Animals.* London: Macmillan, 1876. 2 vols.

———. "On the Value of the 'Nearctic' as one of the Primary Zoological Regions." *Nature*, 27 (1883): 482–483.

———. *Darwinism: An Exposition of the Theory of Natural Selection with Some of Its Applications.* London: Macmillan, 1889.

———. *Natural Selection and Tropical Nature.* New ed. London: Macmillan, 1895.

———. *The World of Life: A Manifestation of Creative Power, Directive Mind, and Ultimate Purpose.* London: Chapman and Hall, 1910.

———. *Island Life: Or the Phenomena and Causes of Insular Faunas and Floras.* 3d ed., rev. London: Macmillan, 1911.

———. *My Life: A Record of Events and Opinions.* New York: Dodd, Mead & Co., 1905. 2 vols.

Waters, C. Kenneth, and Albert Van Helden, eds. *Julian Huxley: Biologist and Statesman of Science.* Houston: Rice University Press, 1992.

Waterschoot van der Gracht, W. A. J. M. van, et al., *Theory of Continental Drift: A Symposium on the Origin and Movement of Land Masses both Inter-Continental and Intra-Continental, as Proposed by Alfred Wegener.* Tulsa, Oklahoma: American Association of Petroleum Gelogists, 1928.

Watson, D. M. S. "On the Primitive Tetrapod Limb." *Anatomische Anzeiger*, 46 (1913): 24–27.

———. "The Deinocephalia, an Order of Mammal-Like Reptiles." *Proceedings of the Zoological Society of London*, 1914: 749–786.

———. "The Evolution of the Tetrapod Shoulder Girdle and Fore-limb." *Journal of Anatomy*, 52 (1917): 1–63.

———. "Structure, Evolution, and Origin of the Amphibia.—The 'Orders' Rachitomi and Sterospondyli." *Philosophical Transactions of the Royal Society*, 209 B (1919): 1–74.

———. Croonian Lecture. "The Evolution and Origin of the Amphibia." *Philosophical Transactions of the Royal Society*, 214 B (1926): 189–257.

———. "The Acanthodian Fishes." *Philosophical Transactions of the Royal Society*, 228 B (1937): 49–146.

Webster, Gerry. "The Relations of Natural Forms." In Mae-Wan Ho and Peter T. Saunders, eds., *Beyond Neo-Darwinism: An Introduction to the New Evolutionary Paradigm.* London: Academic Press, 1984, pp. 193–217.

Wegener, Alfred. *The Origin of Continents and Oceans.* Trans. John Biram. New York: Dover, 1966.

Weindling, Paul J. *Darwinism and Social Darwinism in Imperial Germany: The Contribution of the Cell Biologist Oscar Hertwig (1849–1922).* Stuttgart and New York: Gustav Fischer, 1991.

Wells, Herbert George. *The Outline of History.* 4th revision. London: Cassell, 1925.

———. *The Short Stories of H. G. Wells.* London: Ernest Benn, 1927.

———. *An Experiment in Autobiography.* London: Gollancz, 1934. 2 vols.

———. *H. G. Wells: Early Writings in Science and Science Fiction.* Ed. Robert M. Philmus and David Y. Hughes. Berkeley: University of California Press, 1975.

Wells, H. G., J. Huxley, and G. P. Wells. *The Science of Life.* London: Cassell, 1931.

Westoll, T. S. "The Origin of the Tetrapods." *Biological Reviews of the Cambridge Philosophical Society,* 18 (1943): 78–98.

White, Errol Ivor. "The Ostracodem *Pteraspis* KNER and the Relationships of the Agnathous Vertebrates." *Philosophical Transactions of the Royal Society,* 225 B (1935): 381–457.

Wiedersheim, Robert. *Das Gliedmassen Skelet der Wirbelthiere.* Jena: Gustav Fischer, 1892.

———. *The Structure of Man: An Index to His Past History.* Trans. H. and M. Bernard. London: Macmillan, 1895.

Wilder, Harris Hawthorne. *History of the Human Body.* New York: Henry Holt, 1909.

Willey, Arthur. *Amphioxus and the Ancestry of the Vertebrates: with a Preface by Henry Fairfield Osborn.* New York: Macmillan, 1894.

———. *Convergence in Evolution.* London: John Murray, 1911.

Williams-Ellis, Amabel. *Darwin's Moon: A Biography of Alfred Russel Wallace.* London: Blackie, 1966.

Willis, J. C. *Age and Area: A Study in Geographical Distribution and Origin of Species.* Cambridge: Cambridge University Press, 1922.

Williston, Samuel W. "The Faunal Relations of the Early Vertebrates." *Journal of Geology,* 17 (1909): 389–402.

Willmer, Pat. *Invertebrate Relationships: Patterns in Animal Evolution.* Cambridge: Cambridge University Press, 1990.

Winsor, Mary P. *Reading the Shape of Nature: Comparative Zoology at the Agassiz Museum.* Chicago: University of Chicago Press, 1991.

———. "The Lessons of History." In R. W. Scotland, D. J. Siebert, and D. M.

Williams, eds., *Models in Phylogeny Reconstruction*. Systematics Association Special Volume No. 52. Oxford: Clarendon Press, 1994, pp. 1–9.

Wollaston, A. F. R. *Life of Alfred Newton: Professor of Comparative Anatomy, Cambridge University, 1866–1907*. London: John Murray, 1921.

Woodward, Arthur Smith. *Catalogue of Fossil Fishes in the British Museum (Natural History)*. London: British Museum (Natural History), 1889–1901. 4 vols.

———. "The Evolution of Fins." *Natural Science*, 1 (1892): 28–35.

———. *Outlines of Vertebrate Palaeontology*. Cambridge: Cambridge University Press, 1894.

———. "The Problem of the Primaeval Sharks." *Natural Science*, 6 (1895): 38–43.

———. "Edward Drinker Cope." *Natural Science*, 10 (1897): 377–381.

———. "The Evolution of Mammals in South America." In *The Darwin-Wallace Celebration Held on Thursday, 1st July, 1908*. London: Linnean Society, 1908, pp. 79–80.

———. President's Address, Geological Section. *Report of the British Association for the Advancement of Science*, 1909 meeting: 462–471.

Wortman, J. L. "*Psittacotherium*, a Member of a New and Primitive Suborder of the Edentata." *Bulletin of the American Museum of Natural History*, 8 (1896): 259–262.

———. "Studies of Eocene Mammalia in the Marsh Collection, Peabody Museum: Primates." *American Journal of Science*, ser. 4, 15 (1903): 163–176, 399–414, 419–436.

Young, David. *The Discovery of Evolution*. Cambridge: Cambridge University Press, 1992.

Young, Robert M. *Darwin's Metaphor: Nature's Place in Victorian Culture*. Cambridge: Cambridge University Press, 1985.

Zeigler, Heinrich Ernst. "Über den derzeitigen Stand der Cölomfrage." *Verhandlung der Deutschen zoologischen Gesselschaft*, 8 (1898): 14–78.

Zittel, Karl von. *Handbuch der Palaeontologie*. Munich and Leipzig: R. Oldenbourg, 1876–93. 4 vols.

———. "The Geological Development, Descent, and Distribution of the Mammalia." *Geological Magazine*, ser. 3, 10 (1893): 401–412, 455–468, 501–514.

———. *Textbook of Palaeontology*. Trans. Lucy P. Bush and Marguerite L. Engler; rev. Arthur Smith Woodward. London: Macmillan, 1926. 3 vols.

Abel, O., 275, 317, 349, 358–9, 436–7, 447
acquired characters, inheritance of. *See* Lamarckism
Actinozoa, 181, 183–4
adaptation, 8, 46–7, 57, 62–4, 253–7; in arthropods, 118; in birds, 260–2; and convergence, 71, 127–8; in reptiles, 333. *See also* convergence; environment,as factor in evolution; function
adaptive radiation: of arthropods, 101, 135–6; in Cambrian, 194–5; of mammals, 319, 329, 336–8
Africa, 297, 299–300, 302–3, 382, 389, 391–2, 405, 408
Agassiz, A., 30, 81–2, 321–2, 325, 447
Agassiz, L., 30, 208, 231, 389, 437, 447
Alectoromorphae, 385
Allen, G. E., 19, 82
Allen, J. A., 380, 382
Allis, E. P., 238
Ameghino, F., 409–10, 447
America: biogeography of, 277, 383–4, 397, 403; biology in, 26, 28–9, 30–1, 33–4. *See also* South America

American Museum of Natural History, 28, 30–1, 338, 415
amniote egg, 261–2
amphibians: as ancestors of mammals, 282–4, 294–7, 285, 300–1; derived from crossopterygians, 233–9, 245–6, 247–8; derived from lungfish, 231–3, 239–42, 247–8; diphyletic, 242–4; evolution of, 349; limbs of, 245–52; origin of, 204–5, 220, 229–31, 244–5, 253–8, 261–2, 445
Amphioxus, 55; as ancestral chordate, 144, 146, 148–53, 155–7, 186, 188–91, 199, 206–7, 222, 426; degenerate, 157, 182; derived from annelids, 169
analogy, 46–8, 69, 165, 269
ancestor, common, 56–7, 59–60, 61, 63; of amphibians, 251, of arthropods, 98–106, 124, 130; of mammals, 289; of molluscs, 91–3. *See also* common descent; phylogeny
Anderson, D. T., 139
Andrews, E. A., 167
annelids: ancestral to arthropods, 97, 114–5, 117–20, 121–2, 128–9, 134–5, 138; ancestral to vertebrates, 94, 157–71; origin of, 94, 95

anomodonts, 299–300
Antarctica, 372, 382, 389, 393, 395, 402, 407, 413–5
Apus, 118–9, 136, 137
arachnids: ancestral to vertebrates, 172–7; fossil, 104–5, 132–3; *Limulus* and, 54, 123–6; relation to other arthropods, 103–6, 113, 119–20, 126, 129–30
Archaeopteryx, 60, 263–5, 267, 268, 272, 273, 275–8
archetype, 43–4, 48–9, 54–5, 57, 144, 219. *See also* idealism
archipterygium, 203–5, 224–6, 232, 245
Arctic, as center of evolution, 402. *See also* north, as center of evolution
Argyll, Duke of. *See* Campbell, G. D. H.
Arkell, W. J., 328
Arldt, T., 398, 414, 416
"arms race," 364, 438
arrow-worm, 148
arthropods, 54, 97–140, 365; ancestral to vertebrates, 94–5, 171–80; classification of, 99–101, 103–6; fossil, 100–1, 104–5, 132–8; monophyletic, 106–12; 120, 138; polyphyletic, 68, 104–5, 112–20, 122–32, 139–40
ascidians, 149–54, 156–7, 189, 192–4, 199
Asia, 382–4, 401
Atlantic ocean, 372, 396, 398, 411–2
Atlantis, 396, 397
auditory ossicles, 295, 304–8
Australia: biogeography of, 386, 387, 390–3, 412; and lungfish, 220, 232, 239–4; and marsupials, 290; and platypus, 28, 286–7

Baer, K. E. von, 46, 76, 141, 154, 329, 447
Bain, A., 297

Balanoglossus, 154–5, 181, 182, 185–91, 197–8
Balfour, F. M., 37, 448; on arthropods, 112, 115, 122; on fish, 215, 223, 227, 232; on phylogeny, 78, 79–80; on vertebrate origins, 145, 167, 122
Barbour, T., 410
Barrell, J., 254–5, 448
Bastian, H. C., 85
Bates, H. W., 9
Bateson, W., 13, 34, 448; on origin of vertebrates, 185–7; rejection of morphology by, 74, 81, 146, 187; on saltations, 53, 144, 350
Bather, F., 83, 326, 327, 351–2, 357, 448
"*Bathybius haeckelii*," 85
Baur, G., 238, 268, 298–9, 302, 411, 448
Beard, J., 167–8
Beddard, F. E., 376, 384, 393, 403, 413, 414
Beebe, W., 276–7, 278, 448
Beecher, C. E., 133–4, 347–8, 357, 448
Beeson, R., 142, 150
Belloc, Hilaire, 444
Beneden, E. van, 124
Bensley, B. A., 290, 393
Benton, M., 258, 280, 366
Bergson, H., 429
Bernard, H. M., 21–2, 102, 118–20, 128, 134–5
Bernissart, 265, 316
Berrill, N. J., 201–2
Beurlen, K., 311–2, 358–9
biochemistry, 200–1
biogenetic law. *See* recapitulation theory
biogeography, 318, 319, 354, 361, 438; and centers of origin, 394–418; Darwinism and, 10, 374–9; and "lost worlds," 389–94; and zoological provinces, 379–89
biometry, 9

bipolar theory, 390, 395, 414
birds: biogeography of, 381, 385, 402, 413; origin and evolution of, 259–60, 262–80, 323–5, 364
Blanford, W. T., 408, 413, 448
blastopore, 87, 184, 191
bone, evolution of, 206, 208, 214–5, 216–7
Bourne, G. C., 37
bradytely, 426
brain: of birds and pterosaurs, 271; of early vertebrates, 158, 169, 178; of mammals, 311, 340–2
Branchiostoma. *See Amphioxus*
Britain: biogeography of, 372, 398–9; biology in, 25–8, 31–2, 34, 35–9
British Museum (Natural History), 25, 28, 31–2, 210, 213, 297
Brongniart, A., 362
Brooks, W. K., 33, 135–7, 165, 185, 192–5, 448
Broom, R.: on amphibians, 248, 275, 334, 448–9; on origin of mammals, 292–3, 300, 301–7, 310
Browne, J., 10, 371
Burgess shale, 135, 138, 139

Caldwell, W. H., 28, 61, 286–7, 302
Calman, W. T., 138
Cambrian explosion, 194
Cambridge University, 37–8, 271
camel, evolution of, 349
Campbell, G. D. H., Duke of Argyll, 57
Cannon, H. G., 65–6, 139
carbon dioxide, in atmosphere, 362–3, 395, 397
Carpenter, G. H., 130–1, 400
Cephalaspis, 212, 213, 214
Ceratodus. See lungfish
chain of being. *See* linear evolutionism
Challenger expedition, 28, 182, 389–90, 433
Chamberlin, T. C., 33; on role of

geological change, 137, 177, 218, 254, 362–3, 370, 403–4, 432–3, 449; on origin of vertebrates, 195–6
Chambers, R., 11, 25, 425
Chicago, University of, 33
chordates, 144, 148–57, 185–202. *See also* vertebrates
cladism, 5, 44, 51–2, 60, 68, 230, 323–5
Cladoselache, 216, 228
classification. *See* taxonomy
Claus, C., 35, 102, 110–12, 125–6, 449
cleidiotic egg, 261–2
coelacanth, 241
coelom, 88–9, 182
Colbert, E. H., 3, 280
Coleman, W., 41, 43
common descent, in arthropods, 98–106; as basis for classification, 42–4, 49–55, 58–62, 67, 69, 320–8; in vertebrates, 143–4. *See also* ancestor, common; taxonomy; tree of life
Competenzkonflict, 166, 225–6
Compsognathus, 265–6, 269
Comstock, J. H., 324
continental drift, 372, 375, 395, 405, 416–7
continents: permanence of, 379, 396, 416; sunken, 395, 397–9, 407, 411–6
convergence, 69–74, 317; in arthropods, 99–100, 114, 117–8, 126, 128–9, 139–40; in birds, 266–7, 269, 271–2, 274, 278; in fish, 232–3, 235, 237, 240–1; in human origins, 345–6; and Lamarckism, 66, 70, 100; in mammals, 295, 349; in origin of vertebrates, 143–4, 189; and taxonomy, 44, 315, 322, 324, 327. *See also* parallelism
cooling-earth theory, 368, 394–5, 400, 413–4

Cope, E. D., 316, 449; on amphibians, 207, 235; on birds, 266–7;
on fish, 212–3, 215; on Lamarckism, 252, 343–4, 345, 433; on law
of unspecialized, 207–8, 334, 428;
on mammals, 298, 344, 365; on
orthogenesis, 66, 70, 78, 356
crabs, 110, 386
Craw, R., 323
crayfish, 102, 321, 385–6
creation, 48. See also theistic evolutionism
Cretaceous-Tertiary boundary, 336–7
crocodiles, 333
Croizat, L., 375, 377, 405–6, 417
Croll, J., 395
crossopterygian fish, 229, 233–4,
235, 237, 239, 243, 245–6,
248–51
crustaceans, 103–6, 106–20, 133–5
Cuvier, G., 45–6, 62
cycles, in history of life, 318–9,
347–9, 355–70, 403–5, 429,
433, 435–8
cyclostomes, 178, 206, 208–12
Cyognathus, 305, 308

Dames, W., 268
Darwin, Charles, 1–2, 6–8, 16, 356,
421, 444, 449; on biogeography,
354, 360–1, 372–3, 374, 376,
401, 408–9, 412, 439; on fossil
record, 329–30; on mammals,
289; on origin of life, 85; on origin
of lungs, 239; on origin of vertebrates, 153; on taxonomy 49–51,
53, 63, 67, 75–6, 82, 110, 320
Darwinism, 6–12, 41–4, 421, 441–
6; and biogeography, 10, 372–6,
384–9, 405–6; opposition to,
176–7, 187, 223–4, 271–2, 284,
286–7, 297, 303, 312, 318, 343–
4, 347–50, 356–9, 376, 424,
443–5; and paleontology, 5, 7,
11–2, 254, 257–8, 284–6, 352,
354, 360–2; and taxonomy, 49–

53, 314–5, 317, 320–8. See also
modern synthesis; natural selection; social Darwinism; struggle
for existence
Darwinian revolution, 1–2, 3–4, 10–
1, 15–8, 435
Davidoff, M. von, 224
Dawkins, W. B., 398, 439
Dean, B., 210, 215, 216, 228, 326,
355, 449
De Beer, G., 37, 83, 444
degeneration, 61–2, 65, 96, 146,
153–4, 431–4; of Amphioxus,
161, 163–4, 206–7, 210; of annelids, 167; of ascidians, 152–3,
154, in fossil record, 353–4,
356–8; of wings in birds, 269–
70, 278
Delsman, H., 170
Dendy, A., 325–6, 356
Depéret, C., 317, 335, 358, 449
Desmond, A., 271, 284, 297, 299,
316, 333
developmentalism, 17–8, 63, 68, 76–
7, 442–3, 445. See also embryology; linear evolution; progress;
recapitulation theory
De Vries, H., 375
Deuterostomia, 87, 95, 191
Dimetrodon, 298
Dinocerata, 338, 340
dinosaurs, 316; as ancestors of birds,
262–3, 265–7, 270, 272, 277–
80; evolution of, 334, 349, 357;
extinction of, 338, 354–5, 363–4,
365–8, 436, 438, 440, 446; as rivals to mammals, 310–2
Dipleurula, 96
Diploblastica, 87, 183
Diplodocus, 316
Dipnoi. See lungfish
discontinuity, of fossil record, 353,
361. See also fossil record, cycles
in
dispersal, 373, 375, 377–8, 397–9,
402–6, 409, 417, 437

divergence. *See* adaptive radiation; specialization
Dixon, C., 395–6
Dohrn, A., 29, 449; on arthropods, 112, 114–5; on degeneration, 95, 163–4, 425, 431; on fins, 224–5, 225–6; on function, 64–5, 162–3, 308; on vertebrates, 94, 142–3, 146, 157–8, 160–6, 193, 210
Dollo, L., 449; on amphibians, 236–7, 245; on biogeography, 390, 410, 414; on dinosaurs, 265, 316; on irreversibility of evolution, 278, 318, 325, 339, 358; on marsupials, 289–90
Doyle, A. Conan, 389, 409, 426
Duncan, P. M., 292

ear, evolution of, 295, 303–8
echinoderms: origin of, 89; linked to chordates, 188, 190–1, 197–8, 201
echinoids, 322
edentates, 408–9
Eigenmann, C., 30, 326
Eisig, H., 168
elasmobranches, 176, 214–6, 222–3
elephants, 350
embryology, 8, 53, 74–83, 329, 344; and ancestry of invertebrate phyla, 91–5, 113; and ancestry of mammals, 286–7, 294–5, 301, 304; and ancestry of vertebrates, 148–53, 161–2, 166, 169; of fish, 223, 224–6. *See also* recapitulation theory
endothermy, 260, 271, 280
enteropneusts. *See Balanoglossus*
environment, as factor in evolution, 4–5, 16, 65, 73, 95, 380, 423, 429–33, 440; of amphibians, 206, 235, 237, 254–7; of arthropods, 135–6; of early vertebrates, 163–4, 192, 193–6, 201–2, 218; of mammals, 310–2, 361–3, 365–70, 394–5, 400, 413–4. *See*

also adaptation; convergence; flight; geological change; terrestrial habitat
Eohippus, 333, 342, 363
"*Eozoon canadense*," 84
Eryops, 250
eugenics, 432
Euparkeria, 275
eurypterids, 104–5, 125, 132–3, 137, 173, 176, 179, 218
Eusthenopteron, 235, 248–51
evolution. *See* Darwinism; developmentalism; Lamarckism; linear evolution; orthogenesis; phylogeny; tree of life
extinction, 319, 340, 347, 357–9, 366–70, 436–7
eyes, evolution of, 116–7

Falconer, H., 329
feathers, 263–4, 268, 276–8
Fernald, H. T., 127–8
Fichman, M., 371
fieldwork, 24, 27–9, 204–5, 239–40, 286–7
fins, origin of, 162, 219–29
fin–fold theory, 222–7
fish: evolution of, 176, 207, 209–18, 326, 334, 425; fins of, 219–29. *See also* crossopterygian fish; cyclostomes; elasmobranches; lungfish
Fisher, R. A., 442
flight, 262–3, 266, 268, 273–80
Flower, W. H., 32, 270, 301, 386, 387, 393, 449–50
Flynn, T. T., 290–1
Forbes, E., 372, 381, 398
Forbes, H. O., 413–4
Forrest, H. E., 400
fossils, classification of, 44, 321–2
fossil record: of amphibians, 242–4; ancestral forms in, 59–60, 315; of arthropods, 100–1, 104–5, 132–8; and biogeography, 391, 397, 403–5, 409, 412–6; of

fossil record (*continued*)
  birds, 263–4, 269, 273–80; cycles
  in, 318–9, 347–9, 352–66; of
  echinoids, 322; as evidence for
  evolution, 11–2, 207–8, 329–33;
  of fish, 209–28, 230, 235–8,
  248–51; imperfection of, 41, 60,
  329, 336, 353; of mammals, 317–
  8, 329–33, 335–9, 340–6; of
  molluscs, 315, 346–7; and origin
  of vertebrates, 173, 176; parallelism in, 70–1, 318–9, 327; of reptiles, 261–2, 265–8, 269–72,
  316–7, 333–4; trends in, 339–52.
  See also adaptive radiation; living
  fossils; *names of individual species
  and groups*
Freud, Sigmund, 35, 111
Friedmann, H., 71
function: change of, 57, 64–5, 161–
  3, 167, 304–8; as determinant of
  structure, 62–6, 139, 205–6, 242,
  252–3, 342–3. *See also* adaptation; morphology, functional
Furbringer, M., 225, 226, 248, 269,
  323, 450

Gadow, H., 180, 269, 307; on biogeography, 377–8, 386, 393,
  412, 414, 439
Galapagos islands, 372, 410–1
Garstang, W., 153, 198–9
Gaskell, W. H., 171–2, 177–80
Gastraea theory, 86–7, 156
gastrula, 85, 156, 173
Gaudry, A., 330, 410, 450
Gaupp, E., 35, 227, 260, 307–8,
  450
Gegenbaur, C., 21, 35, 58, 440; on
  arthropods, 120; on fins, 204,
  219–21, 224, 232, 246; on limbs,
  246–7; on vertebrates, 155, 159,
  165–6, 181
Gemmill, J. F., 198
genetics, 2, 3. *See also* modern
  synthesis

Geoffroy Saint–Hilaire, E., 59, 157,
  160, 165
geographical distribution. *See*
  biogeography
geological change, as influence on
  evolution, 367–8, 394–5, 407
geological surveys, 28, 212–3
geology, and biogeography, 377,
  378–9, 392–3, 395, 403–5, 407,
  417–8
germ layers, 59, 87, 148–50, 156,
  173, 178, 183
germ plasm, 9
Germany, 21, 25, 32, 34–5
Giard, A., 154, 450
gill arches, 221, 225, 227
Gill, T., 207, 232, 450
Gilmour, J. S. L., 320
Gislen, T., 198
Goethe, J. W. von, 46, 62
Goldschmidt, R., 312, 359, 432
Gondwanaland, 398
Goode, G. B., 32
Goodrich, E. S., 14, 26, 37, 53, 88,
  444, 450; on amphibians, 240–1,
  248; on fins, 227–8; on mammals,
  344; on recapitulation theory, 80,
  357; on reptiles, 309; on vertebrates, 145
Gould, S. J., 38, 75, 100–1, 103, 135,
  139–40, 339, 419–20, 425
Gray, A., 356, 397,
greenhouse effect, 395
Gregory, J. W., 400
Gregory, W. K., 227, 450; on birds,
  277; on fish, 216, 228–9; on function, 65, 250–1; on limbs, 248–
  52; on mammals, 291, 307–9,
  338, 344; on taxonomy, 326–7,
  425
Grinnell, J., 32
Grobben, K., 87, 191
Gunther, A., 32, 220, 321, 387, 411

Haacke, W., 402
Haeckel, E., 7, 13, 17, 21, 85–6, 451;

on amphibians, 231; on arthropods, 110, 120; on biogeography, 373, 412–3; on birds, 265, 269; on mammals, 282–3, 335; on phylogeny, 57–8, 60, 63–4, 89–94, 425, 426–7; on recapitulation theory, 74–5, 77, 79, 110; on reptiles, 261; on vertebrates, 148, 150–3, 156, 208–9
haemocoel, 89
hair, origin of, 294
Haldane, J. B. S., 83, 360, 364–5, 444
Handlirsch, A., 137, 323
Hansen, H. C., 129, 130
Hartlaub, G., 413
Harvard University. *See* Museum of Comparative Zoology
Hatschek, B., 35, 87, 88, 91–4, 111, 113–4, 451
Haughton, S., 395
head, evolution of: in arthropods, 119–20; in vertebrates, 144–5, 169–70, 172–3, 342, 350. *See also* brain
Heer, O., 397, 412, 451
Heider, K., 35, 94, 111, 113, 117, 126, 197
Heilmann, G., 277–9
Heilprin, A., 336–7, 376, 383–4, 406, 451
Hennig, W., 314, 324
heredity, 9. *See also* genetics, Lamarckism
Herrick, C. J., 365
Hertwig, O., 88, 209, 451
Hertwig, R., 156
Hesperornis, 267
Hilgendorf, F., 315, 451
Hill, J. P., 289, 290
Hipparion, 330
Holarctic province, 384, 401, 403–5
Holmgren, N., 242–3, 248
homology, 46–8, 228; in arthropods, 99–100, 127; between types, 59, 141–2, 165; and convergence, 72;

Darwinism and, 53–4; in early vertebrates, 143–4, 158–60, 172–3; in limbs, 245–7, 250–1, 270–4
homoplasy. *See* convergence
Hooker, J. D., 10, 372, 378, 397, 402, 413
"hopeful monster," 312, 432. *See also* saltations
Hopkins, W., 394
horns, evolution of, 343, 348, 350, 357–8
horse, evolution of, 330–3, 342, 363
horseshoe crab. *See Limulus*
Howes, G. B., 70, 294
Howarth, H., 398
Hubrecht, A. A. W., 164, 181–3, 295–6, 346, 451
human origins, 247, 296, 302, 345–6, 405, 419–20, 424–6, 429, 431, 435
Hutchinson, G. E., 135
Hutchinson, H. N., 272
Hutton, F. W., 128, 413
Huxley, J. S., 37, 197–8, 257–8, 280, 320, 327–8, 339, 390, 444, 451; on competition, 364–5, 438; on progress, 434
Huxley, T. H., 21, 22, 25, 36, 144, 451; on arthropods, 102, 114, 122; on "*Bathybius*," 85; on biogeography, 383, 384–6, 387, 391–2, 400, 412; on birds, 260, 262–7, 273, 428; on crocodiles, 333; on fish, 212, 232, 425; on horse evolution, 331–3; on laws of evolution, 342; on mammals, 260, 282, 283–6, 287, 290, 294, 296, 301, 334–5; on persistence of type, 263–4, 267, 328–9, 331, 334–5; on taxonomy, 321, 334; on vertebrates, 179
Hyatt, A., 66, 78, 82–3, 347, 355–6, 357, 437, 452
Hyperodapedon, 391

ice ages, 369, 372, 388, 395, 398–9
Ichthyornis, 267
Ichthyostega, 244
idealism, 43–4, 48–9, 54–6, 58–9,
    64, 171, 286, 312; in history, 312.
    See also Platonism
ideology. *See* Lamarckism, and soci-
    ety; social Darwinism; race theory;
    imperialism
Iguanodon, 265, 268, 316
Ihering, H. von, 94, 412, 414, 416
Imperial College, London, 21, 36, 38
imperialism, 435–40
insects, 32, 102, 103–4, 314, 323;
    origin of, 110, 113–4, 120–1,
    125, 131
instinct, 365
"invasions," 403–5, 409, 418,
    438–9
irreversibility of evolution, law of,
    236, 318, 325, 339
islands, biogeography of, 373. *See
    also* Australia; Galapagos islands;
    Madagascar; New Zealand

Jarvik, E., 243–4
jaws, evolution of 210, 213–4, 295,
    303–8
Jena, University of, 35
Johns Hopkins University, 33, 193
Johnson, M., 421
Joly, J., 367
Jones, F. Wood, 66, 70, 73, 345

Kerr, J. G., 241–2
Kesteven, H. L., 242
Kingsley, J. S., 127, 128–9, 234,
    294–6, 452
Klaatsch, H., 225, 247, 294
Knight, C. R., 309, 316, 368, 420
Koken, E., 357, 452
Kölliker, A. von, 150
Kowalevsky, A., 142, 148–50, 452
Kowalevsky, V., 330–1, 335, 342–3,
    363, 365, 452
Kuhn, T. S., 15

Kukenthal, W., 288–9, 344
Kuppfer, K. R. von, 155–6

labyrinthodonts, 231
Lackoff, G., 421
Lamarckism, 2, 8, 65–6, 252, 297,
    343, 349, 433, 443; in arthropod
    evolution, 128, 139; and conver-
    gence, 70, 73, 100; and degenera-
    tion, 270; and recapitulation
    theory, 75, 78, 82, 347, 437; and
    society, 423
lamprey. *See* cyclostomes
land. *See* terrestrial habitat
land–bridges, 372, 375, 378–9, 396,
    398–9, 404, 407, 415–6
Lang, A., 117, 452
Lang, W. D., 349, 357
Lankester, E. Ray, 26, 27, 29–30, 32,
    36–7, 443–4, 452; on biogeogra-
    phy, 384, 408, 409; on birds, 268,
    278; on coelom, 89; on degenera-
    tion, 95, 153–4, 164, 431–2; on
    fish, 212, 213, 227; on *Limulus,*
    54, 68, 99, 102, 123–6, 130, 137;
    on molluscs, 91, 94; on phylogeny,
    61–2, 69, 78, 80, 86, 87–8, 425,
    426; on vertebrates, 154, 164,
    171, 180, 182
larvae, of arthropods, 109–13, 116;
    of ascidians, 150–7, 199; evolu-
    tion of, 80
Laurie, M., 129, 132
laws, of evolution, 339–52
Le Conte, J., 357, 367, 394–5
Lemuria, 407, 410, 412–3
Leydig, F., 158–9
life, origin of, 85
limbs: of amphibians, 205, 231–2,
    239, 241, 242–3, 245–52; of ar-
    thropods, 117, 121, 129, 134; of
    birds, 265–6, 268–70, 275–6;
    homologies of, 53, of mammals,
    302, 308–9, 342–3
Limulus, 54, 68, 99, 123–6, 130, 173
linear evolution, 16, 46, 60–1, 75,

88, 91, 95–6, 175–6, 334, 419, 424–8. *See also* parallelism
Linnaeus, C., 45
living fossils, 60–1, 361, 425–6, 428; in arthropods, 104, 121–2, 124; in fish, 215, 220, 240–1; in mammals, 291, 361; in reptiles, 250
Loomis, F. B., 349, 357–8
Lorenz, K., 324
"lost world," 389–94, 409, 426
Lowe, P., 268
Lubbock, J., 25, 131
Lucas, F. A., 278–9
Lull, R. S., 317, 452; on amphibians, 249, 256; on birds, 278; on extinction, 367–8; on fish, 213; on mammals, 296–7, 310, 338; on trends in evolution, 356, 357, 363, 433, 440
lung, evolution of, 244–5, 255
lungfish, 205, 219–21, 222, 229–232, 234–42, 361
Lydekker, R., 25, 301, 357–8, 376, 384–5, 393, 403, 410, 452

MacBride, E. W., 38, 452; on arthropods, 112, 115, 115; on degeneration, 65, 95–6, 432; on Lamarckism, 138; on recapitulation theory, 82, 112, 115, 198–9; on vertebrate origins, 180, 191
MacLeod, J., 126
macromutation. *See* "hopeful monster;" saltations
Madagascar, 382, 392–3, 412–3
Maienschein, J., 13, 19, 20–1, 33
Malthus, T. R., 374
mammals: archaic, 336, 337–8, 339, 340–2; evolution of, 328–33, 334–8, 340–5, 363, 365–6, 376–7; in Mesozoic, 285, 291–3, 344, 392–3; multituberculate, 292–3; origin of, 259–60, 283–4, 294–311; taxonomy of, 281–2, 285–91

mammal-like reptiles, 280, 282, 284, 297–310
Manton, S. M., 66, 68, 103, 139–40
maps, 380, 382, 383, 399, 404, 411, 420
marine biology, 28, 29–30
Marsh, O. C., 316, 453; on birds, 267–8; on brain growth, 340, on Dinocerata, 338, 340; on horse evolution, 333, on Mesozoic mammals, 292, 298, 335, on natural selection, 431
marsupials, 281, 284–5, 289–91, 301–2, 335, 354, 392, 408, 426
mass extinctions, 366–70
materialism, 49–50, 57, 58
Matthew, W. D., 3, 27, 33, 453; on biogeography, 361, 370, 378, 392, 403–6, 414, 415, 416–7, 420; on extinction, 370; on mammals, 310–1, 337, 338; on marsupials, 290, 291; on progress, 434, 436, 438, 439; on trends in evolution, 352, 359
Maurer, F., 294
Mayr, E., 7, 20, 24, 69, 339, 405, 417, 421, 441–2
M'Coy, F., 390
Mehnert, E., 274
Mendelism. *See* genetics
Merostomata. *See* eurypterids
Merriam, C. H., 380
Mesozoic mammals, 285, 291–3, 310, 329, 344, 393
metamerism, 94, 95; in arthropods, 103–4, 112, 127, 136–7; in vertebrates, 142–4, 155, 156, 157, 167, 177, 180, 183, 193
metaphors, 52–3, 96, 338–9, 355–5, 364, 403–5
Metchnikoff, E., 86, 188
Meyerhoff, O., 200
Meyrick, E., 324–5
Miall, L. C., 131
migration. *See* dispersal; "invasions"
Miller, Hugh, 212

Miner, R. W., 250

Minot, C. S., 168–70, 171, 453

missing links, 428; in arthropods, 104, 118, 121–2; between reptiles and birds, 263–6

Mitchell, P. Chalmers, 325, 382, 453

Mitman, G., 177

Mivart, St. G. J., 57, 64, 68, 73, 162, 223–4, 270–1, 286, 287, 322, 387, 453; on birds, 270–1; on fins, 223–4; on origin of mammals, 287–8

moa, 269

modern synthesis, in evolution theory, 5, 328, 352, 441–2

molecular biology, 6, 140, 147, 200–1

molluscs, 78, 91–4, 95, 315

monotremes, 281, 284–8, 292, 294, 302

Moore, J. E. S., 408

Morgan, T. H., 13, 33, 102, 188, 453

morphology: of amphibians, 234, 238, 243–4, 245–53; of arthropods, 98–100, 101–132, 139–40; of birds, 269–73, 274–8. and convergence, 72, 139; Darwinism and, 12–4, 17, 20–3, 41–2, 43–4, 52–62, 141–2, 321, 444; decline of, 13–14, 19, 82, 146–7, 166, 174–5, 187, 227, 297, 443; of fish, 208–10, 214–5, 219–28, 231–8, 240–2; functional, 65–6, 161–3, 205–6, 250–3, 260–1, 307–8; of mammals, 231–91, 294–5, 303–8; origins of, 46–9, 62; of primitive vertebrates, 142, 144–5, 148–52, 157–63, 166–70, 185–7; professional structure of, 26, 29–32, 34–8

Morris, C., 244–5, 255

Moseley, H. N., 27, 37, 121–2, 390, 453

mosquitoes, 32

Moy-Thomas, J. A., 210, 328

Muller, F., 77, 78, 103, 106–10, 113, 322–3, 453–4

Murray, A., 374, 376, 391, 397, 412, 454

Murray, J., 28, 390

Museum of Comparative Zoology, Harvard, 30

museums, 19–20, 23, 27, 30–2, 436

multituberculates, 292–3

mutations (of Waagen), 346–7, 350

mutations (of De Vries), 375. See also saltations

myriapods, 104, 122–3, 131

Naef, A., 42, 43, 55–6, 59, 170–1, 312, 454

Nageli, C. von, 429

Naples. See Stazione Zoologica, Naples

Natural History Museum (London). See British Museum (Natural History)

natural selection, 2, 7–8, 252, 421, 423, 431, 434, 442, 444. See also Darwinism; struggle for existence

natural theology. See creation; theistic evolutionism

Nauplius, 108–12, 114–8, 123, 131, 138

Neal, H. V., 34, 201

Neanderthals, 420, 424, 439

Needham, J., 200

Nelson, G., 375

nemertines, 167, 180–3

neo-Lamarckism. See Lamarckism

neotony, 147, 150, 153, 198–9

nervous system, 158, 168–9, 178, 181

Neumayr, M., 315, 316–7, 411–2, 454

Newton, A., 37, 269, 384, 386–7, 402, 454

New Zealand, 128, 382, 387, 391, 413

Nopsca, F., 273, 278

Normal School of Science, London, 22, 36
north, as center of evolution, 373, 378, 384, 394–407, 439
Nyhart, L., 21, 35, 337
Nuttall, G., 200

occipital condyle, 262
oceans: as barriers to migration, 373, 388–9; evolution in, 389–90; origin of life in, 136, 193, 194–5
Odontornithes, 267–8
O'Hara, R. J., 323
Oliver, D., 397
ontogeny. *See* embryology; recapitulation theory
onychophrans, 104, 120–7. *See also* *Peripatus*
opossum, 291, 393, 408
Ornithorhynchus. *See* platypus
orthogenesis, 8–9, 66, 252, 256, 312, 339–40, 346–52, 357–60, 443
Ortmann, A. E., 410, 412, 414
Osborn, H. F., 17–8, 20, 28, 30–1, 317, 454; on adaptive radiation, 319, 336–8, 429; on amphibians, 239; on arthropods, 137; on biogeography, 384, 400, 408, 414–5; on birds, 274; on convergence, 71, 189; on extinction, 367, 369; on fish, 216; on human origins, 346, 434–5; on limbs, 204; on mammals, 288, 292, 300, 301, 340–1, 344, 350–1, 377; on orthogenesis, 70–1, 340, 346, 350–1, 356, 358; on progress, 436; on vertebrate origins, 189, 197
ostracoderms, 173, 176, 177, 208, 210, 212–5
ostrich, 268. *See also* Ratitae
Owen, Richard, 31, 64, 435–6, 454; on birds, 263–5, 269–70, 271, 272; on crocodiles, 333; on dinosaurs, 316; on homology, 46–8; on lungfish, 231; on origin of

mammals, 284, 297; on sea serpents, 390; on specialization, 329; on vertebrate archetype, 144, 164–5
Oxford University, 36–7

Packard, A. F., 124, 130, 355–6, 367, 382, 454
paedomorphosis. *See* neotony
Palaeospondylus, 210–1
paleontology, 5, 11–2, 217–8, 228–9, 230, 313–4, 317, 328, 442–3; and origin of amphibians, 205–6; and origin of mammals, 282; and recapitulation theory, 82–3. *See also* fossil record
Paludina, 315
Pantotheria, 293
parallelism, 53, 68–9, 71, 317, 322, 327, 345, 352; in arthropods, 99–100, 126; and biogeography, 376, 406; in birds and pterosaurs, 271; in fish, 237–9; in mammals, 281, 284–9, 293, 335, 342–3, 354; and origin of vertebrates, 143–4, 157, 189–91. *See also* convergence
Pareiasaurus, 299
Parker, W. N., 234–5
Patten, W., 94, 143, 146, 164, 171–7, 180, 212, 248, 429, 454
Patterson, C., 230
Pelseneer, P., 94
pelvis, structure of, 283
pelycosaurs, 298
pentadactyle limb, origin of. *See* limbs, of amphibians
Perameles, 289–91
Peripatus, 104, 120–4, 127, 128, 131, 135, 384, 413, 426. *See also* onychophorans
persistence of type, 263–4, 267, 328–9, 331, 334–5
phyllopods, 111–2, 117. *See also* *Apus*

phylogeny, 7–8, 17, 21–2, 40–1, 317, 328–52, 387, 422, 424–35; Darwinism and, 50–5, 56–62; idealism and, 55–6; paleontology and, 11–2. *See also* common descent; linear evolutionism; recapitulation theory; tree of life; *names of species and groups*
placenta, 289–90
placoderms, 213–4
Planorbis, 315
plants, evolution of, 362, 391, 397, 402, 413
plate tectonics. *See* continental drift
Platonism, 271. *See also* idealism
platypus, 281, 283, 286–7, 294
Pocock, R. I., 32, 102, 126, 133
Poebrotherium, 349
Pollard, H. B., 234
polyphyly, 327; of amphibians, 205, 242–4, 248; of arthropods, 68, 104–5, 112–20, 122–32, 139–40; and biogeography, 376–7; of birds, 260, 263, 271; of fish, 217; of humans, 247; of mammals, 260, 281, 287–9, 292–3, 300
Polypterus, 233–4, 240, 247
population expansion, 374–5, 440
prawns, 110
precambrian life, 84, 192, 194
Primates, evolution of, 295–6, 345–6, 426
primitive characters, definitions of, 61, 157–8, 260, 281, 322–5, 389, 431
Proboscidia, 350
professionalization of biology, 19–20, 25–39
progress, 88–9, 95–6, 146, 163, 334, 419, 424–35; and biogeography, 400–6; difficulty of defining, 207; in mammals, 284–5, 291–2; in early vertebrates, 175–7, 253, 255–6
Proterosaurus, 261
Protostomia, 87, 94, 95–6, 191

provinces, in biogeography, 373, 380–9
Pseudosuchians, 275
Pterichthys, 212–3
pterosaurs, 266–7, 270–2
Pycnogonidae, 102, 115
Pycraft, W. P., 275, 277

quadrate-incus theory, 303–8

Rabl, C., 225, 226, 248, 294, 454
race theory, 429, 432, 434–5, 439–40
Ramsay, A. C., 395
Rainger, R., 19–20, 347, 415
Ratitae, 262–3, 265–9, 270, 273–4, 278–80
Raymond, P. E., 137–8
recapitulation theory, 16, 44, 74–80, 82–3, 318, 322, 353–4, 355–7; and arthropods, 106–13, 115–7, 118, 138; and birds, 268, 276–7; and Gastraea theory, 86–7; and origin of vertebrates, 150–3, 161, 178, 215; rejection of, 161, 168, 198–9, 344
Reichert, C., 304
reptiles: and birds, 265–72; evolution of, 333–4; and mammals, 281–3, 288, 297–310; origin of, 259, 261. *See also* crocodiles; dinosaurs; mammal-like reptiles; pterosaurs
revolutions, in science. *See* Darwinian revolution
Richards, R. A., 75
Rolleston, G., 36–7, 128, 454
Romer, A. S., 33, 197, 455; on amphibians, 239, 244, 250, 257, 432; on biogeography, 416; on birds, 280, 434; on extinction, 368; on fish, 210, 213, 214, 217, 218; on mammals, 311, 338, 345, 357; on reptiles, 261–2; on trends in evolution, 352, 360, 364–5
Rosa, D., 358

Roth, S., 410
Rudwick, M. J. S., 420
Ruse, M., 19, 21
Russell, E. S., 13, 21, 41, 43, 142, 157
Rutimeyer, L., 400, 413, 455
Ryder, J. A., 252, 343

sabertoothed tiger, 359–60
Sagitta, 148
Salpa, 192–3
saltations, 53, 144, 187, 312, 342, 344, 350, 358, 432
Saporta, G. de, 362, 397
Sauropsida, 282–3
Save-Soderburgh, G., 217, 243
Scharff, R. F., 374–5, 377, 378, 398–400, 407, 410–1, 455
Schindewolf, O. H., 311–2, 358–9, 368, 370, 455
Schmalhausen, I. I., 244, 248
Schuchert, C., 33, 253, 256, 416, 466
Sclater, P. L., 380, 381–2, 384, 387, 402, 455
scorpions, 132–3. *See also* eurypterids
Scott, D. H., 38
Scott, W. B., 38, 317, 349–50, 351, 359–60, 363, 455
sea. *See* ocean
sea scorpions. *See* eurypterids
sea serpents, 390
sea spiders, 102, 115
Sedgwick, A. (geologist), 271
Sedgwick, A. (morphologist), 37–8, 80–1, 95, 110, 183–4, 455
Seeley, H. G., 455; on origin of mammals, 288, 292, 299–300, 302; on pterosaurs, 267, 271–2, 334
Semon, R., 61, 204–5, 232, 239–40, 247–8, 287, 302, 455
Semper, C., 159–60
senility, racial, 356–7, 360
Seymouria, 262
sharks. *See* elasmobranches
Sharpe, R. B., 323

Shimer, H. W., 83, 357
Simpson, G. G., 5, 21, 280, 284–5, 352, 360, 405, 417, 426, 441–2, 455–6; on Mesozoic mammals, 293–4, 288, 339
Simroth, H., 129, 253
Sinclair, W. J., 414
skull. *See* head, in vertebrates
Smith, G. Elliot, 431
Smith, J. P., 83
Snodgrass, R. E., 131–2
social Darwinism, 3–4, 389, 405, 421–4, 439–40
Sollas, W. J., 136, 439–40
South America, 29, 231, 380, 382, 396, 408–10, 414
southern continents, as centers of evolution, 378, 395, 407–18. *See also* Antarctica
specialization, 48, 207–8, 313, 334, 357–8, 428–9; of mammals, 328–31; of marsupials, 290–1. *See also* adaptive radiation
Spencer, Herbert, 432, 433
Spengel, J. W., 187
Sphenodon, 250, 391
spontaneous generation, 85
Stazione Zoologica, Naples, 29–30, 114, 148, 166, 168, 172, 187
Stehlin, H. G., 408
Stensio, E. A., 214
Stille, H., 367
Stormer, L. 138
stratigraphy, 336–7, 377, 410
struggle for existence, 7–8, 423; between species, 179, 319, 340, 354, 361, 364–6, 374–5, 400, 405, 437–40. *See also,* natural selection
Suess, E., 378, 379, 395, 398, 407, 456
Svertsov, N., 34
systematics. *See* taxonomy

taeniodonts, 338–9
tapirs, 361, 410, 407
taxonomy: of arthropods, 98–108,

taxonomy (*continued*)
124–6, 138–40; and biogeography, 386; Darwinism and, 7, 14, 42–3, 49–54, 67, 314–5, 317, 320–8; of fish, 230–1, 240; origins of, 45–6; parallelism and, 326–7, 351–2. *See also* common descent; cladism; tree of life
teeth, evolution of, 332, 338–9, 344–5, 363
terrestrial habitat, invasion of, 431, 432; by arthropods, 110, 120, 125, 129, 133, 137; by vertebrates, 244–5, 248–50, 253–8, 445
tetrapteryx theory, 276
Thatcher, J. K., 222–3
theistic evolutionism, 48, 64, 68, 303. *See also* Darwinism, opposition to
Theobald, F. V., 32
Theromorpha, 299, 301
Thiele, J., 91
Thiselton-Dyer, W. T., 402
Thompson, D'A. W., 390
Thomson, C. W., 389
Thomson, J. A., 122–3
Thomson, W. (Lord Kelvin), 394
tiger, sabertoothed, 359–60
Titanotheres, 350, 358
Tornaria, 188, 191
Tothill, J. D., 137
tracheae, 104, 125–6, 127, 128–9
Traquair, R., 210, 213–4, 217, 237–8
tree of life, 16–7, 49–53, 75–6, 84, 89–96, 142, 320–8, 420–40; in arthropods, 107–10, 119; in mammals, 282–3, 298, 328–39; reconstruction of, 81–2, 187, 272, 317, 320–2. *See also* common descent; phylogeny; taxonomy
trilobites, 100–1, 104–5, 116, 118–9, 133–5, 137, 390
Triploblastica, 87, 88–9, 183
Tristram, H. B., 402
trituberculate theory, 344–5

trochophore, 91–4
Trouessart, E. L., 414
tuatara (*Sphenodon*), 250, 391
Tullberg, T., 408
tunicates. *See* ascidians
typology, 55, 58. *See also* archetype; idealism; Platonism
typostrophe theory, 311–2, 358–9

Unger, F., 391, 397
ungulates, evolution of, 330–3, 335, 342–3, 363, 391–2
universities, 33–8
University College, London, 36
unspecialized, law of, 207–8, 334, 428

variation, 8, 63, 109, 271. *See also* Lamarckism, orthogenesis, saltations
vertebrates, origin of, 94, 141–7, 180–5; from ascidians, 147–57; from annelids, 157–71; from arthropods, 171–80; linked to echinoderms, 185–202
vicariance biogeography, 375, 405–6
visual reconstructions of past, 309, 316, 420–1
vitalism, 344, 357, 359, 429
Vogt, C., 269, 271, 272

Waagen, W., 346–7, 350, 411, 456
Wachsmuth, C. 326
Walcott, C. D., 100–1, 119, 133, 135–6, 194, 456
Wallace, A. W., 9–10, 25, 362, 369, 438, 456; on biogeography, 373, 374, 378, 380, 382, 383–4, 386–9, 392, 400–2, 402–3, 409
Wallace's line, 380, 386, 388
warm-bloodedness. *See* endothermy
Watson, D. M. S., 36, 202, 456; on amphibians, 238–9, 250, 252, 256–7, 349; on early fish, 214; on origin of mammals, 308; on orthogenesis, 349

Wegener, A., 372, 395, 405, 414–6, 456
Weismann, A., 3, 9
Weldon, W. F. R., 37
Wells, H. G., 22, 197, 257, 280, 338–9, 365, 390, 431–2, 443–4
Westoll, T. S., 239
Whittington, H. B., 139
Wiedersheim, R., 225, 456
Wilder, H. H., 197
Willey, A., 31, 71–2, 126, 188–91, 456
Willis, J. C., 375
Williston, S. W., 317, 342

Winsor, M. P., 320, 326
Woods Hole, 30
Woodward, A. Smith, 317, 456–7; on amphibians, 235–6; on early fish, 177, 210, 212, 215, 216, 228; on South America, 414; on trends in fossil record, 355, 357, 361
Wortman, J. L., 362, 403, 409

Zimmermanm W., 324
Zittel, K. von, 317, 318, 336, 376–7, 414, 457
zoea, 109–12, 113, 115